GCSE PHYSICS

third edition

TOM DUNCAN

JOHN MURRAY

To F.D.B.

Photo acknowledgements

Thanks are due to the following for permission to reproduce copyright photographs:
Cover Science Photo Library **p.vi** Fig. 1a NASA; Fig. 1b Prof. Erwin Mueller/ Science Photo Library; **p.vii** Fig. 2a NASA; Fig. 2b Martin Bond/Science Photo Library; Fig. 2c courtesy of INTELSAT; Fig. 2d photograph by Andrew Beattie, Intelligent Systems Solutions Limited; **p.viii** Christine Boyd; **p.2** Fig. 1.2 Philippe Plailly/Science Photo Library; **p.3** Fig.1.3 (and p.1) Topham Picture Point; **p.6** Fig. 2.4 Zefa; **p.9** Figs. 3.3, 3.4 Last Resort; **p.11** Fig. 4.2 Last Resort; **p.13** Fig. 4.6 Last Resort; **p.15** Fig. 5.4 Last Resort; **p.18** Fig.6.3b Last Resort; Figs. 6.6a,b KeyMed Ltd.; **p.20** Figs. 7.2a,b Last Resort; **p.27** Figs. 9.1b, 9.2c Last Resort; **p.28** Figs. 9.4, 9.5a Last Resort; **p.31** Fig. 10.3 Kodak Limited; **p.32** Fig. 11.2 Fisons; **p.34** Fig. 11.6 California Institute of Technology; Fig. 11.7 NASA; **p.40** Figs. 12.6, 12.7a, 12.8a,b from Llowarch, *Ripple Tank Studies of Wave Motion* (Clarendon Press, Oxford) reproduced by permission of Oxford University Press; Fig. 12.9 HR Wallingford; **p.41** Fig. 12.10a © Peter Gould; **p.43** Fig. 13.1 Last Resort; **p.44** Fig. 13.5a Last Resort; Fig. 13.5b Peter Aprahamian/Science Photo Library; **p.46** Fig. 13.9 Last Resort; Fig. 13.10 (and p.37) Paul Biddle & Tim Malyon/Science Photo Library; Fig. 13.11 AHL Holographics Division, Astor Universal Ltd.; **p.49** Fig. 14.3 Cambridge University Collection of Air Photographs: copyright reserved; Fig. 14.4 Zefa; **p.50** Fig. 14.6 Unilab; Fig. 14.7 Westinghouse*; **p.52** Fig. 15.1 Jonathan Watts/Science Photo Library; **p.55** Fig. 15.9. Susan Leavines/Science Photo Library; **p.56** Fig. 16.1 Last Resort; **p.58** Fig. 16.7 Last Resort; Fig. 16.8 University of Washington Libraries, Special Collections and Preservation, FM-25 (ref. no. 17892-2,8); *Pages* **61** *and* **62** Fig. 17.1 British Airways/Moving Image; **p.64** Fig. 17.7 Salter Weigh-Tronix Limited; **p.67** Fig. 18.3 Israel Government Tourist Office; **p.68** Fig. 19.1 British Amateur Weight Lifting Association; **p.71** Fig. 20.1 (from *PSSC Physics*) reproduced by permission of D.C. Heath and Co., copyright 1965, Education Development Center; **p.72** Fig. 20.6 Salt Union Limited; **p.73** Fig. 20.7 Last Resort; **p.75** Fig. 21.5 Photographed by Oskar Kreisal, from the *Focal Encyclopedia of Photography*, Butterworth Heinemann Ltd.; Fig. 21.8 Last Resort; **p.78** Heather Angel; **p.83** Fig. 22.9 © Grant Smith; **p.84** Fig. 23.3 Popperfoto/Reuter; **p.85** Fig. 23.6a Silsoe Research Institute; Fig. 23.6b Northern Counties Ltd.; **p.88** Fig. 24.2 photograph by British Petroleum; **p.91** Fig. 25.1 BWSF Publications Ltd.; **p.92** Fig. 25.3a Alton Towers; Fig. 25.3b Dimplex (UK) Ltd.; Fig. 25.3c Last Resort; **p.93** Fig. 25.3d Hydro-Electric; **p.97** Fig. 26.3 Last Resort; **p.101** Fig. 27.7 JCB; **p.105** Fig. 28.5 Last Resort; **p.106** Fig. 28.11 NASA; Fig. 28.12 Dr. George Gornacz/Science Photo Library; **p.109** Fig. 29.5 Crown copyright. Reproduced with the permission of the Controller of HMSO; Fig. 29.6 © Ivor Walton; **p.113** Fig. 30.6a Ford Motor Company Limited; Fig. 30.6b (and p.79) © Francois Gohier/Ardea London; **p.115** Fig. 30.13a © Mark Pepper/PPL; **p.118** Fig. 30.22a Condor Ltd.; **p.122** Fig. 31.1 from 'Dynamics of a Golf Swing' by Dr. D. Williams, *Quarterly Journal of Mechanics and Applied Mathematics* (Clarendon Press, Oxford) reproduced by permission of Oxford University Press; **p.129** Fig. 33.2 © Tony Souter/The Hutchison Library; **p.131** Fig. 33.6 (from *PSSC Physics*) reproduced by permission of D.C. Heath and Co., copyright 1965, Education Development Center; **p.132** Fig. 34.1 Hoverspeed Ltd.;**p.135** Fig. 34.5 (and p.121) Sward Sports Ltd.; **p.139** Figs. 35.5a,b Patrick Eagar; **p.141** Fig. 36.1 French Railways - Lafontant; **p.143** Fig. 36.5 Last Resort; Fig. 36.6 Transport Research Laboratory; **p.145** Fig. 37.1 Paul Tomkins STB/Still Moving; **p.146** Fig. 37.5 © Blackpool Pleasure Beach (photograph by Lord Lichfield); **p.150** Fig. 38.1 British Steel; **p.153** Fig. 39.1 photograph courtesy of BOC Gases; Fig. 39.3 Milepost 92 1/2; **p.157** Fig. 40.4 Philippe Plailly/Eurelios/Science Photo Library; **p.164** Fig. 42.3 Last Resort; **p.171** Figs. 44.4a,b Rockwool Limited; **p.172** Fig. 44.5 © Peter Gould; **p.173** Fig. 44.8 Sky Systems Hang Gliding & Paragliding; **p.175** Fig. 45.2 © Gail Goodger/The Hutchison Library; **p.177** Fig. 45.5 James R. Sheppard*; **p.179** Fig. 46.1a ETSU/Department of Trade and Industry; Fig. 46.1b Honda Motor Europe Ltd.; Fig. 46.2 (and p.149) GEC Alsthom; **p.180** Fig. 46.3 ETSU/Department of Trade and Industry; **p.183** Fig. 47.4 © Crown copyright/MOD. Reproduced with the permission of the Controller of HMSO; Fig. 47.5 GEC Alsthom; *Pages* **187** *and* **188** Fig. 48.1 Keith Kent/Science Photo Library; **p.190** Fig. 48.6b Last Resort; **p.201** Figs. 51.2a,b RS Components; **p.206** Figs. 52.3, 52.4a RS Components; **p.210** Fig. 53.3 Ronald R. Read; **p.217** Fig. 55.3 RS Components; **p.218** Fig. 55.5 Landis & Gyr Energy Management (UK) Ltd.; **p.223** Figs. 56.6a,b Tom Duncan; **p.227** Fig. 57.7 Alex Bartel/Science Photo Library; **p.232** Fig. 58.6 Elu Power Tools Ltd.; **p.238** Fig. 60.7 GEC Alsthom; **p.241** Fig. 61.5 GEC Alsthom; **p.250** Fig. 62.10 Unilab; **p.256** Figs. 63.8a,b The Royal Society from C.T.R. Wilson, Proc. Roy. Soc., Lond. A104, (1932), Plate 16, Fig. 1; **p.257** Fig. 63.11 Honeywell/Lippke; **p.264** Fig. 65.1a Casio Electronics Co. Ltd.; Fig. 65.1b S&W Vickers; **p.265** Fig. 65.4a Last Resort; Fig. 65.4b RS Components; **p.267** Figs. 65.8a, 65.9a RS Components; **p.268** Fig. 65.13a John Townson/Creation; **p.271** Fig. 65.21b Unilab; **p.273** Fig. 65.31a RS Components; Fig. 65.31b Unilab; **p.274** Fig. 65.32a Hugh Steeper Limited; Fig. 65.32b CAD image courtesy of Intergraph (UK) Limited; **p.278** Fig. 66.1a Cable & Wireless Visual Resource/ © Cable & Wireless; Fig. 66.1b,c The Science Museum/Science & Society Picture Library; **p.282** Fig. 66.14a John Townson/Creation; **p.284** Fig. 66.17a STC Submarine Systems Ltd.; **p.285** Fig. 66.19 (and p.247) Cable & Wireless Visual Resource/ © Cable & Wireless; **p.293** Fig. 67.6a © R.D. Whyman; Fig. 67.6b © J.A. Walton; Fig. 67.6c © J.F.P. Galvin; **p.294** Fig. 67.8 © University of Dundee; **p.299** Fig. 68.8a © R.F. Saunders; Fig. 68.8b © J.F.P. Galvin; **p.300** Fig. 68.10 © R.N. Hughes; **p.302** Fig. 68.15 Photo ESA; **p.304** Fig. 69.1 Photo ESA; **p.308** Fig. 69.9a California Institute of Technology; Fig. 69.9b V. Shone/Frank Spooner Pictures; **p.309** Fig. 69.9c The J. Allan Cash Photolibrary; Fig. 69.10a reproduced by permission of the Director, British Geological Survey: NERC copyright reserved; Fig. 69.10b courtesy of the Northern Ireland Tourist Board; **p.310** Fig. 69.11 reproduced by permission of the Director, British Geological Survey: NERC copyright reserved; Fig. 69.12 © David W. Jones/Lakeland Life Picture Library; Fig. 69.13 The Natural History Musuem, London; **p.315** Figs. 70.7a,b,c NASA; **p.317** Fig. 70.10 NASA/Science Photo Library; Fig. 70.12 Royal Greenwich Observatory/Science Photo Library; **p.319** Fig. 71.1a Allan Morton/Science Photo Library; Fig. 71.1b NASA; **p.321** Fig. 71.3b (and p.289) Pekka Parviainen/Science Photo Library.

* Every effort has been made to contact these copyright holders, who have given their permission for reproduction in previous editions; the publishers apologise for any breach and will be pleased to rectify this at the earliest opportunity.

First published in 1986 by
John Murray (Publishers) Ltd
50 Albemarle Street, London W1X 4BD

Second edition 1987
Reprinted 1987, 1988 (twice), 1990, 1991, 1992, 1993
Third edition 1995
Reprinted 1995, 1996, 1997, 1998, 2000

Artwork by 1-11 Line Art

Typeset in 11/13 pt Garamond Book uncondensed by Wearset, Boldon, Tyne and Wear

Printed in Great Britain by Butler & Tanner Limited, Frome, Somerset

A catalogue entry for this title may be obtained from the British Library

ISBN 0-7195-5301-6

By the same author

Exploring Physics, Books 1–3
Physics for Today and Tomorrow
Advanced Physics: Fourth Edition
Adventures with Electronics
Adventures with Microelectronics
Adventures with Physics
Adventures with Digital Electronics
Electronics for Today and Tomorrow
Success in Electronics
Physics for the Caribbean (with Deniz Onaç)
Science for Today and Tomorrow (with M. A. Atherton and D. G. Mackean)
Basic Skills: Electronics
GCSE Electronics

CONTENTS

ELECTRICITY AND MAGNETISM

ELECTRONS AND ATOMS

EARTH AND SPACE PHYSICS

Preface

This edition has undergone a major revision to cover the core and extension content of the new GCSE courses and to meet the requirements of National Curriculum Key Stage 4. Amendments to some of the topics have arisen from the restatement and emphasis of certain points more explicitly in the terms required by the syllabuses. Other changes are due to the addition or deletion of subject matter. The topics and items listed below are either new or are much changed.

(a) Core topics

Earth and space physics - a group of five new chapters (67 to 71) has been added to give a basic introduction to meteorology, geology and astronomy.

Energy and energy sources - this topic has been updated and enlarged and includes a new chapter (46) about renewable and non-renewable sources, power stations, and social and environmental considerations. A higher profile is also given to the energy transfer aspect (Chapter 25).

Electronics - greater emphasis has been placed on the systems approach, control and feedback and on social implications (Chapter 65).

(b) Extension topics

Fluids, **fluid flow** including **flight** (Chapter 30), **photons** (Chapter 62) and **telecommunications** (Chapter 66) are now treated more fully to meet the needs of different syllabuses.

(c) Physics investigations

A brief section has been included (p. viii) to assist students with this part of their assessed coursework, and 21 projects suggested.

(d) Questions

Some pruning and reorganization of the 700 or so questions has been undertaken in line with the trend towards the structured variety and away from the multiple-choice type. However, in the banks of both the **Additional questions** (after each group of related topics, for homework) and the **Revision questions** (at the end of the book, for quick, comprehensive revision before examinations), the two-level structure has been retained to meet the needs of a range of students. The first questions in a topic group are 'basic' questions intended for all, and the following 'higher' questions marked with a stripe are for those seeking grades E to A*.

I would like to thank Keith Munnings for his detailed analysis of the various GCSE syllabuses and for the many helpful suggestions he made as a result. Thanks are also due to Neil Duncan, Brian and Malcolm Kennett for advice on certain matters relating to Earth and space physics and to my wife for again undertaking the re-types. I am also much indebted once more to Jane Roth for all her excellent editorial work.

T.D.

Acknowledgement is made to the following boards (answers given being the sole responsibility of the author):

Cambridge Local Examination Syndicate (*C.*)
Joint Matriculation Board (*J.M.B.*)
University of London (*L.*)
Northern Ireland G.C.E. Examination Board (*N.I.*)
Oxford Local Examinations (*O.L.E.*)
Oxford and Cambridge Examination Board (*O. and C.*)
Southern Universities Joint Board (*S.*)
Welsh Joint Education Committee (*W.*)
Associated Lancashire Schools Examining Board (*A.L.*)
East Anglian Examinations Board (*E.A.*)
East Midland Regional Examination Board (*E.M.*)
North West Regional Examinations Board (*N.W.*)
Southern Regional Examinations Board (*S.R.*)
South-East Regional Examinations Board (*S.E.*)
West Midlands Examinations Board (*W.M.*)
Midland Examing Group (*M.E.G.*)

PHYSICS AND TECHNOLOGY

Physicists explore the Universe. Their investigations range from stars that are millions and millions of kilometres away to particles that are smaller than atoms, Figures 1a, b.

Figure 1a Astronomers have found that the many millions of stars in the Universe, of which the Sun is just one, are in widely separated groups called galaxies. The photograph of the spiral galaxy M100 shown here was taken with the Hubble Space Telescope (Figure 11.7) in December 1993. This orbiting telescope will enable astronomers to tackle one of the most fundamental questions in science, i.e. the age and scale of the Universe, by giving much more detailed information about individual stars than is possible with ground-based telescopes

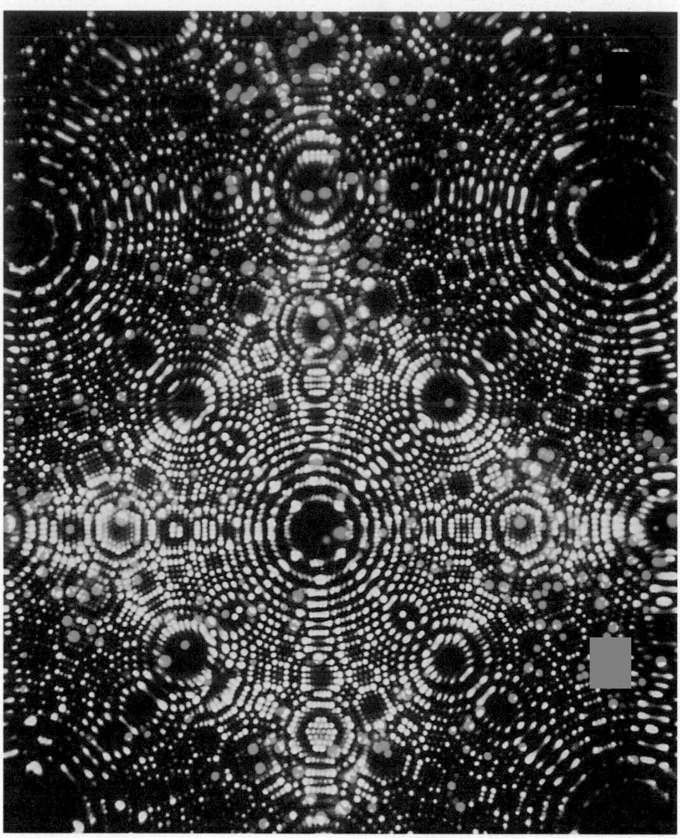

Figure 1b The photograph here shows the atoms in the tip of an iridium needle magnified about 2 million times. It was taken by an instrument called a field ion microscope

As well as having to find the **facts** by observation and experiment, they also must try to discover the **laws** that summarize (often as mathematical equations) these facts. Sense has then to be made of the laws by thinking up and testing **theories** (thought-models) to explain the laws. The reward, apart from a satisfied curiosity, is a better understanding of the physical world. Engineers and technologists use physics to solve **practical problems** for the benefit of people, though in solving them social, environmental and other problems may arise.

In this book we will study the behaviour of **matter** (the stuff things are made of) and the different kinds of **energy** (such as light, sound, heat, electricity). We will also consider the applications of physics in the home, in transport, medicine, research, industry, energy production, meteorology, communications and electronics, Figures 2a, b, c, d.

Mathematics is an essential tool of physics and a 'reference section' of some of the basic mathematics is given at the end of the book along with a suggested procedure for solving physics problems.

Figure 2a The manned exploration of space is such an expensive undertaking that international co-operation would seem to be the way forward. The representation here is of a proposed space station built and operated as a joint venture by the USA and Russia

Figure 2b In the search for alternative energy sources, experimental 'wind farms' of 20 to 30 wind turbines have been set up in suitable locations, such as this in North Wales, to generate at least enough electricity for the local community. There is some objection to their effect on the natural landscape

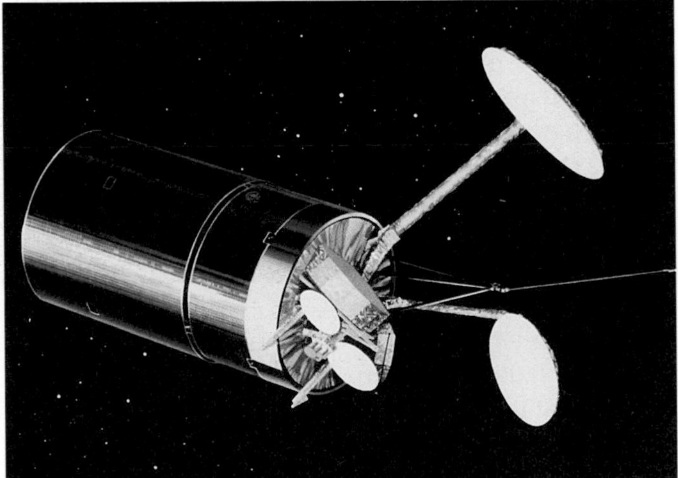

Figure 2c Communications satellites like the INTELSAT VI shown above can handle three television channels plus 30 000 telephone circuits simultaneously. They are in geostationary orbit 36 000 km (22 500 miles) above the equator where they circle the Earth in 24 hours and so appear to be at rest. Microwave signals are sent to the satellite and received from it by Earth stations with large dish aerials like those in Figure 66.19

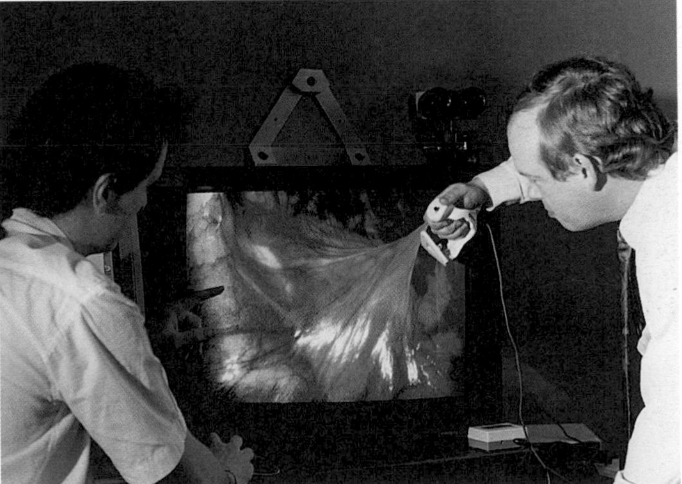

Figure 2d An effect called 'virtual reality' can be produced using computer graphics. In this photograph 'painless surgery' is being carried out on simulated organs and tissues inside a 'virtual body'. A special ultrasonic hand controller is used by the 'surgeon' in conjunction with the computer to carry out the 'operation'

PHYSICS INVESTIGATIONS

During your course you will have to carry out a few scientific investigations aimed at encouraging you to develop some of the **skills** and **abilities** that scientists use to solve real-life problems.

Investigations may arise from the topic you are currently studying in class, or your teacher may provide you with suggestions to choose from, or you may have your own ideas. However an investigation arises, it will probably require at least one hour of laboratory time, but often longer and will involve you in the following three aspects.

1. Planning how you are going to set about finding answers to the questions the problem poses. Making predictions and hypotheses (informed guesses) may help you to focus on what is required at this stage.
2. Performing the necessary experimental work safely, having decided on the equipment needed, what observations and measurements have to be made and what variable quantities need to be manipulated.
3. Presenting and interpreting the results in a way that enables any relationships between quantities to be established, drawing conclusions, giving explanations, assessing the reliability of data gathered and making comparisons with what was expected.

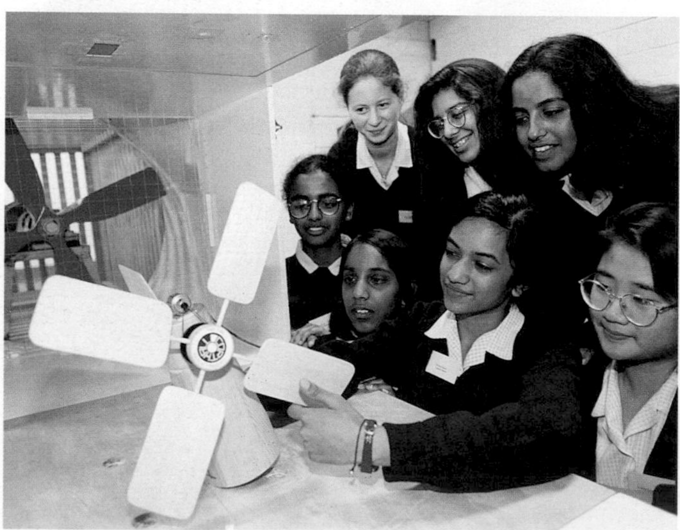

Girls from Copthall School, Mill Hill, London, with their winning entry for a contest to investigate, design and build the most efficient, elegant and cost-effective windmill

A **written report** of the investigation, possibly presented under the above three headings, would normally be made, indicating the aim of the work, giving records of procedures, observations and measurements, and stating the conclusions based on the evidence gathered.

Suggestions for investigations

1 Vibrating of a long steel strip clamped at one end (Chapter 16).
2 Resonance of an air column (Chapter 16).
3 Pitch of note from a vibrating wire (Chapter 16).
4 Stretching of rubber bands (Chapter 19).
5 Stretching of copper wires – **wear safety glasses** (Chapter 19).
6 Compressing of pillars of various sizes and materials (Chapter 21).
7 Strength of bridges made from paper, drinking straws, etc. (Chapter 21).
8 Bending of 'beams' (strips or sheets) of different materials (Chapter 22).
9 Energy values from burning fuel, e.g. firelighter (Chapter 25).
10 Friction – factors affecting (Chapter 26).
11 Fall of ball-bearings in a liquid (Chapter 30).
12 Flow of liquid through tubes (Chapter 30).
13 Viscosity of different liquids (Chapter 30).
14 Speed of a bicycle and its stopping distance (Chapter 36).
15 Circular motion using a bung on a string (Chapter 37).
16 Heat loss using different insulating materials (Chapter 44).
17 Model wind turbine design (Chapter 46).
18 Resistance of a thermistor and temperature (Chapter 51).
19 Heating effect of an electric current (Chapter 54).
20 Strength of an electromagnet (Chapter 57).
21 Efficiency of an electric motor (Chapter 58).

LIGHT AND SIGHT

1 LIGHT RAYS

Sources of light
Rays and beams
Shadows
Speed of light
Practical work
The pinhole camera.

Sources of light

You can see an object only if light from it enters your eyes. Some objects such as the Sun, electric lamps and candles make their own light. We call these **luminous** sources.

Most things you see do not make their own light but reflect it from a luminous source. They are **non-luminous** objects. This page, you and the Moon are examples. Figure 1.1 shows some others.

Luminous sources radiate light when their atoms become 'excited' as a result of receiving energy. In a light bulb, for example, the energy comes from electricity. The 'excited' atoms give off their light haphazardly in most luminous sources.

A light source that works differently is the **laser**, invented in 1960. In it the 'excited' atoms act together and emit a narrow, very bright beam of light which can cut through a thick metal plate with ease, Figure 1.2. Today the laser has a host of applications. It is used in scanners to read the bar code at shop and library check-outs, in compact disc (CD) players, in optical fibre telecommunication systems, in delicate medical operations (e.g. on the eye), in printing, and in surveying and range-finding.

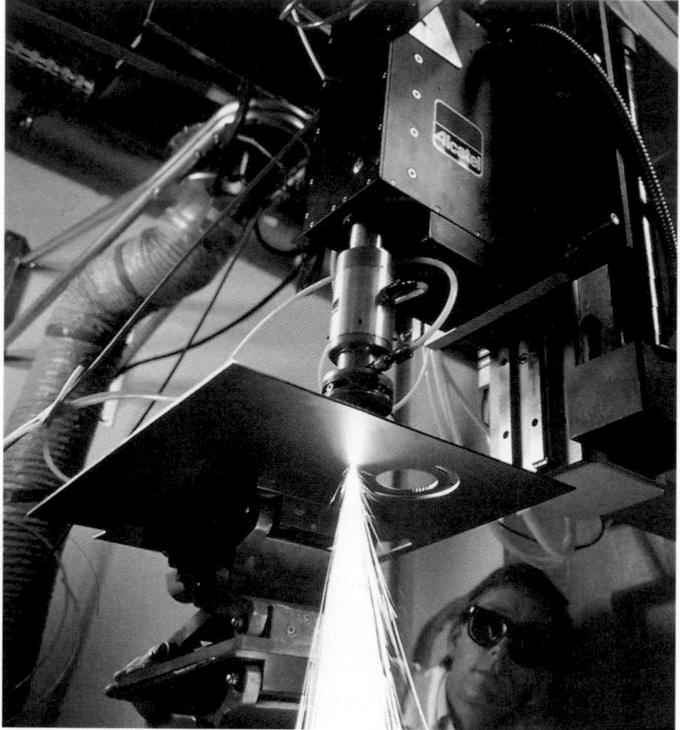
Figure 1.2 Laser beam cutting a metal plate

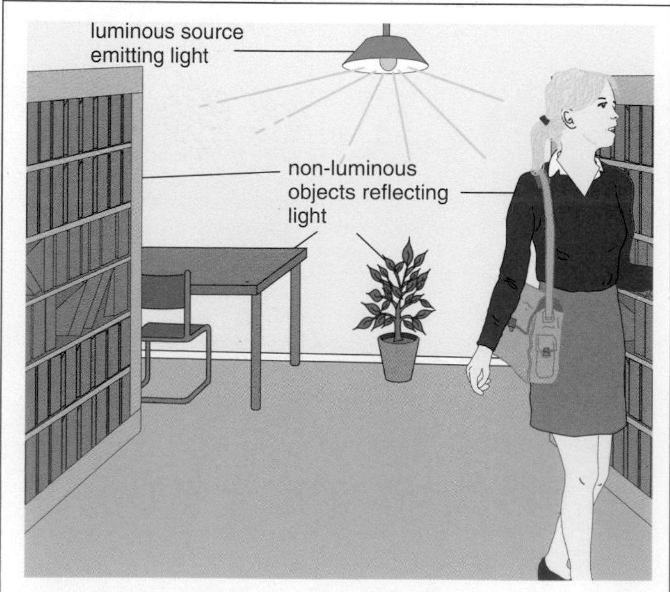

Figure 1.1 Luminous and non-luminous objects

Rays and beams

Sunbeams streaming through trees, Figure 1.3, and light from a cinema projector on its way to the screen both suggest that **light travels in straight lines**. The beams are visible because dust particles in the air reflect light into our eyes.

The direction of the path in which light is travelling is called a **ray** and is represented in diagrams by a straight line with an arrow on it. A **beam** is a stream of light and is shown by a number of rays; it may be parallel, diverging (spreading out) or converging (getting narrower), Figure 1.4.

Figure 1.3 Light travels in straight lines

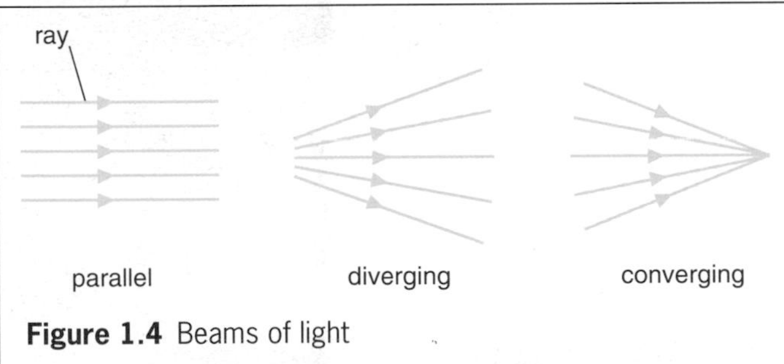

Figure 1.4 Beams of light

Practical work

The pinhole camera

One is shown in Figure 1.5a. Make a small pinhole in the centre of the black paper. Half-darken the room. Hold the box at arm's length so that the pinhole end is nearest to and about 1 metre from a luminous object, e.g. a carbon filament lamp or a candle. Look at the **image** on the screen (an image is a likeness of an object and need not be an exact copy).

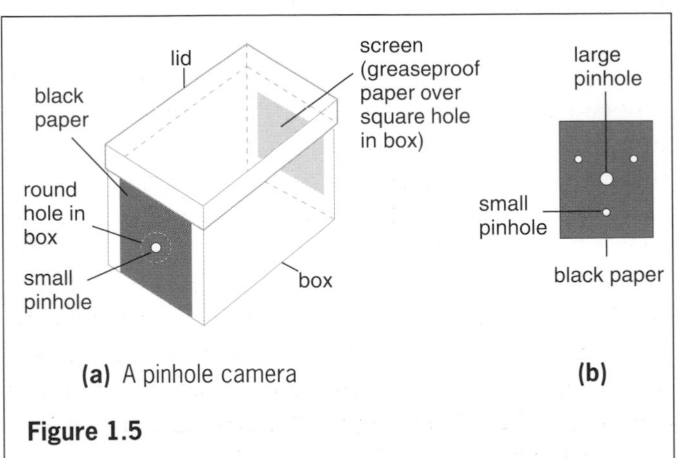

(a) A pinhole camera **(b)**

Figure 1.5

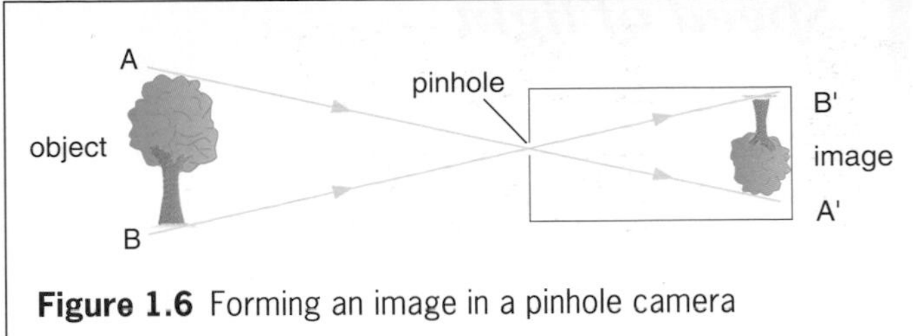

Figure 1.6 Forming an image in a pinhole camera

Can you see *three* ways in which the image differs from the object? What is the effect of moving the camera closer to the object?

Make the pinhole larger. What happens to the (i) brightness, (ii) sharpness, (iii) size of the image?

Make several small pinholes round the large hole, Figure 1.5b, and view the image again.

The forming of an image is shown in Figure 1.6.

Shadows

Shadows are formed for two reasons. First, because some objects, which are said to be **opaque**, do not allow light to pass through them. Second, because light travels in straight lines. The sharpness of the shadow depends on the size of the light source. A very small source of light, called a **point** source, gives a sharp shadow which is equally dark all over. This may be shown as in Figure 1.7a where the small hole in the card acts as a point source.

If the card is removed the lamp acts as a large or **extended** source, Figure 1.7b. The shadow is then larger and has a central dark region, the **umbra**, surrounded by a ring of partial shadow, the **penumbra**. You can see by the rays that some light reaches the penumbra but none reaches the umbra.

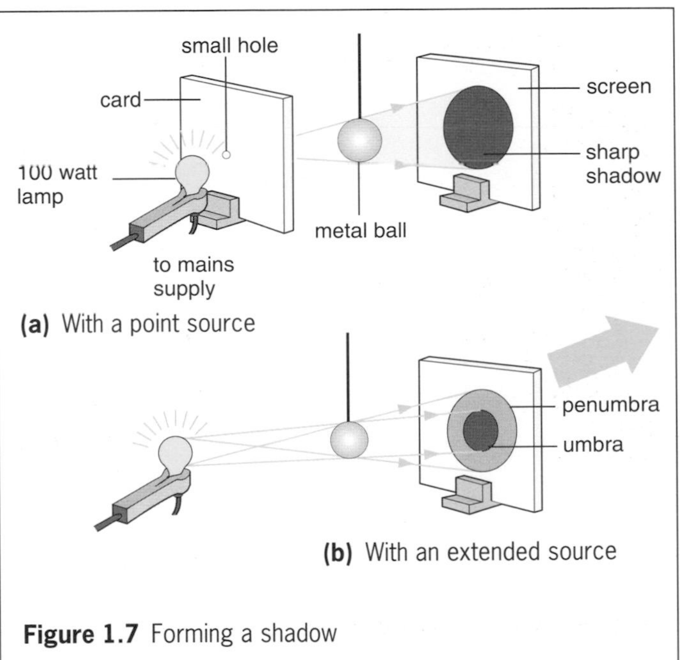

(a) With a point source

(b) With an extended source

Figure 1.7 Forming a shadow

Speed of light

Proof that light travels very much faster than sound is provided by the starter's gun at a sports meeting. The flash from the gun is seen before it is heard. The length of the time lapse is greater the further the observer is from the starter.

The speed of light has a definite value; it does not travel instantaneously from one point to another but takes a certain, very small time. Its speed is about 1 million times greater than that of sound.

QUESTIONS

1 How would the size and brightness of the image formed by a pinhole camera change if the camera was made longer?

2 What changes would occur in the image if the single pinhole in a camera was replaced by **a** four pinholes close together, **b** a hole 1 cm wide?

3 A long narrow bench has two small identical lamps mounted one at each end as in Figure 1.8. A vertical rod is placed on the bench. Copy the diagram and draw the shadows formed, showing the correct size and position. State, giving a reason, which shadow is the darker. *(E.A.)*

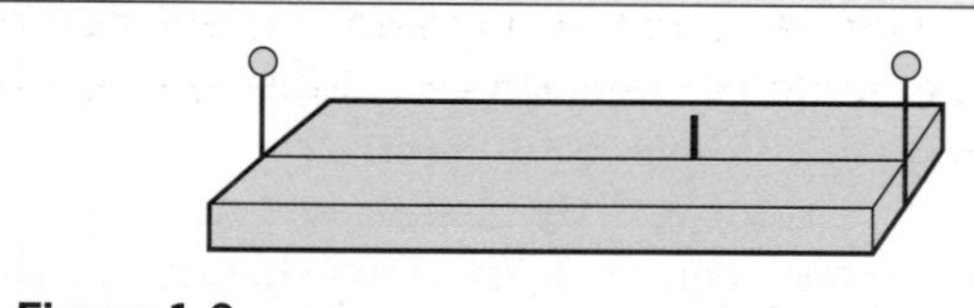

Figure 1.8

4 At a sports meeting should the timekeeper start the clock when he sees the flash from the starter's gun or when he hears it? Explain your answer.

Checklist

After studying this chapter you should be able to

- give examples of effects which show that light travels in a straight line,
- explain the operation of a pinhole camera and draw ray diagrams to show the result of varying the object distance or the length of the camera,
- draw diagrams to show how shadows are formed using point and extended sources, and use the terms **umbra** and **penumbra**,
- recall that light travels much faster than sound.

2 REFLECTION OF LIGHT

Laws of reflection
Periscope
Regular and diffuse reflection

Practical work
Reflection by a plane mirror.

If we know how light behaves when it is reflected we can use a mirror to change the direction in which it is travelling. This happens when a mirror is placed at the entrance of a concealed drive to give warning of approaching traffic.

An ordinary mirror is made by depositing a thin layer of silver on one side of a piece of glass and protecting it with paint. The silver - at the *back* of the glass - acts as the reflecting surface.

PRACTICAL WORK

Reflection by a plane mirror

Draw a line AOB on a sheet of paper and using a protractor mark angles on it. Measure them from the perpendicular ON, which is at right angles to AOB. Set up a plane (flat) mirror with its reflecting surface on AOB.

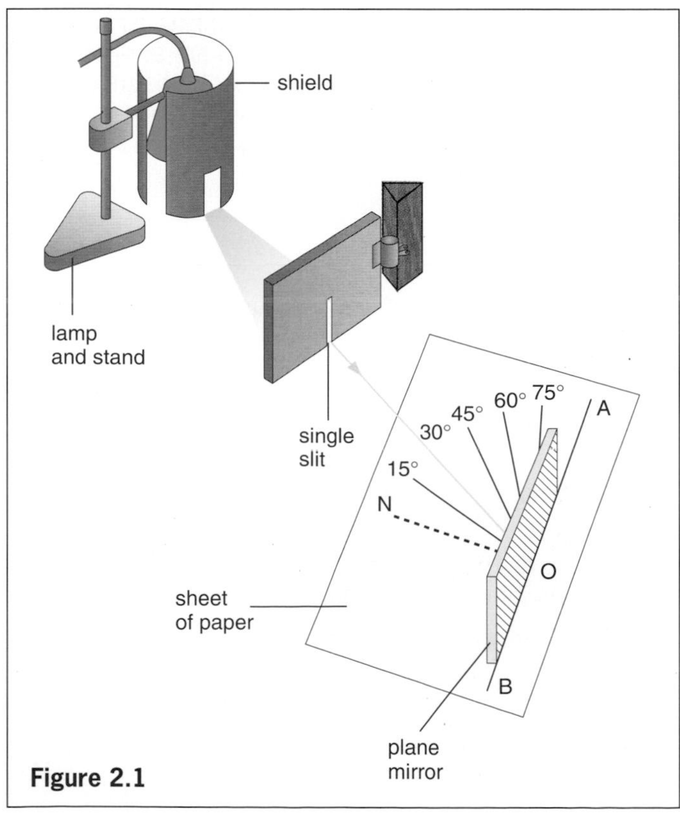

Figure 2.1

Shine a narrow ray of light along say the 30° line onto the mirror, Figure 2.1.

Mark the position of the reflected ray, remove the mirror and measure the angle between the reflected ray and ON. Repeat for rays at other angles. What can you conclude?

Laws of reflection

Terms used in connection with reflection are shown in Figure 2.2. The perpendicular to the mirror at the point where the incident ray strikes it is called the **normal**. Note that the angle of incidence i is the angle between the incident ray and the normal; similarly for the angle of reflection r. There are two laws of reflection.

1 The angle of incidence equals the angle of reflection.
2 The incident ray, the reflected ray and the normal all lie in the same plane. (*This means that they can all be drawn on a flat sheet of paper.*)

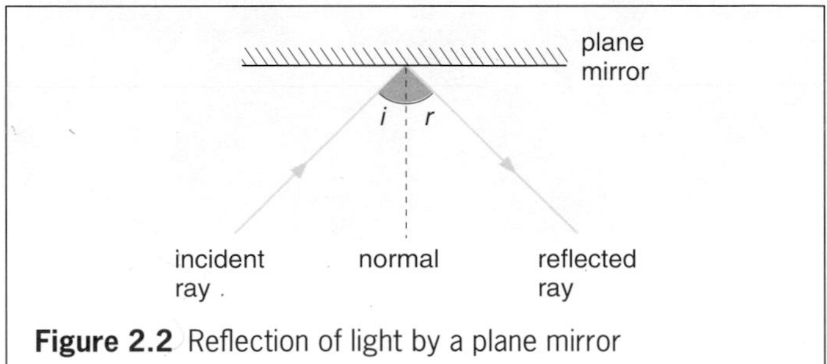

Figure 2.2 Reflection of light by a plane mirror

Periscope

A simple periscope consists of a tube containing two plane mirrors, fixed parallel to and facing one another. Each makes an angle of 45° with the line joining them, Figure 2.3. Light from the object is turned through 90° at each reflection and an observer is able to see over a crowd, for example, Figure 2.4, or over the top of an obstacle.

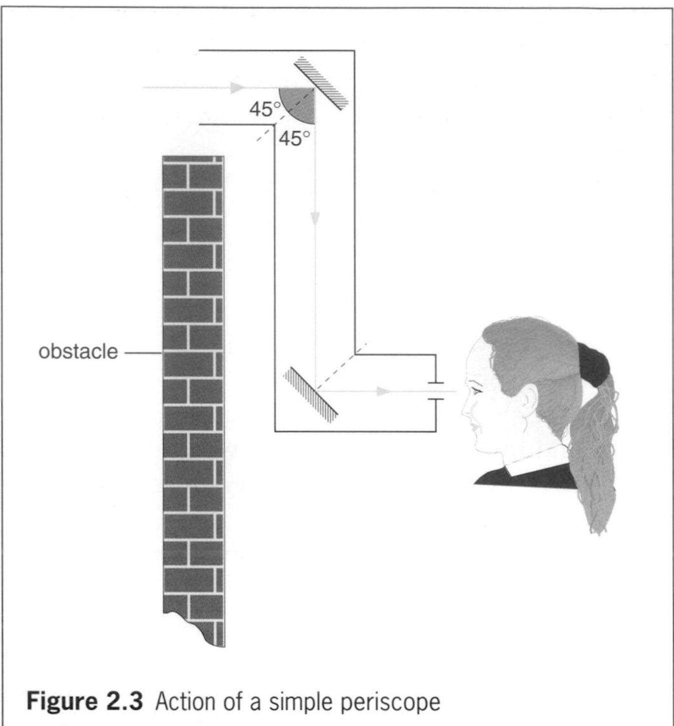

Figure 2.3 Action of a simple periscope

Figure 2.4 Periscopes being used by people in a crowd

In more elaborate periscopes like those used in submarines, prisms replace mirrors (see Chapter 6).

Make your own from a long, narrow cardboard box measuring about 40 cm × 5 cm × 5 cm (e.g. one in which aluminium cooking foil or greaseproof paper is sold), two plane mirrors (7.5 cm × 5 cm) and sticky tape. When you have got it to work, make modifications which turn it into a 'See-back-o-scope' that lets you see what is behind you.

Regular and diffuse reflection

If a parallel beam of light falls on a plane mirror it is reflected as a parallel beam, Figure 2.5a, and **regular** reflection occurs. Most surfaces, however, reflect light irregularly and the rays in an incident parallel beam are reflected in many directions, Figure 2.5b.

Irregular or **diffuse** reflection is due to the surface of the object not being perfectly smooth like a mirror. At each point on the surface the laws of reflection are obeyed but the angle of incidence and so the angle of reflection varies from point to point. The reflected rays are scattered haphazardly. Most objects, being rough, are seen by diffuse reflection and this helps to reduce glare often caused by regular reflection from a mirror.

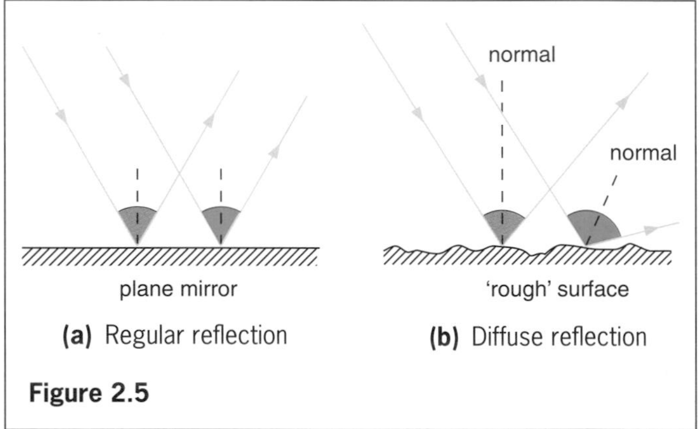

(a) Regular reflection **(b)** Diffuse reflection

Figure 2.5

QUESTIONS

1 Figure 2.6 shows a ray of light PQ striking a mirror AB. The mirror AB and the mirror CD are at right angles to each other. QN is a normal to the mirror AB.

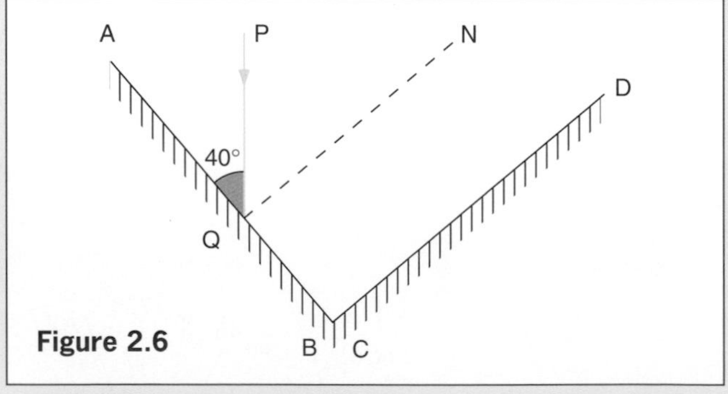

Figure 2.6

a What is the value of the angle of incidence of the ray PQ on the mirror AB?

b Copy the diagram and continue the ray PQ to show the path it takes after reflection at both mirrors.

c Mark on your diagram the value of the angle of reflection on AB, the angle of incidence on CD and the angle of reflection on CD.

d What do you notice about the path of the ray PQ and the final reflected ray? *(E.M.)*

2 When a ray of light incident on a plane mirror at an angle of incidence of 70° is reflected from the mirror it subsequently strikes a second plane mirror placed so that the angle between the mirrors is 45°. The angle of reflection at the second mirror, in degrees, is

A 20 **B** 25 **C** 45 **D** 65 **E** 70 *(N.I.)*

3 A person is looking into a tall mirror in front, Figure 2.7. What part of the mirror is actually needed to see from eye to toes? *(S.R.)*

Figure 2.7

Checklist

After studying this chapter you should be able to

- describe an experiment to show that the angle of incidence equals the angle of reflection,
- state the laws of reflection and use them to solve problems,
- draw a ray diagram to show how a periscope works.

3 PLANE MIRRORS

Real and virtual images	***Kaleidoscope***
Lateral inversion	***Practical work*** Position of image.
Properties of the image	

When you look into a plane mirror on the wall of a room you see an image of the room behind the mirror; it is as if there were another room. Restaurants sometimes have a large mirror on one wall just to make them look larger. You may be able to say how much larger after the next experiment.

The position of the image formed by a mirror depends on the position of the object.

PRACTICAL WORK

Position of image

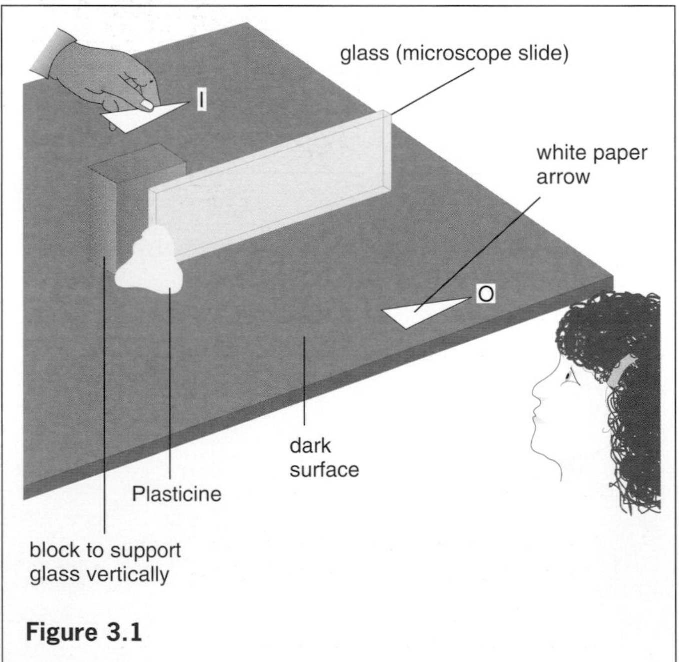

Figure 3.1

Support a piece of thin glass on the bench, as in Figure 3.1. It must be *vertical* (at 90° to the bench). Place a small paper arrow, O, about 10 cm from the glass. The glass acts as a poor mirror and an image of O will be seen in it; the darker the bench top the brighter is the image. How do the sizes of O and its image compare? Imagine a line joining them. What can you say about it?

Lay another identical arrow, I, on the bench behind the glass; move it until it coincides with the image of O. Measure the distances of the points of O and I from the glass along the line joining them. How do they compare? Try O at other distances.

Real and virtual images

A **real** image is one which can be produced on a screen (as in a pinhole camera) and is formed by rays that actually pass through it.

A **virtual** image cannot be formed on a screen and is produced by rays which seem to come from it but do not pass through it. The image in a plane mirror is virtual. Rays from a point on an object are reflected at the mirror and appear to come from the point behind the mirror where the eye imagines the rays intersect when produced backwards, Figure 3.2. IA and IB are construction lines and are shown by broken lines.

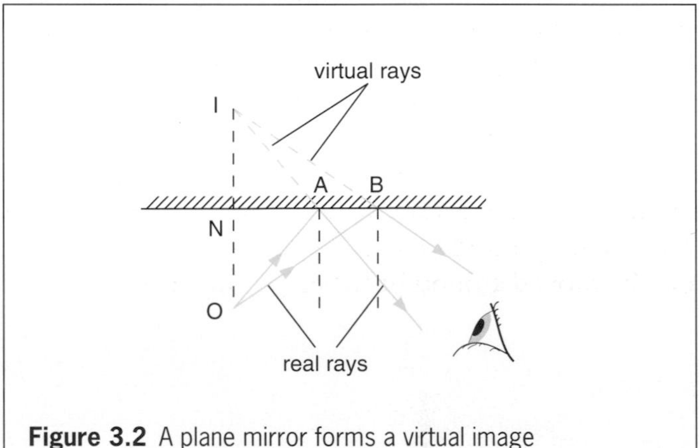

Figure 3.2 A plane mirror forms a virtual image

Lateral inversion

If you close your left eye your image in a plane mirror seems to close the right eye. In a mirror image, left and right are interchanged and the image is said to be **laterally inverted**. The effect occurs whenever an image is formed by one reflection and is very evident if print is viewed in a mirror, Figure 3.3. What happens if two reflections occur as in a periscope?

Figure 3.3 The image in a plane mirror appears laterally inverted

Properties of the image

The image in a plane mirror is

(i) as far behind the mirror as the object is in front and the line joining the object and image is perpendicular to the mirror,
(ii) the same size as the object,
(iii) virtual,
(iv) laterally inverted.

Kaleidoscope

To see how a kaleidoscope works, draw on a sheet of paper two lines at right angles to one another. Using different coloured pens or pencils draw a design between them, Figure 3.5a. Place a small mirror along each line and look into the mirrors, Figure 3.5b. You will see three reflections which join up to give a circular pattern. If you make the angle between the mirrors smaller, more reflections appear but you always get a complete design.

In a kaleidoscope the two mirrors are usually kept at the same angle (about 60°) and different designs are made by hundreds of tiny coloured beads which can be moved around between the mirrors.

Figure 3.4 A kaleidoscope produces hundreds of ever-changing coloured designs using the images formed by two plane mirrors

Make a kaleidoscope using a cardboard tube (e.g. from a toilet roll), some thin card, greaseproof paper, clear sticky tape, small pieces of different coloured cellophane and two mirrors (10 cm × 3 cm) or polished tinplate (10 cm × 6 cm, bent to form two mirrors at 60° to each other).

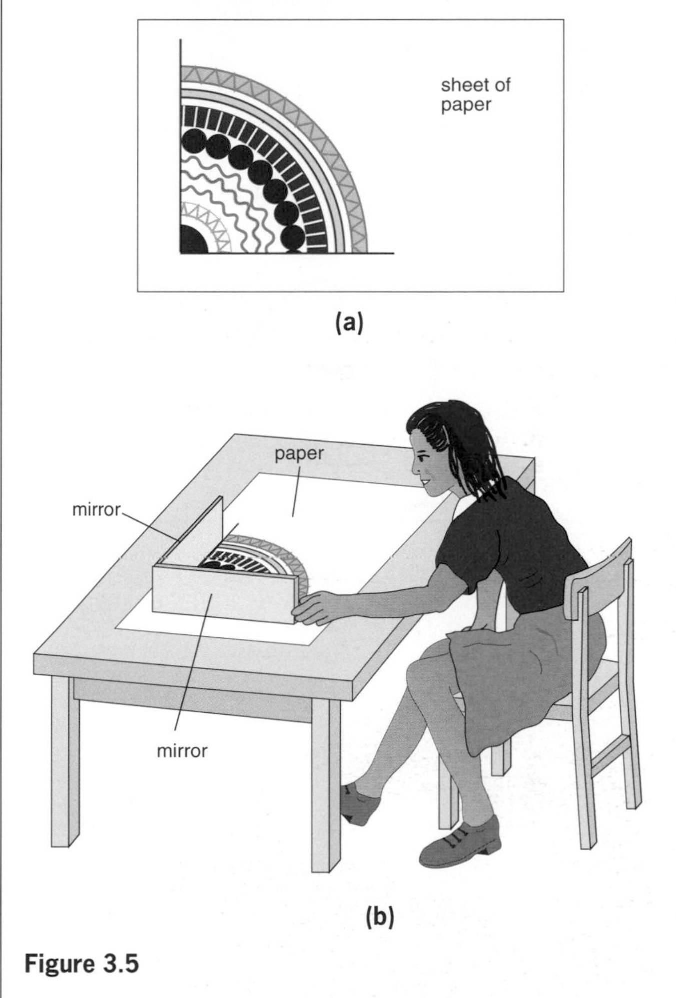

Figure 3.5

QUESTIONS

1 Figure 3.6 shows a plan view of a jar of water and a vertical sheet of glass in a box, the inside of which is painted black.

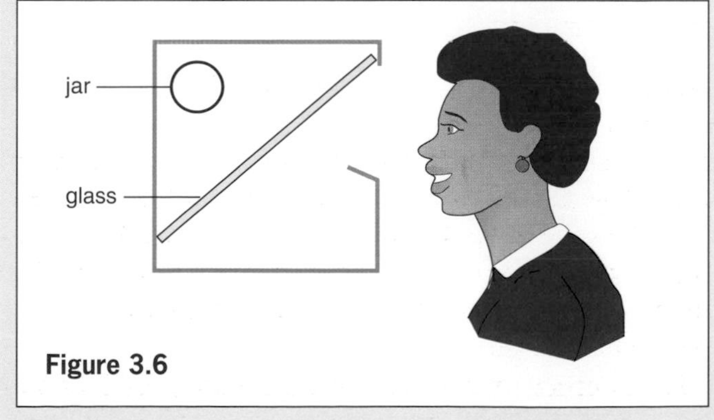

Figure 3.6

a Copy the diagram and show where a candle might be placed so that it appears to the viewer to burn in the jar of water.

b Trace the path of two rays of light from the candle to the eye and show with dotted lines how they appear to come from inside the jar.

c What does the glass do to the light to get this effect? *(S.E.)*

2 The image in a plane mirror of a modern clock (with lines instead of numbers) looks as in Figure 3.7.The correct time is

A 2.25 **B** 2.35 **C** 8.35 **D** 9.25 *(W.M.)*

Figure 3.7

3 A girl stands 5 m away from a large plane mirror. How far must she walk to be 2 m away from her image?

Checklist

After studying this chapter you should be able to

- describe an experiment to show that the image in a plane mirror is as far behind the mirror as the object is in front, and that the line joining the object and image is at right angles to the mirror,
- draw a diagram to explain the formation of a virtual image by a plane mirror,
- explain the term **lateral inversion**,
- explain how a kaleidoscope works.

4 CURVED MIRRORS

Concave and convex mirrors
Principal focus
Ray diagrams
Images in a concave mirror
Uses of curved mirrors
Practical work
f of a concave mirror.

For some purposes curved mirrors are more useful than plane mirrors. They can make things look larger or smaller depending on which way they curve and how far the object is from them. You can see the images they give by looking into both sides of a large, well-polished spoon. There are two main types.

Concave and convex mirrors

A **concave** mirror curves inwards like a cave, Figure 4.1a; a **convex** one curves outwards, Figure 4.1b. Many curved mirrors have spherical surfaces and the **principal axis** of such a mirror is the line joining the **pole** P or centre of the mirror to the **centre of curvature** C. C is the centre of the sphere of which the mirror is a part; it is in front of a concave mirror and behind a convex one.

The **radius of curvature** r is the distance CP.

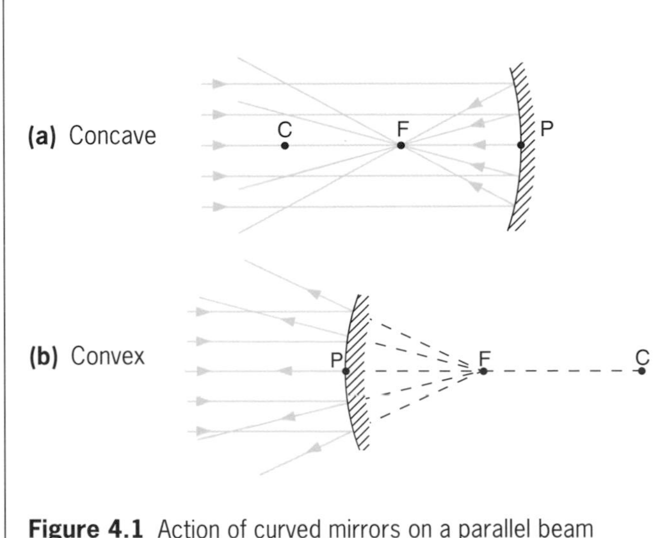

Figure 4.1 Action of curved mirrors on a parallel beam of light

Principal focus

When a beam of light parallel to the principal axis is reflected from a concave mirror, the laws of reflection hold and it converges to a point on the axis called the **principal focus** F. Since light actually passes through it, it is a **real** focus and can be obtained on a screen. A convex mirror has a **virtual** principal focus behind the mirror, from which the reflected beam appears to diverge.

The **focal length** f is the distance FP and experiment and theory show that FP = CP/2 or $f = r/2$, i.e.

focal length = half the radius of curvature

These statements apply only to small mirrors, or to large mirrors if the beam is close to the axis. A large mirror does not form a point focus of a wide parallel beam but causes the reflected rays to form a curve called a **caustic**. One may be seen on the surface of tea in a cup, Figure 4.2. Why?

Figure 4.2 Caustic curve in a cup of tea

Ray diagrams

Facts about the images formed by spherical mirrors can be found by drawing *two* of the following rays.

1. A ray parallel to the principal axis which is reflected through the principal focus F.
2. A ray through the centre of curvature C which hits the mirror normally, is reflected back along its own path (the radius of a sphere is perpendicular to the surface where it meets the surface).
3. A ray through the principal focus F which is reflected parallel to the principal axis.

In diagrams a curved mirror is represented by a straight line. A good-sized object can then be shown and rays from it regarded as forming a point focus, i.e. the mirror behaves as a small one. In numerical questions horizontal distances have sometimes to be scaled down.

Images in a concave mirror

The ray diagrams in Figure 4.3 show the images for four object positions. In each case two rays are drawn from the top A of an object OA and where they intersect after reflection gives the top B of the image IB. The foot I of each image is on the axis since ray OP hits the mirror normally and is reflected back along the axis. In (d) the broken rays and the image are virtual (not real).

PRACTICAL WORK

f of a concave mirror

We use the fact that when an object is placed at the centre of curvature C of a concave mirror, a real image is formed also at C, Figures 4.3c and 4.4a.

Move the mirror, arranged as in Figure 4.4b, until a sharp image of the cross-wire is obtained on the screen. Measure the distance r from the screen to the mirror, then $f = r/2$.

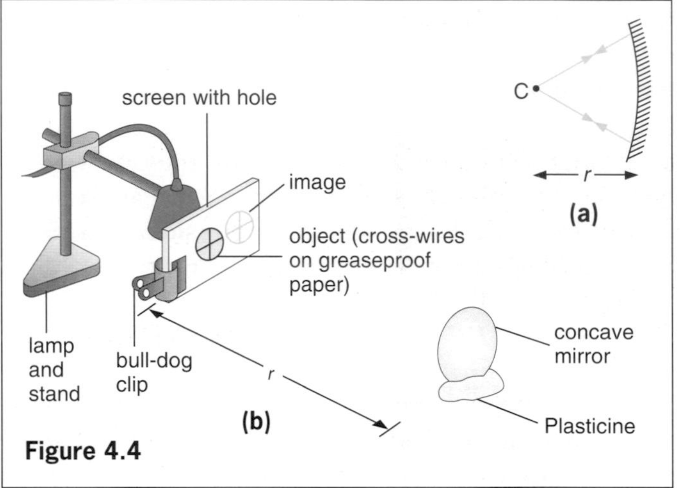

Figure 4.4

Uses of curved mirrors

(a) Reflectors

Concave mirrors are used as reflectors in, for example, car headlamps and torches, because a small lamp at their focus gives a **parallel** reflected beam. This is only strictly true if the mirror has a parabolic shape (rather than spherical) like that in Figure 4.5.

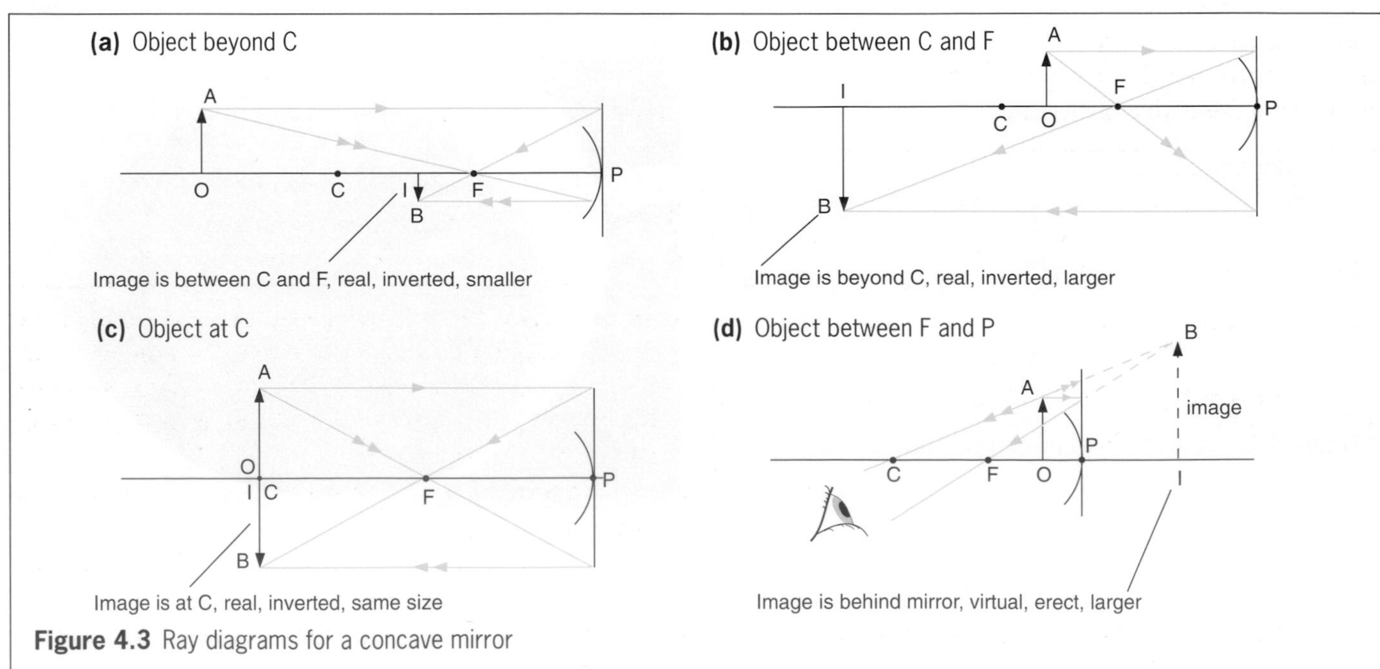

Figure 4.3 Ray diagrams for a concave mirror

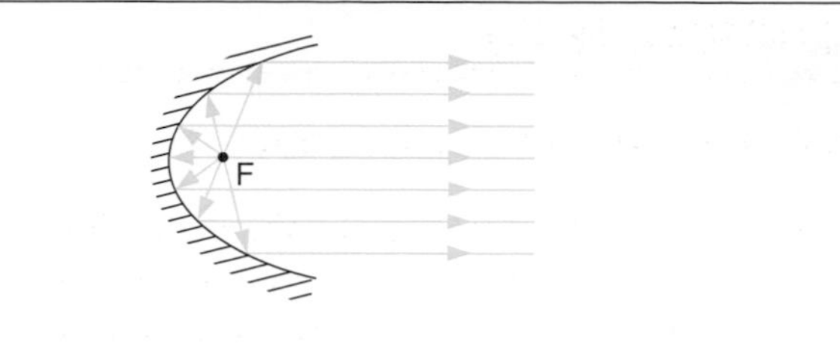

Figure 4.5 A parabolic mirror produces a parallel beam

(b) Make-up, shaving and dental mirrors

A concave mirror forms a **magnified**, **upright** image of an object **inside** its focus, Figure 4.6. This accounts for these uses. The image appears to be behind the mirror and is virtual.

Figure 4.6 A concave mirror forms a magnified image of a close object

(c) Driving mirrors

A convex mirror gives a wider field of view than a plane mirror of the same size, Figures 4.7a, b. For this reason and because it always gives an upright (but smaller) image, it is used as a car driving mirror and on dangerous bends of roads. However it does give the driver a false idea of distance.

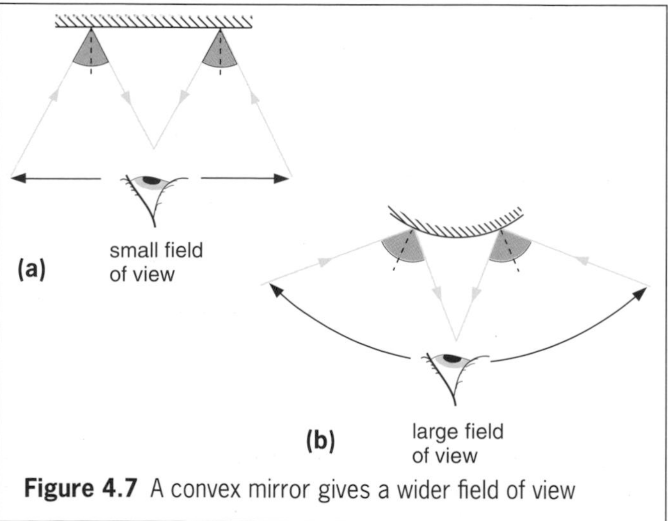

Figure 4.7 A convex mirror gives a wider field of view

QUESTIONS

1 Draw diagrams to represent **a** a convex mirror and **b** a concave mirror. Mark on each the principal axis, the principal focus F and the centre of curvature C if the focal length of **a** is 3 cm and of **b** 4 cm.

2 A concave mirror has a focal length of 4 cm and a real object 2 cm tall is placed 9 cm away from it. By means of an accurate full-size diagram find where the image would be and measure its length.

3 A communications satellite in orbit sends a parallel beam of signals down to Earth. If they obey the same laws of reflection as light and are to be focused on to a small receiving aerial, what is the best shape for the metal 'dish' used to collect them?

4 Account for the use of a convex mirror on the stairs of a double-decker bus.

Checklist

After studying this chapter you should be able to

- draw diagrams to show the action of concave and convex mirrors on a parallel beam of light, and use the terms **principal focus**, **focal length**, **radius** and **centre of curvature**,
- explain why concave mirrors are used as reflectors in car headlamps,
- explain why concave mirrors are used as make-up, shaving and dental mirrors,
- explain why convex mirrors are used as driving mirrors.

5 REFRACTION OF LIGHT

Facts about refraction	***Refraction by a prism***
Real and apparent depth	***Practical work*** Refraction in glass.
Refractive index	

If you place a coin in an empty cup and move back until you *just* cannot see it, the result is surprising if someone *gently* pours in water. Try it.

Although light travels in straight lines in one transparent material, e.g. air, if it passes into a different material, e.g. water, it changes direction at the boundary between the two, i.e. it is bent. The **bending of light** when it passes from one material (called a medium) to another is called **refraction**. It causes effects like the coin trick.

PRACTICAL WORK

Refraction in glass

Shine a ray of light at an angle on to a glass block (which has its lower face painted white or frosted), Figure 5.1. Draw the outline ABCD of the block on the sheet of paper under it. Mark the positions of the various rays in air and in glass.

Remove the block and draw the normals on the paper at the points where the ray enters AB (see Figure 5.1) and where it leaves CD.

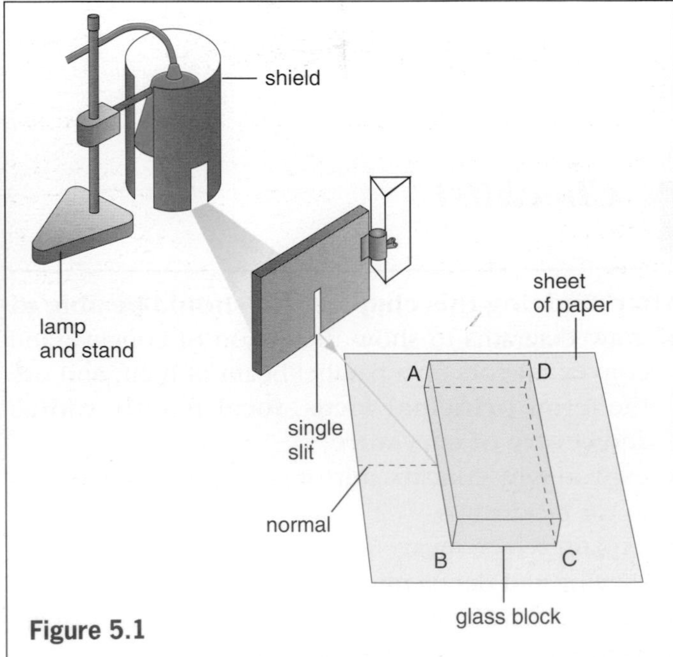

Figure 5.1

What *two* things happen to the light falling on AB? When the ray enters the glass at AB is it bent towards or away from the part of the normal in the block? How is it bent at CD? What can you say about the direction of the ray falling on AB and the direction of the ray leaving CD?

What happens if the ray hits AB at right angles?

Facts about refraction

The previous experiment shows that:

(i) a ray of light is bent **towards** the normal when it enters an optically denser medium at an angle (e.g. from air to glass), i.e. the angle of refraction r is less than the angle of incidence i, Figure 5.2a,
(ii) a ray of light is bent **away from** the normal when it enters an optically less dense medium (e.g. from glass to air),
(iii) a ray emerging from a parallel-sided block is **parallel** to the ray entering, but is displaced sideways,
(iv) a ray travelling along the normal is **not refracted**, Figure 5.2b.

Note. Optically denser means having a greater refraction effect; the actual density may or may not be greater.

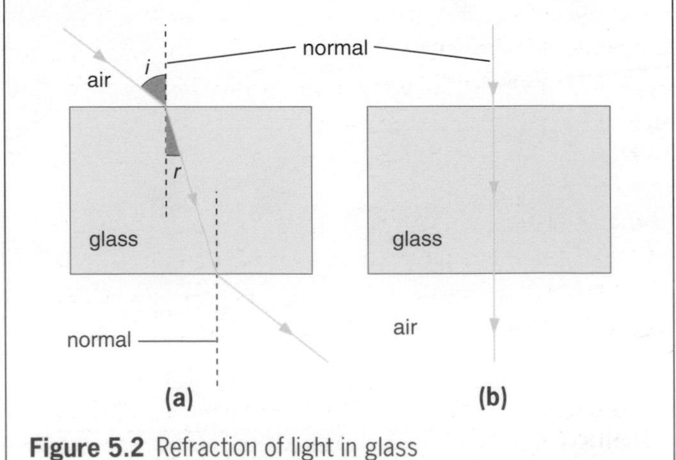

Figure 5.2 Refraction of light in glass

Real and apparent depth

Rays of light from a point O on the bottom of a pool are refracted away from the normal at the water surface since they are passing into an optically less dense medium, i.e. air, Figure 5.3. On entering the eye they appear to come from a point I *above* O; I is the virtual image of O formed by refraction. The apparent depth of the pool is less than its real depth.

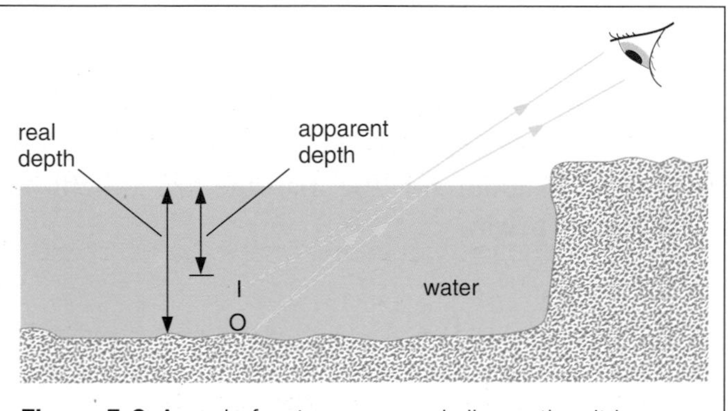

Figure 5.3 A pool of water appears shallower than it is

Figure 5.4 A straight stick seems bent in water. Why?

Refractive index

Light is refracted because its speed changes when it enters another medium. An analogy helps to explain why.

Suppose three people A, B, C are marching in line, with hands linked, on a good road surface. If they approach marshy ground at an angle, Figure 5.5a, A is slowed down first, followed by B and then C. This causes the whole line to swing round and change its direction of motion.

In air (and a vacuum) light travels at 300 000 km/s (3×10^8 m/s), in glass its speed falls to 200 000 km/s (2×10^8 m/s), Figure 5.5b. The **refractive index** n of the medium, i.e. glass, is defined by the equation

$$\text{refractive index } n = \frac{\text{speed of light in air (or a vacuum)}}{\text{speed of light in medium}}$$

$$\therefore \quad n = \frac{300\,000 \text{ km/s}}{200\,000 \text{ km/s}} = \frac{3}{2}$$

Experiments also show that

$$n = \frac{\text{sine of angle of incidence}}{\text{sine of angle of refraction}}$$

$$= \frac{\sin i}{\sin r} \text{ (see Figure 5.2a)}$$

The more light is slowed down when it enters a medium from air, the greater is the refractive index of the medium and the more it is bent.

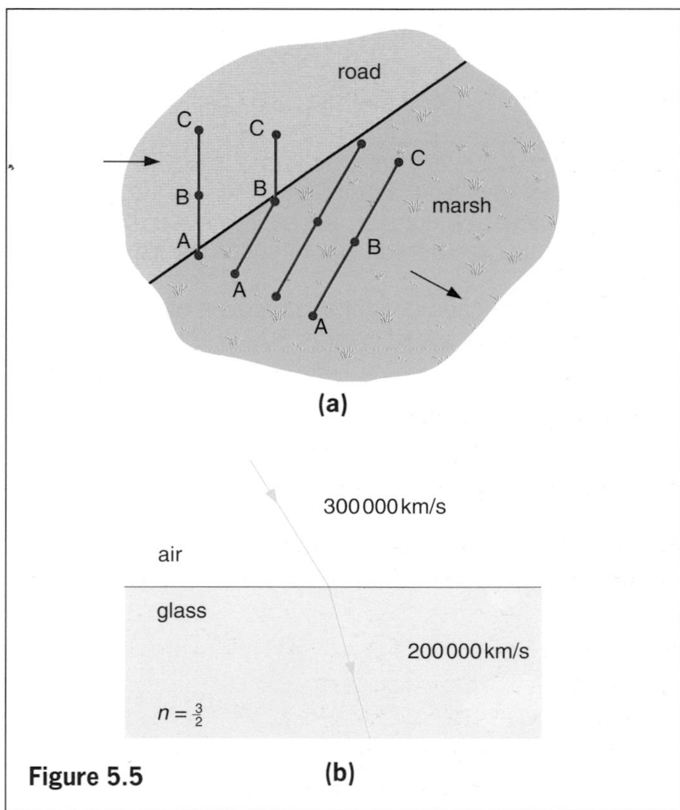

Figure 5.5

Refraction by a prism

In a triangular glass prism, Figure 5.6a, the deviation (bending) of a ray due to refraction at the first surface is added to the deviation at the second surface, Figure 5.6b. The deviations do not cancel out as in a parallel-sided block where the emergent ray, although displaced, is parallel to the incident ray.

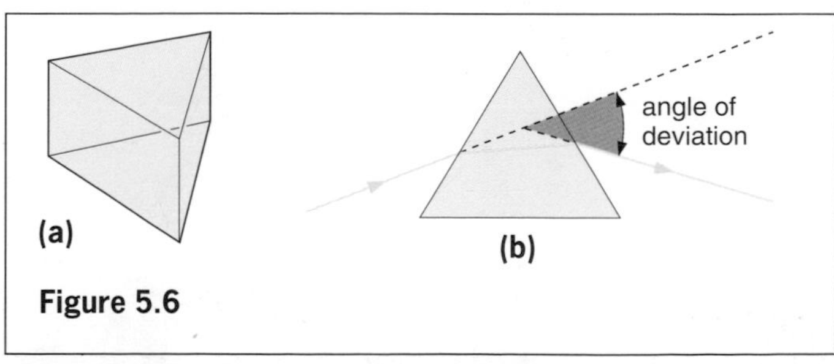

Figure 5.6

QUESTIONS

1 Figure 5.7 shows a ray of light entering a rectangular block of glass.

a Copy the diagram and draw the normal at the point of entry.

b Sketch the approximate path of the ray through the block and out of the other side.

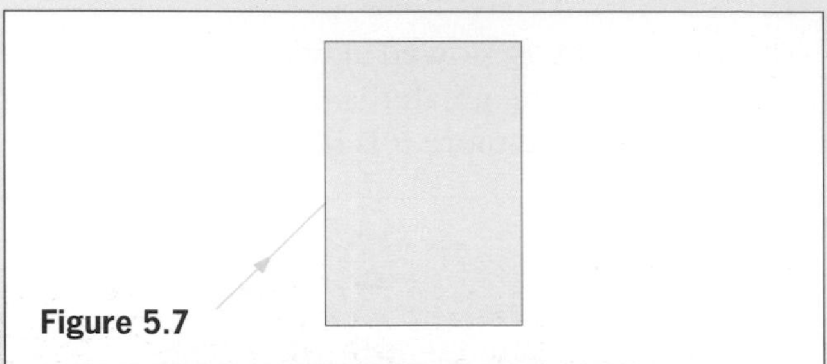

Figure 5.7

2 Draw two rays from a point on a fish in a stream to show where someone on the bank will see the fish. Where must the person aim to spear the fish?

3 What is the speed of light in a medium of refractive index 6/5 if its speed in air is 300 000 km/s?

4 Figure 5.8 shows a ray of light XY striking a glass prism and then passing through it. Which of the rays A to E is the correct representation of the emerging ray? *(J.M.B./A.L./N.W. 16+)*

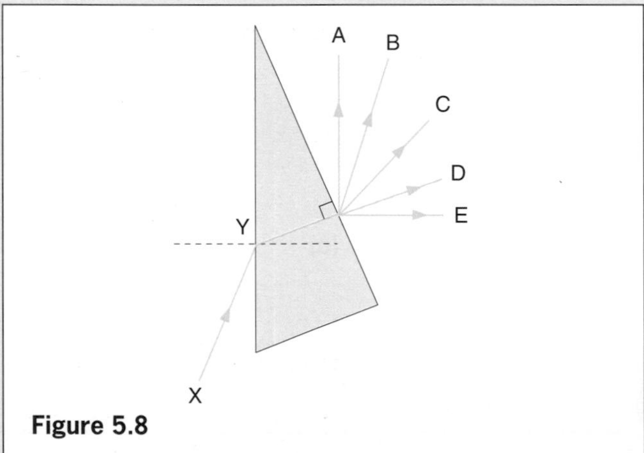

Figure 5.8

Checklist

After studying this chapter you should be able to

- state what the term **refraction** means,
- give examples of effects that show light can be refracted,
- describe an experiment to study refraction,
- draw diagrams of the passage of light rays through rectangular blocks and recall that lateral displacement occurs for a parallel-sided block,
- recall that light is refracted because it changes speed when it enters another medium,
- recall the definition of refractive index as n = speed in air/speed in medium,
- draw a diagram for the passage of a light ray through a prism.

6 TOTAL INTERNAL REFLECTION

Critical angle	***Light pipes and optical fibres***
Multiple images in a mirror	***Practical work*** Critical angle of glass.
Totally reflecting prisms	

Critical angle

When light passes at small angles of incidence from an optically denser to a less dense medium, e.g. from glass to air, there is a strong refracted ray and a weak ray reflected back into the denser medium, Figure 6.1a. Increasing the angle of incidence increases the angle of refraction.

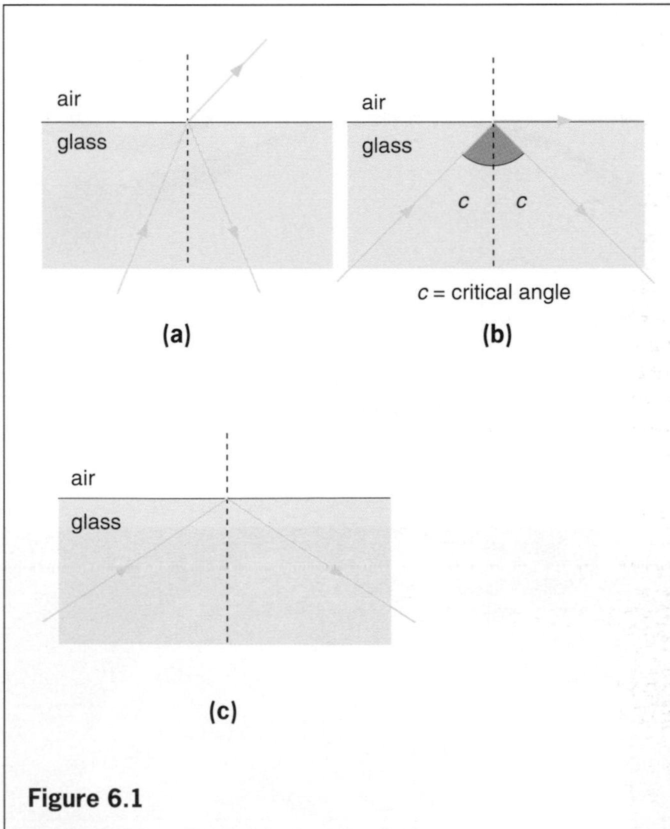

Figure 6.1

At a certain angle of incidence, called the **critical angle** *c*, the angle of refraction is 90°, Figure 6.1b. For angles of incidence greater than *c*, the refracted ray disappears and *all* the incident light is reflected inside the denser medium, Figure 6.1c. The light does not cross the boundary and is said to suffer **total internal reflection**.

PRACTICAL WORK

Critical angle of glass

Place a semicircular glass block on a sheet of paper, Figure 6.2, and draw the outline LOMN where O is the centre and ON the normal at O to LOM. Direct a narrow ray (at an angle of about 30°) *along a radius towards* O. The ray is not refracted at the curved surface. Why? Note the refracted ray in the air beyond LOM and also the weak internally reflected ray in the glass.

Slowly rotate the paper so that the angle of incidence increases until total internal reflection *just* occurs. Mark the incident ray. Measure the angle of incidence; it equals the critical angle.

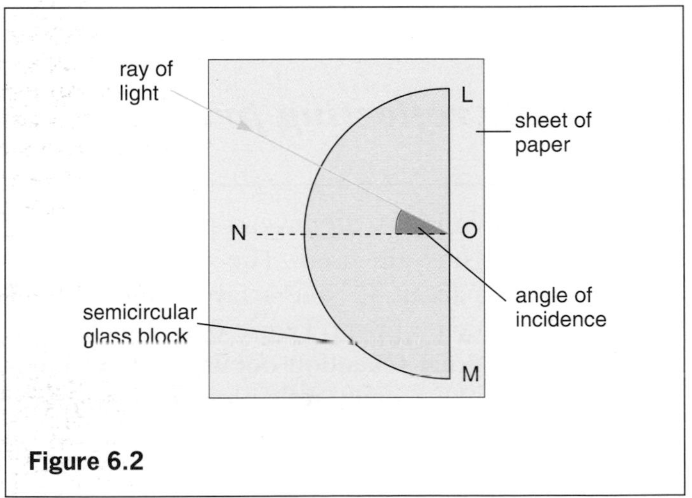

Figure 6.2

Multiple images in a mirror

An ordinary mirror silvered at the back forms several images of one object, due to multiple reflection inside the glass, Figures 6.3a, b (overleaf). These blur the main image I (which is formed by one reflection at the silvering), especially if the glass is thick. The trouble is absent in front-silvered mirrors but they are easily damaged.

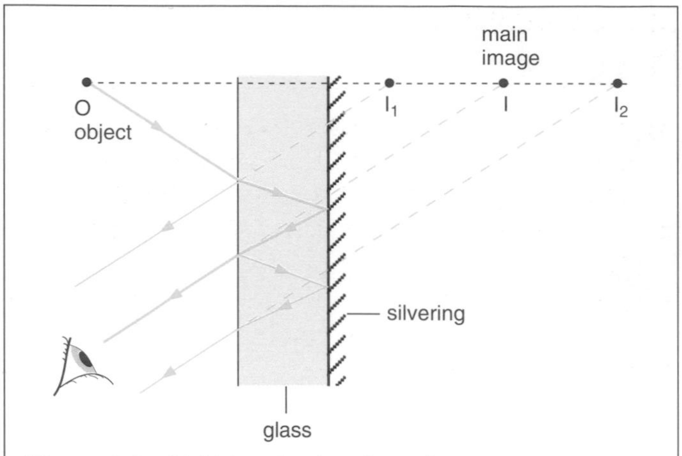

Figure 6.3a Multiple reflections in a mirror

Figure 6.3b The multiple images cause blurring

Totally reflecting prisms

The defects of mirrors are overcome if 45° right-angled glass prisms are used. The critical angle of ordinary glass is about 42° and a ray falling normally on face PQ of such a prism, Figure 6.4a, hits face PR at 45°. Total internal reflection occurs and the ray is turned through 90°. Totally reflecting prisms replace mirrors in good periscopes.

Light can also be reflected through 180° by a prism, Figure 6.4b; this happens in binoculars.

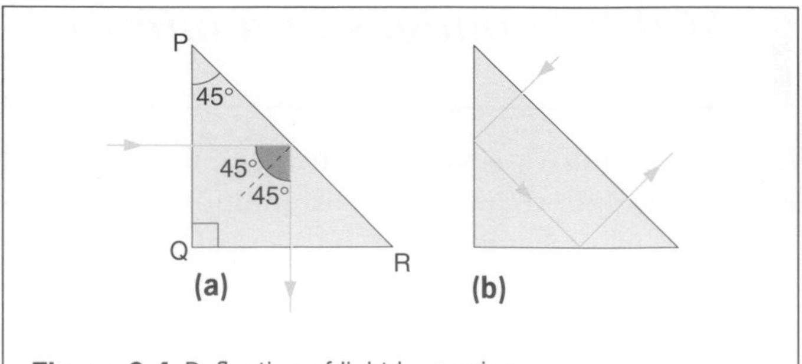

Figure 6.4 Reflection of light by a prism

Light pipes and optical fibres

Light can be trapped by total internal reflection inside a bent glass rod and 'piped' along a curved path, Figure 6.5. A single, very thin glass fibre behaves in the same way. If several thousand such fibres are taped together a flexible light pipe is obtained that can be used, for example, by doctors as an 'endoscope', Figure 6.6a, to obtain an image of an internal organ in the body, Figure 6.6b, or by engineers to light up some awkward spot for inspection. The latest telephone 'cables' are optical (very pure glass) fibres carrying information as pulses of laser light.

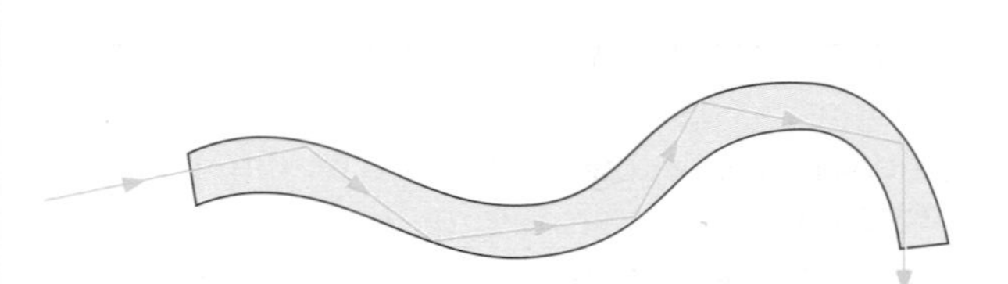

Figure 6.5 Light travels through a curved glass rod or fibre by total internal reflection

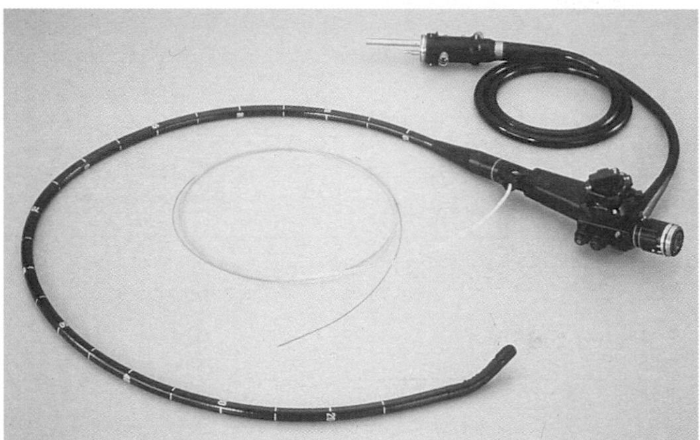

Figure 6.6a Endoscope

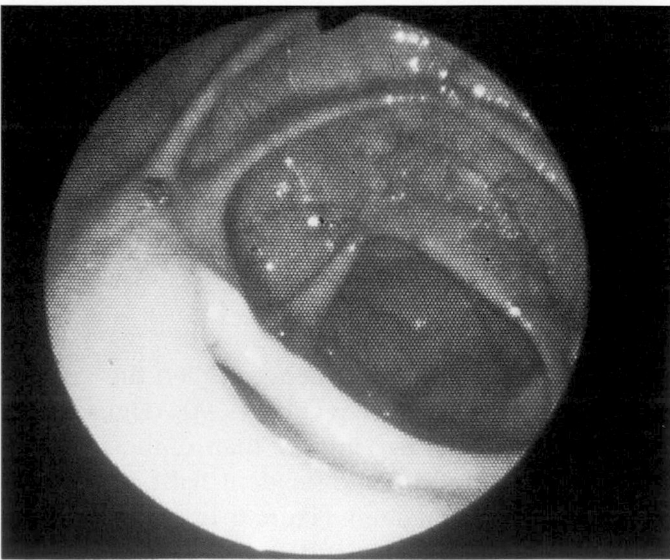

Figure 6.6b Intestine viewed by an endoscope

QUESTIONS

1 Figure 6.7 shows rays of light in a semicircular glass block.

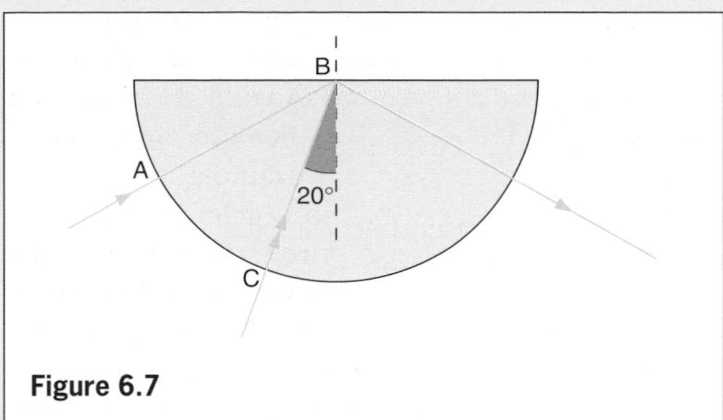

Figure 6.7

a Explain why the ray entering the glass at A is not bent.

b Explain why the ray AB is reflected at B and not refracted.

c Ray CB does not stop at B. Copy the diagram and draw its approximate path after it leaves B.

2 Copy Figures 6.8a and b and complete the paths of the rays.

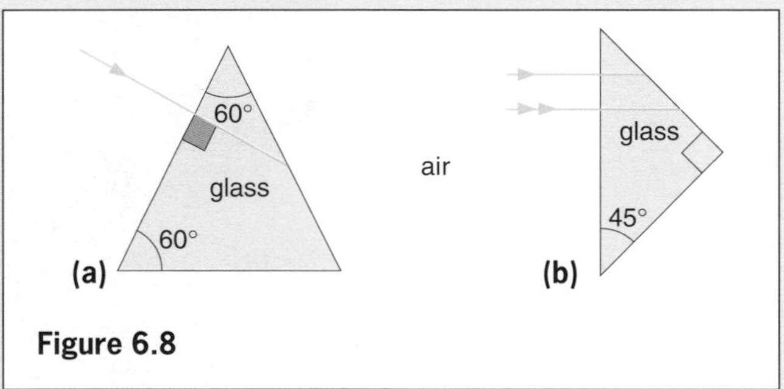

Figure 6.8

Checklist

After studying this chapter you should be able to

- explain with the aid of diagrams what is meant by **critical angle** and **total internal reflection**,
- describe an experiment to find the critical angle of glass or Perspex,
- draw diagrams to show the action of totally reflecting prisms in periscopes and binoculars,
- explain the action and state some uses of optical fibres.

7 LENSES

Convex and concave lenses	***Magnification***
Principal focus	***Practical work*** *f* of a convex lens by distant object method. Images formed by a convex lens.
Ray diagrams	
The lens formula	

Convex and concave lenses

Lenses are used in many optical instruments (Chapters 10 and 11); they often have spherical surfaces and there are two types. A **convex** lens is thickest in the centre and is also called a **converging** lens because it bends light inwards, Figure 7.1a. You may have used one as a magnifying glass, Figure 7.2a, or as a burning glass. A **concave** or **diverging** lens is thinnest in the centre and spreads light out, Figure 7.1b; it always gives a diminished image, Figure 7.2b.

The centre of a lens is its **optical centre** C; the line through C at right angles to the lens is the **principal axis**.

The action of a lens can be understood by treating it as a number of prisms (most cut-off), each of which bends the ray towards its base, as in Figure 7.1c, d. The centre acts as a parallel-sided block.

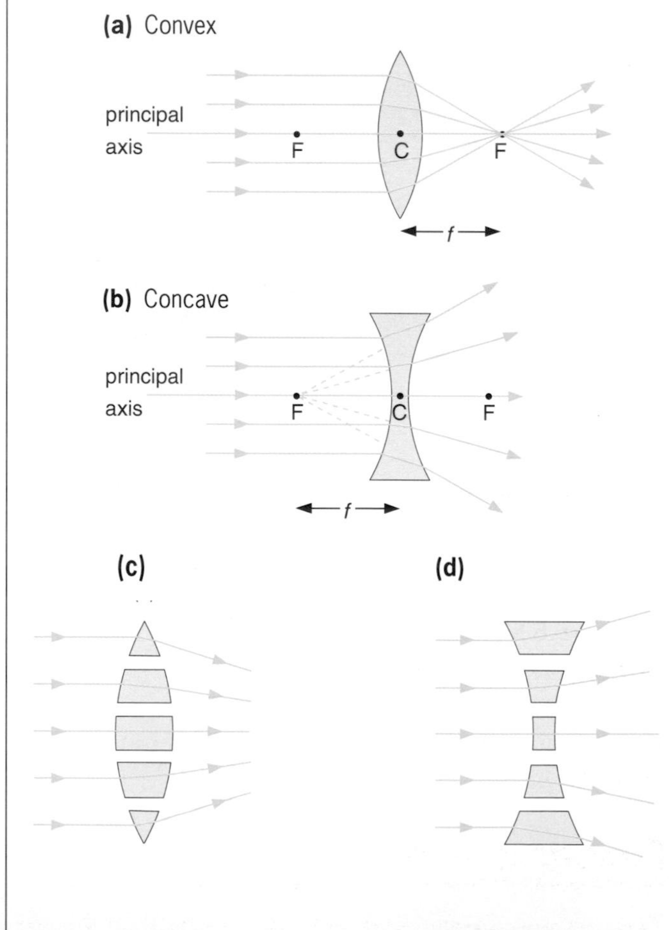

Figure 7.1 Action of lenses on a parallel beam of light

Principal focus

When a beam of light parallel to the principal axis passes through a convex lens it is refracted so as to converge to a point on the axis called the **principal focus** F. It is a real focus. A concave lens has a virtual principal focus behind the lens, from which the refracted beam seems to diverge.

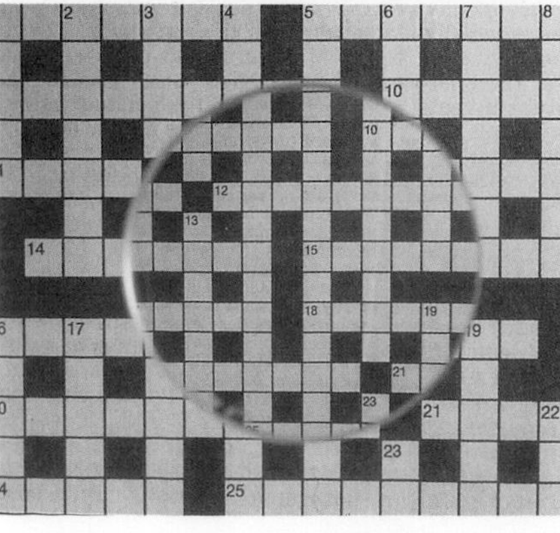

Figure 7.2a A convex lens (left) forms a magnified image of a close object

Figure 7.2b A concave lens (right) always forms a diminished image

Since light can fall on both faces of a lens it has two principal foci, one on each side, equidistant from C. The distance CF is the **focal length** *f* of the lens; it is an important property of a lens. The more curved the lens faces are, the smaller is *f* and the more powerful is the lens

Practical work

f of a convex lens by distant object method

We use the fact that rays from a **point** on a very distant object, i.e. at infinity, are nearly parallel, Figure 7.3a.

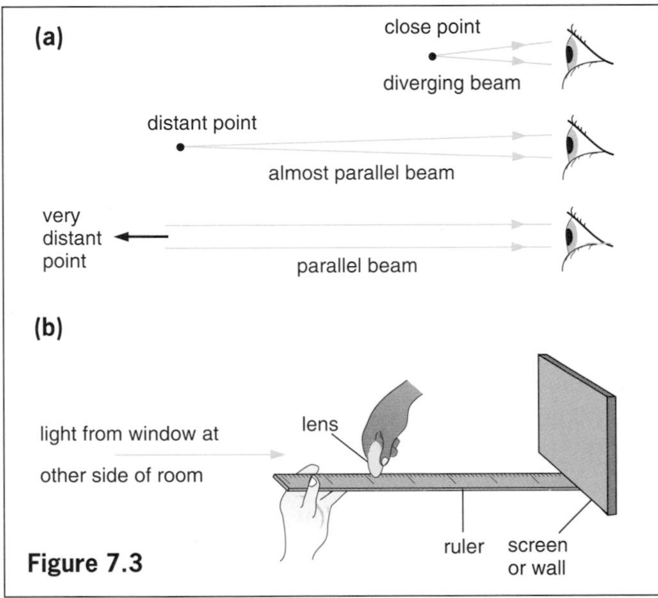

Figure 7.3

Move the lens, arranged as in Figure 7.3b, until a **sharp** image of a window at the other side of the room is obtained on the screen. The distance between the lens and the screen is *f* roughly. Why?

Practical work

Images formed by a convex lens

In the formation of images by lenses two important points on the principal axis are F and 2F; 2F is at a distance of twice the focal length from C.

First find the focal length of the lens by the 'distant object method', then fix the lens upright with Plasticine at the centre of a metre rule. Place small pieces of Plasticine at the points F and 2F on both sides of the lens, as in Figure 7.4.

Place a small lighted candle (or other light source) as the object on the rule beyond 2F and move a white card, on the other side of the lens from the candle, until a sharp image of the candle flame is obtained on the card.

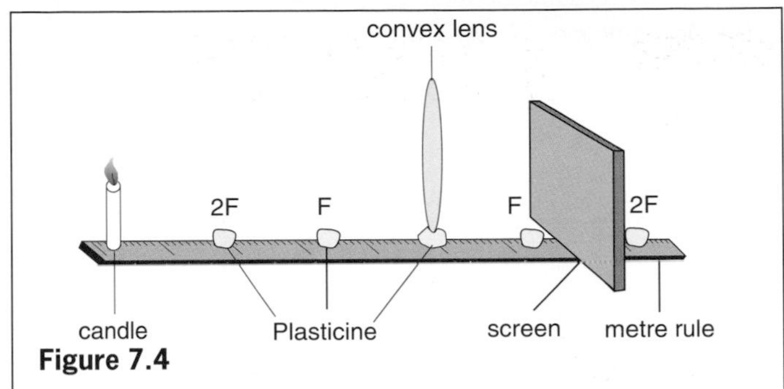

Figure 7.4

Note and record, in a table like the one below, the image position as 'beyond 2F', 'between 2F and F' or 'between F and lens'. Also note whether the image is 'larger' or 'smaller' than the actual flame or 'same size' and if it is 'upright' or 'inverted'. Now repeat with the candle at 2F, then between 2F and F.

Object position	Image position	Larger, smaller or same size?	Upright or inverted?
Beyond 2F			
At 2F			
Between 2F and F			
Between F and lens			

So far all the images have been real since they can be obtained on a screen. When the candle is between F and the lens, the image is **virtual** and is seen by *looking through the lens* at the candle. Do this. Is the virtual image larger or smaller than the object? Is it upright or inverted? Record your findings in the table.

Ray diagrams

Information about the images formed by a lens can be obtained by drawing *two* of the following rays.

1. A ray parallel to the principal axis which is refracted through the principal focus F.
2. A ray through the optical centre C which is undeviated for a thin lens.
3. A ray through the principal focus F which is refracted parallel to the principal axis.

In diagrams a thin lens is represented by a straight line at which all the refraction is considered to occur.

In each ray diagram in Figure 7.5 (overleaf) two rays are drawn from the top A of an object OA and where they intersect after refraction gives the top B of the image IB. The foot I of each image is on the axis since ray OC passes through the lens undeviated. In (d) the broken rays, and the image, are virtual.

(a) Object beyond 2F

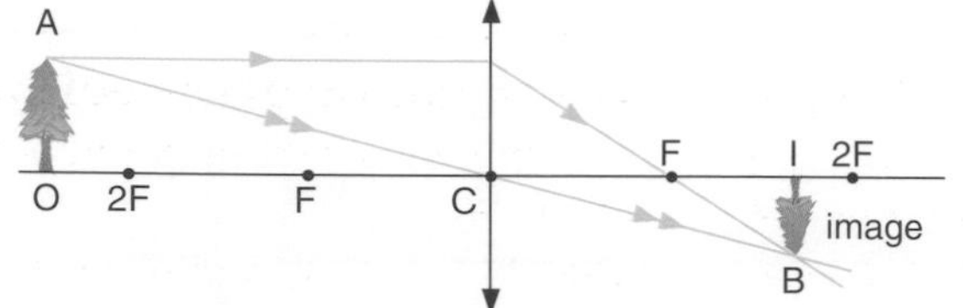

Image is between F and 2F, real, inverted, smaller

(b) Object between 2F and F

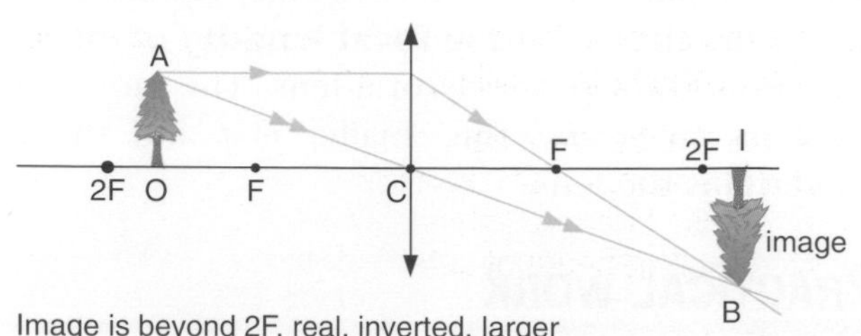

Image is beyond 2F, real, inverted, larger

(c) Object at 2F

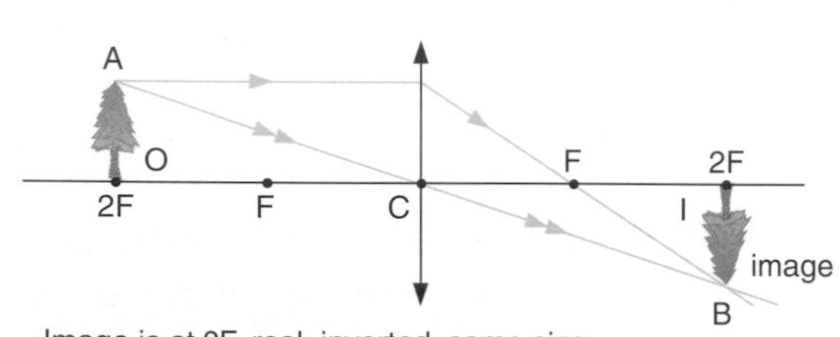

Image is at 2F, real, inverted, same size

(d) Object between F and C

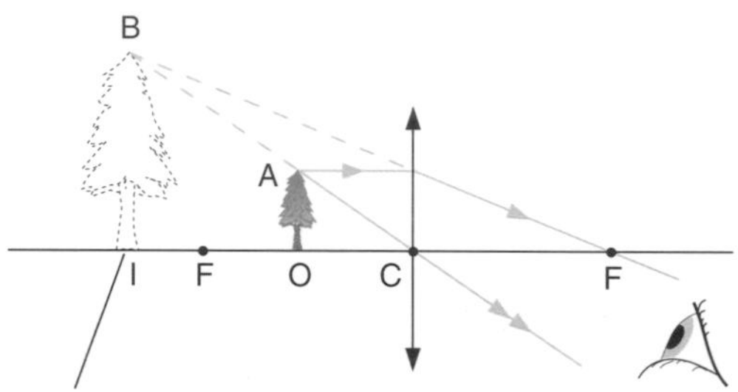

Image is behind object, virtual, erect, larger

Figure 7.5 Ray diagrams for a convex lens

The lens formula

The position and nature of the image formed by a lens can be obtained by calculation. It can be shown that if

u = distance of object from lens
v = distance of image from lens
and f = focal length of lens

then

$$\frac{1}{f} = \frac{1}{u} + \frac{1}{v}$$

For this equation to be true for both convex and concave lenses, real and virtual images, we must use the following sign rule.

Distances to real images are positive.
Distances to virtual images are negative.

The focal length of a convex lens is therefore given a + sign, and a concave lens a − sign.

Worked examples

1 Calculate the position and nature of the image formed by a convex lens of focal length 15 cm, of an object placed 20 cm from the lens.

We have

$$f = +15 \text{ cm} \quad \text{and} \quad u = +20 \text{ cm}$$

Rearranging the lens equation and substituting in it,

$$\frac{1}{v} = \frac{1}{f} - \frac{1}{u} = \frac{1}{(+15)} - \frac{1}{(+20)} = +\frac{4}{60} - \frac{3}{60} = +\frac{1}{60}$$

$$\therefore \quad v = +60 \text{ cm}$$

The image is real, since v is positive, and is 60 cm from the lens.

2 An object is placed 30 cm in front of a concave lens of focal length 20 cm. Where is the image formed and is it real or virtual?

We have

$$u = +30 \text{ cm} \quad \text{and} \quad f = -20 \text{ cm}$$

As before,

$$\frac{1}{v} = \frac{1}{f} - \frac{1}{u} = \frac{1}{(-20)} - \frac{1}{(+30)} = -\frac{3}{60} - \frac{2}{60} = -\frac{5}{60}$$

$$\therefore \quad v = -\frac{60}{5} = -12 \text{ cm}$$

The image is virtual, since v is negative, and is 12 cm from the lens.

Magnification

The **linear magnification** m is given by

$$m = \frac{\text{height of image}}{\text{height of object}}$$

It can be shown that in all cases

$$m = \frac{\text{distance of image from lens}}{\text{distance of object from lens}} = \frac{v}{u}$$

QUESTIONS

1 A small electric lamp when placed at the focal point of a converging lens will produce a

A parallel beam of light
B converging beam of light
C diffuse beam of light
D diverging beam of light. (*W.M.*)

2 **a** What kind of lens is shown in Figure 7.6?

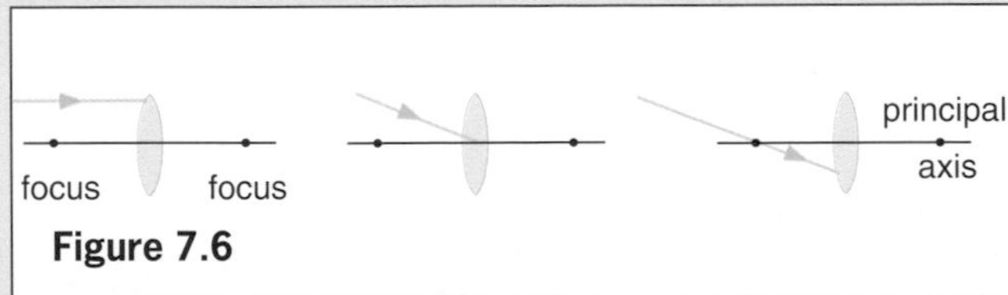

Figure 7.6

b Copy the diagrams and complete them to show the path of the light after passing through the lens.

c Figure 7.7 shows an object AB 6 cm high placed 18 cm in front of a lens of focal length 6 cm. Draw the diagram to scale and by tracing the paths of rays from A find the position and size of the image formed.

d Check your answer to **c** by calculation.

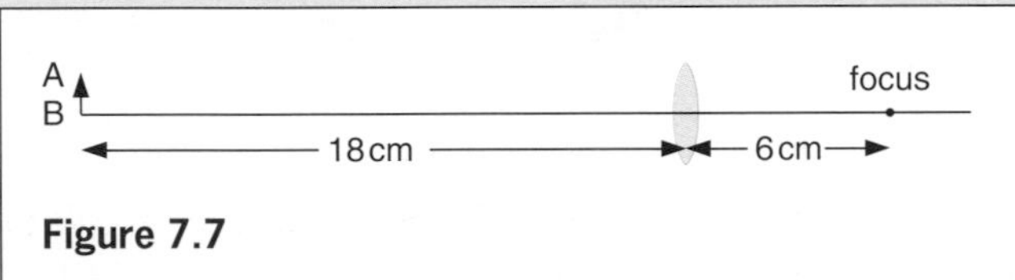

Figure 7.7

3 Where must the object be placed for the image formed by a convex lens to be

a real, inverted and smaller than the object,

b real, inverted and same size as the object,

c real, inverted and larger than the object,

d virtual, upright and larger than the object?

Checklist

After studying this chapter you should be able to

- explain the action of a lens in terms of refraction by a number of small prisms,
- draw diagrams showing the effects of convex and concave lenses on a beam of parallel rays,
- recall the meaning of **optical centre**, **principal axis**, **principal focus** and **focal length**,
- describe an experiment to measure the focal length of a convex lens,
- draw ray diagrams to show image formation by a convex lens,
- draw scale diagrams to solve problems on convex lenses,
- use the lens formula to solve problems on lenses,
- recall the meaning of the term **linear magnification**.

8 THE EYE

Structure of the eye
Defects of vision
Persistence of vision

Practical work
The eyes: blind spot, binocular vision, inverted image on retina.

1 *Structure of the eye*

An image is formed on the **retina** of the eye, Figure 8.1, by successive refraction at the **cornea**, the **aqueous humour**, the **lens** and the **vitreous humour**. Electrical signals then travel along the **optic nerve** to the brain to be interpreted. In good light, the **yellow spot (fovea)** is most sensitive to detail and the image is automatically formed there.

Objects at different distances are focused by the **ciliary muscles** changing the shape and so the focal length of the lens – a process called **accommodation**. The lens fattens to view near objects. The **iris** has a central hole, the **pupil**, whose size it decreases in bright light and increases in dim light. Excessive brightness can damage the eye.

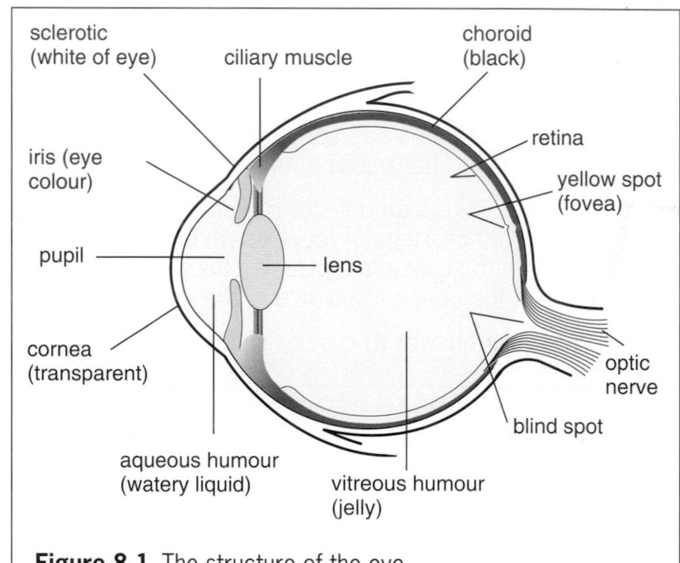

Figure 8.1 The structure of the eye

PRACTICAL WORK

The eyes

(a) Blind spot

This is the small area of the retina where the optic nerve leaves the eye. It has no light-sensitive nerve endings and in each eye it is closer to the nose than the yellow spot.

To show its existence, close your left eye and look at the cross in Figure 8.2. You will also see the black dot.

Slowly bring the book towards you. At a certain distance the dot disappears; its image has fallen on the blind spot of your right eye.

Figure 8.2

(b) Binocular vision

Your two eyes see an object from slightly different angles, giving two slightly different images which the brain combines to give a three-dimensional impression. This also helps you to judge distances.

Roll a sheet of paper into a tube and hold it to your right eye with your right hand, Figure 8.3. Close your left eye and place your left hand halfway along the tube. Open your left eye. Is there a 'hole' in your hand to 'see' through? Explain.

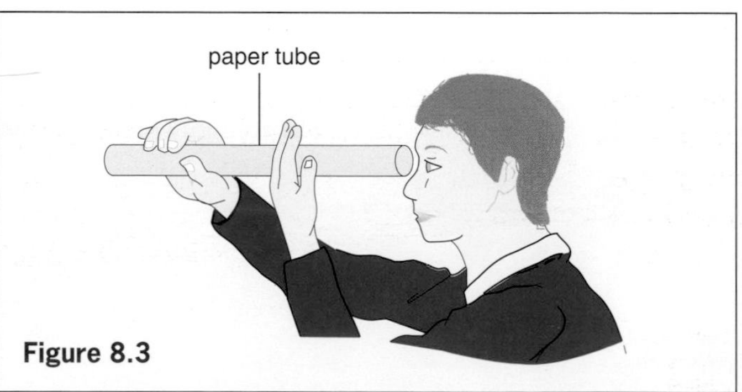

Figure 8.3

(c) Image inverted on retina, Figure 8.4a

To show this make a pinhole in a piece of paper and hold it about 10 cm away. Close one eye and look at the hole against the sky or something bright. Hold a pin, head up, very close to your eye, Figure 8.4b.

Keep looking at the hole and move the pin about slowly until you 'see' it inverted in the hole. Make several pinholes round the first and look again. What do you see?

The pin is too close to the eye for a real image (except a large blur) to be formed on the retina, but a sharp **upright shadow** is produced by light from the pinhole and this falls directly (i.e. still upright) on the retina. But as you 'see' it upside down you know that the brain must turn it upside down. If the brain does that with the shadow it will do the same with any image falling on the retina. So, as you normally see an upright object as upright, the image must be upside down on the retina.

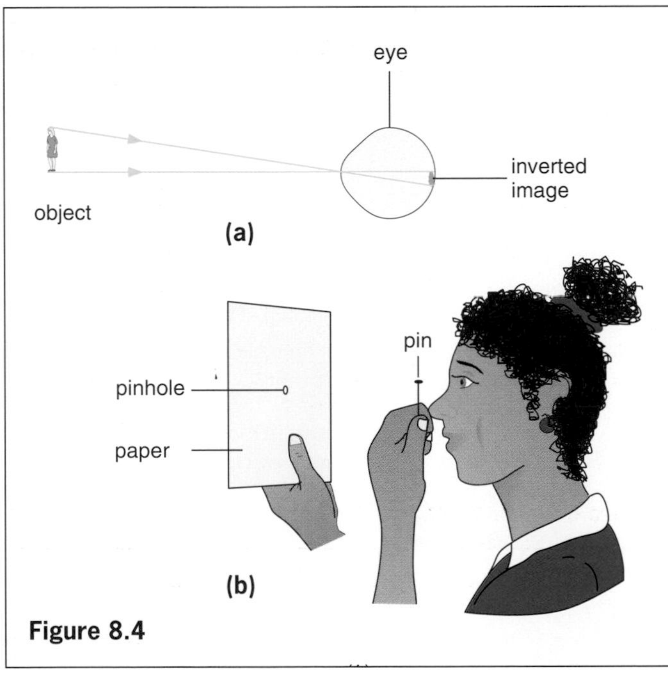

Figure 8.4

Defects of vision

The average adult eye can focus objects comfortably from about 25 cm (the **near point**) to infinity (the **far point**). Your near point may be less than 25 cm; it gets farther away with age.

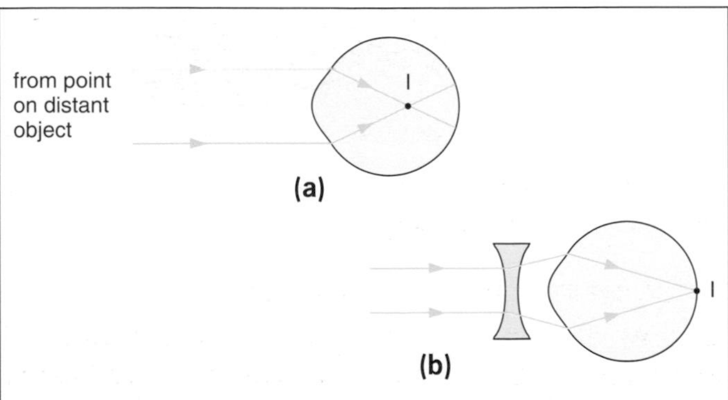

Figure 8.5 Short sight and its correction by a concave lens

(a) Short sight

A short-sighted person sees near objects clearly but the far point is closer than infinity. The image of a distant object is formed in front of the retina because the eyeball is too long or because the eye lens cannot be made thin enough, Figure 8.5a. The defect is corrected by a concave spectacle lens (or contact lens) which diverges the light before it enters the eye, to give an image on the retina, Figure 8.5b.

(b) Long sight

A long-sighted person sees distant objects clearly but the near point is beyond 25 cm. The image of a near object is focused behind the retina because the eyeball is too short or because the eye lens cannot be made thick enough, Figure 8.6a. A convex spectacle lens (or contact lens) corrects the defect, Figure 8.6b.

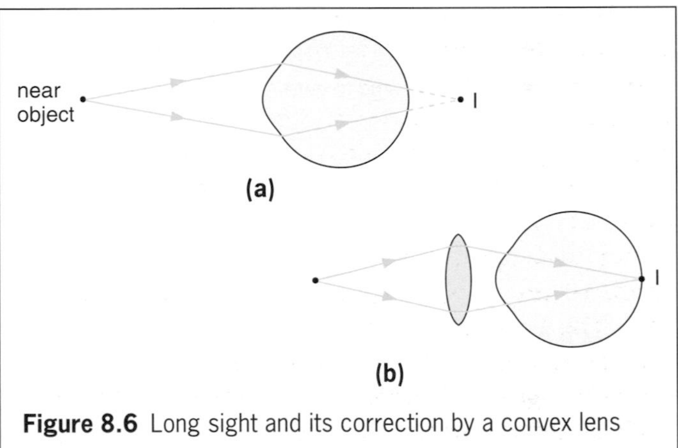

Figure 8.6 Long sight and its correction by a convex lens

(c) Astigmatism

A person having this defect will see one set of lines in Figure 8.7 more sharply than the others. It is caused by the cornea being more curved in some directions than others. Correction may be achieved with a non-spherical spectacle lens whose curvature has the effect of increasing or decreasing that of the cornea in the required directions.

Figure 8.7

Persistence of vision

An image lasts on the retina for about one-tenth of a second after the object has disappeared, as can be shown by spinning a card like that in Figure 8.8. The effect makes possible the production of motion pictures. In a TV receiver 25 complete pictures are produced every second.

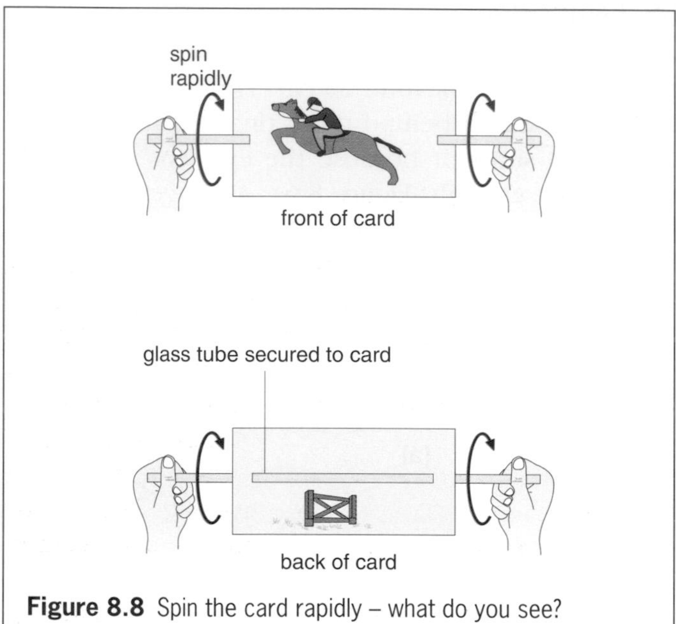

Figure 8.8 Spin the card rapidly – what do you see?

QUESTIONS

1 Name the part of the eye

a which controls how much light enters it,

b on which the image is formed,

c which changes the focal length of the lens.

2 Refraction of light in the eye occurs at

A the lens only **B** the iris **C** the cornea only
D the pupil **E** both the cornea and the lens.

3 A short-sighted person has a near point of 15 cm and a far point of 40 cm.

a Can he see clearly an object at a distance of (i) 5 cm, (ii) 25 cm, (iii) 50 cm?

b To see clearly an object at infinity what kind of spectacle lens does he need?

4 The near point of a long-sighted person is 50 cm from the eye.

a Can she see clearly an object at (i) a distance of 20 cm, (ii) infinity?

b To read a book held at a distance of 25 cm will she need a convex or a concave spectacle lens?

Checklist

After studying this chapter you should be able to

- draw and label the main parts of the eye,
- explain how refraction at the cornea and lens produces an image on the retina,
- explain how the lens allows accommodation to occur,
- recall the meaning of **short sight** and describe how it is corrected,
- recall the meaning of **long sight** and describe how it is corrected,
- recall the meaning of **astigmatism** and describe how it is corrected,
- explain the terms **binocular vision** and **persistence of vision**.

9 COLOUR

Dispersion	***Mixing coloured lights***
Recombining the spectrum	***Practical work*** A pure spectrum.
Colour of an object	

Dispersion

Colourful clothes, colour television and the flashing coloured lights in a disco all help to make life brighter. It was Newton who, in 1666, set us on the road to understanding how colours may arise. He produced them by allowing sunlight (which is white) to fall on a triangular glass prism, Figure 9.1a. The band of colours obtained, Figure 9.1b, is a **spectrum** and the effect is called **dispersion**. He concluded that (i) white light is a mixture of many colours of light which the prism separates out because (ii) the refractive index of glass is different for each colour, being greatest for violet light.

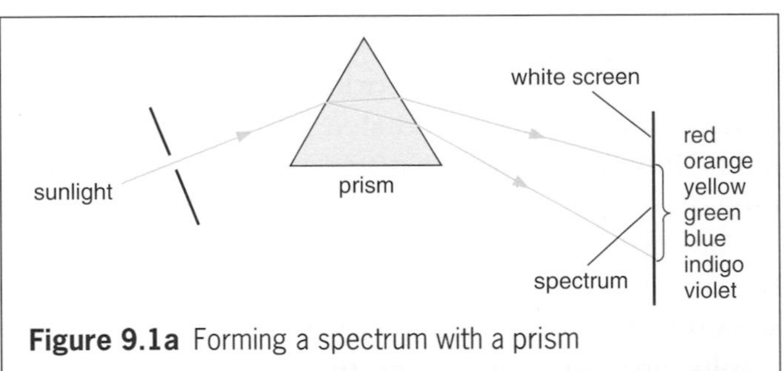

Figure 9.1a Forming a spectrum with a prism

Figure 9.1b White light shining through cut crystal can produce a spectrum

PRACTICAL WORK

A pure spectrum

A pure spectrum is one in which the colours do not overlap, as they do when a prism alone is used. It needs a lens to focus each colour as in Figure 9.2a.

Arrange a lens L, Figure 9.2b, so that it forms an image of the vertical filament of a lamp on a screen at S_1, 1 m away. The filament acts as a narrow source of white light. Insert a 60° prism P and move the screen to S_2, keeping it at the same distance from L, to receive the spectrum; rotate P until the spectrum is pure, Figure 9.2c.

Place different colour filters between P and S_2. What happens to the spectrum?

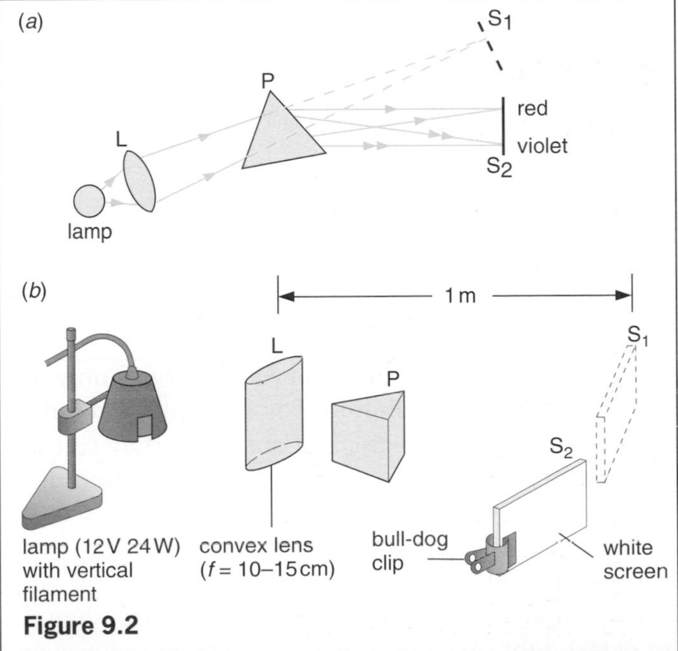

Figure 9.2

Figure 9.2c A pure spectrum

Recombining the spectrum

The colours of the spectrum can be recombined to form white light by

(i) arranging a second prism so that the light is deviated in the opposite direction, Figure 9.3a, or
(ii) using an electric motor to rotate at high speed a disc with the spectral colours painted on its sectors, Figure 9.3b. (The whiteness obtained is slightly grey because paints are not pure colours.)

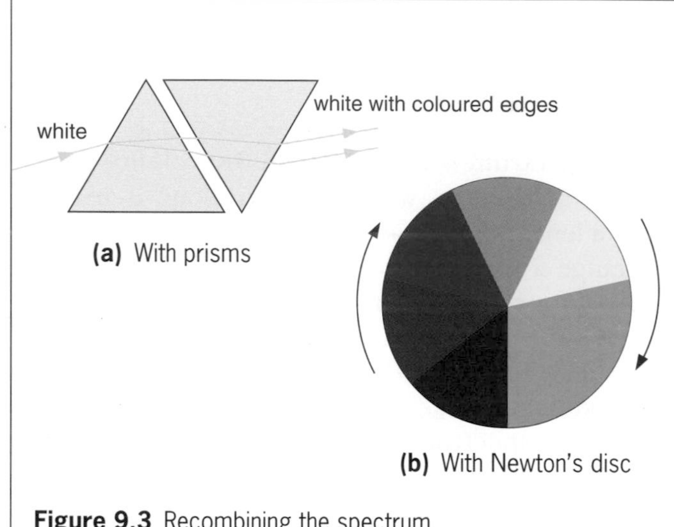

Figure 9.3 Recombining the spectrum

Colour of an object

The colour of an object depends on (i) the colour of the light falling on it and (ii) the colour(s) it transmits or reflects.

(a) Filters

A filter lets through light of certain colours only and is made of glass or celluloid. For example, a red filter transmits mostly red light and absorbs other colours; it therefore produces red light when white light shines through it.

(b) Opaque objects

These do not allow light to pass but are seen by reflected light. A white object reflects all colours and appears white in white light, red in red light, blue in blue light, etc. A blue object appears blue in white light because the red, orange, yellow, green and violet colours in white light are absorbed and only blue reflected. It also looks blue in blue light but in red light it appears black since no light is reflected and blackness indicates the absence of colour.

Figure 9.4 A blue object does not appear blue in red light. (If it were pure blue, it would appear black.) Why?

Mixing coloured lights

In science red, green and blue are **primary** colours (they are not the artist's primary colours) because none of them can be produced from other colours of light. However, they give other colours when suitably mixed.

The primary colours can be mixed by shining beams of red, green and blue light onto a white screen so that they partially overlap, Figure 9.5a. The results are summarized in the 'colour triangle' of Figure 9.5b.

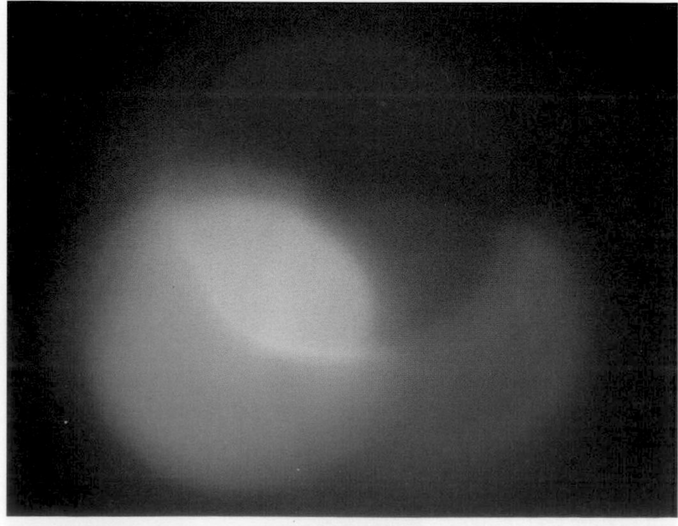

Figure 9.5a Mixing the three primary colours

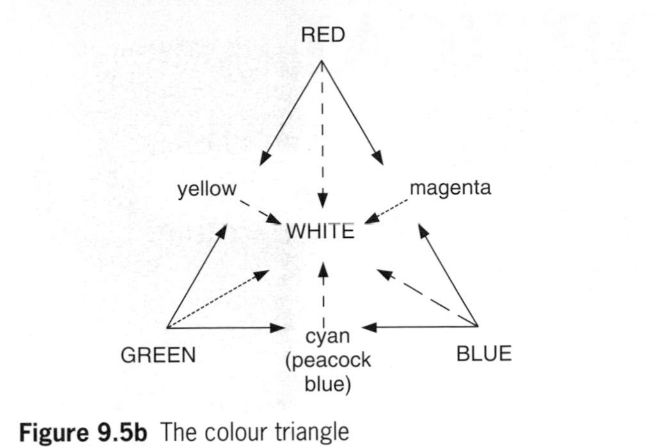

Figure 9.5b The colour triangle

The colours formed by adding two primaries are called **secondary** colours; they are yellow, cyan (turquoise or peacock blue) and magenta. The three primary colours give white light, as do the three secondaries. We would also expect a primary colour and the secondary opposite it in the colour triangle to give white. Why? Any *two* colours producing white light are **complementary** colours, e.g. blue and yellow.

QUESTIONS

1 Copy Figure 9.6, mark and label

a the reflected ray of light at the first surface,

b the path followed by the light through and out of the prism,

c what is seen on the screen AB. *(E.M.)*

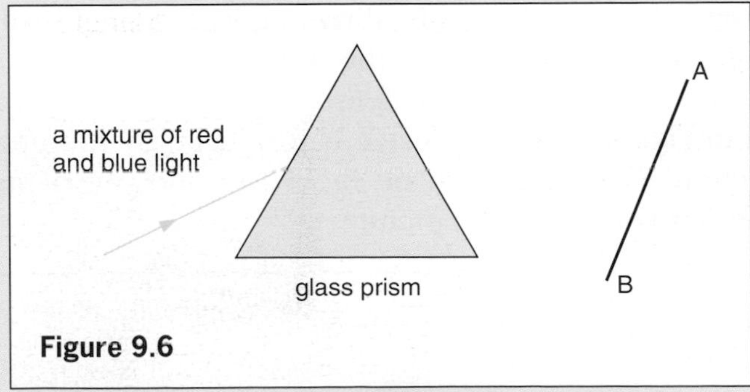

Figure 9.6

2 A sheet of white paper is viewed through a piece of blue glass and the paper looks blue. This is because

A the colour of the glass is reflected onto the paper
B blue light is absorbed by the glass
C blue light travels faster than red through glass
D the glass absorbs all colours except blue
E the glass adds blue light to the light coming from the paper. *(O.L.E./S.R. 16+)*

3 a What colour is formed when the following beams of equally bright light are shone onto a white screen and completely overlap: (i) yellow and blue, (ii) cyan (turquoise) and red?

b A book which looks red in white light is viewed in magenta light: what colour does it appear?

c White light is viewed through a piece of yellow filter and a piece of red filter held in contact. What colour light is seen? *(N.W.)*

Checklist

After studying this chapter you should be able to

- explain the terms **spectrum** and **dispersion**,
- describe how a prism is used to produce a spectrum from white light,
- describe how the colours of the spectrum can be recombined to give white light,
- recall the factors affecting the colour of an object,
- recall the results of mixing coloured lights.

10 SIMPLE OPTICAL INSTRUMENTS

Camera	*Magnifying glass*
Projector	

Camera

A camera is a light-tight box in which a convex lens forms a real image on a film, Figure 10.1. The image is smaller than the object and nearer to the lens. The film contains chemicals that change on exposure to light; it is **developed** to give a negative. From the negative a photograph is made by **printing**.

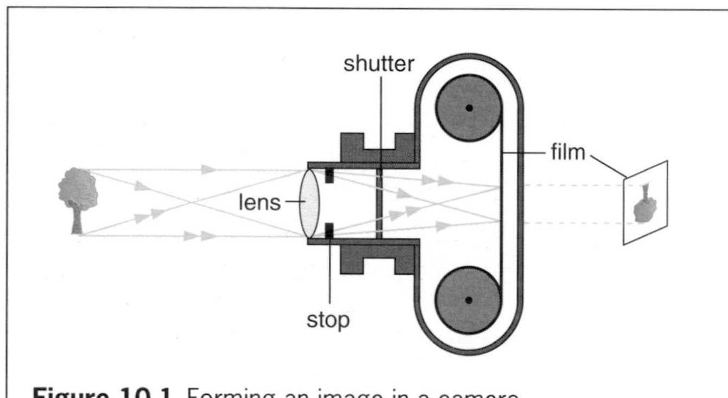

Figure 10.1 Forming an image in a camera

(a) Focusing

In simple cameras the lens is fixed and all distant objects, i.e. beyond about 2 metres, are in reasonable focus. Roughly how far from the film will the lens be if its focal length is 5 cm?

In other cameras exact focusing of an object at a certain distance is done by altering the lens position. For near objects it is moved away from the film, the correct setting being shown by a scale on the focusing ring.

(b) Shutter

When a photograph is taken, the shutter is opened for a certain time and exposes the film to light entering the camera. Sometimes exposure times can be varied and are given in fractions of a second, e.g. 1/1000, 1/60, etc. Fast-moving objects require short exposures.

(c) Stop

The brightness of the image on the film depends on the amount of light passing through the lens when the shutter is opened and is controlled by the size of the hole (aperture) in the stop. In some cameras this is fixed but in others it can be made larger for a dull scene and smaller for a bright one.

The aperture may be marked in ***f*-numbers**. The diameter of an aperture with *f*-number 8 is $\frac{1}{8}$ of the focal length of the lens and so the *larger* the *f*-number the *smaller* the aperture. The numbers are chosen so that on passing from one to the next higher, e.g. from 8 to 11, the area of the aperture is halved.

Projector

A projector forms a real image on a screen of a slide in a slide projector, or of a film in a cine-projector. The image is larger than the slide or frame of film and is further away from the lens. It is usually so highly magnified that very strong but even illumination of the slide or film is needed if the image is also to be bright. This is achieved by directing light from a small but powerful lamp on to the 'object' by means of a concave mirror and a condenser lens system arranged as in Figure 10.2. The image is produced by the projection lens which can be moved in and out of its mounting to focus the picture.

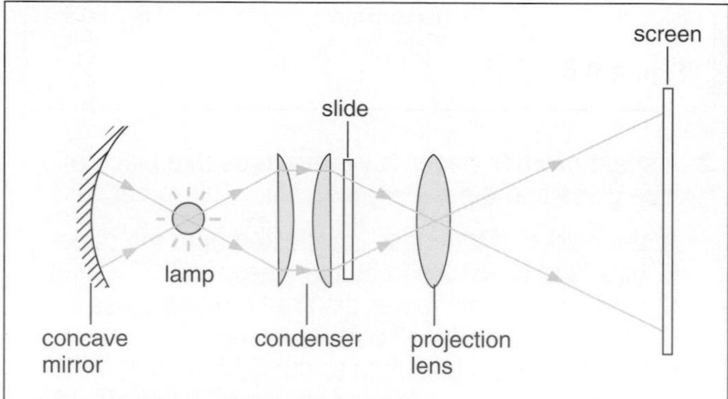

Figure 10.2 Illuminating the slide in a projector

In a projector like that in Figure 10.3 the slide must be inverted to give an upright image, and must be between 2F and F from the lens (see Figure 7.5b).

Figure 10.3 Modern slide projector

Magnifying glass

The apparent size of an object depends on its actual size and on its distance from the eye. The sleepers on a railway track are all the same length but those nearby seem longer. This is because they enclose a larger angle at your eye than more distant ones: as a result their image on the retina is larger so making them appear bigger.

A convex lens gives an enlarged, upright virtual image of an object placed inside its principal focus F, Figure 10.4a. It acts as a magnifying glass since the angle β made at the eye by the image, formed at the near point, is greater than the angle α made by the object when it is viewed directly at the near point without the magnifying glass, Figure 10.4b.

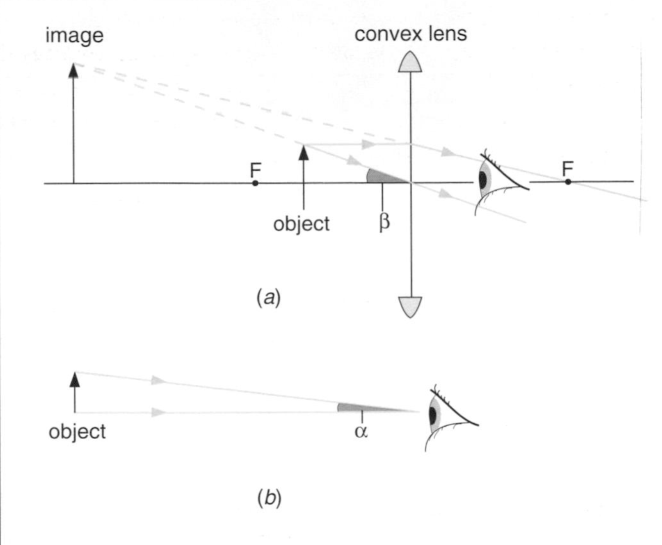

Figure 10.4 Magnification by a convex lens: $\beta > \alpha$

The fatter (more curved) a convex lens is, the shorter is its focal length and the more it magnifies. Too much curvature however distorts the image.

QUESTIONS

1 Figure 10.5a shows a camera focused on an object in the middle distance. Is the camera shown in Figure 10.5b focused on a distant or a close object? Give a reason. *(E.A.)*

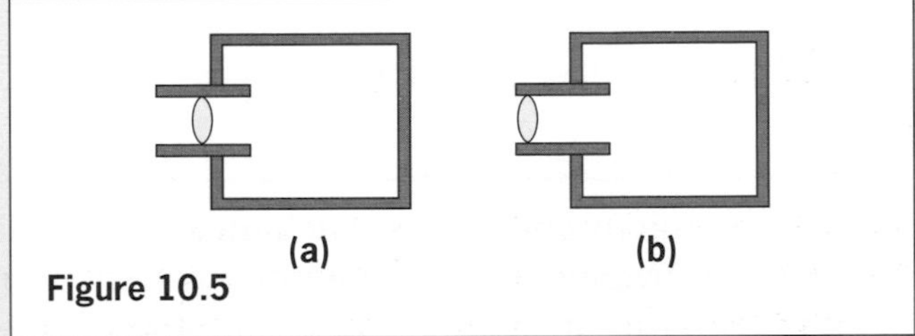

Figure 10.5

2 If a projector is moved farther away from the screen on which it was giving a sharp image of a slide, state

a *three* changes that occur in the image,

b how the projection lens must be adjusted to re-focus the image.

3 **a** Three converging lenses are available having focal lengths of 4 cm, 40 cm and 4 m respectively. Which one would you choose as a magnifying glass?

b An object 2 cm high is viewed through a converging lens of focal length 8 cm. The object is 4 cm from the lens. By means of a ray diagram find the position, nature and magnification of the image.

Checklist

After studying this chapter you should be able to describe with the aid of diagrams how a single lens is used

- in a simple camera,
- in a projector,
- in a magnifying glass.

11 MICROSCOPES AND TELESCOPES

Compound microscope	***Practical work***
Refracting astronomical telescope	Simple lens telescopes.
Reflecting astronomical telescope	

Compound microscope

A compound microscope gives much greater magnification than a magnifying glass and less distortion. In its simplest form it consists of two **short-focus** convex lenses arranged as in Figure 11.1. The one nearer the object, called the **objective**, forms a real, enlarged, inverted image I_1 of a small object O placed just outside its principal focus F_o. I_1 is just inside the principal focus F_e of the second lens, the **eyepiece**, which treats I_1 as an object and acts as a magnifying glass to give a further enlarged, virtual image I_2.

The yellow arrowed lines are actual rays from O, by which the eye sees the top of I_2. The thin black lines without arrows are construction lines drawn to find the position of I_2.

To reduce distortion in a microscope, the objective and eyepiece each consist of several lenses, Figure 11.2. The object is seen inverted.

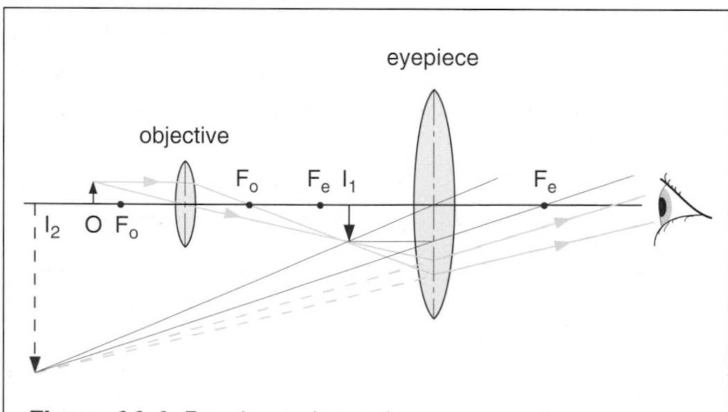

Figure 11.1 Forming an image in a compound microscope

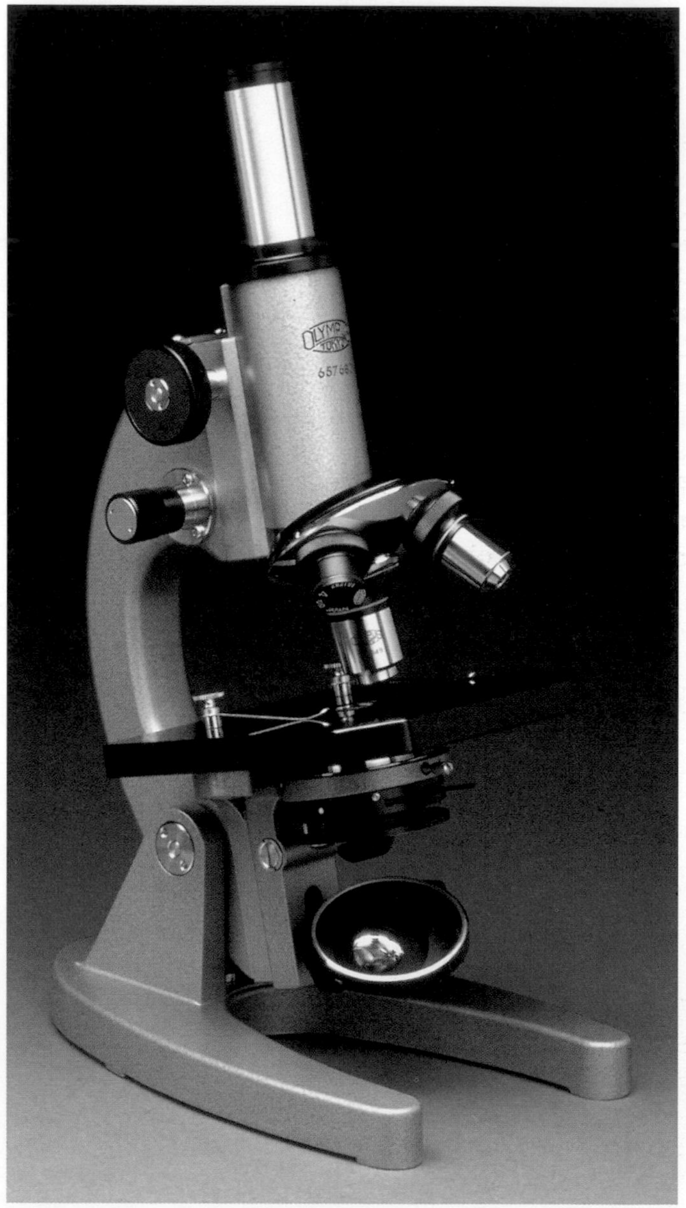

Figure 11.2 Compound microscope

Refracting astronomical telescope

This consists of two convex lenses, a **long-focus** objective and a **short-focus** eyepiece. Rays from a **point** on a distant object, e.g. a star, are nearly parallel on reaching the telescope (see Chapter 7). The objective forms a real, inverted, diminished image I_1 of the object at its principal focus F_o, Figure 11.3. The eyepiece acts as a magnifying glass, treating I_1 as an object and forming a magnified, virtual image.

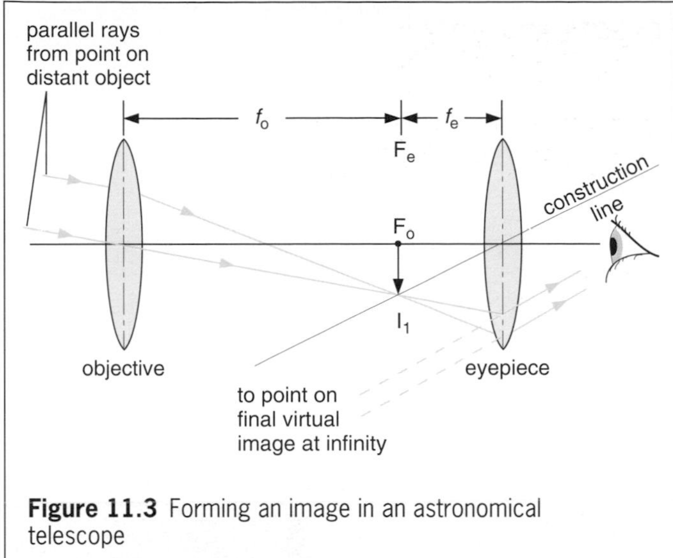

Figure 11.3 Forming an image in an astronomical telescope

Normally the eyepiece is adjusted to give this final image at infinity, i.e. a long way off, and so I_1 will be at the principal focus F_e of the eyepiece. That is, F_o and F_e coincide. The separation of the lenses is $f_o + f_e$. The object is seen inverted.

A telescope magnifies because the final image it forms subtends a much greater angle at the eye than does the distant object viewed without the telescope. Figure 11.3 shows that the longer the focal length of the objective, the larger is I_1, and so for greatest magnification the objective should have a long focal length and the eyepiece a short one (like any magnifying glass).

PRACTICAL WORK

Simple lens telescopes

Arrange a carbon filament lamp at the far end of the room. Mount a long-focus convex lens (e.g. 50 cm) at eye-level, Figure 11.4a. Point it at the lamp and find its image on a piece of greaseproof paper.

(a) Astronomical

Insert a short-focus convex lens (e.g. 5 cm) as the eyepiece and adjust it so that you can see the image on the greaseproof paper clearly, Figure 11.4b. Remove the paper and view the image of the lamp. Look at a book beside the lamp and other distant objects.

(b) Galilean

In an astronomical telescope the final image is inverted. Replace the convex eyepiece with a **short-focus concave** lens and adjust its distance from the objective till you see distant objects clearly. Are their images inverted or upright? This lens combination is used in Galilean telescopes and in opera glasses.

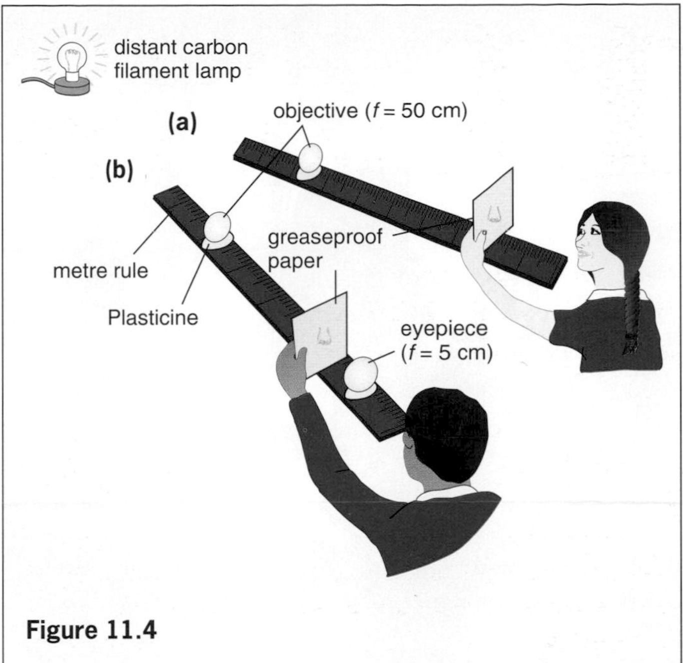

Figure 11.4

Reflecting astronomical telescope

The objective of a telescope must have a large diameter (aperture) as well as a large focal length. This (i) gives it good light-gathering power so that faint objects can be seen and (ii) enables it to reveal detail.

The largest **lens** telescope has an objective of diameter 1 metre; anything more would sag under its own weight. The biggest astronomical telescopes use long-focus **concave mirrors** as objectives; they can be supported at the back. Figure 11.5 shows how parallel rays from a distant point object are reflected at the objective and then intercepted by a small plane mirror before they form a real image I_1. This image is magnified by the eyepiece.

The Mount Palomar telescope in California, Figure 11.6 (overleaf), has a mirror of diameter 5 metres.

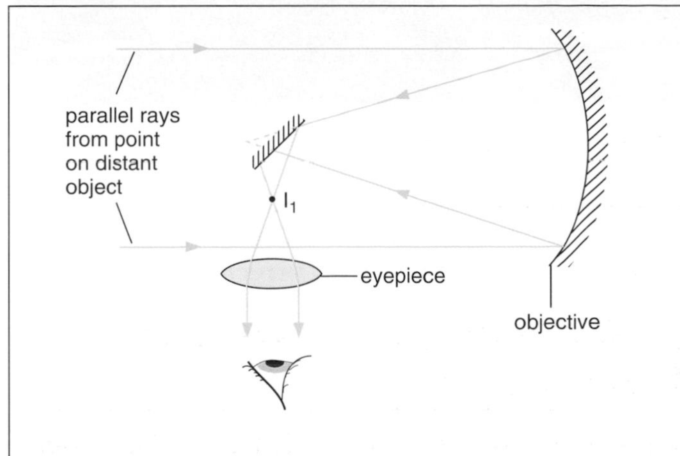

Figure 11.5 Forming an image in a reflecting telescope

Figure 11.6 The 200 inch (5 metre) reflecting telescope on Mount Palomar, California

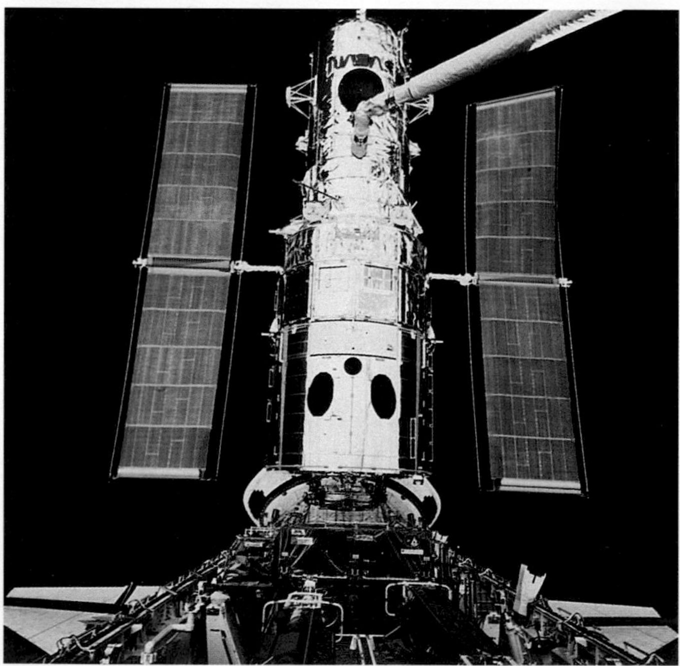

Figure 11.7 The Hubble Space Telescope

The Hubble Space Telescope, Figure 11.7 (launched in 1990 with a faulty main mirror, of diameter 2.4 metres, which was corrected by astronauts during space walks in 1993) is intended to send back to Earth pictures of distant galaxies (large collections of stars) that will enable more reliable estimates to be made of the age and size of the Universe (see Chapter 71) than are possible from Earth-based observations. It orbits the Earth at a height of 590 km and is expected to do so for about 15 years.

QUESTIONS

1 Select from the list **A** to **E** the pairs of lenses most suitable for **a** a compound microscope, **b** an astronomical telescope.

A Two concave lenses of focal lengths 100 cm and 5 cm.
B Two convex lenses of focal lengths 100 cm and 5 cm.
C Two concave lenses of focal lengths 5 cm and 3 cm.
D Two convex lenses of focal lengths 5 cm and 3 cm.
E A convex lens of focal length 5 cm and a concave lens of focal length 3 cm. *(J.M.B.)*

2 A and B are two convex lenses correctly set up as a telescope to view a distant object, Figure 11.8. One has focal length 5 cm and the other 100 cm.

a What is A called and what is its focal length?

b How far from A is the first image of the distant object?

c What is B called?

d What acts as the 'object' for B and how far must B be from it if someone looking through the telescope is to see the final image at the same distance as the distant object?

e What is the distance between A and B with the telescope set up as in **d**?

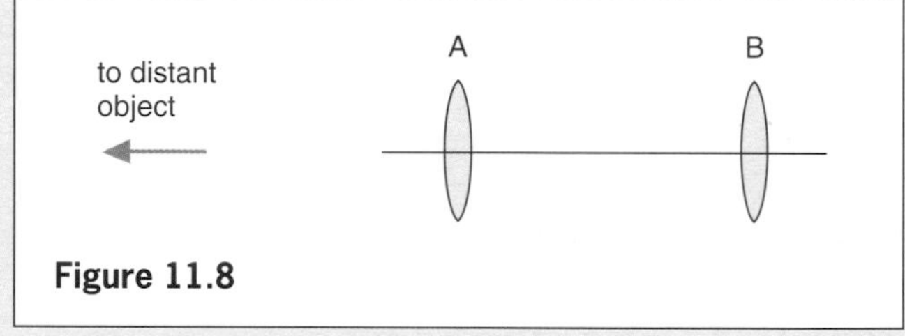

Figure 11.8

3 **a** What is used as the objective in (i) an optical refracting telescope, (ii) an optical reflecting telescope?

b Give *two* reasons for an optical telescope having an objective of large diameter. *(J.M.B./A.L./N.W.16+)*

Checklist

After studying this chapter you should be able to describe with the aid of diagrams

- a compound microscope,
- a refracting astronomical and a Galilean telescope,
- a reflecting astronomical telescope.

ADDITIONAL QUESTIONS

The first questions under each topic heading are 'basic' questions intended for all students; the following questions marked with a stripe are 'higher' questions for those seeking grades E to A*.

Light rays; reflection of light; mirrors

1 Which one of the following statements about the image produced in a pinhole camera is correct?

A The image is bigger if the object is further away.
B The image is smaller if the screen is nearer the pinhole.
C The image is brighter if the object is further away.
D The image is sharper if the pinhole is made bigger.
E The image is bigger if the object is brighter.
(O.L.E./S.R.16+)

2 The image in a plane mirror is

A upright, real with a magnification of 2
B upright, virtual with a magnification of 1
C inverted, real with a magnification of $\frac{1}{2}$
D inverted, virtual with a magnification of 1
E inverted, real with a magnification of 2.

3 A parallel beam of light is produced by reflection from a mirror when a point source of light is

A at the focus of a concave mirror
B at the focus of a convex mirror
C inside the focus of a concave mirror
D inside the focus of a convex mirror
E in front of a plane mirror.

4 A boy finds that by holding an opaque circular disc, 8.0 mm in diameter, at a distance of 90 cm from his eye he can just cover the full moon. Calculate the diameter of the moon if it is at a distance of 380 000 km. Upon what property of light does this test depend? *(S.)*

5 Show by a ray diagram how a suitably placed eye sees an image of a point object which is placed 10 cm in front of a plane mirror. Show clearly the position of the image and give two reasons why it is described as virtual. *(L.)*

6 By how far does the distance between a girl and her image decrease if she walks from a position where she is 10 m away from a mirror to one where she is 3 m away?

Refraction of light; total internal reflection

7 Light travels up through a pond of water of critical angle 49°. What happens at the surface if the angle of incidence is **a** 30°, **b** 60°?

8 Which diagram shows the ray of light refracted correctly?

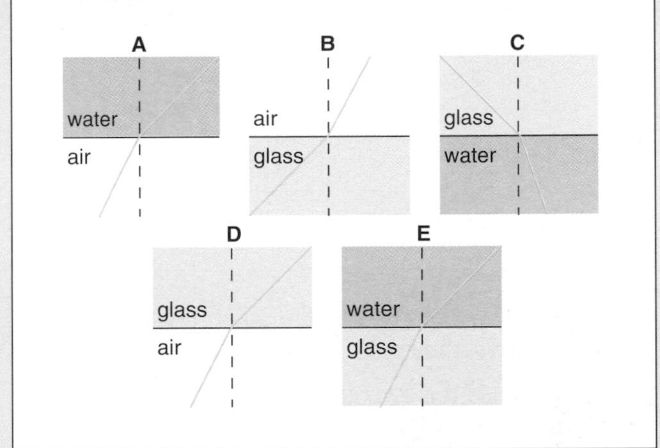

9 The diagram below shows the end view of a glass prism whose angles are 30°, 60° and 90°. Light passes into the prism in the direction AB and some of it after reflection in the face XY emerges in the direction BA. What is the angle of refraction of the ray AB at B? *(O. and C. part qn.)*

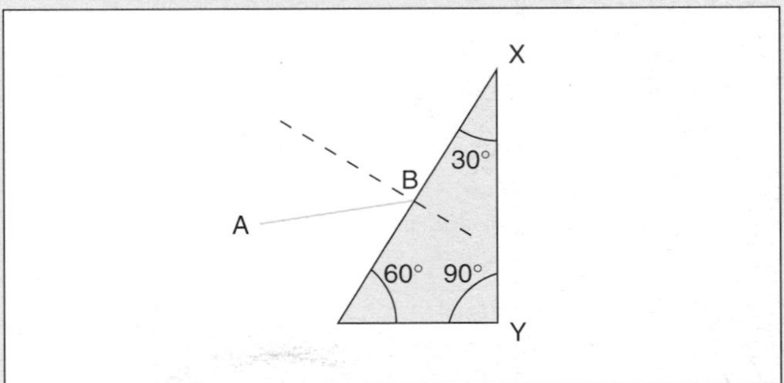

Lenses; the eye

10 An illuminated object is placed at a distance of 15 cm from a convex lens of focal length 10 cm. The image obtained on a screen is

A upright and magnified
B upright and the same size
C inverted and magnified
D inverted and the same size
E inverted and diminished. *(N.I.)*

11 An object is placed 4 cm in front of a convex lens. A real image is produced 16 cm from the lens.

a What is the magnification produced by the lens?

b By means of a full size diagram determine the focal length of the lens. Mark the principal focus. *(E.M.)*

12 The diagram shows a simplified section of a human eye.

a Name the parts labelled A, B and C and explain what happens there.

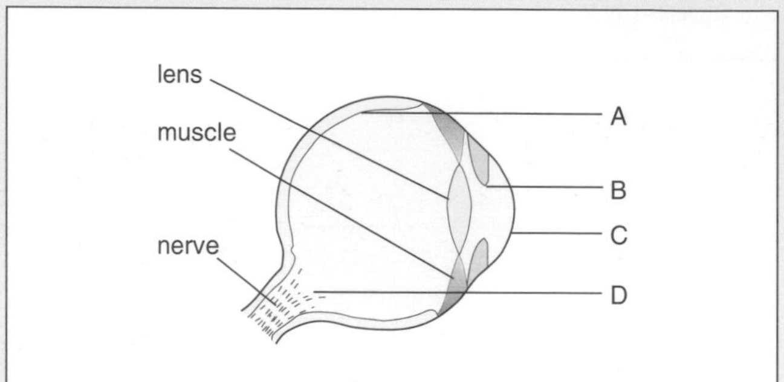

b What is the effect if light goes through the lens on to the nerve ending D?

c Why does the eye automatically turn to look straight at an object if possible? *(S.R.)*

13 A man is obliged to hold a newspaper at arm's length in order to read it.

a State a defect of vision which would cause this.

b What type of spectacle lenses would he require to correct this defect?

14 An object is set up 0.20 m (20 cm) in front of lens A and the details of the image are noted and are shown below. The process is repeated for a different lens B.

Lens A. Real, inverted, magnified and at a great distance.
Lens B. Real, inverted and same size as the object.

State the type of each lens and give as definite a value as possible for each focal length, explaining how you arrive at the values given. *(J.M.B.)*

Colour

15 Draw a diagram showing an arrangement of apparatus to project a pure spectrum from white light onto a screen and indicate the paths of rays that produce the blue and red parts of the spectrum. Account for the separation of the colours. *(L.)*

16 Three lamphouses A, B and C, each with a filter of one of the primary colours as shown in the diagram, are set up equal distances apart on a line parallel to a white screen so as to direct beams of equal intensity onto the centre of the screen. An opaque rectangular object painted white is placed halfway between the line of the lamps and the screen as shown. If the width of this object is slightly less than the distance between adjacent lamps, describe and explain the appearance of the object and the screen, using a diagram to indicate the parts of the screen referred to in your description. *(J.M.B.)*

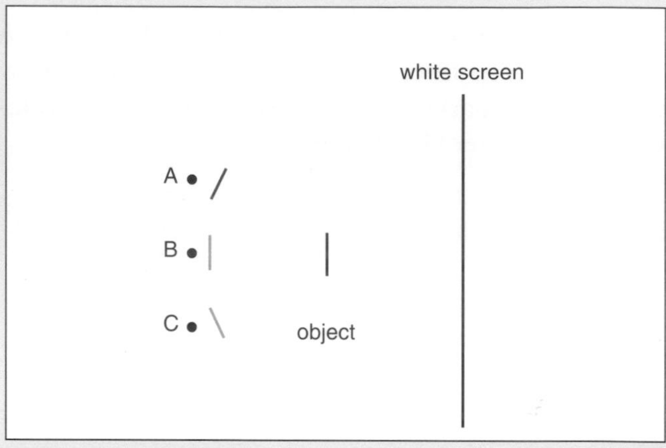

Optical instruments

17 Compare the optical system of the eye with that of a camera. Discuss in particular the focusing of objects at different distances, the regulation of the amount of light entering the system and the nature of the images formed. *(N.I.)*

18 Draw a diagram of a slide projector and use it to explain the function of the condenser lens and the concave mirror placed behind the light source.

What advantage has a matt white surface for the screen over a glossy surface?

A slide projector using a slide 5 cm × 5 cm produces a picture 3 m × 3 m on a screen placed at a distance of 24 m from the projection lens. How far from the lens must the slide be? Make an approximate estimate of the focal length of the projection lens. If this lens gets broken and the only substitute available is one of about half its focal length, what would you do to arrange that the picture is still the same size? *(O. and C.)*

19 Draw a ray diagram to show how two lenses may be used to construct a compound microscope. State the type of each lens and its approximate focal length. Show on the diagram the positions of the principal foci of the lenses in relation to the object and image positions. *(J.M.B.)*

WAVES AND SOUND

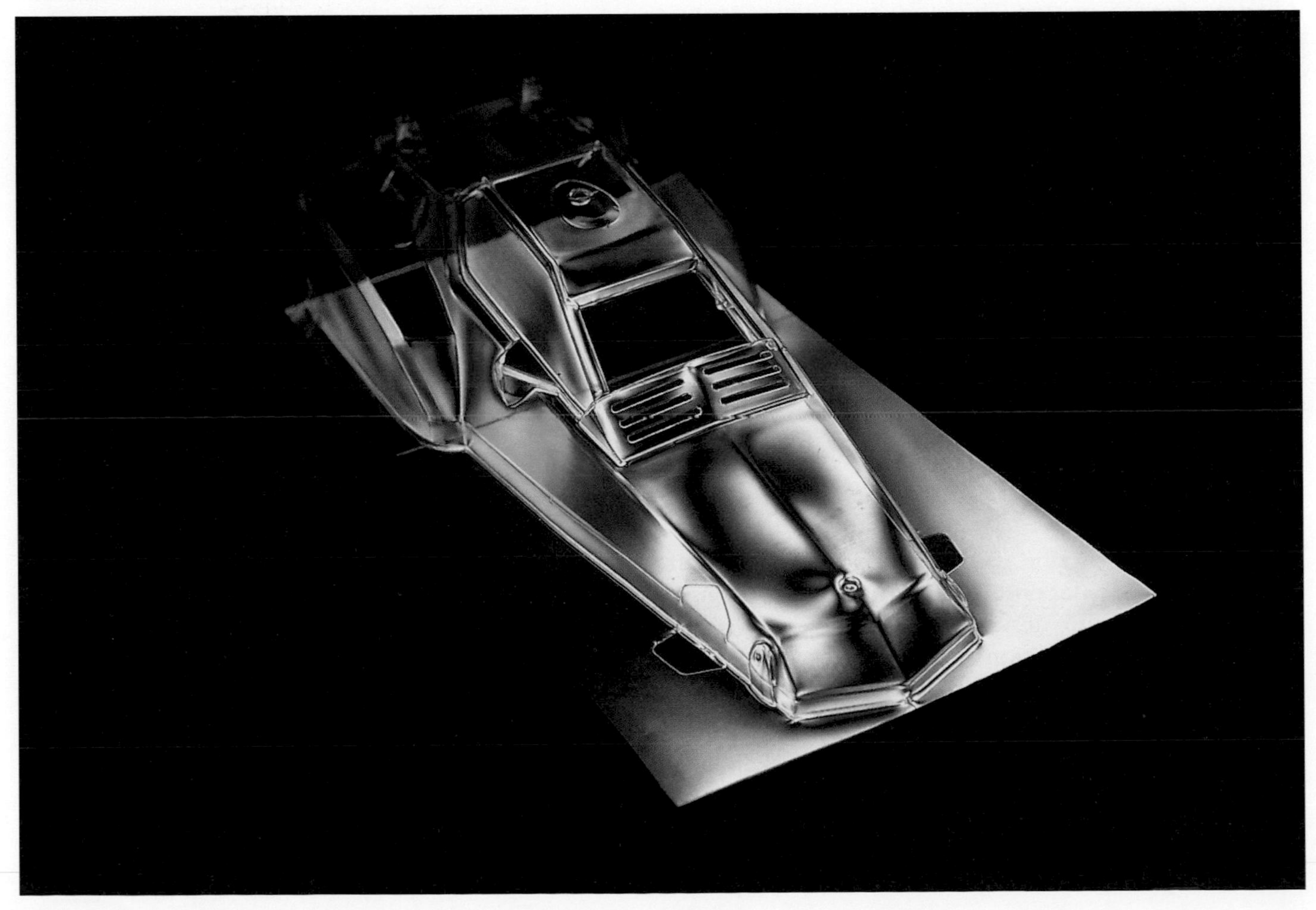

12 MECHANICAL WAVES

Types of wave	***Diffraction***
Describing waves Wavelength, frequency, speed, amplitude, phase.	***Interference***
The wave equation	***Polarization***
Reflection	***Practical work*** The ripple tank.
Refraction	

Types of wave

Several kinds of wave occur in physics. **Mechanical waves** are produced by a disturbance, e.g. a vibrating object, in a material medium and are transmitted by the particles of the medium vibrating to and fro. Such waves can be seen or felt and include waves on a rope or spring, water waves and sound waves in air or in other materials.

A **progressive** or travelling wave is a disturbance which carries energy from one place to another without transferring matter. There are two types, **transverse** and **longitudinal** (Chapter 15).

In a transverse wave, the direction of the disturbance is at **right angles** to the direction of travel of the wave. One can be sent along a rope (or a spring) by fixing one end and moving the other rapidly up and down, Figure 12.1. The disturbance generated by the hand is passed on from one part of the rope to the next which performs the same motion but slightly later. The humps and hollows of the wave travel along the rope as each part of the rope vibrates transversely about its undisturbed position.

Water waves are transverse waves.

Figure 12.1 A transverse wave

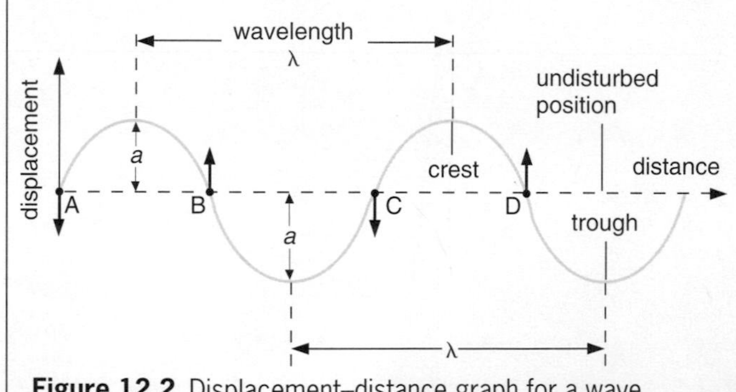

Figure 12.2 Displacement–distance graph for a wave at a particular instant

Describing waves

Terms used to describe waves can be explained with the aid of a **displacement–distance** graph, Figure 12.2. It shows the distance moved sideways from their undisturbed positions, of the parts vibrating at different distances from the cause of the wave, at a **certain time**.

(a) Wavelength, represented by the Greek letter λ (lambda), is the distance between successive crests.

(b) Frequency f is the number of complete waves generated per second. If the end of a rope is jerked up and down twice in a second, two waves are produced in this time. The frequency of the wave is 2 vibrations per second or 2 **hertz** (2 Hz; the hertz being the unit of frequency) as is the frequency of jerking of the end of the rope. That is, the frequencies of the wave and its source are equal.

The frequency of a wave is also the number of crests passing a chosen point per second.

(c) Speed v of the wave is the distance moved by a crest or any point on the wave in 1 second.

(d) Amplitude a is the height of a crest or the depth of a trough measured from the undisturbed position of what is carrying the wave, e.g. a rope.

(e) Phase. The arrows at A, B, C, D show the directions of vibration of the parts of the rope at these points. The parts at A and C have the same speed in the same direction and are **in phase**. At B and D the parts are also in phase but they are **out of phase** with those at A and C because their directions of vibration are opposite.

The wave equation

The faster the end of a rope is waggled, the shorter the wavelength of the wave produced. That is, the higher the frequency of a wave the smaller its wavelength. There is a useful connection between f, λ and v which is true for all types of wave.

Suppose waves of wavelength $\lambda = 20$ cm travel on a long rope and three crests pass a certain point every second. The frequency $f = 3$ Hz. If Figure 12.3 represents this wave motion then if crest A is at P at a particular time, 1 second later it will be at Q, a distance from P of three wavelengths, i.e. $3 \times 20 = 60$ cm. The speed of the wave $v = 60$ cm per second (60 cm/s), obtained by multiplying f by λ. Hence

speed of wave = frequency × wavelength

or $$v = f\lambda$$

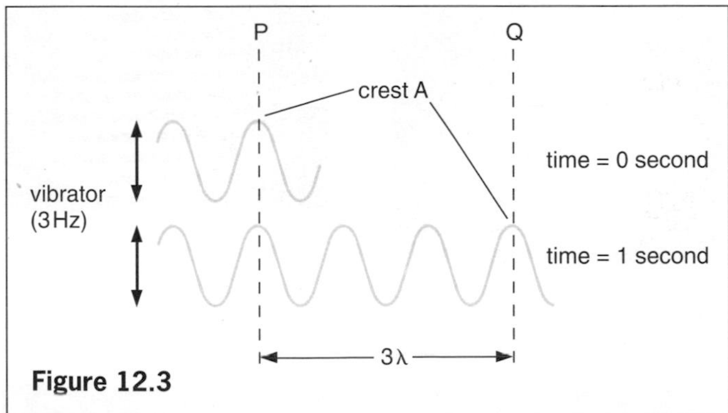

Figure 12.3

Practical work

The ripple tank

The behaviour of water waves can be studied in a ripple tank. It consists of a transparent tray containing water, having a lamp above and a white screen below to receive the wave images, Figure 12.4.

Pulses (i.e. short bursts) of ripples are obtained by dipping a finger in the water for circular ones and a ruler for straight ones. **Continuous ripples** are generated by an electric motor on a bar which gives straight ripples if it just touches the water and circular ripples if the bar is raised and a small ball fitted to it so as to be in the water.

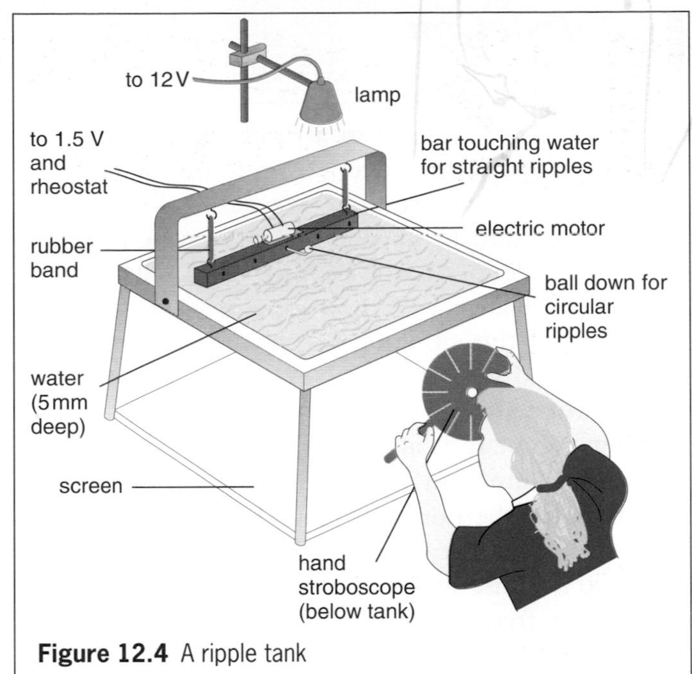

Figure 12.4 A ripple tank

Continuous ripples are studied more easily if they are *apparently* stopped ('frozen') by viewing the screen through a disc with equally spaced slits, that can be spun by hand, i.e. a stroboscope. If the disc speed is such that the waves have advanced one wavelength each time a slit passes your eye, they appear at rest.

Reflection

In Figure 12.5 **straight** water waves are represented falling on a metal strip placed in a ripple tank at an angle of 60°, i.e. the angle i between the direction of travel of the waves and the normal to the strip is 60°, as is the angle between the wavefront and the strip. The **wavefronts** are represented by straight lines and can be thought of as the crests of the waves. They are at right angles to the direction of travel, i.e. to the **rays**. The angle of reflection r is 60°. Incidence at other angles shows that the angles of reflection and incidence are always equal.

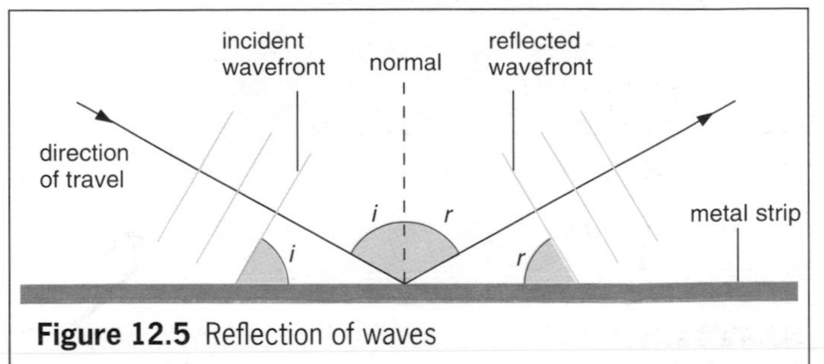

Figure 12.5 Reflection of waves

Refraction

If a glass plate is placed in a ripple tank so that the water is about 1 mm deep over it but 5 mm elsewhere, continuous straight waves in the shallow region are found to have a shorter wavelength than those in the deeper parts, i.e. the wavefronts are closer together, Figure 12.6. Both sets of waves have the frequency of the vibrating bar and since $v = f\lambda$, if λ has decreased so has v, since f is fixed. Hence **waves travel more slowly in shallow water**.

When the plate is at an angle to the waves, Figure 12.7a, their direction of travel in the shallow region is bent towards the normal, Figure 12.7b, i.e. refraction occurs. We saw earlier (Chapter 5) that light is refracted because its speed (and wavelength) changes (but not its frequency) when it enters another medium. The refraction of water waves for the same reason seems to suggest that light itself may be a kind of wave motion.

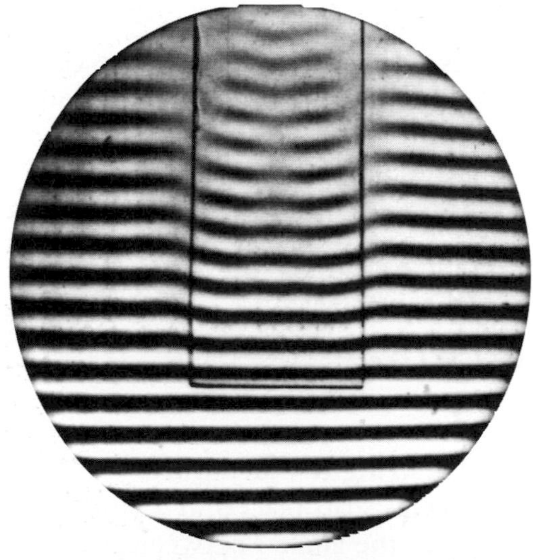

Figure 12.6 Waves in shallower water have a shorter wavelength

Figure 12.7a Waves are refracted at the boundary between deep and shallow regions

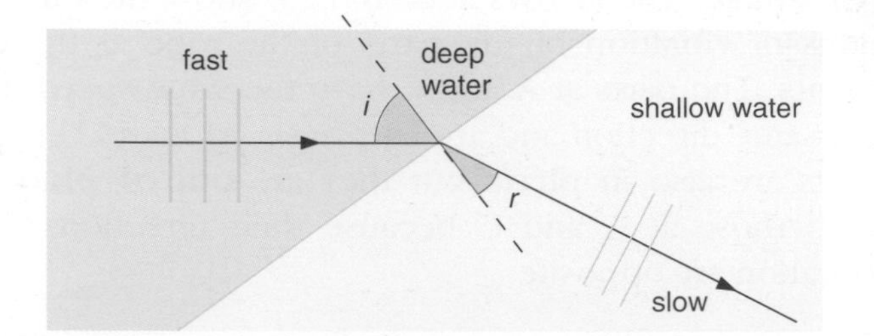

Figure 12.7b The direction of travel is bent towards the normal in the shallow region

Diffraction

In Figures 12.8a, b, straight water waves in a ripple tank are falling on gaps formed by obstacles. In (a) the gap width is about the same as the wavelength of the waves (1 cm); those passing through are circular and spread out in all directions. In (b) the gap is wide (10 cm) compared with the wavelength and the waves continue straight on; some spreading occurs but it is less obvious.

The spreading of waves at the edges of obstacles is called **diffraction**; when designing harbours, engineers use models like that in Figure 12.9 to study it.

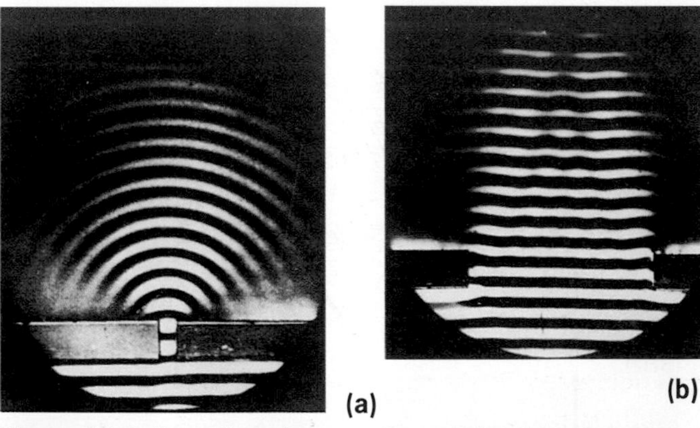

Figure 12.8 Spreading of waves after passing through a narrow and a wide gap

Figure 12.9 Model of a harbour used to study wave behaviour

Interference

When two sets of continuous **circular** waves cross in a ripple tank, a pattern like that in Figure 12.10a is obtained.

At points where a crest from S_1 arrives at the same time as a crest from S_2, a bigger crest is formed and the waves are said to be **in phase**. At points where a crest and a trough arrive together, they cancel out (if their amplitudes are equal); the waves are exactly **out of phase** (due to travelling different distances from S_1 and S_2) and the water is undisturbed, Figure 12.10b. The dark 'spokes' radiating from S_1 and S_2 join such points.

Interference or **superposition** is the combination of waves to give a larger or a smaller wave. Figure 12.10c shows how the pattern in Figure 12.10a is formed. All points on AB are equidistant from S_1 and S_2 and since these vibrate in phase, crests (or troughs) from S_1 arrive at the same time as crests (or troughs) from S_2. Hence along AB reinforcement occurs by superposition and a wave of double amplitude is obtained. Points on CD are half a wavelength nearer to S_1 than to S_2, i.e. there is a path difference of half a wavelength. Therefore crests (or troughs) from S_1 arrive simultaneously with troughs (or crests) from S_2 and the waves cancel. Along EF the difference of distances from S_1 and S_2 to any point is one wavelength, making EF a line of reinforcement.

Study this effect with the two ball 'dippers' about 3 cm apart on the bar. Also observe the effect of changing (i) the frequency and (ii) the separation of the 'dippers'; use a stroboscope when necessary. You will find that if the frequency is increased, i.e. the wavelength decreased, the 'spokes' are closer together. Increasing the separation has the same effect.

Similar patterns are obtained if straight waves fall on two small gaps: interference occurs between the emerging (circular) diffracted waves.

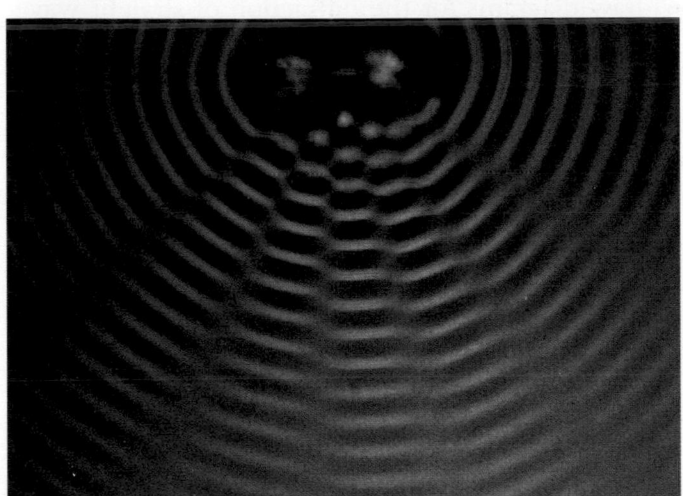

Figure 12.10a Interference of circular waves

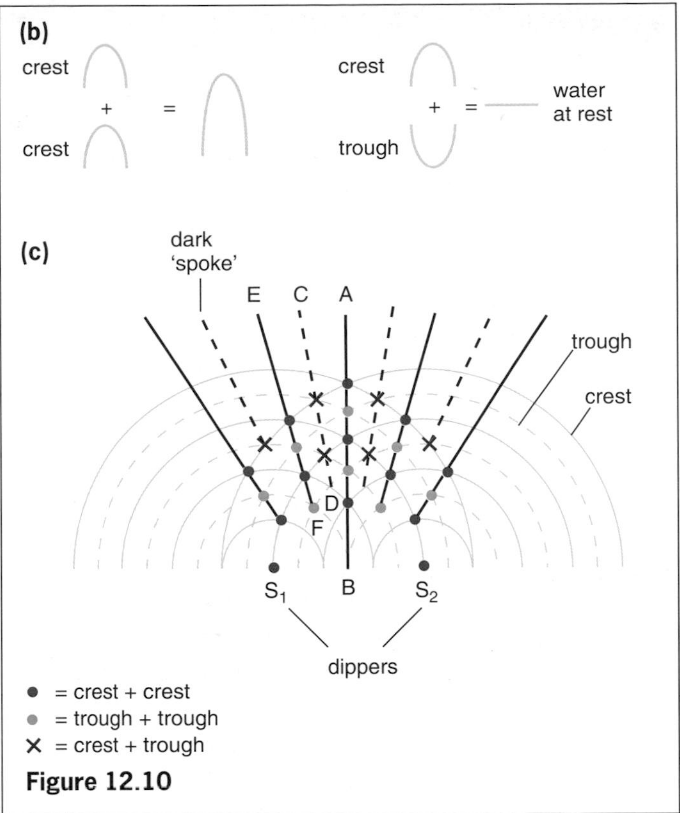

Figure 12.10

Polarization

This effect occurs only with transverse waves. It can be shown by fixing a rope at one end, A in Figure 12.11, and passing it through two slits B and C. If end A is moved to and fro in all directions (as shown by the short arrowed lines), vibrations of the rope occur in every plane and transverse waves travel towards B.

At B only waves due to vibrations in a vertical plane can emerge from the vertical slit. The wave between B and C is said to be **plane polarized** (in the vertical plane containing the slit at B). By contrast the waves between A and B are unpolarized. If the slit at C is vertical, the wave travels on, but if it is horizontal as shown, the wave is stopped and the slits are said to be 'crossed'.

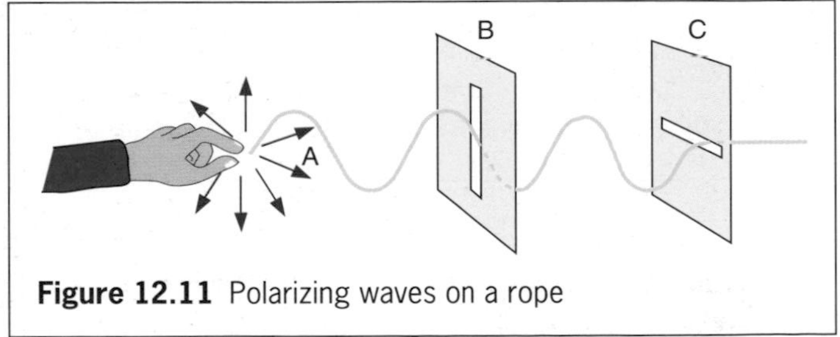

Figure 12.11 Polarizing waves on a rope

QUESTIONS

1 The lines in Figure 12.12 are crests of straight ripples.

a What is the wavelength of the ripples?

b If ripple A occupied 5 seconds ago the position now occupied by ripple F, what is the frequency of the ripples?

c What is the speed of the ripples?

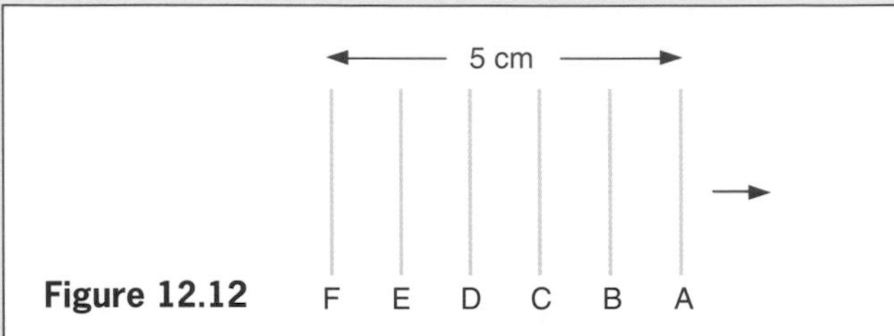

Figure 12.12

2 During the refraction of a wave which of the following properties change: **a** the speed, **b** the frequency, **c** the wavelength?

3 One side of a ripple tank ABCD is raised slightly, Figure 12.13, and a ripple started at P by a finger. After a second the shape of the ripple is as shown.

a Why is it not circular?

b Which side of the tank has been raised?

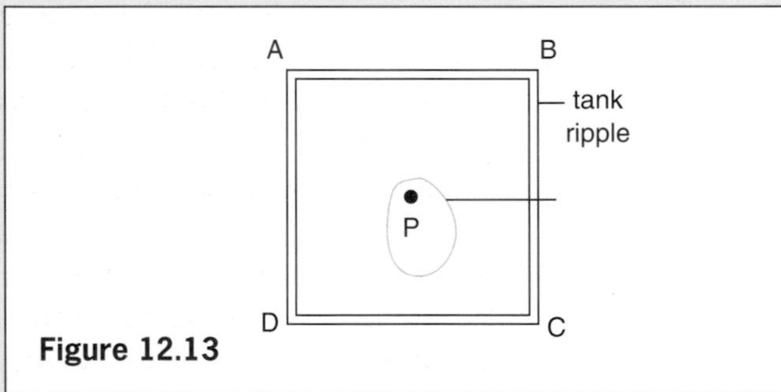

Figure 12.13

4 Figure 12.14 gives a full-scale representation of the water in a ripple tank 1 second after the vibrator was started. The coloured lines represent crests.

a What is represented at A at this instant?

b Estimate (i) the wavelength, (ii) the speed of the waves, and (iii) the frequency of the vibrator.

c Sketch a suitable attachment which could have been vibrated up and down to produce this wave pattern.

d Explain how the waves combine (i) at B and (ii) at C.

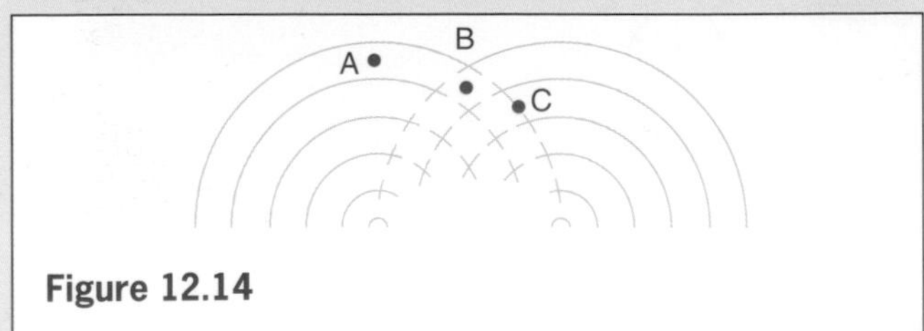

Figure 12.14

5 Copy Figure 12.15 and show on it what happens to the waves as they pass through the gap when the water is much shallower on the right-hand side than on the left.

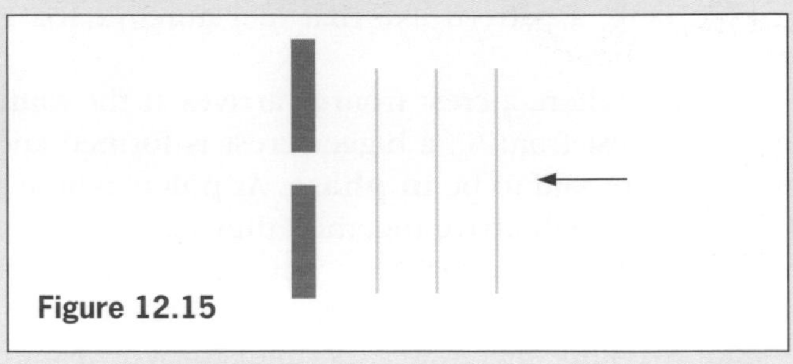
Figure 12.15

Checklist

After studying this chapter you should be able to

- describe the production of pulses and progressive transverse waves on ropes, springs and ripple tanks,
- recall the meaning of **wavelength**, **frequency**, **speed**, **amplitude** and **phase**,
- represent a transverse wave on a displacement–distance graph and extract information from it,
- recall the wave equation $v = f\lambda$ and use it to solve problems,
- describe an experiment to show reflection of waves,
- recall that the angle of reflection equals the angle of incidence and draw a diagram for the reflection of straight wavefronts at a plane surface,
- describe experiments to show refraction of waves,
- recall that refraction at a straight boundary is due to change of wave speed but *not* of frequency,
- draw a diagram for the refraction of straight wavefronts at a straight boundary,
- explain the term **diffraction**,
- describe experiments to show diffraction of waves,
- draw diagrams for the diffraction of straight wavefronts at single slits of different widths,
- predict the effect of changing the wavelength or the size of the gap on diffraction of waves at a single slit,
- describe an experiment to show interference of water waves using two point sources,
- draw interference patterns for waves from two point sources,
- explain interference as the superposition of crests or troughs at points where waves arrive **in phase** or **out of phase**,
- predict the effect on the interference pattern of changing the separation of the two sources or the wavelength of the waves,
- recall the meaning of a **plane polarized** wave.

13 LIGHT WAVES

Diffraction	***Polarization*** Using polarizing material. By reflection.
Interference	***Using polarization*** Reducing glare. Stress analysis.
Everyday interference effects	***Lasers and holograms***
Colour, wavelength and frequency	
Diffraction grating	
Red sunsets; blue sky	

Although we cannot see how light travels, it displays the properties of waves.

Diffraction

Diffraction occurs when light passes the edge of an object, but it is not easy to detect. This suggests that light has a very small wavelength since we saw that diffraction of water waves is most obvious at a gap when its width is comparable with the wavelength of the waves. Figure 13.1 is the diffraction pattern of a very narrow vertical slit (1/100 mm or less) and shows how light has spread into regions that would be in shadow if it went exactly in straight lines.

Interference

A steady interference pattern is obtained with water waves in a ripple tank because both sets of waves have the same frequency and wavelength and are exactly in phase when they leave the sources S_1 and S_2. They are said to be **coherent**. It is impossible to obtain a steady interference pattern with light using two separate lamps since most light sources emit light waves of many wavelengths in short erratic bursts, each out of phase with the next. The two sets of waves are not coherent.

These difficulties were overcome by Young in 1801 by allowing light from *one* lamp in a darkened room to fall on two narrow parallel slits very close together (about 0.5 mm separation), as shown in the modern arrangement of Figure 13.2. A pattern of equally spaced bright and dark bands, called **fringes**, is obtained on a screen. The waves leaving the slits are coherent since any phase changes in the bursts of light from the lamp affect both sets of waves at the same time.

The bright bands are coloured, except for the centre one which is white. If a red filter is placed in front of the lamp, the bright bands are red and are spaced farther apart than those given by a blue filter, Figures 13.3a, b (overleaf).

The fringes can be explained by assuming that diffraction occurs at each slit, and in the region where the two diffracted beams cross there is interference,

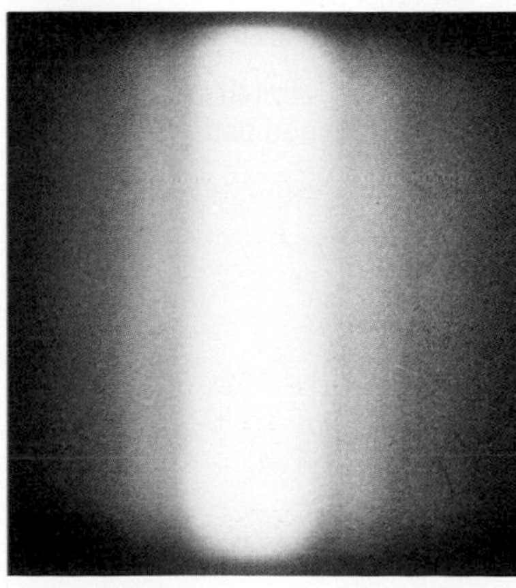

Figure 13.1 Diffraction of white light at a narrow slit

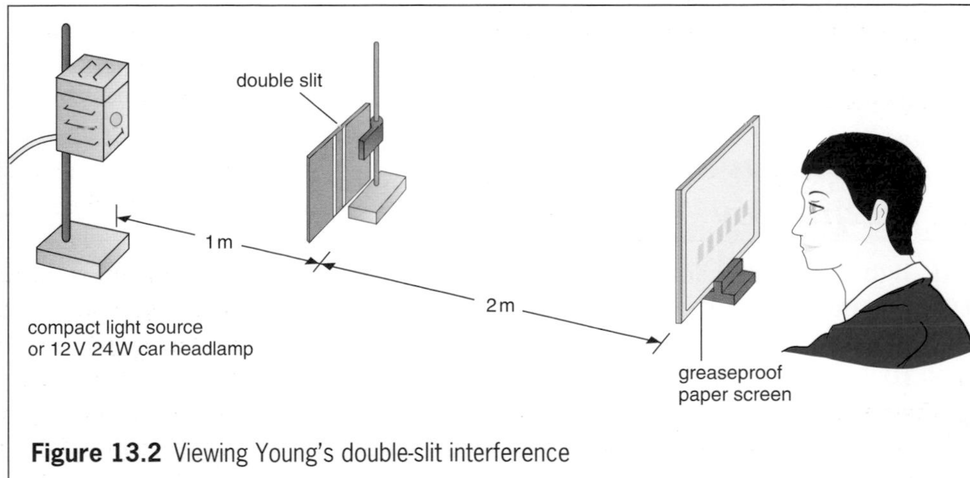

Figure 13.2 Viewing Young's double-slit interference

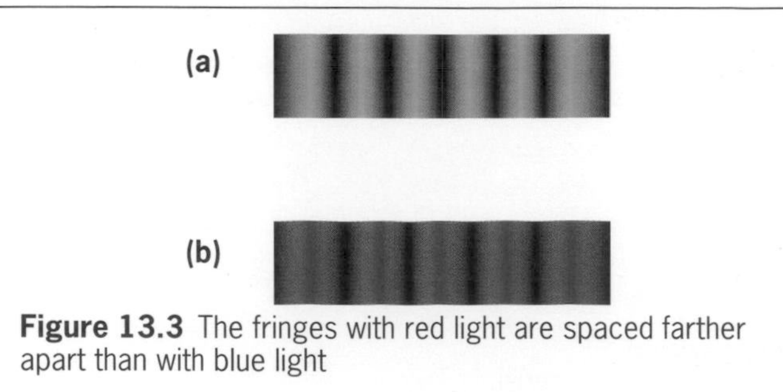

Figure 13.3 The fringes with red light are spaced farther apart than with blue light

Figure 13.4. At points on the screen where a 'crest' from one slit arrives at the same time as a 'crest' from the other, the waves are in phase and there are bright bands. Dark bands occur where 'crests' and 'troughs' arrive simultaneously and the waves cancel. We then have: light + light = darkness. This makes sense only if we regard light as having a wave nature. Figure 12.10c, which we used to explain the water wave interference pattern, is also a help when thinking about how light produces interference effects.

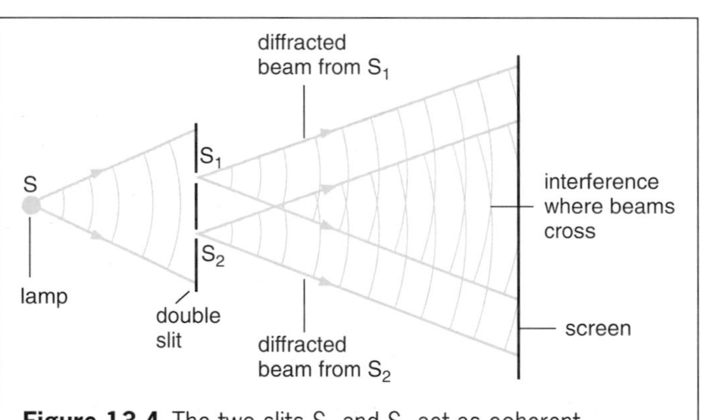

Figure 13.4 The two slits S_1 and S_2 act as coherent sources

Everyday interference effects

The colour effects produced by thin oil films on a wet road, Figure 13.5a, and in soap bubbles, Figure 13.5b, are due to interference between light reflected from the two surfaces of the film or bubble. Figure 13.6 shows this for a film of oil on water.

Colour, wavelength and frequency

Red light has the longest wavelength of about 0.000 7 mm (7×10^{-7} m = 0.7 μm), while violet light has the shortest of about 0.000 4 mm (4×10^{-7} m = 0.4 μm). Colours between these in the spectrum of white light have intermediate values. Light of one colour and so of one wavelength is called **monochromatic** light.

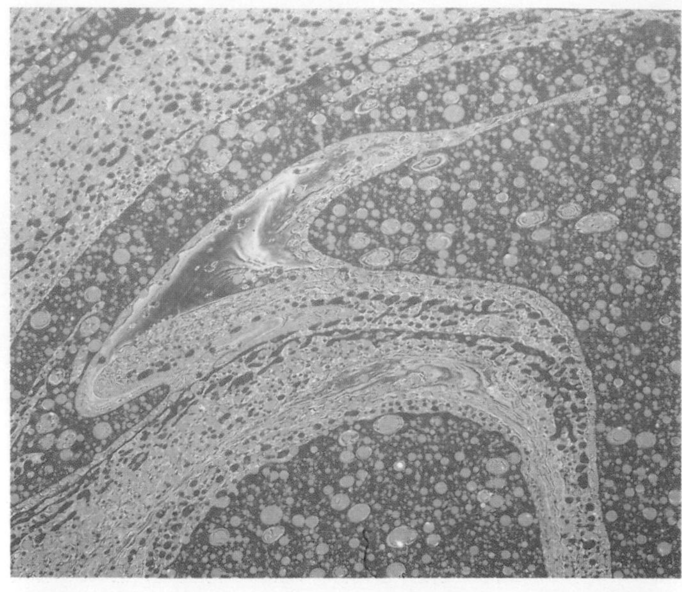

(a)

(b)

Figure 13.5 Interference of light reflected from thin oil films and soap bubbles can give rise to beautiful coloured patterns

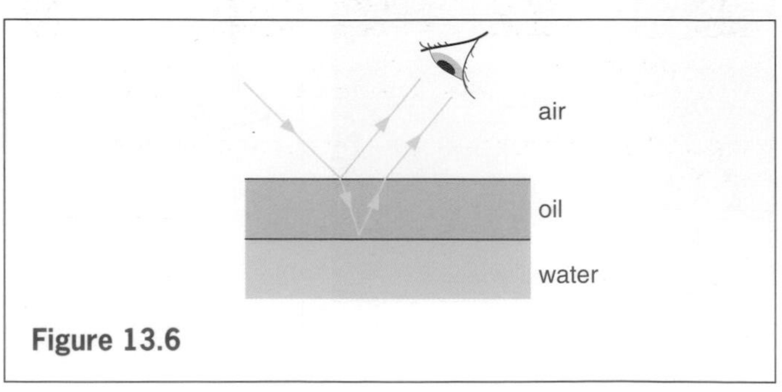

Figure 13.6

Remembering that $v = f\lambda$ for all waves including light, it follows that red light has a lower frequency (f) than violet light since (i) the wavelength (λ) of red light is greater and (ii) all colours travel with the same speed (v) of 3×10^8 m/s in air (strictly a vacuum). It is the **frequency** of light which decides its colour, rather than its wavelength which is different in different media, as is its speed (Chapter 5).

Different frequencies of light travel at different speeds through a transparent medium and so are refracted by different amounts. This explains dispersion (Chapter 9), i.e. why the refractive index of a material depends on the wavelength of the light.

The **amplitude** of a light (or any other) wave is greater the greater the **intensity** of the source, i.e. the brighter it is in the case of light.

Diffraction grating

A diffraction grating consists of a piece of glass or plastic with a large number of parallel lines marked on it. The lines scatter light and in effect are opaque. The thin clear strips between the lines transmit light and act as slits.

When white light from a straight filament lamp is viewed through a grating a number of spectra (plural of spectrum) are seen, each like that given by a prism. The spectra are said to be of various 'orders'. A 'fine' grating with say 300 lines per mm gives the zero, first and second orders, Figure 13.7. A 'coarse' grating, e.g. 100 lines per mm, gives more orders closer together and the 'coarser' the grating the more the pattern resembles the two-slits one produced by the arrangement in Figure 13.2.

The slits are only a few wavelengths wide and when light falls on them, diffraction occurs causing cylindrical waves to spread out from each one. In certain directions the diffracted waves are in phase and if brought to a focus by a lens or the eye, they reinforce. In other directions they more or less cancel, i.e. destructive interference occurs.

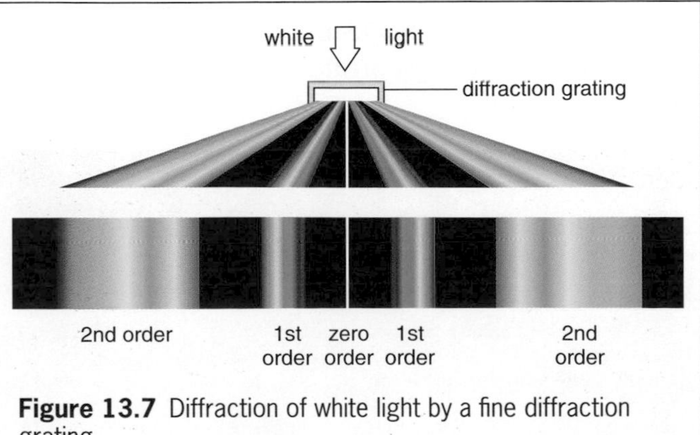

Figure 13.7 Diffraction of white light by a fine diffraction grating

Red sunsets; blue sky

The wavelength of red light is about twice that of blue light. The longer the wavelength the more penetrating the light, while the shorter the wavelength the more easily it is scattered.

When the sun is setting the light from it has to travel through a greater thickness of the Earth's atmosphere and only the longer wavelength red light is able to get through. Sunsets are therefore red.

The shorter wavelengths, like blue, are scattered in all directions by the atmosphere, which is why the sky looks blue.

Polarization

Interference and diffraction need a wave model to explain the effects they produce but they do not show whether the waves are transverse or longitudinal. Polarization suggests that light waves are transverse.

(a) Using polarizing material

If a lamp is viewed through a piece of polarizing material such as Polaroid (used in sunglasses), apart from it seeming slightly less bright, there is no effect when the Polaroid is rotated. However, using two pieces, one of which is kept at rest and the other rotated slowly, Figure 13.8, the light is cut off more or less completely in one position. The Polaroids are then 'crossed'. Rotation through a further 90° allows maximum light through.

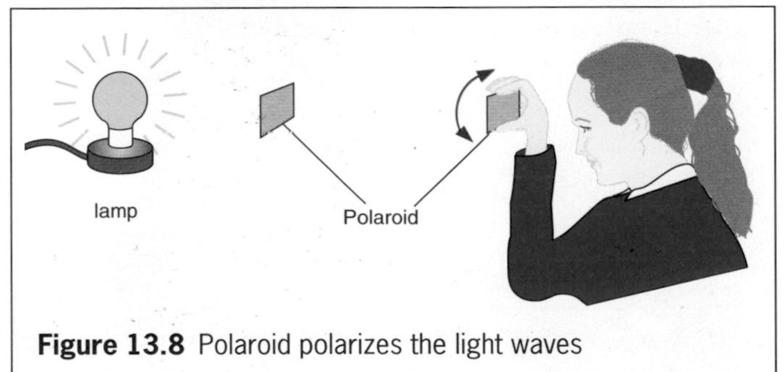

Figure 13.8 Polaroid polarizes the light waves

This experiment is similar to that shown in Figure 12.11 where transverse waves are sent along a rope to two slits B and C. Slit B produces a wave that is **polarized** in a vertical plane. This wave cannot pass through horizontal slit C. In this position slits B and C are 'crossed'. A longitudinal wave, e.g. vibrations along a coiled spring, would emerge from both slits whatever their positions, i.e. it cannot be polarized.

The experiment with the Polaroids can be explained if we regard light as a transverse wave motion. Light from the Sun and other sources is unpolarized and consists of 'vibrations' in every plane at right angles to the direction of travel of the wave. The Polaroid nearest the lamp polarizes the light by transmitting only the 'vibrations' in one particular plane. The other Polaroid transmits or absorbs the plane-polarized light falling on it depending on its orientation with respect to the first Polaroid.

(b) By reflection

When unpolarized light falls on glass, water or a polished surface, the reflected ray is, in general, partly plane polarized. At a certain angle of incidence the polarization is complete.

Using polarization

(a) Reducing glare

Glare caused by light reflected from a smooth surface can be reduced by using polarizing material such as Polaroid since the reflected light is partly or completely polarized. Thus polarizing discs, suitably orientated, are used in sunglasses and also in photography as 'filters' in front of the camera lens, enabling detail to be seen that would otherwise be hidden by glare, Figure 13.9.

Figure 13.9 Glare can be reduced with a polarizing filter in front of the camera

(b) Stress analysis

If glass, Perspex and some other plastics are under stress (e.g. by bending or uneven heating) and are viewed in white light between two 'crossed' sheets of polarizing material, coloured fringes are seen round the regions of stress. The technique is used to study stress in plastic models of structures, Figure 13.10.

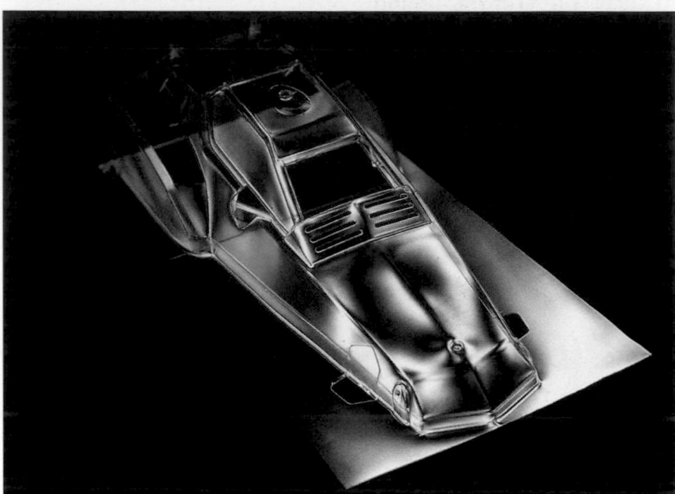

Figure 13.10 Photoelastic stress patterns in a plastic model car

Lasers and holograms

A laser produces a very intense beam of light which is monochromatic, coherent and parallel. A hologram is a three-dimensional image of an object recorded on a special photographic plate. The image is formed by the interference of two beams of light from a laser, one of which has been reflected from the object. When developed and illuminated, the image looks 'real', appears to float in space and moves when the viewer does.

Phonecards contain a hologram which keeps a check on the number of units used. Credit cards contain holograms to prevent forgery. Holograms are also used to make safety checks on car tyres and engines, Figure 13.11, to design artificial limbs and to display and market goods.

Figure 13.11 Hologram of an engine

QUESTIONS

1 In Figure 13.12, L is a source of light, wavelength λ, S_1S_2 a double slit and O, O_1 and O_2 are points on a screen.

a If the central bright band is formed at O, how do the distances S_1O and S_2O compare?

b If the first dark band next to the central bright band is formed at O_1, how do S_1O_1 and S_2O_1 compare?

c If the first bright band occurs at O_2 next to the first dark band, how do S_1O_2 and S_2O_2 compare?

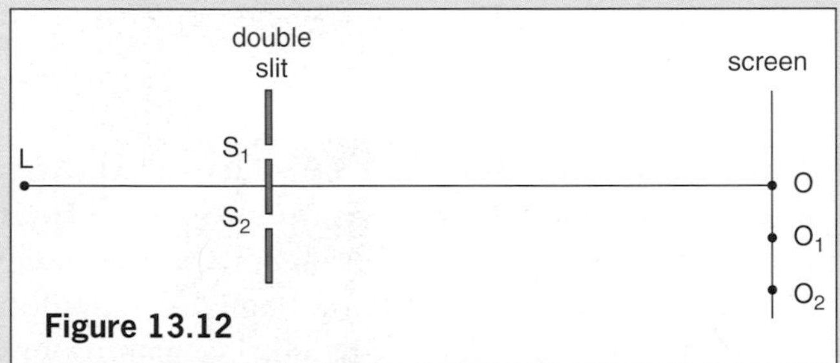

Figure 13.12

2 a Why is the diffraction of light not easy to detect?

b Why is it not possible to produce a steady interference pattern using two light sources?

3 Give the approximate wavelength in micrometres (μm) of **a** red light, **b** violet light.

4 a State *two* ways of producing polarized light.

b State *two* uses of polarized light.

c Is it possible to polarize longitudinal waves?

Checklist

After studying this chapter you should be able to

- explain why diffraction of light is not normally observed,
- describe a simple Young's double-slit experiment to show light has a wave-like nature,
- recall that the colour of light depends on its frequency, that red light has a lower frequency (but longer wavelength) than blue light and that all colours travel at the same speed in air,
- describe a diffraction grating and describe what is seen when white light is viewed through a grating,
- describe some evidence that light is a transverse wave,
- give some uses of the polarization of light,
- explain briefly how a hologram is made.

14 ELECTROMAGNETIC RADIATION

Properties	***X-rays***
Infrared radiation	***Practical work*** Wave nature of microwaves.
Ultraviolet radiation	
Radio waves	

Light is one member of a family of electromagnetic radiation, which forms a continuous spectrum well beyond both ends of the visible (light) spectrum, Figure 14.1. While each type of radiation has a different source, all result from electrons in atoms undergoing an energy change and all have certain properties in common.

Properties

1 They **travel through space at 300 000 km/s** (3×10^8 m/s), i.e. with the speed of light.
2 They **exhibit interference, diffraction and polarization**, which suggests they have a transverse wave nature.
3 They **obey the wave equation** $v = f\lambda$ where v is the speed of light, f is the frequency of the waves and λ is the wavelength. Since v is constant for a particular medium, it follows that large f means small λ.
4 They **carry energy from one place to another and can be absorbed by matter to cause heating and other effects**. The higher the frequency and the smaller the wavelength of the radiation, the greater is the energy carried, i.e. gamma rays are more 'energetic' than radio waves. This is shown by the **photoelectric effect** in which electrons are ejected from metal surfaces when electromagnetic waves fall on them. As the frequency of the waves increases so too does the speed (and energy) with which electrons are emitted.

Because of its electrical origin, its ability to travel in a vacuum (e.g. from the Sun to the Earth) and its wave-like properties (i.e. **2** above), electromagnetic radiation is regarded as a **progressive transverse wave**. The wave is a combination of travelling electric and magnetic forces. The forces vary in value and are directed at right angles to each other and to the direction of travel of the wave, as shown by the representation in Figure 14.2.

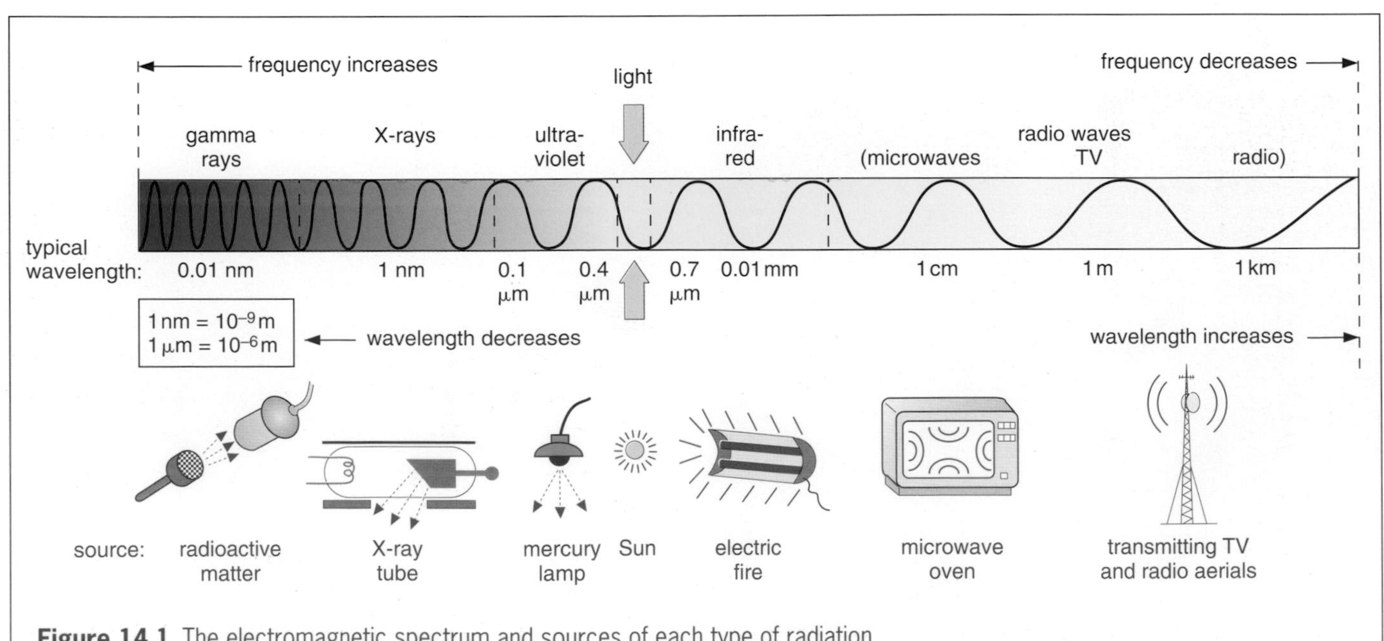

Figure 14.1 The electromagnetic spectrum and sources of each type of radiation

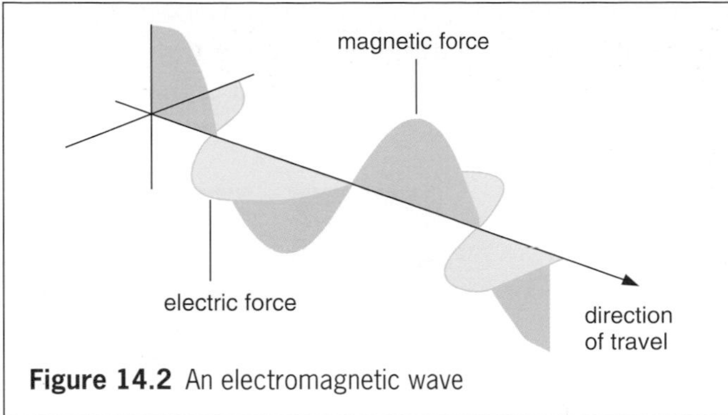

Figure 14.2 An electromagnetic wave

Infrared radiation

Our bodies detect infrared radiation (IR) by its heating effect on the skin. It is sometimes called 'radiant heat' or 'heat radiation'.

Anything which is hot but not glowing, i.e. below 500 °C, emits IR alone. At about 500 °C a body becomes red-hot and emits red light as well as IR – the heating element of an electric fire, a toaster or a grill are examples. At about 1000 °C things such as lamp filaments are white-hot and radiate IR and white light, i.e. all the colours of the visible spectrum.

Infrared is also detected by special temperature-sensitive photographic films which allow pictures to be taken in the dark. Infrared sensors are used on satellites and aircraft for weather forecasting, agricultural monitoring and assessing heat loss from buildings, Figure 14.3.

Infrared lamps are used to dry the paint on cars during manufacture and in the treatment of muscular complaints. Remote control keypads for televisions contain a small infrared transmitter for changing programmes.

Figure 14.3 Infrared aerial photograph showing contrasting vegetation.

Ultraviolet radiation

Ultraviolet (UV) rays have shorter wavelengths than light. They cause sun-tan and produce vitamins in the skin but can penetrate deeper causing skin cancer. Dark skin is able to absorb more UV so reducing the amount reaching deeper tissues.

Ultraviolet causes fluorescent paints and clothes washed in some detergents to fluoresce, i.e. they glow by re-radiating as light the energy they absorb as UV.

Figure 14.4 White clothes fluorescing at a disco

A UV lamp used for scientific or medical purposes contains mercury vapour and this emits UV when an electric current passes through it. Fluorescent tubes also contain mercury vapour and their inner surfaces are coated with special powders called phosphors which radiate light.

Radio waves

Radio waves have the longest wavelengths in the electromagnetic spectrum. They are radiated from aerials and used to 'carry' sound, pictures and other information over long distances.

(a) Long, medium and short waves (2 km to 10 m)

These can bend (diffract) round obstacles so they can be received when hills etc. are in their way, Figure 14.5a (overleaf). They are also reflected by layers of electrically charged particles in the upper atmosphere (the **ionosphere**), which makes long-distance radio reception possible, Figure 14.5b.

(b) VHF (very high frequency) and UHF (ultra high frequency) waves

These have shorter wavelengths and need a clear, straight-line path to the receiver. They are not reflected by the ionosphere. They are used for local radio and for television.

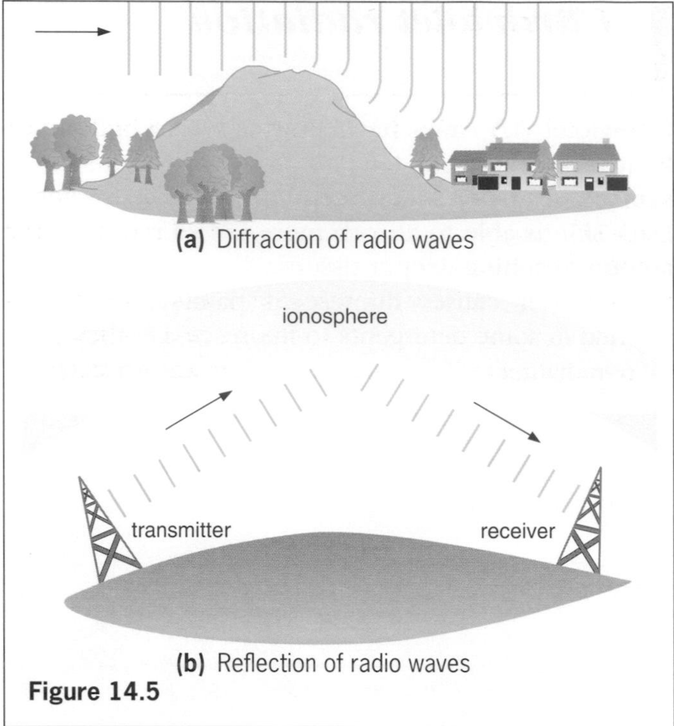

(a) Diffraction of radio waves

(b) Reflection of radio waves

Figure 14.5

(c) Microwaves (wavelengths of a few cm)

These are used for international (as well as national) telephone and television links via geostationary communications satellites, which go round the equator at the same rate as the Earth spins and so appear at rest. Signals are beamed through the ionosphere by large dish aerials, like those in Figure 66.19, to the satellite where they are amplified and sent back to a dish aerial in another part of the world. Microwaves are also used for **radar** detection of ships and aircraft.

Microwaves can be used for cooking since they cause water molecules in the moisture of the food to vibrate vigorously at the frequency of the microwaves. As a result, heating occurs inside the food which cooks itself.

Living cells can be damaged or killed by the heat produced when microwaves are absorbed by water in the cells.

PRACTICAL WORK

Wave nature of microwaves

The 3 cm microwave transmitter and receiver of Figure 14.6 can be used. The arrangement shown is set up for double-slit interference; the reading on the meter connected to the horn receiver rises and falls as it is moved across behind the three metal plates.

Diffraction can be shown similarly using the two wide metal plates to form a single slit.

If the grid of vertical metal wires is placed in front of the transmitter, the signal is absorbed but transmission occurs when the wires are horizontal, showing that the microwaves are vertically polarized. This can also be shown by rotating the receiver through 90° in a vertical plane from the maximum signal position, when the signal decreases to a minimum.

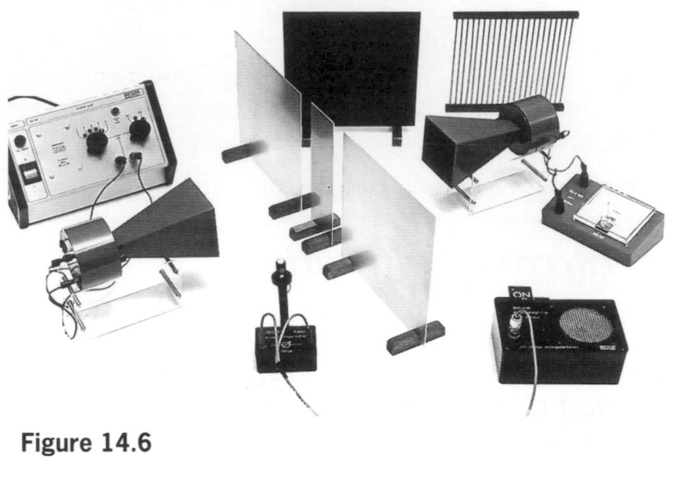

Figure 14.6

X-rays

These are produced when high-speed electrons are stopped by a metal target in an X-ray tube. X-rays have smaller wavelengths than UV.

They are absorbed to some extent by living cells and can penetrate some solid objects and affect a film. With materials like bones, teeth and metals which they do not pass through easily, shadow pictures can be taken, like that in Figure 14.7 of someone shaving. In industry X-ray photography is used to inspect welded joints. X-ray machines need to be shielded with lead since normal body cells can be killed by high doses and made cancerous by lower doses.

Gamma rays (Chapter 63) are more penetrating and dangerous than X-rays. They are used to kill cancer cells, and also harmful bacteria in food and on surgical instruments.

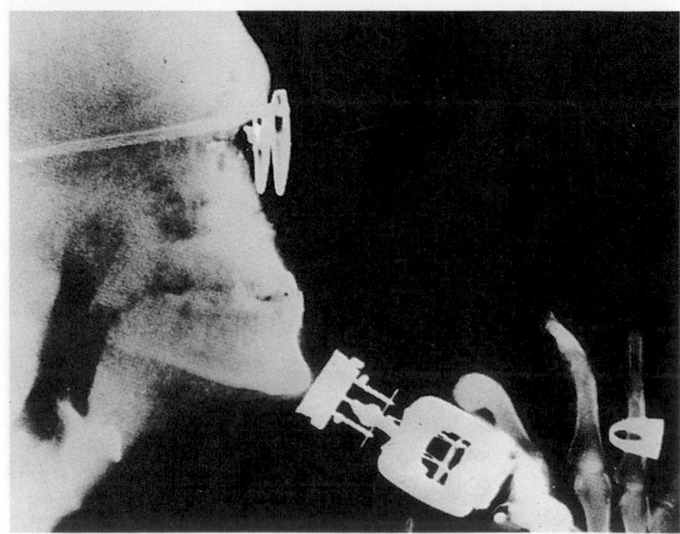

Figure 14.7 X-rays cannot penetrate bone and metal

QUESTIONS

1 Name *four* properties common to all electromagnetic waves.

2 Which of the following types of radiation has the largest frequency?

A UV **B** radio waves **C** light **D** X-rays **E** IR

3 Name one type of electromagnetic radiation which

a causes sun-tan,

b is used for satellite communication,

c passes through a thin sheet of lead,

d is used to take photographs in haze.

4 A VHF radio station transmits on a frequency of 100 MHz (1 MHz = 10^6 Hz). If the speed of radio waves is 3×10^8 m/s

a what is the wavelength of the waves,

b how long does the transmission take to travel 60 km?

Checklist

After studying this chapter you should be able to

- recall the types of electromagnetic radiation,
- recall that all electromagnetic waves have the same speed in space and are progressive transverse waves,
- distinguish between **infrared radiation**, **ultraviolet radiation**, **radio waves** and **X-rays** in terms of their wavelengths, properties and uses.

15 SOUND WAVES

Origin and transmission of sound
Longitudinal waves
The ear
Reflection and echoes
Speed of sound
Diffraction and interference
Ultrasonics

Origin and transmission of sound

Sources of sound all have some part which **vibrates**. A guitar has strings, Figure 15.1, a drum has a stretched skin and the human voice has vocal cords. The sound travels through the air to our ears and we hear it. That the air is necessary may be shown by pumping out a glass jar containing a ringing electric bell, Figure 15.2; the sound disappears though the striker can still be seen hitting the gong. Evidently sound cannot travel in a vacuum as light can.

Other materials, including solids and liquids, transmit sound. 'Cathedral chimes' may be heard if you jingle together some spoons tied to a piece of string with its ends in your ears, Figure 15.3. Not only does string (a solid) transmit sound but it does so better than air.

Sound also gives interference and diffraction effects. Because of this and its other properties, we believe it is a form of energy (as the damage from supersonic booms shows) which travels as a progressive wave, but of a type called **longitudinal**.

Figure 15.1 A guitar string vibrating. The sound waves produced are amplified when they pass through the circular hole into the guitar's sound box

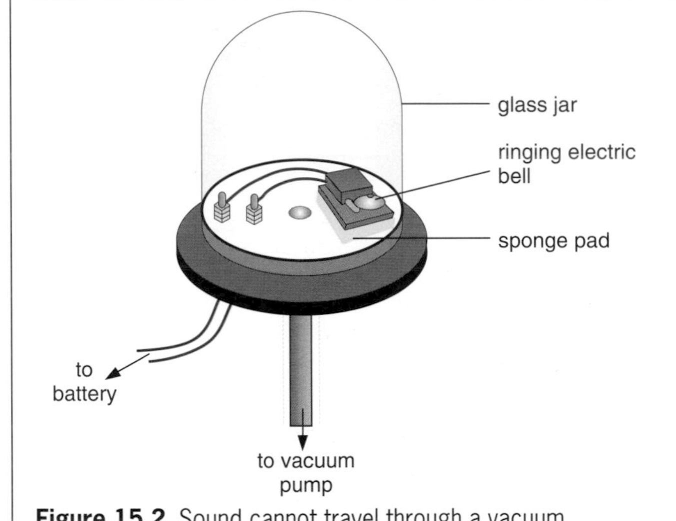

Figure 15.2 Sound cannot travel through a vacuum

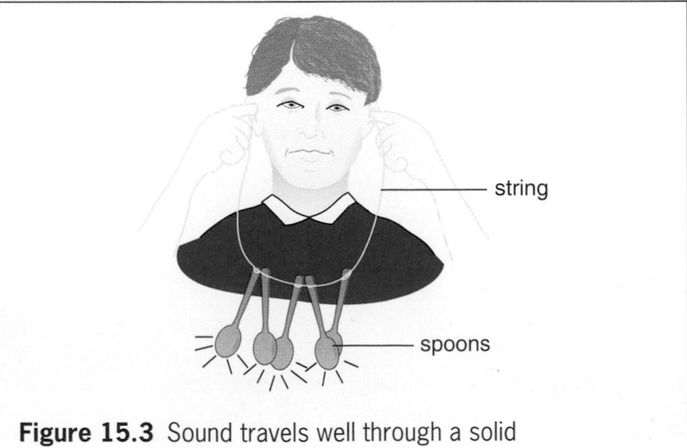

Figure 15.3 Sound travels well through a solid

Longitudinal waves

(a) Waves on a spring

In a progressive longitudinal wave the particles of the transmitting medium vibrate to and fro along the same line as that in which the wave is travelling and not at right angles to it as in a transverse wave. A longitudinal wave can be sent along a spring, stretched out on the bench and fixed at one end, if the free end is repeatedly pushed and pulled sharply. Compressions C (where the coils are closer together) and rarefactions R (where the coils are farther apart), Figure 15.4, travel along the spring.

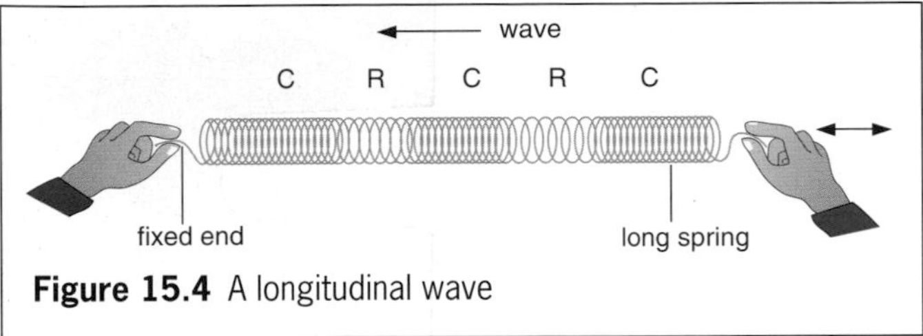

Figure 15.4 A longitudinal wave

(b) Sound waves

A sound wave produced for example by a loudspeaker consists of a train of compressions ('squashes') and rarefactions ('stretches') in the air, Figure 15.5.

The speaker has a cone which is made to vibrate in and out by an electric current. When the cone moves out the air in front is compressed; when it moves in the air is rarefied (goes 'thinner'). The wave progresses through the air but the air as a whole does not move. The air particles (molecules) vibrate backwards and forwards a little as the wave passes. When the wave enters your ear the compressions and rarefactions cause small, rapid pressure changes on the eardrum and you experience the sensation of sound.

The number of compressions produced per second is the frequency f of the sound wave (and equals the frequency of the vibrating cone); the distance between successive compressions is the wavelength λ. As for transverse waves the speed $v = f\lambda$.

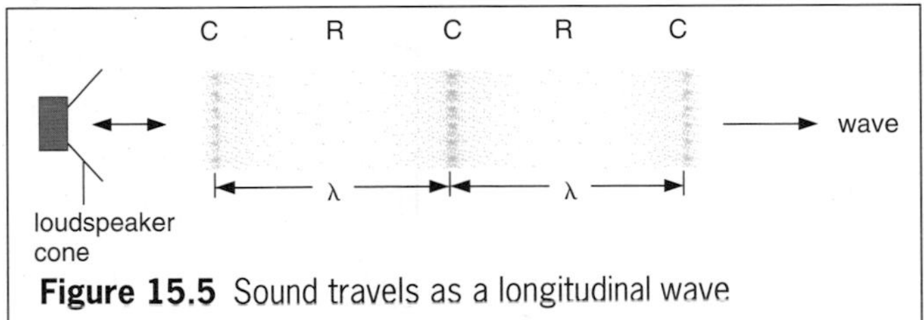

Figure 15.5 Sound travels as a longitudinal wave

The ear

The ear has three main parts, Figure 15.6.

(a) Outer ear

The **ear canal** collects and directs the sound waves on to a thin membrane called the **eardrum**. This is made to vibrate at the same frequency by the air vibration of the sound waves.

(b) Middle ear

This contains three tiny bones, the **ossicles**, the outermost being joined to the eardrum and the innermost fitting into a small hole in the skull called the **oval window**. When the eardrum vibrates so too do the ossicles.

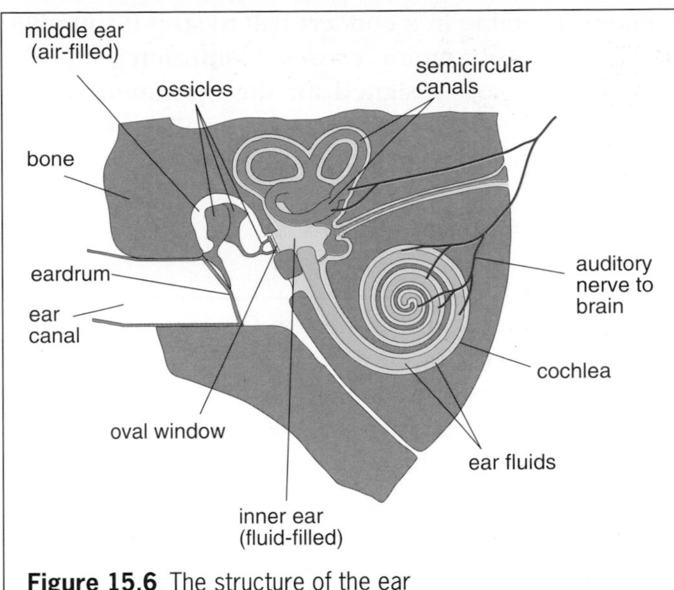

Figure 15.6 The structure of the ear

(c) Inner ear

The vibrations of the ossicles are passed on to the sensitive part of the inner ear. This is a fluid in a coiled tube known as the **cochlea** containing the **auditory nerve endings** which send off impulses to the brain when the fluid in the cochlea vibrates. Nerve endings in the first part of the cochlea respond to high frequency vibrations and those in the last part to low frequencies, enabling the brain to distinguish between high- and low-pitched sounds.

Human beings hear only sounds with frequencies from about 20 Hz to 20 000 Hz. These are the **limits of audibility**; the upper limit decreases with age.

The **semicircular canals** are part of the inner ear but are not concerned with hearing. The information received by the brain from their sense organs enables us to stand upright when still and to keep our balance when moving around.

A hearing aid can be used to send an amplified sound to the eardrum when someone can hear only quite loud sounds. If the eardrum or middle ear has worse damage, an aid can pass on vibrations to the cochlea directly through the skull bones. Deafness occurs if the cochlea or auditory nerve is damaged.

Reflection and echoes

Sound waves are reflected well from hard, flat surfaces such as walls or cliffs and obey the same laws of reflection as light. The reflected sound forms an **echo**.

If the reflecting surface is nearer than 15 m, the echo joins up with the original sound which then seems to be prolonged. This is called **reverberation**.

Some is desirable in a concert hall to stop it sounding 'dead', but too much causes 'confusion'. Modern concert halls are designed for the optimum amount of reverberation. Seats and some wall surfaces are covered with sound-absorbing material.

Speed of sound

The speed of sound depends on the material through which it is passing. It is greater in solids than in liquids or gases because the molecules in a solid are closer together than in a liquid or a gas. Some values are given in the table below.

Material	air (0 °C)	water	concrete	steel
Speed (metres/second)	330	1400	5000	6000

In air the speed **increases with temperature** and at high altitudes, where the temperature is lower, it is less than at sea-level. Changes of atmospheric pressure do not affect it.

An estimate of the speed of sound can be made directly if you stand about 100 metres from a high wall or building and clap your hands. Echoes are produced. When the clapping rate is such that each clap coincides with the echo of the previous one, the sound has travelled to the wall and back in the time between two claps, i.e. an interval. By timing 30 intervals with a stopwatch, the time t for one interval can be found. Also, knowing the distance d to the wall, a rough value is obtained from

$$\text{speed of sound in air} = \frac{2d}{t}$$

Diffraction and interference

(a) Diffraction

Audible sounds have wavelengths from about 1.5 centimetres (frequency 20 kHz) up to 15 metres (20 Hz) and so suffer diffraction by objects of similar size, e.g. a doorway 1 metre wide. This explains why we hear sound round corners. Low frequency (longer wavelength) notes are diffracted more than higher frequencies (shorter wavelengths) such as those caused by traffic noise.

(b) Interference

In Figure 15.7 sound waves of the same frequency from two loudspeakers (supplied by one signal generator) produce a steady interference pattern. The resulting variations in loudness of the sound, due to the waves reinforcing and cancelling one another, can be heard when you walk past the loudspeakers.

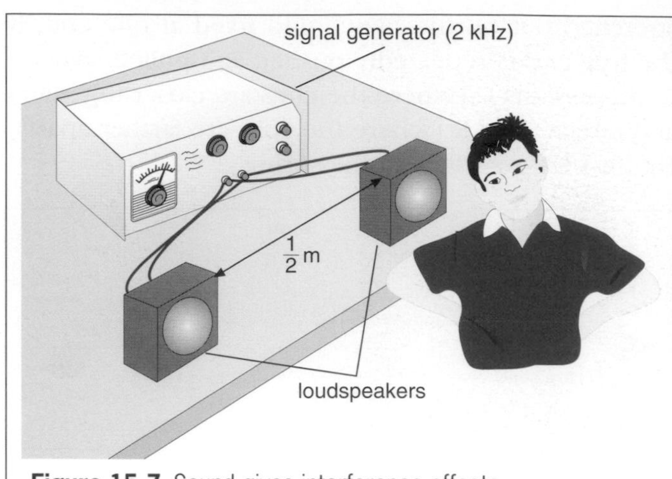

Figure 15.7 Sound gives interference effects

Ultrasonics

Sound waves with frequencies above 20 kHz are called **ultrasonic** waves. They are emitted by bats and enable them to judge the distance of an object from the time taken by the reflected wave to return.

In the echo-sounding system called **sonar**, ships use ultrasonic waves to measure the depth of the sea and to detect shoals of fish, Figure 15.8. In industry they are used to reveal flaws in welded joints; also holes of any shape or size can be cut in glass and steel by ultrasonic drills. Objects such as street lamp covers can be cleaned by immersion in a tank of water which has an ultrasonic vibrator in the base. Ultrasonic scans are used as a check on expectant mothers, and other medical applications of ultrasonic waves are being developed, Figure 15.9.

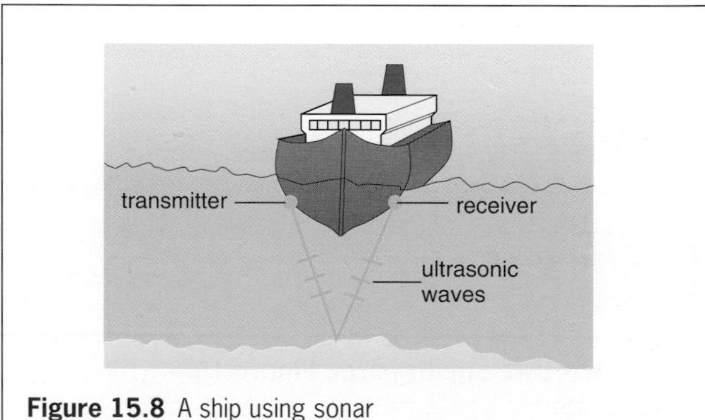

Figure 15.8 A ship using sonar

Ultrasonic waves are produced by ultrasonic vibrators. These contain a quartz crystal which is made to vibrate electrically at the required frequency. An ultrasonic receiver also consists of a quartz crystal but it works in reverse, i.e. when it is set into vibration by ultrasonic waves it generates an electrical signal which is then amplified.

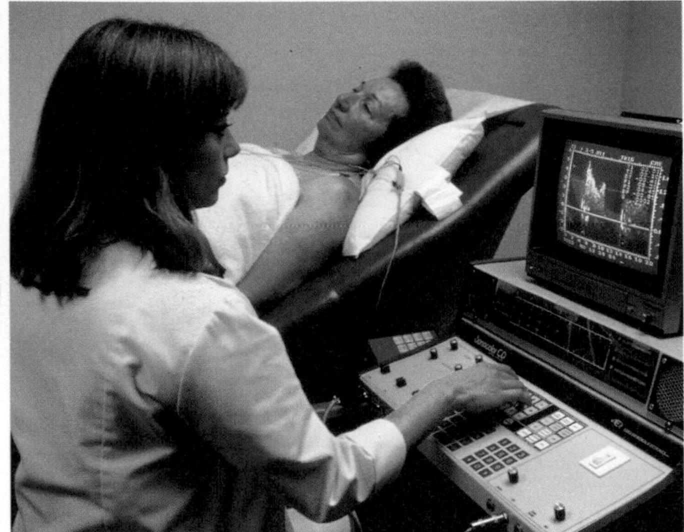

Figure 15.9 Patient undergoing an 'echocardiography' examination, in which reflections of ultrasonic waves are analysed to build up a picture of the heart on a screen and aid detection of abnormalities

QUESTIONS

1 A coastguard sees a distress rocket burst in the sky and 5 seconds later he hears the bang. If the speed of sound is 330 m/s, how far is the exploding rocket from the coastguard?

A 66 m **B** 335 m **C** 1100 m **D** 1650 m **E** 3300 m *(S.E.)*

2 The time-keeper of a 110 m race stands near the finishing tape and starts his stop-watch on hearing the bang from the starting pistol.

a When should he have started his stop-watch?

b Give a reason for your answer to **a**.

c Calculate the error in his timing assuming he makes no further errors. (Speed of sound = 330 m/s) *(E.M.)*

3 a What is the relationship connecting the frequency of a sound source with its wavelength and the speed of sound in air?

b An echo sounder in a ship produces a sound pulse and an echo is received from the sea bed after 0.4 s. Assuming the speed of sound in sea water to be 1500 m/s calculate the depth of the sea bed.

c If the echo sounder had produced continuous waves of frequency 6000 Hz (6 kHz), what would have been their wavelength in the sea water? Give your value in cm. *(N.W.)*

4 If 5 s elapse between a lightning flash and the clap of thunder how far away is the storm? Speed of sound = 330 m/s.

5 a A girl stands 160 m away from a high wall and claps her hands at a steady rate so that each clap coincides with the echo of the one before. If she makes 60 claps in 1 minute, what value does this give for the speed of sound?

b If she moves 40 m closer to the wall she finds the clapping rate has to be 80 per minute. What value do these measurements give for the speed of sound?

c If she moves again and finds the clapping rate becomes 30 per minute, how far is she from the wall if the speed of sound is the value you found in **a**?

6 a What properties of sound suggest it is a wave motion?

b How does a progressive transverse wave differ from a longitudinal one? Which type is sound?

Checklist

After studying this chapter you should be able to

- recall that sound is produced by vibrations,
- describe an experiment to show that sound is not transmitted through a vacuum,
- describe how sound travels in a medium as progressive longitudinal waves,
- draw and label the structure of the human ear and describe the functions of its parts,
- recall the limits of audibility (i.e. the range of frequencies) for the normal human ear,
- explain echoes and reverberation,
- describe a simple method of estimating the speed of sound and recall its approximate value,
- solve problems using the speed of sound, e.g. thundercloud proximity,
- describe experiments to show diffraction and interference of sound waves,
- recall some uses of ultrasonics.

16 MUSICAL NOTES

Pitch	*Vibrating strings; stationary waves*
Loudness	*Resonance*
Quality	*Noise pollution*

Irregular vibrations such as those of motor engines cause **noise**; regular vibrations such as occur in the instruments of a jazz band, Figure 16.1, produce **musical notes** which have three properties - pitch, loudness and quality.

Figure 16.1 Musical instruments produce regular sound vibrations

Pitch

The pitch of a note depends on the frequency of the sound wave reaching the ear, i.e. on the frequency of the source of sound. A high-pitched note has a high frequency and a short wavelength. The frequency of middle C (in 'scientific pitch') is 256 vibrations per second or 256 Hz and that of upper C is 512 Hz. Notes are an **octave** apart if the frequency of one is twice that of the other. Pitch is like colour in light; both depend on the frequency.

Notes of known frequency can be produced in the laboratory by a signal generator supplying alternating electric current (a.c.) to a loudspeaker. The cone of the speaker vibrates at the frequency of the a.c. which can be varied and read off a scale on the generator. A set of tuning forks with frequencies marked on them can also be used. A tuning fork, Figure 16.2, has two steel prongs which vibrate when struck; the prongs move in and out *together*, generating compressions and rarefactions.

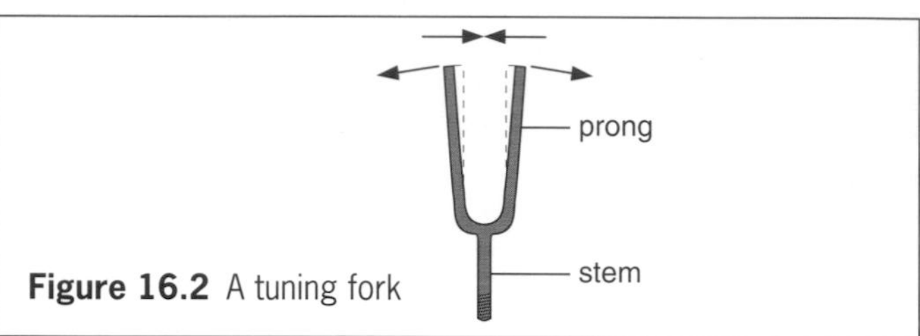

Figure 16.2 A tuning fork

Loudness

A note is louder when more sound energy enters our ears per second than before and is caused by the source vibrating with a larger amplitude. If a violin string is bowed more strongly, its amplitude of vibration increases as does that of the resulting sound wave and the note heard is louder because more energy has been used to produce it.

Quality

The same note on different instruments sounds different; we say the notes differ in **quality** or **timbre**. The difference arises because no instruments (except a tuning fork and a signal generator) emit a 'pure' note, i.e. of one frequency. Notes consist of a main or **fundamental** frequency mixed with others, called **overtones**, which are usually weaker and have frequencies that are exact multiples of the fundamental. The number and strength of the overtones decides the quality of a note. A violin has more and stronger higher overtones than a piano. Overtones of 256 Hz (middle C) are 512 Hz, 768 Hz and so on.

The **waveform** of a note played near a microphone connected to a CRO can be displayed on the CRO screen. Those for the *same* note on three instruments are given in Figure 16.3. Their different shapes show that while they have the same fundamental frequency, their quality differs. The 'pure' note of a tuning fork has a **sine** waveform and is the simplest kind of sound wave. **Note.** Although the waveform on the CRO screen is transverse it represents a longitudinal sound wave.

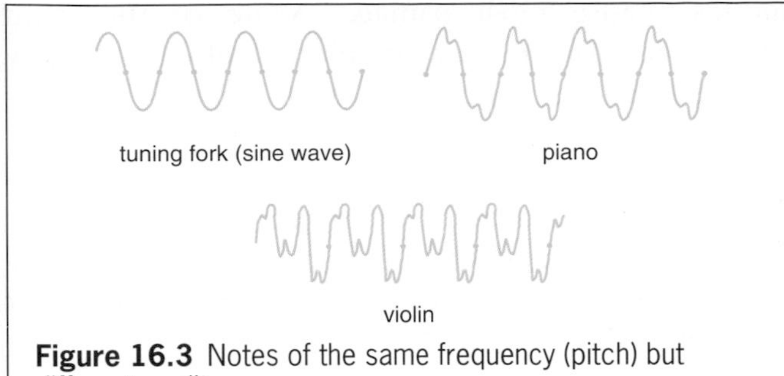

Figure 16.3 Notes of the same frequency (pitch) but different quality

Vibrating strings; stationary waves

In a string instrument such as a guitar the 'string' is a tightly stretched wire or length of gut. When it is plucked transverse waves travel to both ends, which are fixed, and are reflected. Interference occurs between the incident and reflected waves and a **stationary** or **standing** wave pattern is formed. In this, certain points on the string, called **nodes**, are at rest whilst points midway between each pair of nodes are in continuous vibration with maximum amplitude; they are called **antinodes**.

A string can vibrate in various ways. If the standing wave has one loop, Figure 16.4a, the fundamental note is emitted. If there is more than one loop, Figures 16.4b, c, overtones are produced, but in all cases the **separation of the nodes N and the antinodes A is one-quarter of the wavelength** of the wave on the string causing the note. In an instrument, a string vibrates in several ways at the same time depending on where it is plucked; this decides the quality of the note.

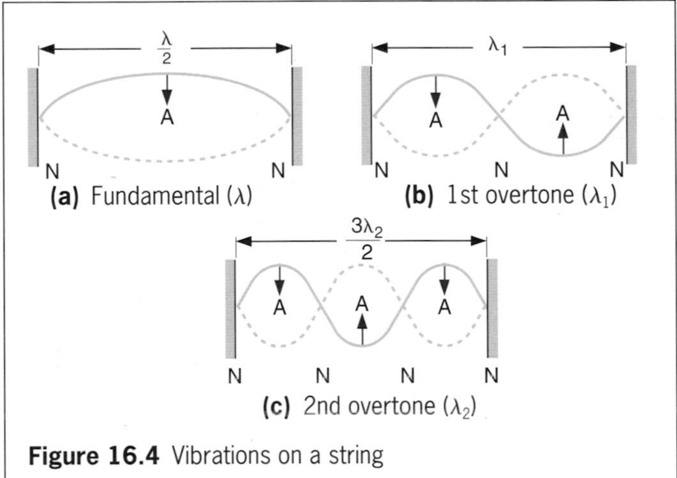

Figure 16.4 Vibrations on a string

The standing wave patterns of a vibrating string (or rubber cord) may be viewed as in Figure 16.5.

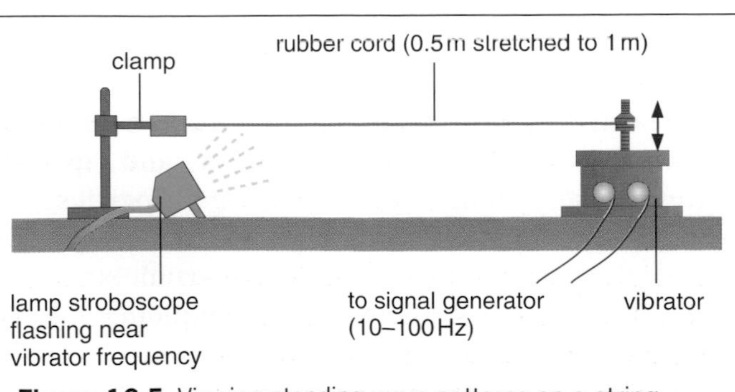

Figure 16.5 Viewing standing wave patterns on a string

The frequency of the fundamental note emitted by a vibrating string depends on its

(i) **length**: short strings emit high notes and halving the length doubles the frequency,
(ii) **tension**: tight wires produce high notes, and
(iii) **mass per unit length**: thin strings give high notes.

Resonance

All objects have a natural frequency of vibration. The vibration can be started and increased by another object vibrating at the same frequency. The effect is called **resonance**. For example, when the heavy pendulum X in Figure 16.6 is set swinging, it forces all the light ones to swing at the same frequency, but D, which has the same length as X, does so with a much larger amplitude, i.e. D **resonates** with X.

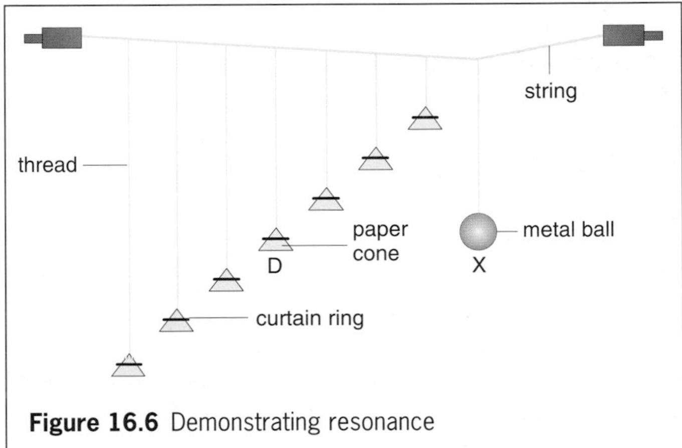

Figure 16.6 Demonstrating resonance

(a) Advantages

A playground swing can be made to swing high by someone pushing in time with the free swinging, Figure 16.7 (overleaf).

Resonance occurs in sound when, for example, the column of air in a wind instrument or the air in the hollow body of a string instrument is made to vibrate. The air resonates and this creates the volume of sound.

Electrical resonance is used in radio and television tuners so that only transmissions of a particular frequency are picked up.

Figure 16.7 Pushing at the right time makes the swing go higher

'Stones' of calcium compounds can grow in the kidneys and cause severe pain. One treatment focuses beams of ultrasonic waves on them at their resonant frequency in the hope that this will break them up.

(b) Disadvantages

If a large structure such as a tall tower or a suspension bridge starts vibrating at its natural frequency, the result can be disastrous. The collapse of the Tacoma Narrows Bridge, USA, in 1940 may have been due to a cross-wind causing resonant vibrations, Figure 16.8. Models of new bridges are now tested in wind tunnels to check that their natural frequencies are outside the range of vibrations caused by wind.

Some singers who can produce very high frequency notes are said to be able to break wine glasses when the notes have the same frequency as the natural frequency of the glass.

Figure 16.8 The Tacoma Narrows Bridge after its collapse

Noise pollution

Unpleasant sounds are called noises. High-pitched noises are usually more annoying than low-pitched ones. Noise can damage the ears, cause tiredness and loss of concentration and, if it is very loud, result in sickness and temporary deafness. Sudden increases in loudness cause most damage. Some of the main 'noise polluters' are aircraft, motor vehicles, greatly amplified music and many types of machinery including domestic appliances.

Ways of reducing unwanted noise include designing quieter engines and better exhaust systems. For example, rotating shafts in machinery can be balanced better so that they do not cause vibration. Car engines are often mounted on metal brackets via rubber blocks which absorb vibrations and do not pass them on to the car body.

The use in the home of sound-insulating materials, such as carpets and curtains, and of double-glazed windows also helps. The farther away the noise originates the weaker it is, so distance is a natural barrier, as are trees between houses and a noisy road. Tractor drivers, factory workers, pneumatic drill operators and others exposed regularly to noise often have to wear ear protectors.

Noise levels are measured in **decibels** (dB) by a noise meter. The sound level the average human ear can just detect, called the **threshold of hearing**, is taken as 0 dB. Normal conversation is about 60 dB, a jet plane overhead is 100 dB and the **threshold of pain** (which explains itself) is 120 dB.

QUESTIONS

1 a Draw the waveform of (i) a loud, low-pitched note and (ii) a soft, high-pitched note.

b If the speed of sound is 340 m/s what is the wavelength of a note of frequency (i) 340 Hz, (ii) 170 Hz?

2 a What change does your ear detect when you are listening to a sound if (i) the **amplitude** is raised, (ii) the **frequency** is raised?

b A certain length of guitar string gives the note middle C. (i) What note do you get with half the length? (ii) How could you raise the pitch of the note without changing the length? *(S.R. part qn.)*

Checklist

After studying this chapter you should be able to

- use the terms **pitch**, **loudness** and **quality** (timbre) and connect them to wave properties,
- describe stationary (standing) waves and explain how they are produced in string instruments,
- recall the factors affecting the frequency of the note emitted by a vibrating string,
- describe resonance and its effects,
- discuss noise pollution constructively.

ADDITIONAL QUESTIONS

The first questions under each topic heading are 'basic' questions intended for all students; the following questions marked with a stripe are 'higher' questions for those seeking grades E to A*.

Mechanical waves

1 A straight vibrator causes water ripples to travel across the surface of a shallow tank. The waves travel a distance of 33 cm in 1.5 s and the distance between successive wave crests is 4.0 cm. Calculate the frequency of the vibrator. *(W.)*

2 The diagram shows a series of waves as might be produced in a laboratory experiment with a ripple tank. The waves are travelling from region A to region B.

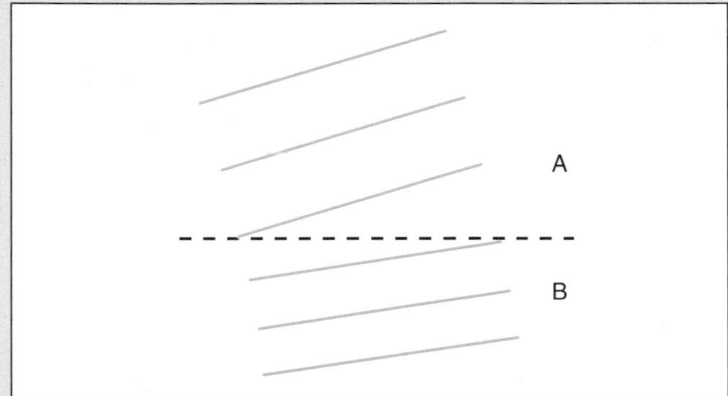

a Describe how these waves are produced.

b What is happening along the dotted line?

c By measurement, determine the wavelength of the waves in the region A and the region B.

d What can you say about the frequency of the waves in region A and region B?

e What can you say about the speed of waves in region A and region B?

f Where is the water deepest – at A or B? *(E.M.)*

3 The diagrams (a) and (b) show ripples approaching metal plates with gaps in them. Gap (a) is narrow compared with the wavelength and gap (b) is wide compared with the wavelength.

Copy the diagrams and draw the shapes of the ripples after passing through the gaps.

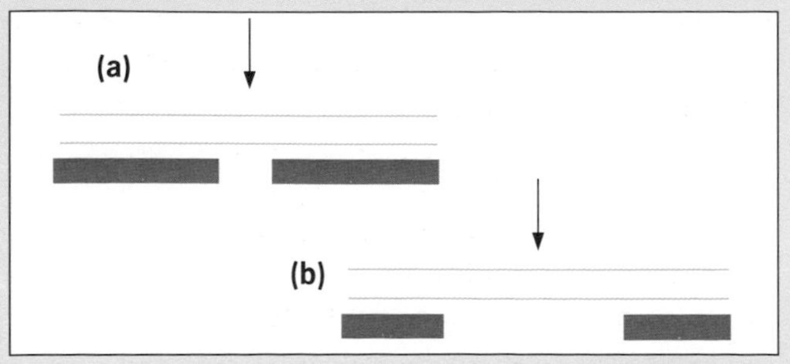

4 The coloured line shows the position of a water wave travelling to the right. X, Y and Z are corks floating on the surface and the black dotted line is the undisturbed water surface. As the wave moves forward does (i) X, (ii) Y, (iii) Z, move up or down or stay where it is?

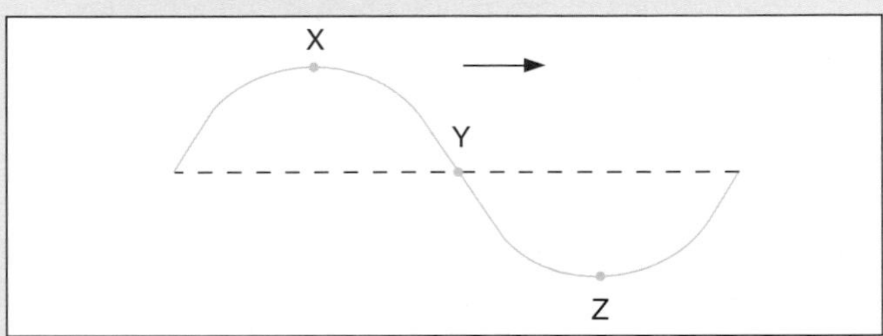

Light waves

5 In the diagram light waves are incident on an air–glass boundary. Some are reflected and some are refracted in the glass. One of the following is the same for the incident wave and the refracted wave. Which?

A speed **B** wavelength **C** direction
D brightness **E** frequency

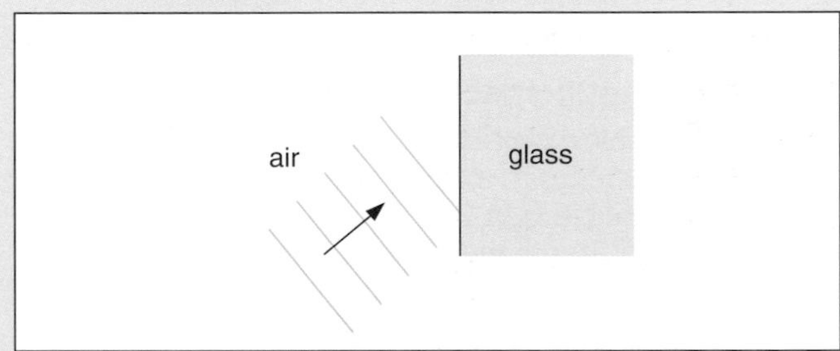

6 A double slit is formed on a blackened glass plate by drawing two lines close together. When a blue light is viewed through the double slit, an interference pattern like that in diagram (a) is seen.

a What can be concluded about the nature of light from this pattern?

b If the blue light is replaced by one giving a different colour of light, the pattern in diagram (b) is seen. Account for its being more spread out.

c What could be the colour of the light in the second case?

7 If you held a diffraction grating with its lines vertical close to the eye in a darkened room and looked through the grating at a small, bright source of white light some distance away, describe and explain what you might expect to see.

What is meant by the **order of a spectrum**? *(O. and C. part qn.)*

8 In the diagram plane wavefronts of light, i.e. a parallel beam, are falling normally on a convex lens.

a Use the wave theory to explain why they are changed to curved wavefronts on passing through the lens.

b What happens at I?

c What does the distance EI equal?

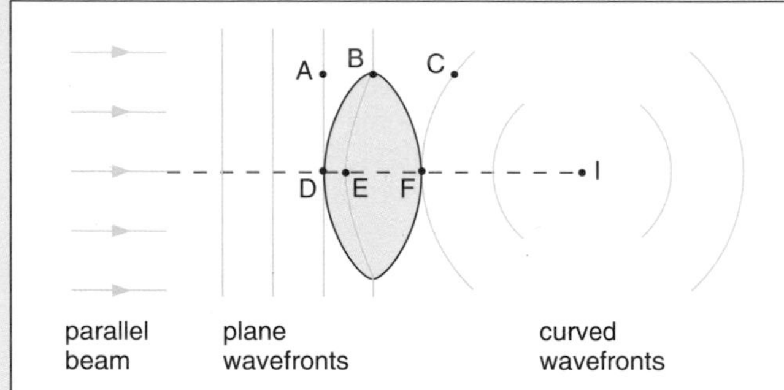

Electromagnetic spectrum

9 The strip below represents part of the electromagnetic spectrum extending on both sides of the region of visible light C. If A is the region of shortest wavelengths and F the region of longest wavelengths, name which section will represent the **a** UV region, **b** X-rays, **c** IR region.

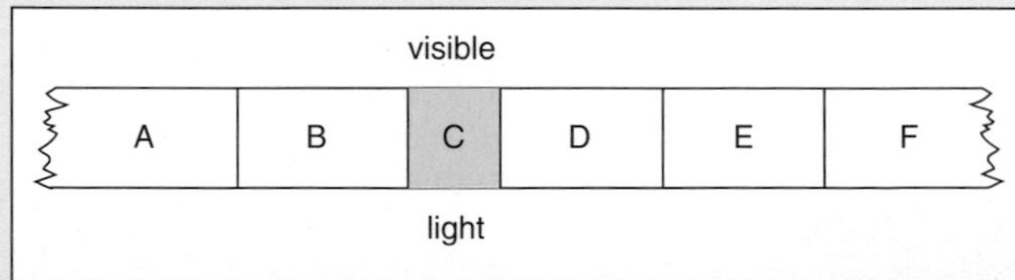

State one difference between the behaviour of ultraviolet and X-rays. *(L.)*

10 Light travels at 3×10^8 m/s in air (300 000 000 m/s).

a How fast do radio waves travel in air?

b What is the wavelength of the 80 MHz waves used for broadcasting on VHF? (1 MHz = 1 000 000 Hz)

c What is the frequency of the 1500 m radio waves used on the Long Wave Band? *(S.R.)*

Sound waves

11 Two loudspeakers play the same note from a signal generator. A person moving across the line between them (see diagram) finds the note is loud in some places and soft in others.

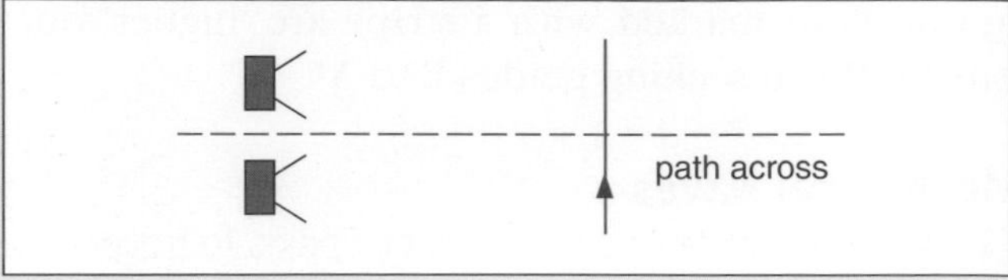

Why is the note loud in some positions and soft in others? *(O.L.E./S.R.16+ part qn.)*

12 The property which distinguishes longitudinal waves from transverse waves is the

A wavelength **B** velocity **C** ability to be refracted
D need for a material medium
E relative directions of oscillation and propagation. *(J.M.B.)*

13 An echo-sounder in a trawler receives an echo from a shoal of fish 0.4 s after it was sent. If the speed of sound in water is 1500 m/s, how deep is the shoal?

A 150 m **B** 300 m **C** 600 m **D** 7500 m
E 10 000 m

14 a State three differences between sound waves and radio waves.

b Explain the following: (i) echo, (ii) reverberation.

c A girl claps her hands at a steady rate of 30 claps a minute and hears the echoes from a high wall. When she is 170 m away from the wall, she hears the echo of each clap midway between it and the next clap. What is the speed of sound from these observations?

Musical notes

15 a On what does (i) the loudness, (ii) the pitch of a sound depend?

b Why does middle C on the piano sound different from the same note on the violin?

16 How is a standing wave produced? How does it differ from a progressive wave?

17 What is meant by resonance? Give an example.

MATTER AND MOLECULES

17 MEASUREMENTS

Units and basic quantities	*Volume*
Powers of ten shorthand	*Mass*
Length	*Time*
Significant figures	*Practical work* Oscillations of a mass–spring system.
Area	

Units and basic quantities

Before a measurement can be made, a standard or **unit** must be chosen. The size of the quantity to be measured is then found with an instrument having a scale marked in the unit.

Three basic quantities we have to measure in physics are **length**, **mass** and **time**. Units for other quantities are based on them. The SI (Système International d'Unités) system is a set of metric units now used in many countries. It is a decimal system in which units are divided or multiplied by 10 to give smaller or larger units.

Powers of ten shorthand

This is a neat way of writing numbers especially if they are large or small. It works like this:

$$
\begin{aligned}
4000 &= 4 \times 10 \times 10 \times 10 &&= 4 \times 10^3 \\
400 &= 4 \times 10 \times 10 &&= 4 \times 10^2 \\
40 &= 4 \times 10 &&= 4 \times 10^1 \\
4 &= 4 \times 1 &&= 4 \times 10^0 \\
0.4 &= 4/10 = 4/10^1 &&= 4 \times 10^{-1} \\
0.04 &= 4/100 = 4/10^2 &&= 4 \times 10^{-2} \\
0.004 &= 4/1000 = 4/10^3 &&= 4 \times 10^{-3}
\end{aligned}
$$

The small figures 1, 2, 3, etc., are called **powers of ten** and give the number of times the number has to be multiplied by 10 if it is greater than 1 or divided by 10 if it is less than 1. Note that 1 is also written as 10^0 and for numbers less than 1 the power has a negative sign.

Length

The unit of length is the **metre** (m) and is the distance, believed never to alter, occupied by a certain number of wavelengths of a particular colour of light in a vacuum. Previously it was the distance between two marks on a certain metal bar. Submultiples are:

$$
\begin{aligned}
1 \text{ centimetre (cm)} &= 10^{-2} \text{ m} \\
1 \text{ millimetre (mm)} &= 10^{-3} \text{ m} \\
1 \text{ micrometre } (\mu\text{m}) &= 10^{-6} \text{ m} \\
1 \text{ nanometre (nm)} &= 10^{-9} \text{ m}
\end{aligned}
$$

A multiple for large distances is

$$1 \text{ kilometre (km)} = 10^3 \text{ m } (\tfrac{5}{8} \text{ mile approx.})$$

Many length measurements are made with rulers; the correct way to read one is shown in Figure 17.2. The reading is 76 mm or 7.6 cm. Your eye must be right over the mark on the scale or the thickness of the ruler causes errors.

Figure 17.1 Measuring instruments on the flight deck of Concorde provide the pilot with information about the performance of the aircraft

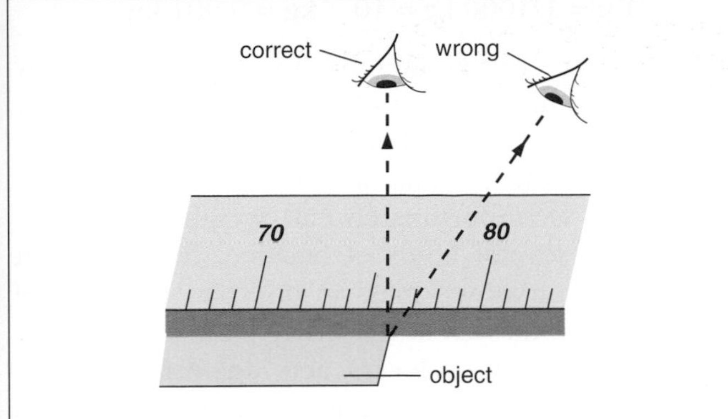

Figure 17.2 The correct way to measure with a ruler

Significant figures

Every measurement of a quantity is an attempt to find its true value and is subject to errors arising from limitations of the apparatus and the experimenter. The number of figures, called **significant** figures, given for a measurement indicates how accurate we think it is and more figures should not be given than is justified.

For example, a value of 4.5 for a measurement has two significant figures; 0.0385 has three significant figures, 3 being the most significant and 5 the least, i.e. it is the one we are least sure about since it might be 4 or it might be 6. Perhaps it had to be estimated by the experimenter because the reading was between two marks on a scale.

When doing a calculation your answer should have the same number of significant figures as the measurements used in the calculation. For example, if your calculator gave an answer of 3.4185062, this would be written as 3.4 if the measurements had two significant figures. It would be written as 3.42 for three significant figures. Note that in deciding the least significant figure you look at the next figure. If it is less than 5 you leave the least significant figure as it is (hence 3.41 becomes 3.4) but if it equals or is greater than 5 you increase the least significant figure by 1 (hence 3.418 becomes 3.42).

If a number is expressed in powers of ten form (also called **standard form**), the number of significant figures is the number of digits before the power of ten. For example, 2.73×10^3 has three significant figures.

Area

The area of the square in Figure 17.3a with sides 1 cm long is 1 square centimetre (1 cm^2). In Figure 17.3b the rectangle measures 4 cm by 3 cm and has an area of $4 \times 3 = 12$ cm^2 since it has the same area as twelve squares each of area 1 cm^2. The **area of a square** or **rectangle** is given by

$$\text{area} = \text{length} \times \text{breadth}$$

The SI unit of area is the square metre (m^2) which is the area of a square with sides 1 m long. Note that

$$1\ \text{cm}^2 = \frac{1}{100}\ \text{m} \times \frac{1}{100}\ \text{m} = \frac{1}{10\ 000}\ \text{m}^2 = 10^{-4}\ \text{m}^2$$

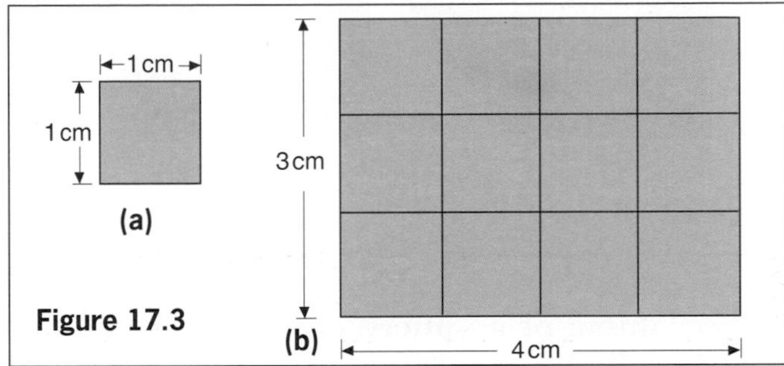

Figure 17.3

Sometimes we need to know the **area of a triangle** (Chapter 32). It is given by

$$\text{area of triangle} = \tfrac{1}{2} \times \text{base} \times \text{height}$$

For example in Figure 17.4

$$\text{area } \triangle ABC = \tfrac{1}{2} \times AB \times AC$$
$$= \tfrac{1}{2} \times 4\ \text{cm} \times 6\ \text{cm} = 12\ \text{cm}^2$$

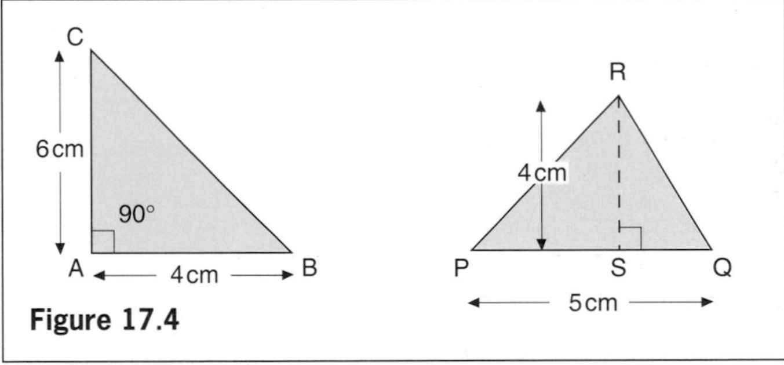

Figure 17.4

$$\text{area } \triangle PQR = \tfrac{1}{2} \times PQ \times SR$$
$$= \tfrac{1}{2} \times 5\ \text{cm} \times 4\ \text{cm} = 10\ \text{cm}^2$$

The **area of a circle** of radius r is πr^2 where $\pi = 22/7$ or 3.14; its **circumference** is $2\pi r$.

Volume

Volume is the amount of space occupied. The unit of volume is the **cubic metre** (m^3) but as this is rather large, for most purposes the **cubic centimetre** (cm^3) is used. The volume of a cube with 1 cm edges is 1 cm^3. Note that

$$1\ \text{cm}^3 = \frac{1}{100}\ \text{m} \times \frac{1}{100}\ \text{m} \times \frac{1}{100}\ \text{m}$$

$$= \frac{1}{1\ 000\ 000}\ \text{m}^3 = 10^{-6}\ \text{m}^3$$

For a regularly shaped object such as a rectangular block, Figure 17.5 shows that

volume = length × breadth × height

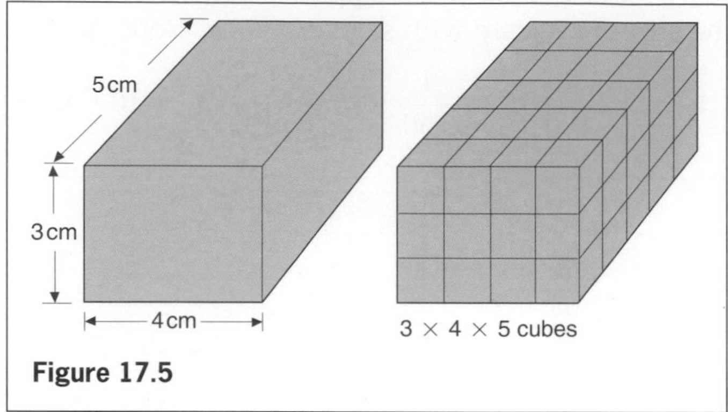

Figure 17.5

The **volume of a sphere** of radius r is $\frac{4}{3}\pi r^3$ and that of a **cylinder** of radius r and height h is $\pi r^2 h$.

The **volume of a liquid** may be obtained by pouring it into a measuring cylinder, Figure 17.6a. A known volume can be run off accurately from a burette, Figure 17.6b. When making a reading both vessels must be upright and the eye level with the bottom of the curved liquid surface, i.e. the meniscus. The meniscus formed by mercury is curved oppositely to that of other liquids and the top is read.

Liquid volumes are also expressed in litres (l); 1 litre = 1000 cm^3. One millilitre (1 ml) = 1 cm^3.

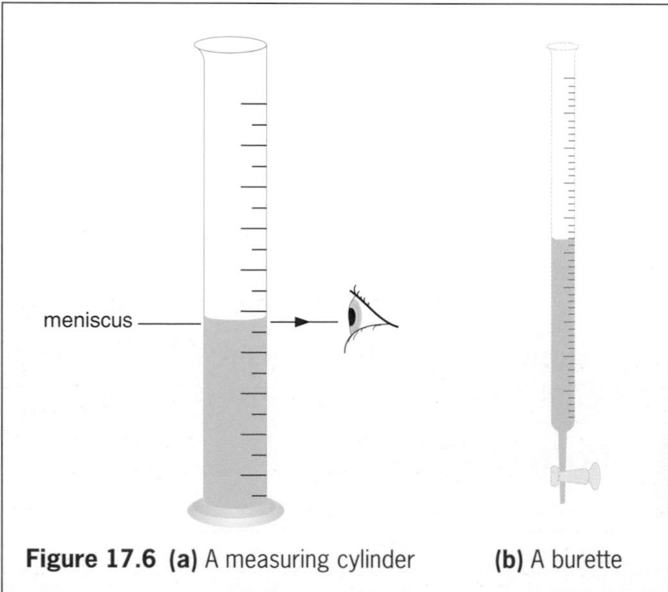

Figure 17.6 (a) A measuring cylinder **(b)** A burette

Mass

The mass of an object is the measure of the amount of matter in it. The unit of mass is the **kilogram** (kg) and is the mass of a piece of platinum-iridium alloy at the Office of Weights and Measures in Paris. The gram (g) is one-thousandth of a kilogram.

$$1 \text{ g} = 1/1000 \text{ kg} = 10^{-3} \text{ kg} = 0.001 \text{ kg}$$

The term **weight** is often used when mass is really meant. In science the two ideas are distinct and have different units as we shall see later. The confusion is not helped by the fact that mass is found on a balance by a process we unfortunately call 'weighing'!

There are several kinds of balance. In the **beam balance** the unknown mass in one pan is balanced against known masses in the other pan. In the **lever balance** a system of levers acts against the mass when it is placed in the pan. A direct reading is obtained from the position on a scale of a pointer joined to the lever system. A digital **top-pan balance** is shown in Figure 17.7.

Figure 17.7 A digital top-pan balance

Time

The unit of time is the second which until 1968 was based on the length of a day, this being the time for the Earth to revolve once on its axis. However, days are not all of exactly the same duration and the second is now defined as the time interval for a certain number of energy changes to occur in the caesium atom.

Time-measuring devices rely on some kind of constantly repeating oscillations. In traditional clocks and watches a small wheel (the balance wheel) oscillates to and fro; in digital clocks and watches the oscillations are produced by a tiny quartz crystal. A swinging pendulum controls a pendulum clock.

PRACTICAL WORK

Oscillations of a mass–spring system

In this investigation you have to make time measurements using a stopwatch or clock.

Support a spiral steel spring as in Figure 17.8a. Hang a mass (e.g. 50 g) from its lower end so that it is stretched several centimetres. Pull the mass *vertically*

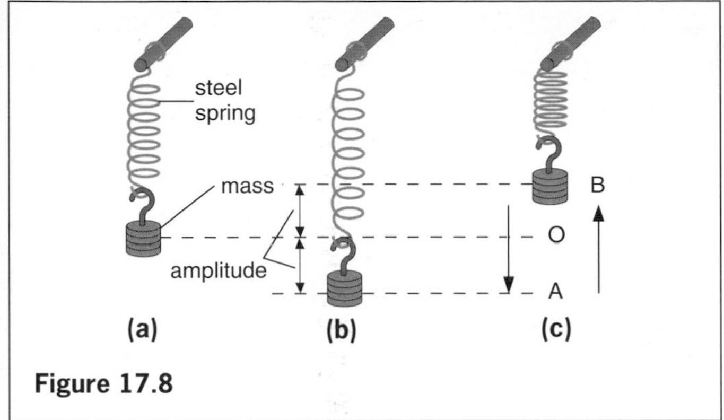

Figure 17.8

downwards a few centimetres, Figure 17.8b, and release it so that it oscillates up and down above and below its rest position.

Find the time for the mass to make several complete oscillations; one oscillation is from A to O to B to O to A, Figure 17.8c. Repeat the timing a few times for the same number of oscillations and work out the average. The time for one oscillation is the **period** T. What is it for your system? The **frequency** f of the oscillations is the number of complete oscillations per second and equals $1/T$. Calculate f.

How does the amplitude of the oscillations, Figure 17.8b, change with time? Investigate the effect on T of (i) a greater mass, (ii) a smaller mass, (iii) a different spring.

Mechanical oscillations (vibrations) have their uses (e.g. to produce sound) but if they build up in a system (e.g. in a bridge, Chapter 16) they can cause damage.

QUESTIONS

1 How many millimetres are there in **a** 1 cm, **b** 4 cm, **c** 0.5 cm, **d** 6.7 cm, **e** 1 m?

2 What are these lengths in metres: **a** 300 cm, **b** 550 cm, **c** 870 cm, **d** 43 cm, **e** 100 mm?

3 **a** Write the following in powers of ten form with one figure before the decimal point:

100 000; 3500; 428 000 000; 504; 27 056

b Write out the following in full:

10^3; 2×10^6; 6.9×10^4; 1.34×10^2; 10^9

4 **a** Write these fractions in powers of ten form:

1/1000; 7/100 000; 1/10 000 000; 3/60 000

b Express the following decimals in powers of ten form with one figure before the decimal point:

0.5; 0.084; 0.000 36; 0.001 04

5 The pages of a book are numbered 1 to 200 and each leaf is 0.10 mm thick. If each cover is 0.20 mm thick, what is the thickness of the book?

6 How many significant figures are there in a length measurement of **a** 2.5 cm, **b** 5.32 cm, **c** 7.180 cm, **d** 0.042 cm?

7 A rectangular block measures 4.1 cm by 2.8 cm by 2.1 cm. Calculate its volume giving your answer to an appropriate number of significant figures.

8 A metal block measures 10 cm × 2 cm × 2 cm. What is its volume? How many blocks each 2 cm × 2 cm × 2 cm have the same total volume?

9 How many blocks of ice cream each 10 cm × 10 cm × 4 cm can be stored in the compartment of a freezer measuring 40 cm × 40 cm × 20 cm?

10 A Perspex box has a 6 cm square base and contains water to a height of 7 cm, Figure 17.9.

a What is the volume of the water?

b A stone is lowered into the water so as to be completely covered and the water rises to a height of 9 cm. What is the volume of the stone?

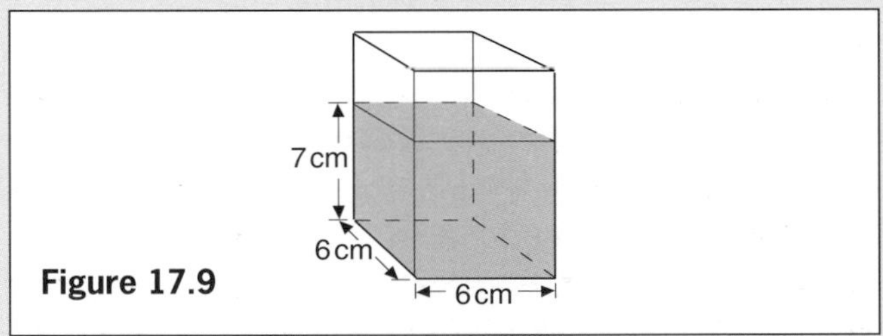

Figure 17.9

Checklist

After studying this chapter you should be able to

- recall three basic quantities in physics,
- write a number in powers of ten (standard) form,
- recall the unit of length and the meaning of the prefixes **kilo**, **centi**, **milli**, **micro**, **nano**,
- use a ruler to measure length so as to minimize errors,
- give a result to an appropriate number of significant figures,
- measure areas of squares, rectangles, triangles and circles,
- measure the volume of regular solids and of liquids,
- recall the unit of mass and how mass is measured,
- recall the unit of time and how time is measured,
- describe an experiment to find the period of a mass–spring system.

18 DENSITY

Calculations	*Floating and sinking*
Simple density measurements	

In everyday language lead is said to be 'heavier' than wood. By this it is meant that a certain volume of lead is heavier than the same volume of wood. In science such comparisons are made by using the term **density**. This is the **mass per unit volume** of a substance and is calculated from

$$\text{density} = \frac{\text{mass}}{\text{volume}}$$

The density of lead is 11 grams per cubic centimetre (11 g/cm^3) and this means that a piece of lead of volume 1 cm^3 has mass 11 g. A volume of 5 cm^3 of lead would have mass 55 g. Knowing the density of a substance the mass of *any* volume can be calculated. This enables engineers to work out the weight of a structure if they know from the plans the volumes of the materials to be used and their densities. Strong enough foundations can then be made.

The SI unit of density is the **kilogram per cubic metre**. To convert a density from g/cm^3, normally the most suitable unit for the size of sample we use, to kg/m^3, we multiply by 10^3. For example the density of water is 1.0 g/cm^3 or 1.0×10^3 kg/m^3.

The approximate densities of some common substances are given in the table.

Table 18.1 Densities of some common substances

Solids	Density g/cm^3	Liquids	Density g/cm^3
aluminium	2.7	meths	0.80
copper	8.9	paraffin	0.80
iron	7.9	petrol	0.80
gold	19.3	pure water	1.0
glass	2.5	mercury	13.6
wood (teak)	0.80	**Gases**	kg/m^3
ice	0.92	air	1.3
polythene	0.90	hydrogen	0.09

Calculations

Using the symbols d for density, m for mass and V for volume, the expression for density is

$$d = \frac{m}{V}$$

Rearranging the expression gives

$$m = V \times d \quad \text{and} \quad V = \frac{m}{d}$$

These are useful if d is known and m or V have to be calculated. If you do not see how they are obtained refer to the *Mathematics for physics* section on p. 344. The triangle in Figure 18.1 is an aid to remembering them. If you cover the quantity you want to know with a finger, e.g. m, it equals what you can still see, i.e. $d \times V$. To find V, cover V and you get $V = m/d$.

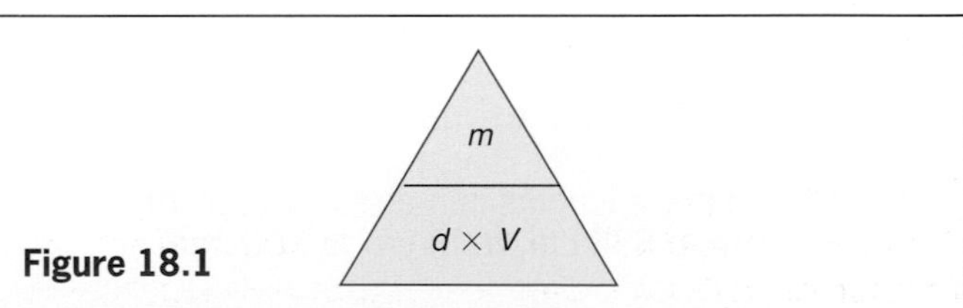

Figure 18.1

Worked example

Taking the density of copper as 9 g/cm^3, find **a** the mass of 5 cm^3 and **b** the volume of 63 g.

a $d = 9$ g/cm^3, $V = 5$ cm^3 and m is to be found.

$$\therefore \quad m = V \times d = 5\text{ cm}^3 \times 9\text{ g/cm}^3 = 45\text{ g}$$

b $d = 9$ g/cm^3, $m = 63$ g and V is to be found.

$$\therefore \quad V = \frac{m}{d} = \frac{63\text{ g}}{9\text{ g/cm}^3} = 7\text{ cm}^3$$

Simple density measurements

If the mass m and volume V of a substance are known its density can be found from $d = m/V$.

(a) Regularly shaped solid

The mass is found on a balance and the volume by measuring its dimensions with a ruler.

(b) Irregularly shaped solid, e.g. pebble or glass stopper

The solid is weighed and its volume measured by one of the methods shown in Figures 18.2a, b. In (a) the volume is the difference between the first and second readings. In (b) it is the volume of water collected in the measuring cylinder.

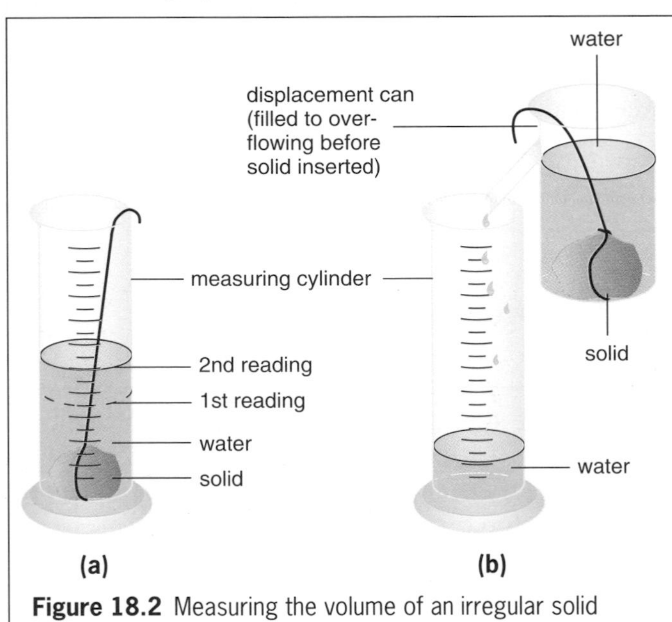

Figure 18.2 Measuring the volume of an irregular solid

(c) Liquid

A known volume is transferred from a burette or a measuring cylinder into a weighed beaker which is reweighed to give the mass of liquid.

(d) Air

A 500 cm^3 round-bottomed flask is weighed full of air and then after removing the air with a vacuum pump; the difference gives the mass of air in the flask. The volume of air is found by filling the flask with water and pouring it into a measuring cylinder.

Floating and sinking

An object sinks in a liquid of smaller density than its own; otherwise it floats, partly or wholly submerged. For example, a piece of glass of density 2.5 g/cm^3 sinks in water (density 1.0 g/cm^3) but floats in mercury (density 13.6 g/cm^3). An iron nail sinks in water but an iron ship floats because its average density is less than that of water.

Figure 18.3 Why is it easy to float in the Dead Sea?

QUESTIONS

1 **a** If the density of wood is 0.5 g/cm^3 what is the mass of (i) 1 cm^3, (ii) 2 cm^3, (iii) 10 cm^3?

b What is the density of a substance of (i) mass 100 g and volume 10 cm^3, (ii) volume 3 m^3 and mass 9 kg?

c The density of gold is 19 g/cm^3. Find the volume of (i) 38 g, (ii) 95 g of gold.

2 A piece of steel has a volume of 12 cm^3 and a mass of 96 g. What is its density in **a** g/cm^3, **b** kg/m^3?

3 What is the mass of 5 m^3 of cement of density 3000 kg/m^3?

4 What is the mass of air in a room measuring 10 m $\times$ 5.0 m $\times$ 2.0 m if the density of air is 1.3 kg/m^3?

5 A perspex box has a 10 cm square base and contains water to a height of 10 cm. A piece of rock of mass 600 g is lowered into the water and the level rises to 12 cm.

a What is the volume of water displaced by the rock?

b What is the volume of the rock?

c Calculate the density of the rock. *(E.M.)*

Checklist

After studying this chapter you should be able to

- define **density** and perform calculations using $d = m/V$,
- describe experiments to measure the density of solids, liquids and air,
- relate floating and sinking to density.

19 WEIGHT AND SPRINGS

Force	***Hooke's law***
Weight	***Practical work*** Stretching a spring.
The newton	

Force

A **force** is a push or a pull. It can cause a body at rest to move, or if the body is already moving it can change its speed or direction of motion. It can also change its shape or size.

Weight

We all constantly experience the force of gravity, i.e. the pull of the Earth. It causes an unsupported body to fall from rest to the ground.

The weight of a body is the force of gravity on it.

The nearer a body is to the centre of the Earth the more the Earth attracts it. Since the Earth is not a perfect sphere but is flatter at the poles, the weight of a body varies over the Earth's surface. It is greater at the poles than at the equator.

Gravity is a force which can act through space, i.e. there does not need to be contact between the Earth and the object on which it acts as there does when we push or pull something. Other action-at-a-distance forces which, like gravity, decrease with distance are:

(i) **magnetic** forces between magnets, and
(ii) **electric** forces between electric charges.

The newton

The unit of force is the **newton** (N). It will be defined later (Chapter 34); the definition is based on the change of speed a force can produce on a body. Weight is a force and therefore should be measured in newtons.

Figure 19.1 A weightlifter in action exerts both a pull and a push

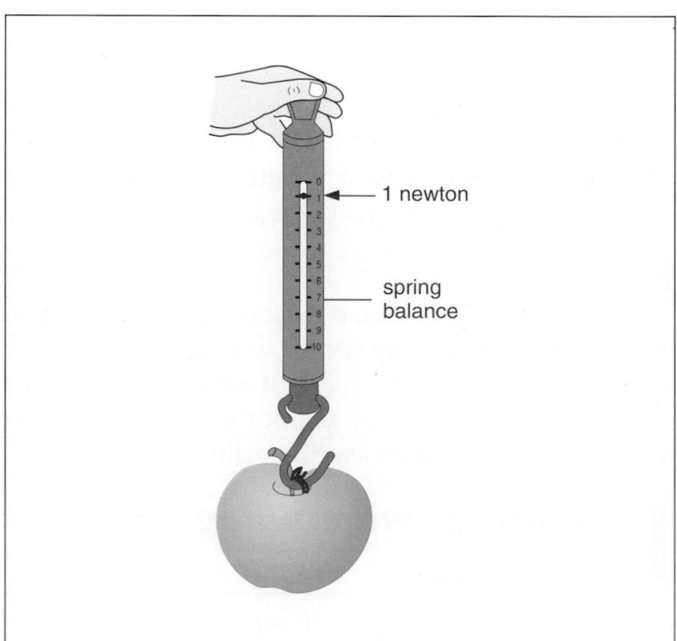

Figure 19.2 The weight of an average-sized apple is about 1 newton

The weight of a body can be measured by hanging it on a **spring balance** marked in newtons, Figure 19.2, and letting the pull of gravity stretch the spring in the balance. The greater the pull the more the spring stretches. On most of the Earth's surface:

> The weight of a body of mass 1 kg is 9.8 N.

Often this is taken as 10 N. A mass of 2 kg has a weight of 20 N and so on. The mass of a body is the same wherever it is and does not depend on the presence of the Earth as does weight.

Mass is measured by a lever, beam or top-pan balance; weight is measured by a spring balance.

PRACTICAL WORK

Stretching a spring

Arrange a steel spring as in Figure 19.3. Read the scale opposite the bottom of the hanger. Add 100 g loads one at a time (thereby increasing the stretching force by steps of 1 N) and take the readings after each one. Enter the readings in a table for loads up to 500 g.

Stretching force (N)	**Scale reading** (mm)	**Total extension** (mm)

Do the results suggest any rule about how the spring behaves when it is stretched?

Sometimes it is easier to discover laws by displaying the results on a graph. Do this on graph paper by plotting **stretching force** readings along the *y*-axis (vertical axis) and **total extension** readings along the *x*-axis (horizontal axis). Every pair of readings will give a point; mark them by small crosses and draw a smooth line through them. What is its shape?

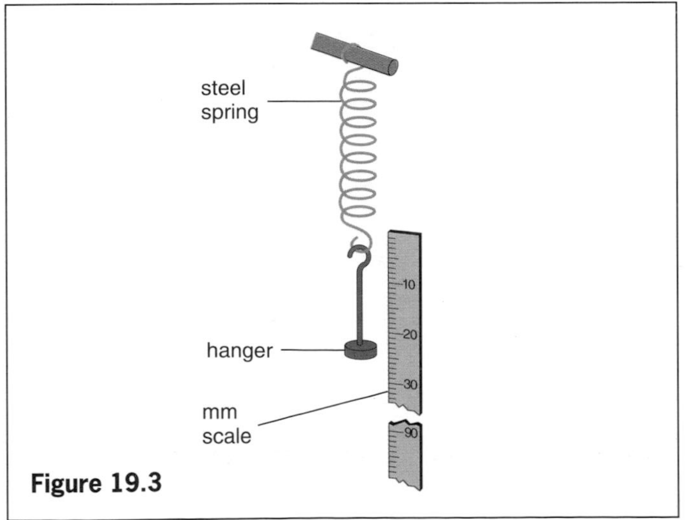

Figure 19.3

Hooke's law

Springs were investigated by Hooke about 300 years ago. He found that the extension was proportional to the stretching force so long as the spring was not permanently stretched. This means that doubling the force doubles the extension, trebling the force trebles the extension and so on. Using the sign for proportionality we can write **Hooke's law** as

> extension $\propto$ stretching force

It is true only if the **elastic limit** of the spring is not exceeded. The spring returns to its original length when the force is removed.

The graph of Figure 19.4 is for a spring stretched beyond its elastic limit E. OE is a straight line passing through the origin O and is graphical proof that Hooke's law holds over this range. If the force for point A on the graph, called the **yield point**, is applied to the spring, the elastic limit is passed and on removing the force some of the extension (OS) remains. Over which part of the graph does a spring balance work?

The **force constant** *k* of a spring is the force needed to cause unit extension, e.g. 1 m. If a force *F* produces extension *e* then

$$k = \frac{F}{e}$$

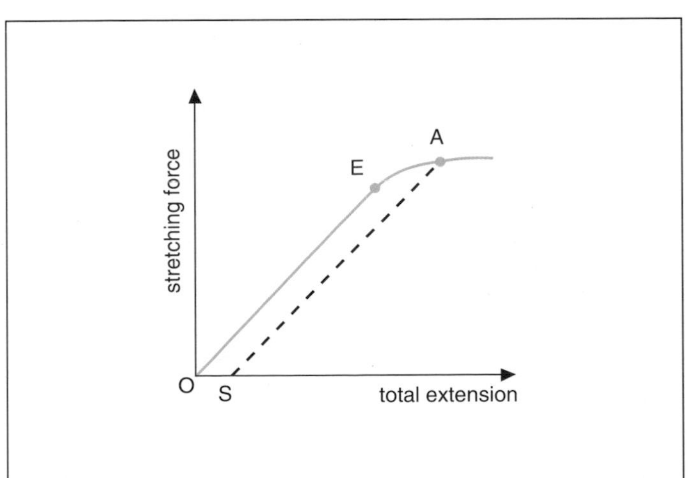

Figure 19.4 Force–extension graph for a spring

Worked example

A spring is stretched 10 mm (0.01 m) by a weight of 2.0 N. Calculate **a** the force constant k and **b** the weight W of an object which causes an extension of 80 mm (0.08 m).

a We have $k = \frac{F}{e} = \frac{2.0\ \text{N}}{0.01\ \text{m}} = 200\ \text{N/m}$

b W = stretching force $F = k \times e$
$= 200\ \text{N/m} \times 0.08\ \text{m} = 16\ \text{N}$

QUESTIONS

1 A body of mass 1 kg has weight 10 N at a certain place. What is the weight of **a** 100 g, **b** 5 kg, **c** 50 g?

2 The force of gravity on the Moon is said to be one-sixth of that on the Earth. What would a mass of 12 kg weigh **a** on the Earth and **b** on the Moon?

3 What is the force constant of a spring which is stretched

a 2 mm by a force of 4 N,

b 4 cm by a mass of 200 g?

4 Figure 19.5 shows four diagrams, not to scale, of the same spring which obeys Hooke's law.

a What is the length *x*?

b What is the mass *M*? *(E.A.)*

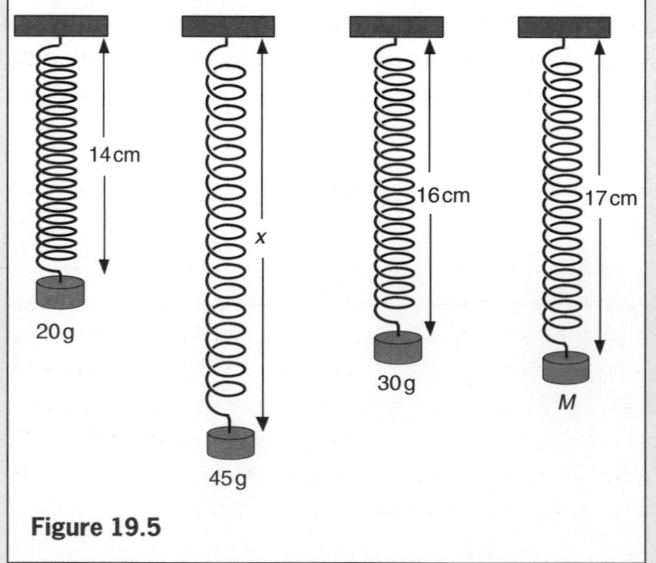

Figure 19.5

Checklist

After studying this chapter you should be able to

- recall that a force can cause a change in the motion, size or shape of a body,
- recall that the weight of a body is the force of gravity on it,
- recall the unit of force and how force is measured,
- describe an experiment to study the relation between force and extension for springs,
- draw conclusions from force–extension graphs,
- recall Hooke's law and solve problems using it,
- explain the term **force constant**.

20 MOLECULES

Kinetic theory of matter	*Practical work*
Solids, liquids, gases.	Brownian motion.
Crystals	

Matter is made up of tiny particles or **molecules** which are too small for us to see directly. But they can be 'seen' by scientific 'eyes'. One of these is the electron microscope: Figure 20.1 is a photograph taken with such an instrument showing molecules of a protein. Molecules consist of even smaller particles called **atoms** and are in continuous motion.

Figure 20.1 Protein molecules

PRACTICAL WORK

Brownian motion

The apparatus is shown in Figure 20.2a. First fill the glass cell with smoke using a burning drinking straw (made of waxed paper), Figure 20.2b. Replace the lid on the apparatus and set it on the microscope platform. Connect the lamp to a 12 V supply; the glass rod acts as a lens and focuses light on the smoke.

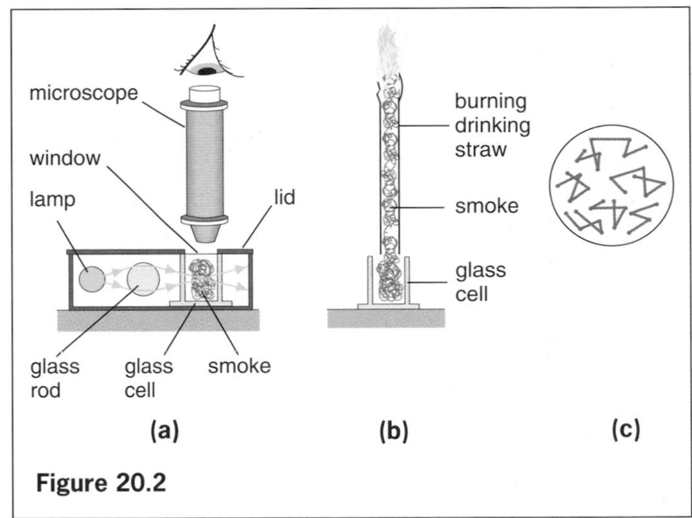

Figure 20.2

Carefully adjust the microscope until you see bright specks dancing around haphazardly, Figure 20.2c. The specks are smoke particles seen by reflected light; their random motion is due to collisions with fast-moving air molecules in the cell. The effect is called **Brownian motion**.

Kinetic theory of matter

As well as being in continuous motion, molecules also exert strong electric forces on one another when they are close together. The forces are both attractive and repulsive. The former hold molecules together and the latter cause matter to resist compression. The **kinetic theory** can explain the existence of the solid, liquid and gaseous states.

(a) Solids

The theory states that in solids the molecules are close together and the attractive and repulsive forces between neighbouring molecules balance. Also each molecule vibrates to and fro about a fixed position.

It is just as if springs, representing the electric forces between molecules, held the molecules together, Figure 20.3. This enables the solid to keep a

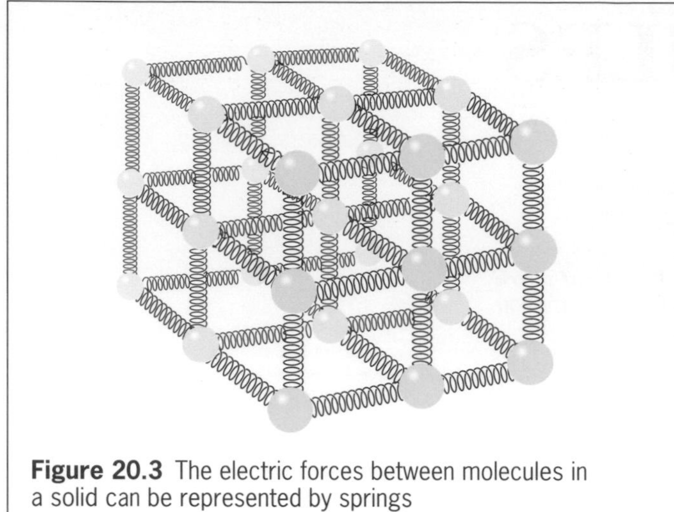

Figure 20.3 The electric forces between molecules in a solid can be represented by springs

definite shape and volume, while still allowing the individual molecules to vibrate backwards and forwards. The theory shows that the molecules in a solid could be arranged in a regular, repeating pattern like those formed by crystalline substances.

(b) Liquids

The theory considers that in liquids the molecules are slightly farther apart than in solids but still close enough together to have a definite volume. As well as vibrating, they can at the same time move rapidly over short distances, slipping past each other in all directions. They are never near another molecule long enough to get trapped in a regular pattern which would stop them flowing and taking the shape of the vessel containing them.

A model to represent the liquid state can be made by covering about a third of a *tilted* tray with marbles ('molecules'), Figure 20.4. It is then rotated to and fro erratically and the motion of the marbles observed. They are able to move around but most stay in the lower half of the tray, so the liquid has a fairly definite volume. A few energetic ones 'escape' from the 'liquid' into the space above. They represent molecules that have 'evaporated' from the 'liquid' surface and become 'gas' or 'vapour' molecules. The thinning out of the marbles near the 'liquid' surface can also be seen.

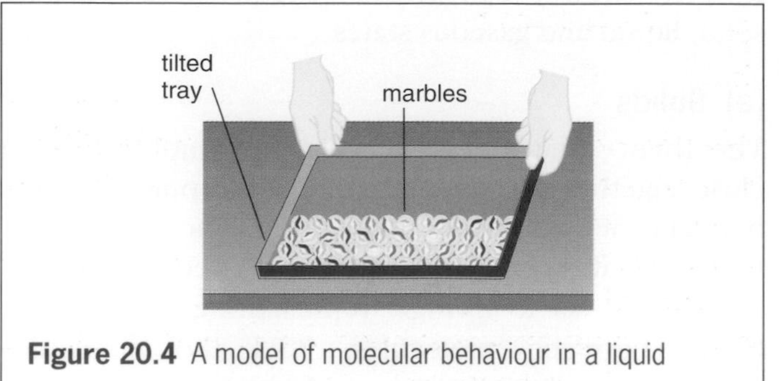

Figure 20.4 A model of molecular behaviour in a liquid

(c) Gases

The molecules in gases are much farther apart than in solids or liquids (about ten times) and so gases are much less dense and can be squeezed into a smaller space. The molecules dash around at very high speed (500 m/s for air molecules) in all the space available. It is only during the brief spells when they collide with other molecules or with the walls of the container that the molecular forces act.

A model of a gas is shown in Figure 20.5. The faster the vibrator works the more often the ball-bearings have collisions with the lid, the tube and with each other, representing a gas at a higher temperature. Adding more ball-bearings is like pumping more air into a tyre; it increases the pressure. If a polystyrene ball (1 cm diameter) is dropped into the tube its irregular motion represents Brownian motion.

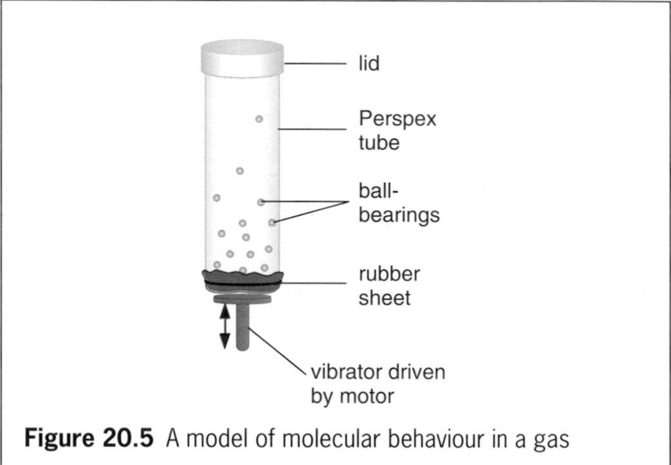

Figure 20.5 A model of molecular behaviour in a gas

Crystals

Crystals have hard, flat sides and straight edges. Whatever their size, crystals of the same substance have the same shape. This can be seen by observing through a microscope very small cubic salt crystals growing as water evaporates from salt solution on a glass slide, Figure 20.6.

Figure 20.6 Salt crystals

Figure 20.7 Splitting a calcite crystal

A calcite crystal will split cleanly if a trimming knife, held exactly parallel to one side of the crystal, is struck by a hammer, Figure 20.7.

These facts suggest crystals are made of small particles (e.g. atoms) arranged in an orderly way in planes. Metals have crystalline structures, but many other common solids such as glass, plastics and wood do not.

QUESTIONS

1 When viewing Brownian motion in a smoke cell the observer sees moving specks of light which are

A molecules moving in random motion
B molecules vibrating regularly
C molecules colliding with each other
D smoke particles vibrating regularly
E smoke particles in random motion. *(J.M.B.)*

2 Using what you know about the compressibility (squeezability) of the different states of matter, explain why

a air is used to inflate tyres,

b steel is used to make railway lines.

3 Explain in terms of molecular behaviour why bromine molecules which travel at about 200 m/s in a vacuum take several minutes to travel 40 cm inside a tube of air, Figure 20.8.

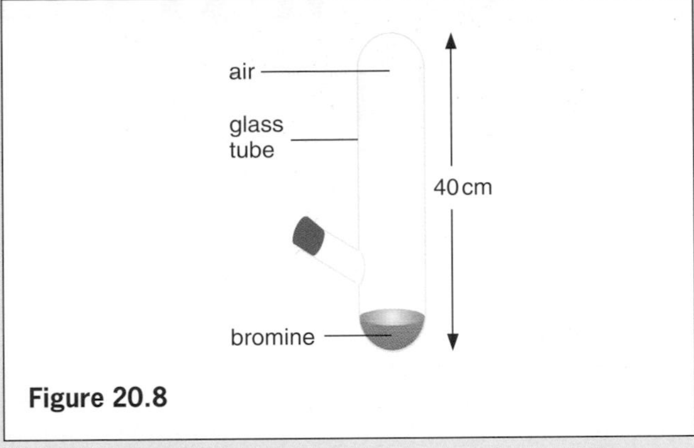

Figure 20.8

4 Graphite consists of **layers** of carbon atoms that are $2\frac{1}{2}$ times farther apart than the atoms in the layers. Why is it soft and flaky and used as a lubricant and in pencils?

Checklist

After studying this chapter you should be able to

- describe and explain an experiment to show Brownian motion,
- use the kinetic theory to explain the physical properties of solids, liquids and gases,
- recall how crystals support the view that matter is made of small particles.

21 PROPERTIES OF MATTER

Diffusion

Smells, pleasant or otherwise, travel quickly and are caused by rapidly moving molecules. The spreading of a substance of its own accord is called **diffusion** and is due to molecular motion.

Diffusion of gases can be shown if some brown nitrogen dioxide gas is made by pouring a mixture of equal volumes of concentrated nitric acid and water on copper turnings in a gas jar. When the action has stopped, a gas jar of air is inverted over the bottom jar, Figure 21.1. The brown colour spreads into the upper jar showing that nitrogen dioxide molecules diffuse upwards against gravity. Air molecules also diffuse into the lower jar.

The speed of diffusion of a gas depends on the speed of its molecules and is greater for light molecules. The apparatus of Figure 21.2 shows this. When hydrogen surrounds the porous pot, the liquid in the U-tube moves in the direction of the arrows. This is due to the lighter, faster molecules of hydrogen diffusing into the pot faster than the heavier, slower molecules of air diffuse out. The opposite happens when carbon dioxide surrounds the pot. Why?

Diffusion in liquids can be seen using the arrangement in Figure 21.3. After 24 hours the blue copper sulphate solution has diffused upwards into the water.

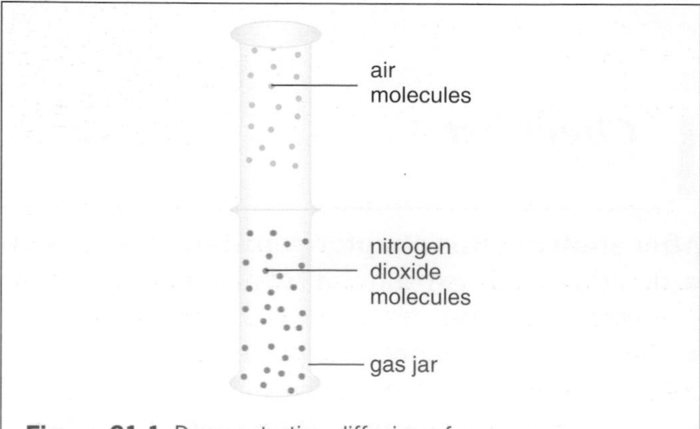

Figure 21.1 Demonstrating diffusion of a gas

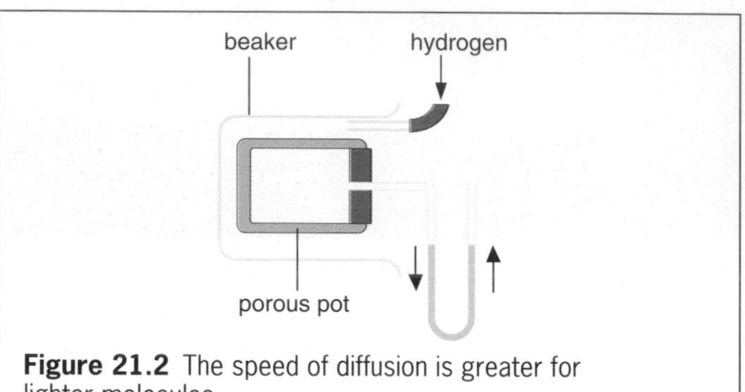

Figure 21.2 The speed of diffusion is greater for lighter molecules

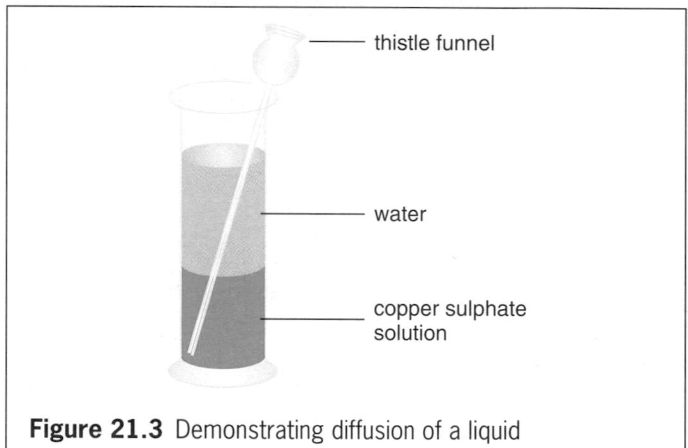

Figure 21.3 Demonstrating diffusion of a liquid

Surface tension

(a) Some effects

A needle, though made of steel which is denser than water, will float on a **clean** water surface. A film, formed by dipping an inverted funnel in a detergent solution, **rises up** the funnel, Figure 21.4a. When the film inside the cotton loop in Figure 21.4b, is broken, the loop forms a **circle**.

These facts suggest that the surface of a liquid (in a vessel or on a film) behaves as if covered with an elastic skin that is trying to shrink. The effect is called **surface tension** and it also explains why small liquid drops are nearly spherical, as can be seen when water drips from a tap, Figure 21.5. They have this

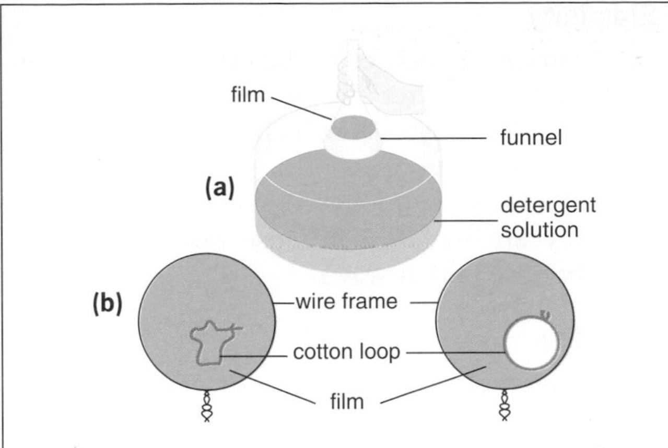

Figure 21.4 A film of liquid tries to shrink because of surface tension

Figure 21.5
Small liquid drops are almost spherical due to surface tension shrinking the surface to a minimum

shape because a sphere has the minimum surface area for a given volume. Surface tension can be reduced if the liquid is 'contaminated' by adding, for example, detergent: a needle floating in water then sinks.

(b) Molecular explanation

In the body of a liquid the attractive and repulsive forces between the molecules balance. If the molecules come closer together, the repulsive forces become stronger and push them apart again. If the separation of the molecules increases, the attractive force becomes greater and pulls them back.

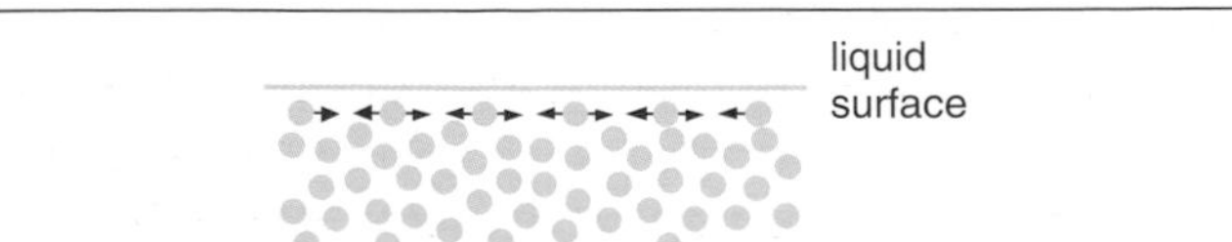

Figure 21.6 The molecules in the surface layer are more widely spaced and in a state of tension

At the surface of a liquid, as our 'tilted tray of marbles' model showed (Chapter 20), the molecules are further apart than inside the liquid and so experience attractive forces on either side from their neighbours, Figure 21.6. This puts them in a state of **tension**, causing the surface to behave like a stretched elastic skin.

Adhesion and cohesion

The force of attraction between molecules of the same substance is known as **cohesion**, that between molecules of different substances is **adhesion**. The adhesion of water to glass is greater than the cohesion of water, and water spilt on **clean** glass 'wets' it by spreading to a thin film, Figure 21.7a. By contrast, mercury on glass forms small spherical drops (or large flattened ones) because cohesion of mercury is stronger than its adhesion to glass, Figure 21.7b. Water on a wax surface also forms small drops because the cohesion of water is greater than the adhesion of water to wax, Figure 21.7c. Leaves have a waxy surface, Figure 21.8.

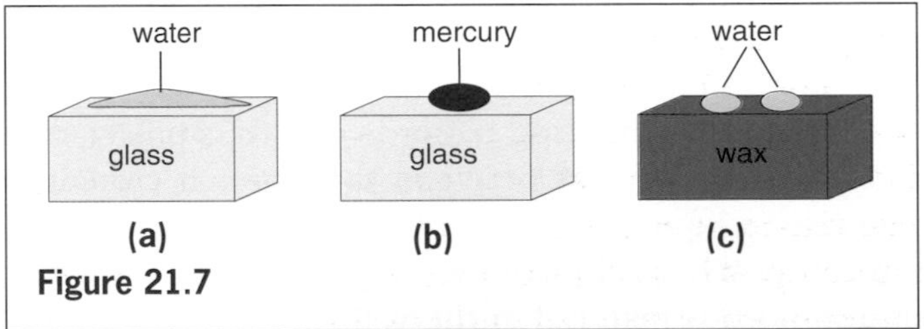

Figure 21.7

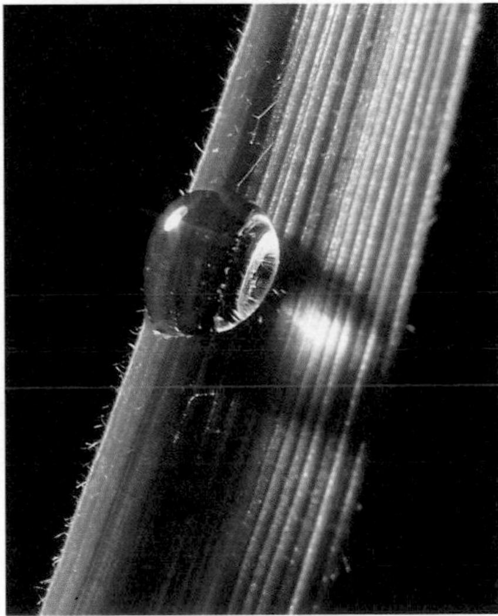

Figure 21.8
Water does not wet a waxy surface because the cohesion of the water molecules is greater than the adhesion of water to wax

Use is made of the behaviour of water on a wax surface when a garment is treated with a waxy substance to waterproof it. Rain collects in droplets on the garment but does not spread out and wet it.

The cleaning action of detergents depends on their ability to weaken the cohesion of water. Instead of forming drops on greasy clothes, the water penetrates the fabric and releases dirt.

Capillarity

If a glass tube of small bore (a capillary tube) is dipped into water, the water rises up the tube a few centimetres, Figure 21.9a. The narrower the tube the greater the rise. Adhesion between water and glass exceeds cohesion between water molecules, the meniscus curves up, and surface tension forces cause the water to rise. The effect is called **capillarity** or **capillary action**.

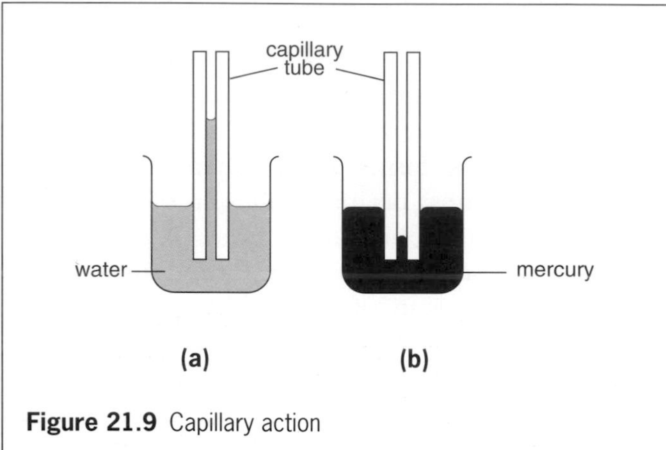

Figure 21.9 Capillary action

The action of blotting paper is due to capillary rise in the narrow spaces between the fibres it contains. The rise of liquid wax up a candle wick occurs in the same way. The damp-course in a house, i.e. the layer of non-porous material in the walls just above ground level but below the floor, prevents water rising up the pores in the bricks by capillary action from the ground and thereby prevents dampness.

Mercury is depressed in a capillary tube, Figure 21.9b. Why?

Mechanical properties of materials

When selecting a material for a particular job we need to know how it behaves when forces act on it, i.e. what its mechanical properties are.

(a) Strength

A strong material requires a large force to break it. The strength of some materials depends on how the force is applied. For example, concrete is strong when compressed but weak when stretched, i.e. in tension.

(b) Stiffness

A stiff material resists forces which try to change its shape or size. It is not flexible. All materials 'give' to some extent although the change may be very small. Steel is strong and stiff, putty is neither. Rope is not stiff but can be strong in tension.

(c) Elasticity

An elastic material such as rubber recovers its original shape and size after the force deforming it has been removed. A material which does not recover but is deformed permanently, like Plasticine, is **plastic**. (Note that synthetic 'plastics' are not always plastic; they are so-called because during manufacture they behave in that way.)

(d) Ductility

Materials which can be rolled into sheets, drawn into wires or worked into other useful shapes without breaking, are ductile. Metals owe much of their usefulness to this property.

(e) Brittleness

A brittle material is fragile and breaks suddenly. Bricks, cast iron and glass are brittle.

QUESTIONS

1 Explain why

a diffusion occurs more quickly in a gas than in a liquid,

b diffusion is still quite slow even in a gas,

c an inflated balloon gradually goes down even when tied.

2 Explain the following observations as fully as you can.

a A small needle may be floated on the surface of water, but if a drop of detergent is added to the water the needle sinks.

b Damp-courses are used in modern houses.

(E.A. part qn.)

Checklist

After studying this chapter you should be able to

- describe and explain simple experiments on diffusion,
- state evidence for the existence of surface tension and give an explanation in terms of intermolecular forces,
- recall the meanings of **adhesion**, **cohesion** and **capillarity**,
- recall the meanings of **strength**, **stiffness**, **elasticity**, **ductility** and **brittleness**.

ADDITIONAL QUESTIONS

The first questions under each topic heading are 'basic' questions intended for all students; the following questions marked with a stripe are 'higher' questions for those seeking grades E to A*.

Measurements; density

1 If 200 cm^3 of water (density 1.0 g/cm^3) is mixed with 300 cm^3 of methylated spirit (density 0.80 g/cm^3), what is the density of the mixture?

2 **a** Why does a piece of wood float and a piece of lead sink in water?

b A wooden block, whose volume is 16 cm^3, has a hole with a volume of 1.0 cm^3 drilled in it. The hole is filled with lead. Will the block sink or float in water? (Give reasons for your answer and show any calculations you make.) Density of lead = 11 g/cm^3; density of wood = 0.50 g/cm^3; density of water = 1.0 g/cm^3.

c Which is denser, milk or cream? *(S.E.)*

3 The diagram shows a measuring cylinder containing some sugar (mass 32.0 g) in paraffin oil (mass 56.4 g, density 0.80 g/cm^3) which does not dissolve the sugar. The total volume of the substances in the cylinder is 90.5 cm^3. Calculate **a** the volume of the paraffin oil, **b** the volume of the sugar, **c** the density of the sugar. *(C.)*

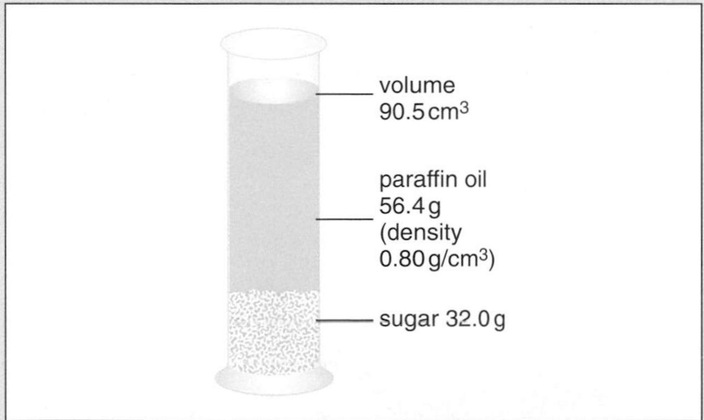

4 **a** Name the basic units of: length, mass, time.

b What is the difference between two measurements with values of 3.4 and 3.42?

c Write expressions for (i) the area of a circle, (ii) the volume of a sphere, (iii) the volume of a cylinder.

Weight and springs

5 The springs in the diagram are identical. If the extension produced in (a) is 4 cm, what are the extensions in (b) and (c)?

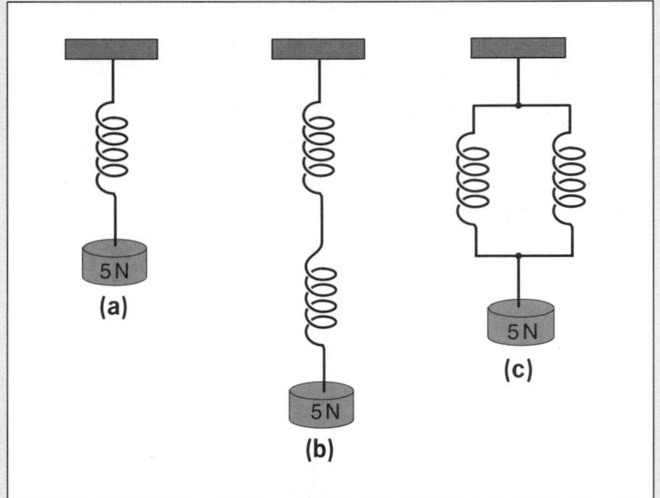

6 A vertical spring of unstretched length 30 cm is rigidly clamped at its upper end. When an object of mass 100 g is placed in a pan attached to the lower end of the spring, its length becomes 36 cm. For an object of mass 200 g in the pan, the length becomes 40 cm. Calculate the mass of the pan. Name and state clearly the law you have assumed. *(S.)*

7 A light helical spring hangs vertically with its upper end fixed. A light pointer, attached to its lower end, can be used to take readings on a vertical scale when masses of various sizes are attached to the lower end of the spring. The following table gives the scale readings of the pointer for different attached masses.

Mass attached in kg	0	0.2	0.4	0.6	0.8
Scale reading in mm	120	126	132	138	144

a Make a table showing corresponding values of the force on the spring in newtons and its resulting extension in metres.

b Plot a graph of force (y-axis) against extension (x-axis).

c State Hooke's law. Are the readings for this spring consistent with it? Explain.

d Use your graph to find the force that produces an extension of 15 mm.
(Weight of 1 kg is 10 N). *(O.L.E.)*

Molecules

8 Describe the differences between solids, liquids and gases in terms of

a the arrangement of the molecules throughout the bulk of the material,

b the separation of the molecules, and

c the motion of the molecules. *(J.M.B.)*

9 Smoke particles in an air cell viewed under a microscope exhibit Brownian motion.

a Copy the diagram below and draw a suitable arrangement to show how the particles in the air can be illuminated.

b How is the motion of the smoke particles best described?

c What accounts for the motion of the smoke particles?

d The motion is viewed using bigger smoke particles. What difference in the motion would this lead to? Give the reason for the difference. *(O.L.E./S.R. 16+)*

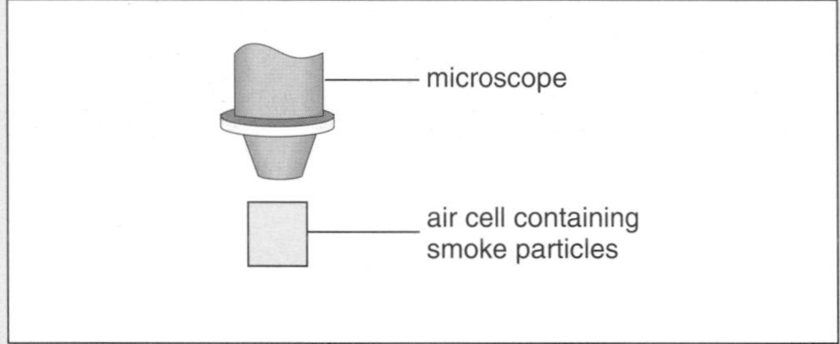

Properties of matter

10 A porous pot holding air is fitted with a tube dipping into water.

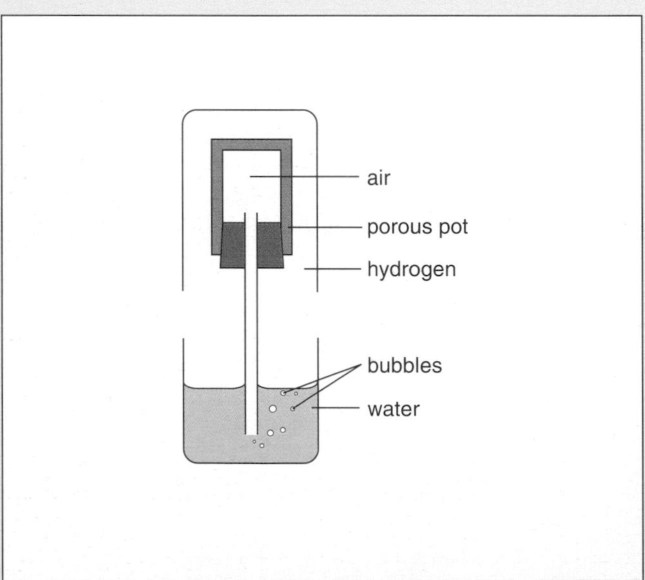

a When the pot is surrounded by hydrogen, air bubbles rapidly out of the water, but after a short time the bubbles slow down and stop. Explain these stages.

b If now the jar of hydrogen is removed, water rises rapidly up the tube, slows down and stops, and very slowly returns. Explain these stages.

c What difference does it make when the experiment is repeated using carbon dioxide instead of hydrogen? Why? *(S.R.)*

11 a A piece of thread is carefully placed in a soap film which has been formed on a metal ring, diagram (a). Copy diagram (b) and show clearly what happens if the soap film inside the thread is pierced by a needle.

b What name is given to the force acting on the thread? *(E.M.)*

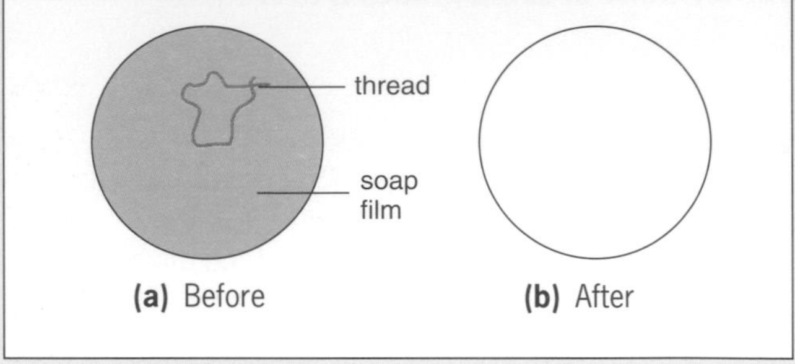

12 a Name the effect shown in tube A below.

b Copy the diagram and mark the water levels in B and C and the mercury levels in D and E.

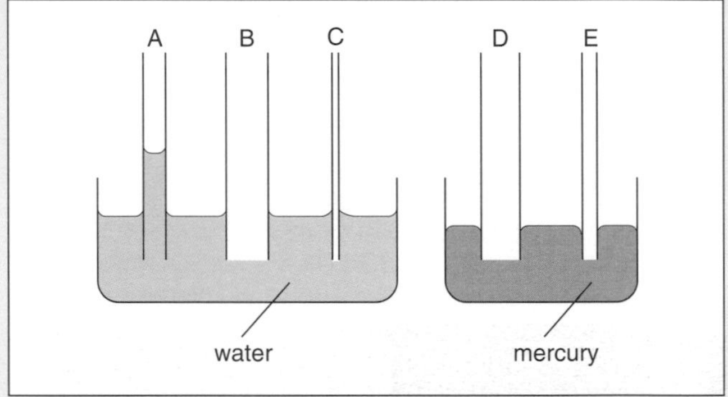

13 How can the pond skater in the photograph walk over the water surface without sinking?

FORCES AND PRESSURE

22 MOMENTS AND LEVERS

Moment of a force

Levers

Conditions for equilibrium

Beams and structures

Practical work
Law of moments.

Moment of a force

The handle on a door is at the outside edge so that it opens and closes easily. A much larger force would be needed if the handle were near the hinge. Similarly it is easier to loosen a nut with a long spanner than with a short one.

The **turning effect** or **moment of a force** depends on both the size of the force and how far it is applied from the pivot or **fulcrum**. It is measured by multiplying the force by the **perpendicular** distance of the line of action of the force from the fulcrum. The unit is the **newton metre** (N m).

moment of a force = force × perpendicular distance from fulcrum

In Figure 22.1 a force F acts on a gate in (a) at the edge, in (b) at the centre.

In (a)

moment of F about O = 5 N × 3 m = 15 N m

In (b)

moment of F about O = 5 N × 1.5 m = 7.5 N m

The turning effect of F is greater in (a); this agrees with the fact that a gate opens most easily when it is pushed or pulled at the edge.

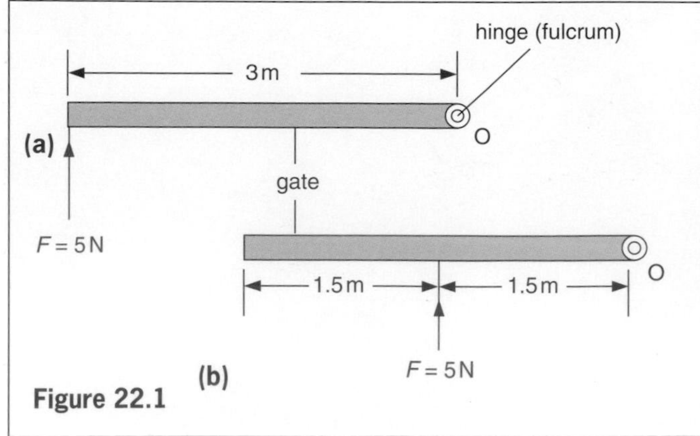

Figure 22.1

PRACTICAL WORK

Law of moments

Balance a half-metre ruler at its centre, adding Plasticine to one side or the other until it is horizontal.

Hang unequal loads m_1 and m_2 from either side of the ruler and alter their distances d_1 and d_2 from the centre until the ruler is again balanced, Figure 22.2. Forces F_1 and F_2 are exerted by gravity on m_1 and m_2 and so on the ruler; on 100 g the force is 1 N. Record the results in a table and repeat for other loads and distances.

m_1 (g)	F_1 (N)	d_1 (cm)	$F_1 \times d_1$ (N cm)	m_2 (g)	F_2 (N)	d_2 (cm)	$F_2 \times d_2$ (N cm)

F_1 is trying to turn the ruler anticlockwise and $F_1 \times d_1$ is its moment. F_2 is trying to cause clockwise turning and its moment is $F_2 \times d_2$. When the ruler is balanced or, as we say, **in equilibrium**, the results should show that the anticlockwise moment $F_1 \times d_1$ equals the clockwise moment $F_2 \times d_2$.

The **law of moments** (also called the **law of the lever**) is stated as follows.

When a body is in equilibrium the sum of the clockwise moments about any point equals the sum of the anticlockwise moments about the same point.

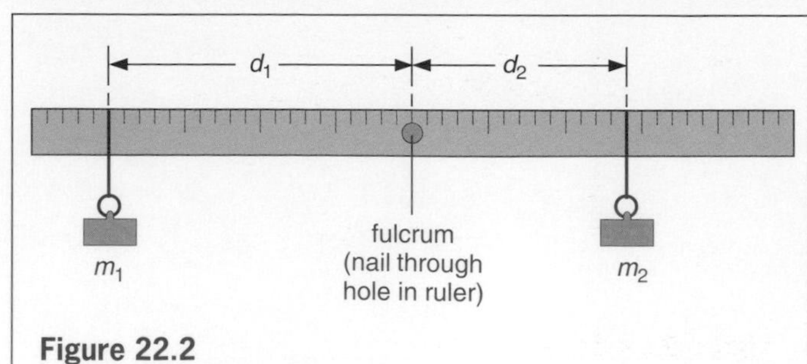

Figure 22.2

Worked example

The see-saw in Figure 22.3 balances when Sue of weight 320 N is at A, Tom of weight 540 N is at B and Harry of weight W is at C. Find W.

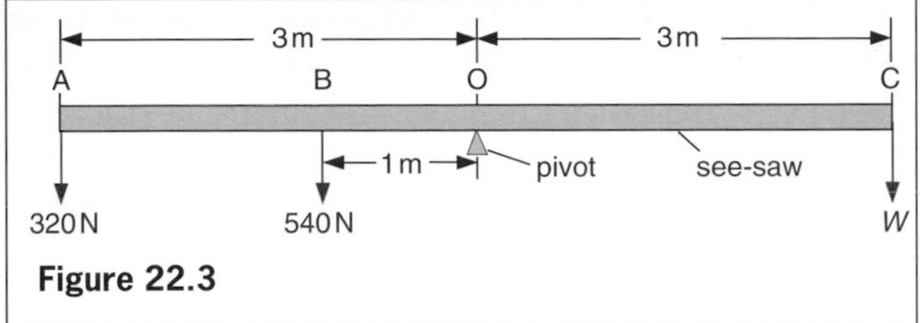

Figure 22.3

Taking moments about the fulcrum O:

anticlockwise moment
$$= 320 \text{ N} \times 3 \text{ m} + 540 \text{ N} \times 1 \text{ m} = 1500 \text{ N m}$$

$$\text{clockwise moment} = W \times 3 \text{ m}$$

By the law of moments,

$$\text{clockwise moments} = \text{anticlockwise moments}$$

$$\therefore \quad W \times 3 \text{ m} = 1500 \text{ N m}$$

$$\therefore \quad W = \frac{1500 \text{ N m}}{3 \text{ m}} = 500 \text{ N}$$

Levers

A lever is any device which can turn about a pivot. In a working lever a force called the **effort** is used to overcome a resisting force called the **load**. The pivotal point is called the **fulcrum**.

When we use a crowbar to move a heavy boulder, Figure 22.4a, our hands apply the effort at one end of the bar and the load is the force exerted by the boulder on the other end. If distances from the fulcrum O are as shown and the load is 1000 N (i.e. the part of the weight of the boulder supported by the crowbar), the effort can be calculated from the law of moments. As the boulder *just begins* to move we can say, taking moments about O, that

$$\text{clockwise moment} = \text{anticlockwise moment}$$

$$\text{effort} \times 200 \text{ cm} = 1000 \text{ N} \times 10 \text{ cm}$$

$$\text{effort} = \frac{10\,000 \text{ N cm}}{200 \text{ cm}} = 50 \text{ N}$$

The crowbar in effect magnifies the effort 20 times but the effort must move farther than the load.

Other examples of levers are shown in Figures 22.4b to e. In (b) the load is between the effort and the fulcrum; in this case as in (a) the effort is less than the load. In (c) the effort (applied by the biceps muscle) is between the load and the fulcrum and is greater than the load, which moves farther than the effort. How does the effort compare with the load for scissors and a spanner in (d) and (e)?

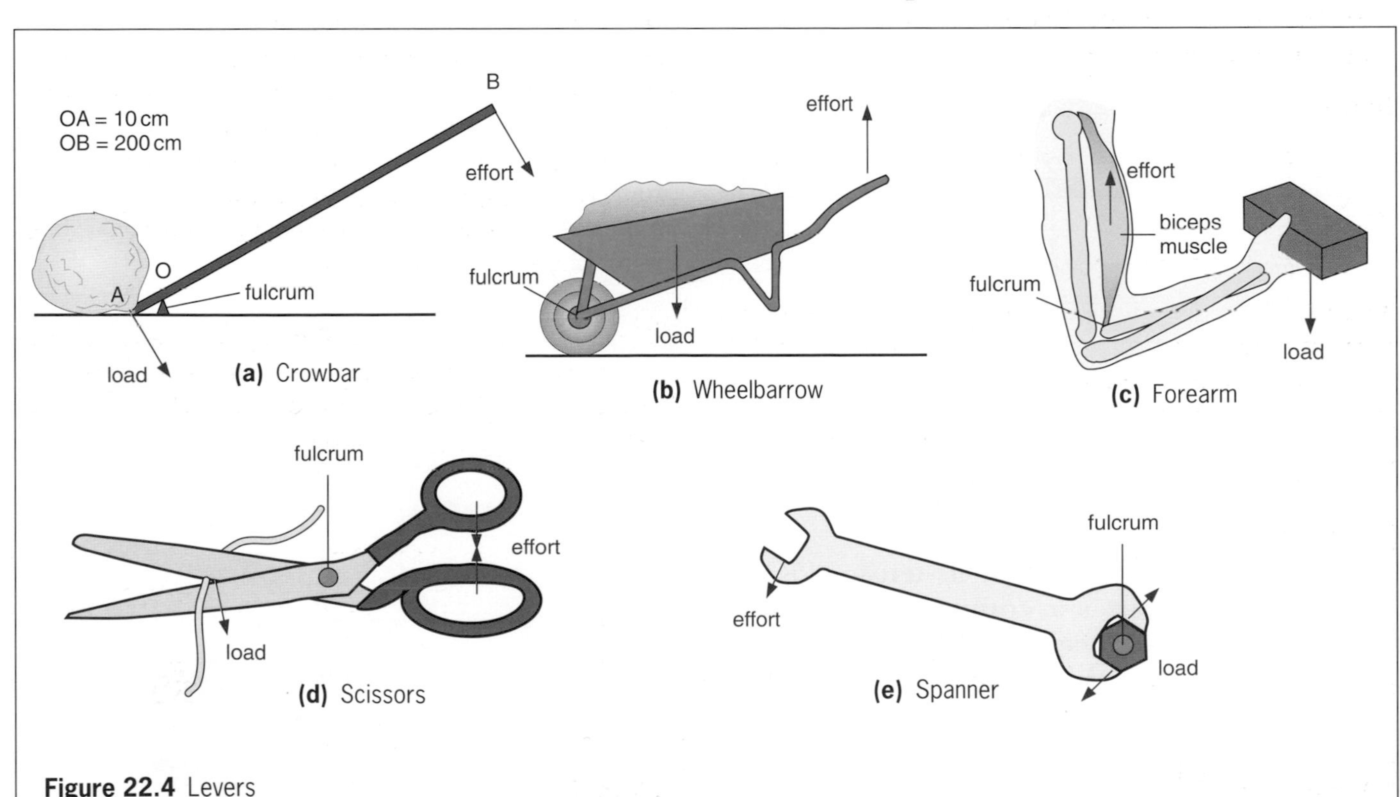

Figure 22.4 Levers

Conditions for equilibrium

Sometimes a number of parallel forces act on a body so that it is in equilibrium. We can then say:

(i) The sum of the forces in one direction equals the sum of the forces in the opposite direction.
(ii) The law of moments must apply.

As an example consider two decorators of weights 500 N and 700 N standing at A and B on a plank resting on two trestles, Figure 22.5. In the next chapter we will see that the whole weight of the plank (400 N) may be taken to act vertically downwards at its centre, O. If P and Q are the upward forces exerted by the trestles on the plank (called **reactions**) then we have from (i)

$$P + Q = 500\text{ N} + 400\text{ N} + 700\text{ N} = 1600\text{ N} \qquad (1)$$

Moments can be taken about any point but taking them about C eliminates the moments due to Q.

$$\text{clockwise moment} = P \times 4\text{ m}$$

anticlockwise moments

$$= (700\text{ N} \times 1\text{ m} + 400\text{ N} \times 2\text{ m} + 500\text{ N} \times 5\text{ m})$$
$$= 700\text{ N m} + 800\text{ N m} + 2500\text{ N m} = 4000\text{ N m}$$

Since the plank is in equilibrium we have from (ii)

$$P \times 4\text{ m} = 4000\text{ N m}$$

$$\therefore \qquad P = \frac{4000\text{ N m}}{4\text{ m}} = 1000\text{ N}$$

From (1), $\qquad Q = 600\text{ N}$

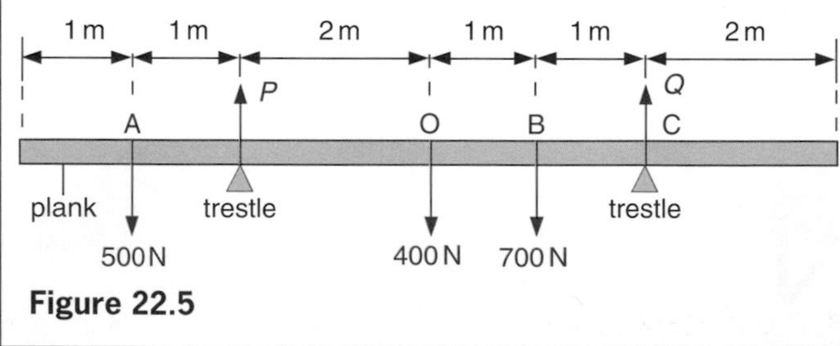

Figure 22.5

Beams and structures

(a) Beams

Stretching a rod or bar puts it in **tension**, Figure 22.6a; squeezing it puts it in **compression** (b); twisting it subjects it to **shearing** (c). The greater the cross-section area of a rod, the greater are the balancing internal resisting forces it sets up to tensile, compressive and shearing forces. Resistance to such forces also depends on shape, so different beams are made for different purposes.

When a beam bends, either under its own weight or because it is loaded, one side is compressed, the other is stretched and the centre is unstressed (a neutral plane), Figure 22.7a. I-shaped steel girders, Figure 22.7b, are used in large structures. They are, in effect, beams that have had material removed from the neutral plane and so weigh less. The top and bottom flanges withstand the compression and tension forces due to loading. In a hollow tube the removal of unstressed material gives similar gains. Concrete, though strong in compression, is weak in tension but can be made stronger if it is reinforced by steel rods. Steel is strong in tension as well as compression.

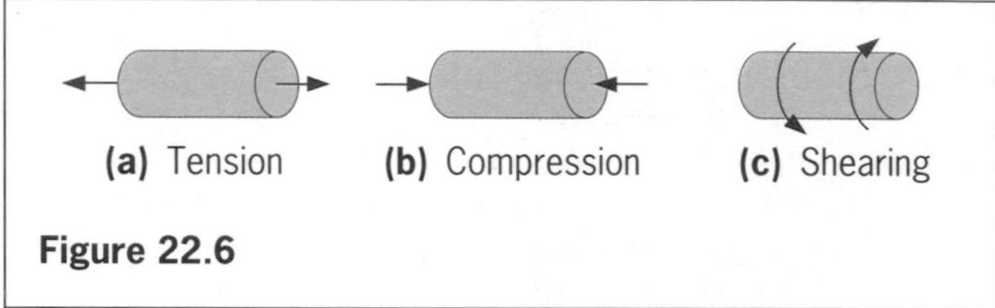
(a) Tension **(b)** Compression **(c)** Shearing

Figure 22.6

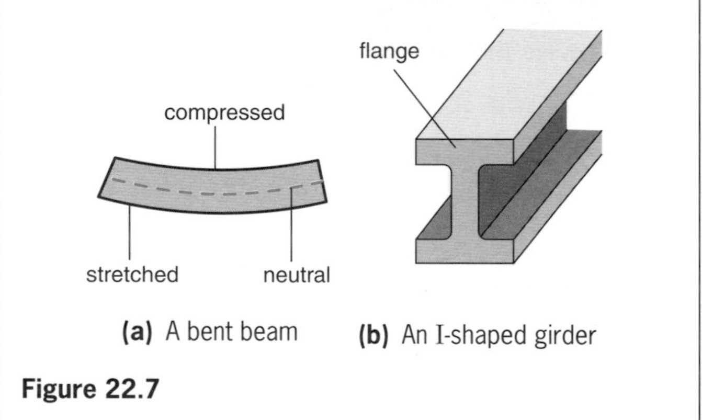

(a) A bent beam **(b)** An I-shaped girder

Figure 22.7

(b) Bridges

In an arched stone bridge, the stone is in compression, since it is weak in tension, Figure 22.8a. A steel girder bridge has no material in the neutral plane, Figure 22.8b; it is strengthened by diagonal bars, Figure 22.9.

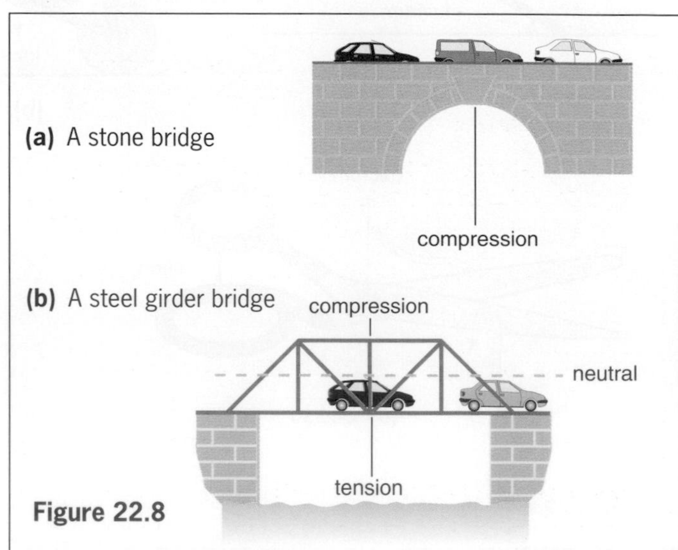

Figure 22.8

Figure 22.9 A steel girder bridge under construction

QUESTIONS

1 Figure 22.10 shows half-metre rules which are marked off at intervals of 5 cm. Identical metal discs are placed on the rules as shown. In each case state whether the rules will turn clockwise, anticlockwise or remain in the horizontal position. Show your working. *(E.A.)*

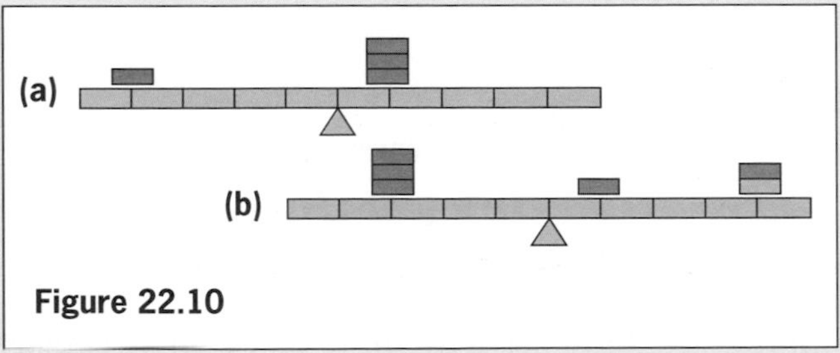

Figure 22.10

2 The metre rule in Figure 22.11 is supported at its centre. If the rule is balanced, the values of *x* and *y* respectively are

A 3 cm (*x*) and 5 cm (*y*) **B** 5 cm (*x*) and 3 cm (*y*)
C 6 cm (*x*) and 10 cm (*y*) **D** 12 cm (*x*) and 20 cm (*y*) *(W.M.)*

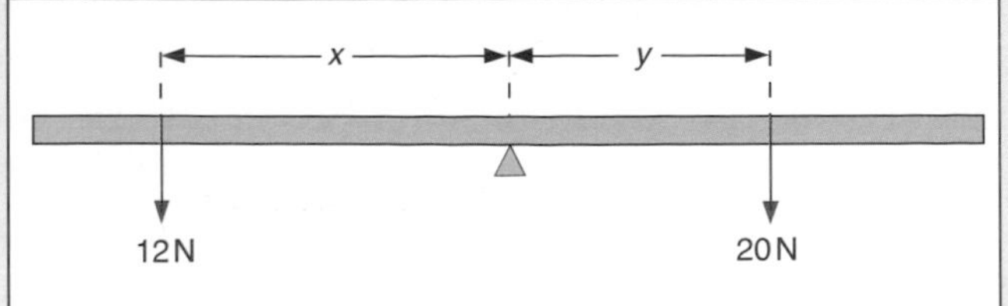

Figure 22.11 Diagram not to scale

3 In Figure 22.12, the distance AC = CB. Calculate in each case the force X which will keep the system stationary. *(E.M.)*

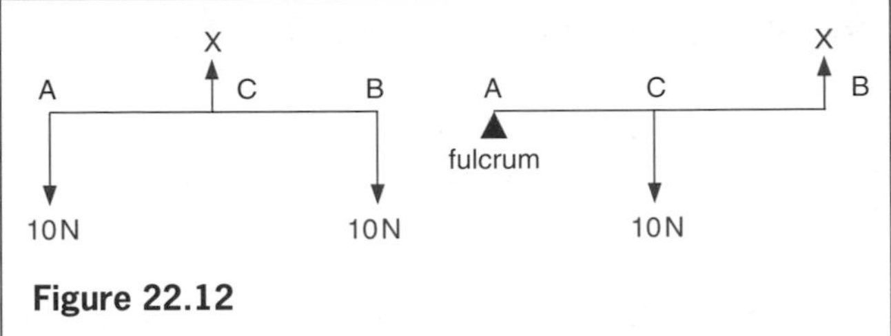

Figure 22.12

4 Figure 22.13 shows three positions of the pedal on a bicycle which has a crank 0.20 m long. If the cyclist exerts the same vertically downward push of 25 N with his foot, in which case is the turning effect (i) $25 \times 0.2 = 5$ N m, (ii) 0, (iii) between 0 and 5 N m? Explain your answers.

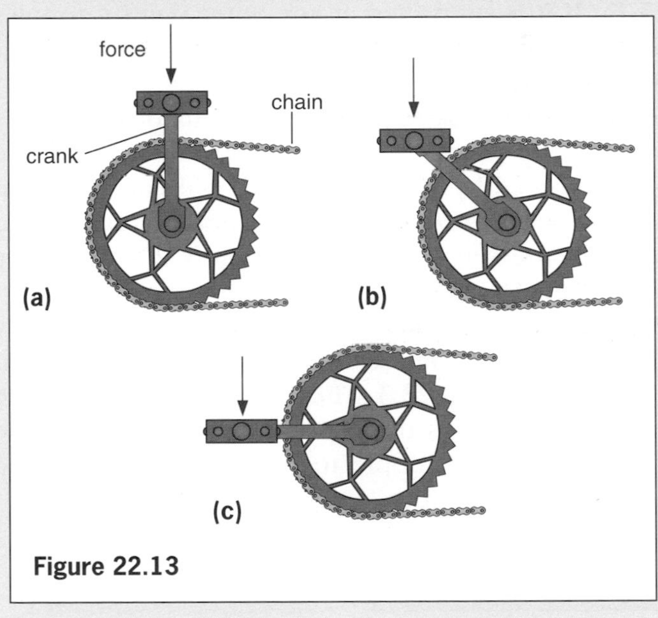

Figure 22.13

Checklist

After studying this chapter you should be able to

- define the **moment** of a force about a point,
- describe an experiment to study turning effects on a body in equilibrium,
- state the law of moments and use it to solve problems,
- explain the action of common tools and devices as levers,
- state the conditions for equilibrium when parallel forces act on a body,
- describe the behaviour of beams and bridges when loaded.

23 CENTRES OF GRAVITY

Toppling	***Practical work***
Stability	Centre of gravity using a plumb line.
Balancing tricks and toys	

A body behaves as if its whole weight were concentrated at one point, called its **centre of gravity** (c.g.) or **centre of mass**, even though the Earth attracts every part of it. The c.g. of a uniform ruler is at its centre and when supported there it balances, Figure 23.1a. If it is supported at any other point it topples because the moment of its weight *W* about the point of support is not zero, Figure 23.1b.

Your c.g. is near the centre of your body and the vertical line from it to the floor must be within the area enclosed by your feet or you will fall over. You can test this by standing with one arm and the side of one foot pressed against a wall, Figure 23.2. Now try to raise the other leg sideways.

A tight-rope walker has to keep his c.g. exactly above the rope. Some carry a long pole to help them to balance, Figure 23.3. The combined weight of the walker and pole is then spread out more and if the walker begins to topple to one side he moves the pole to the other side.

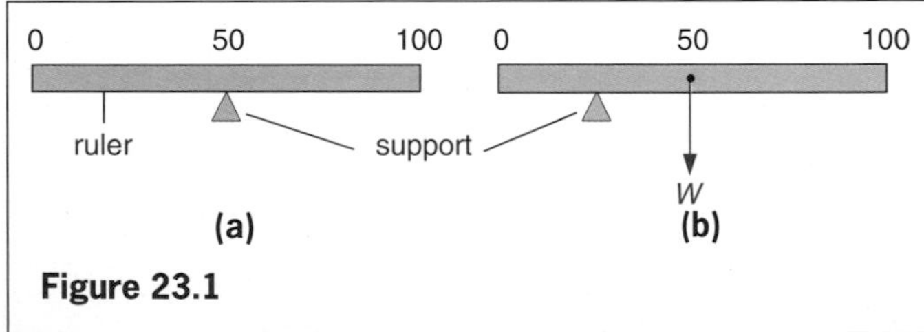

Figure 23.1

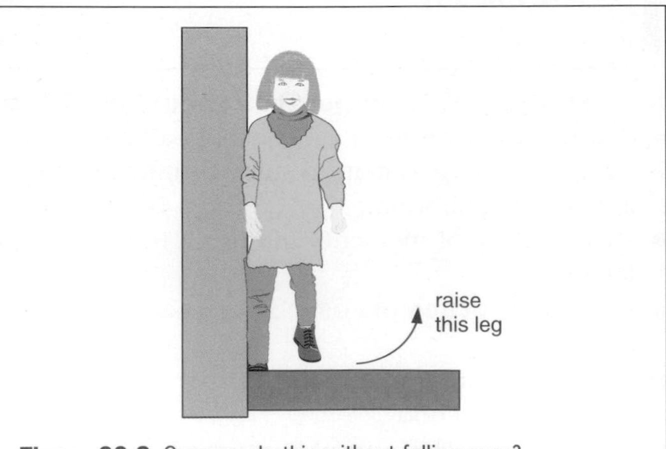

Figure 23.2 Can you do this without falling over?

Figure 23.3 A tight-rope walker using a long pole

The c.g. of a regularly shaped body of the same density all over is at its centre. In other cases it can be found by experiment.

PRACTICAL WORK

Centre of gravity using a plumb line

Suppose we have to find the c.g. of an irregularly shaped lamina (a thin sheet) of cardboard.

Make a hole A in the lamina and hang it so that it can *swing freely* on a nail clamped in a stand. It will come to rest with its c.g. vertically below A. To locate the vertical line through A tie a plumb line (a thread and a weight) to the nail, Figure 23.4, and mark its position AB on the lamina. The c.g. lies on AB.

Hang the lamina from another position C and mark the plumb line position CD. The c.g. lies on CD and must be at the point of intersection of AB and CD. Check this by hanging the lamina from a third hole. Also try balancing it at its c.g. on the tip of your forefinger.

Devise a method using a plumb line of finding the c.g. of a tripod.

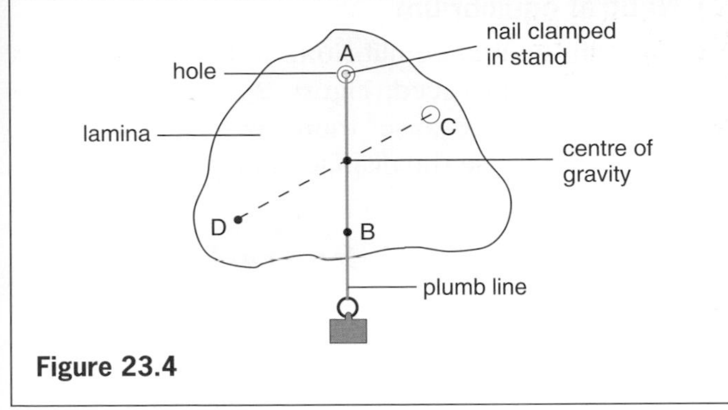

Figure 23.4

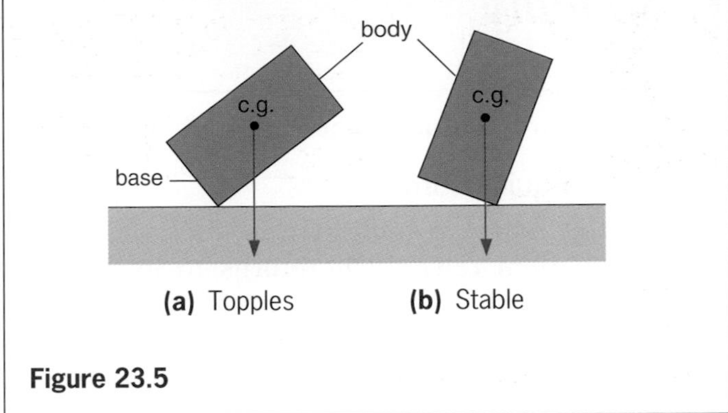

Figure 23.5

Toppling

The position of the c.g. of a body affects whether or not it topples over easily. This is important in the design of such things as tall vehicles (which tend to overturn when rounding a corner), racing cars, reading lamps and even drinking glasses.

A body topples when the vertical line through its c.g. falls outside its base, Figure 23.5a. Otherwise it remains stable, Figure 23.5b.

Toppling can be investigated by placing an empty can on a plank (with a rough surface to prevent slipping) which is slowly tilted. The angle of tilt is noted when the can falls over. This is repeated with 1 kg in the can. How does this affect the position of the c.g.? The same procedure is followed with a second can of the same height as the first but of greater width. It will be found that the second can with the weight in it can be tilted through the greater angle.

The stability of a body is therefore increased by

(i) lowering its c.g.
(ii) increasing the area of its base.

In Figure 23.6a the c.g. of a tractor is being found. It is necessary to do this when testing a new design since tractors are often driven over sloping surfaces and any tendency to overturn must be discovered.

The stability of double-decker buses is being tested in Figure 23.6b. When the top deck only is fully laden with passengers (represented by sand bags in the test), it must not topple if tilted through an angle of 28°.

Racing cars have a low c.g. and a wide wheel base.

Figure 23.6a A tractor under test to find its c.g.

Figure 23.6b A double-decker bus being tilted to test its stability

Stability

Three terms are used in connection with stability.

(a) Stable equilibrium

A body is in 'stable equilibrium' if when slightly displaced and then released it returns to its previous position. The ball at the bottom of the dish in Figure 23.7a is an example. Its c.g. rises when it is displaced. It rolls back because its weight has a moment about the point of contact.

(b) Unstable equilibrium

A body is in 'unstable equilibrium' if it moves farther away from its previous position when slightly displaced. The ball in Figure 23.7b behaves in this way. Its c.g. falls when it is displaced slightly because there is a moment which increases the displacement.

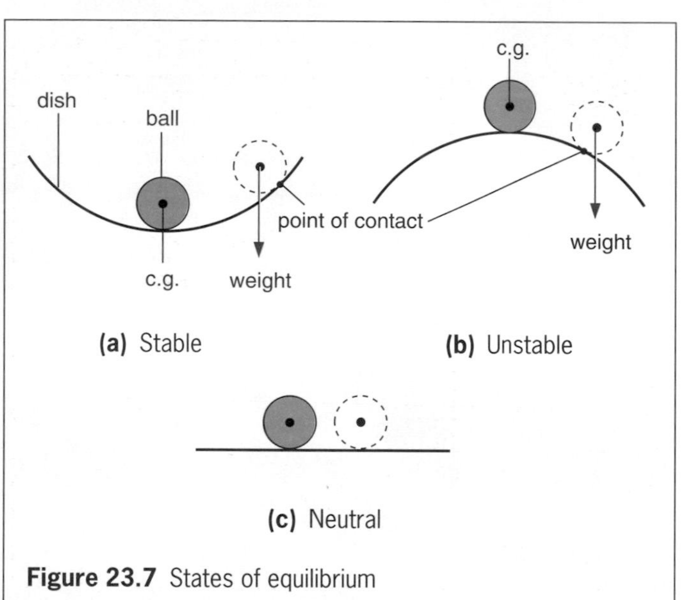

Figure 23.7 States of equilibrium

(c) Neutral equilibrium

A body is in 'neutral equilibrium' if it stays in its new position when displaced, Figure 23.7c. Its c.g. does not rise or fall because there is no moment to increase or decrease the displacement.

Balancing tricks and toys

Some tricks that you can try or toys you can make are shown in Figure 23.8. In each case the c.g. is vertically below the point of support and equilibrium is stable.

A self-righting toy, Figure 23.9, has a heavy base and when tilted, the weight acting through the c.g. has a moment about the point of contact. This restores it to the upright position.

Figure 23.9 A self-righting toy

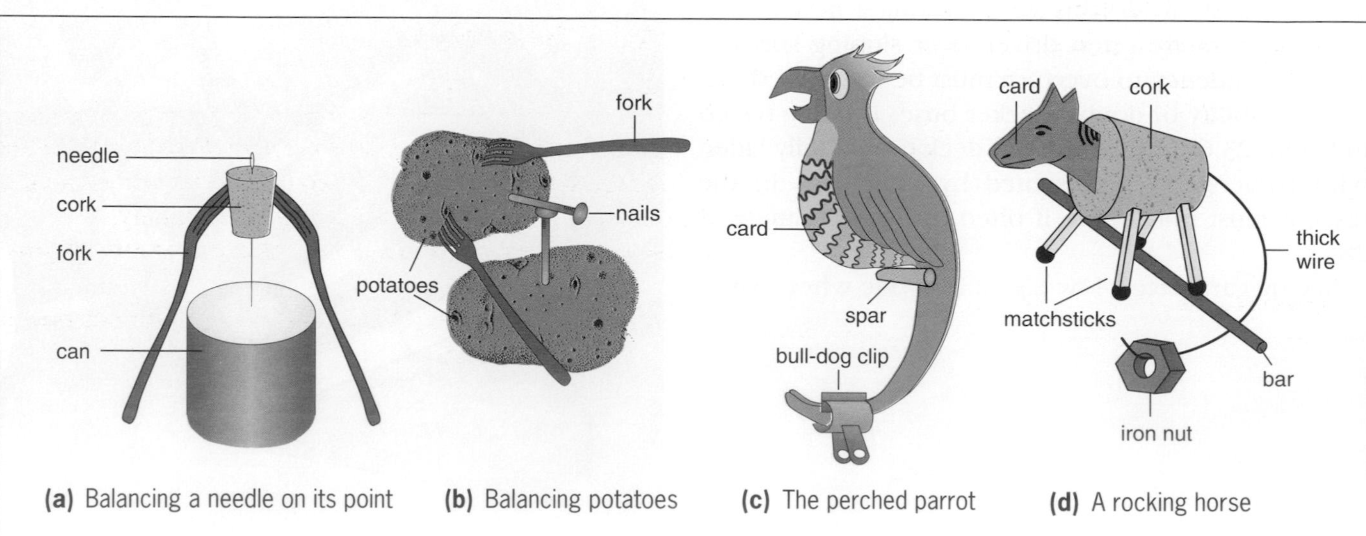

Figure 23.8 Balancing tricks

QUESTIONS

1 Figure 23.10 shows an irregular shape of plywood suspended by a thin thread at two different points, A and B. Copy diagram (b). Mark and label the centre of mass of the plywood. Draw and label across diagram (b) a line which would be vertical if the plywood were suspended at point C. *(E.A.)*

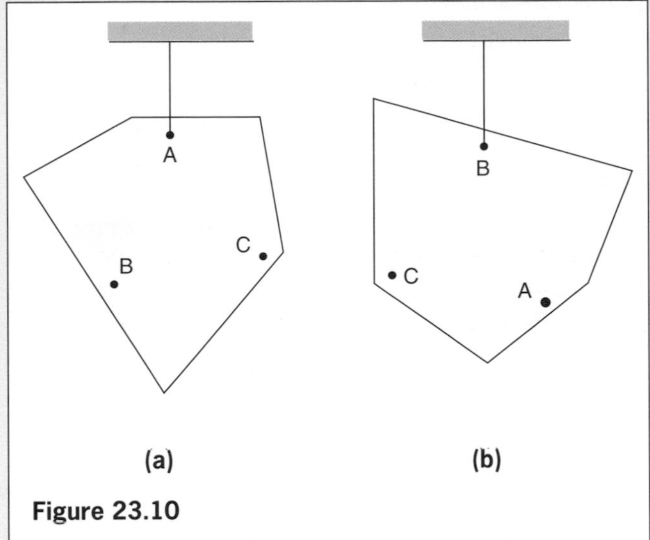

Figure 23.10

2 Figure 23.11 shows a Bunsen burner in three different positions. State in which position it is in **a** stable equilibrium, **b** unstable equilibrium, **c** neutral equilibrium.

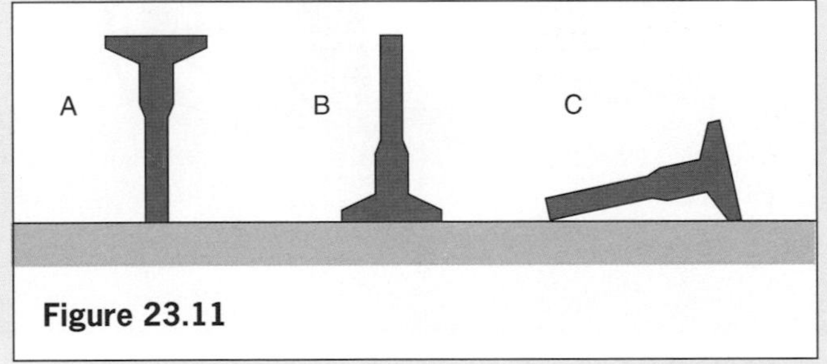

Figure 23.11

3 The weight of the uniform bar in Figure 23.12 is 10 N. Does it balance, tip to the right or tip to the left?

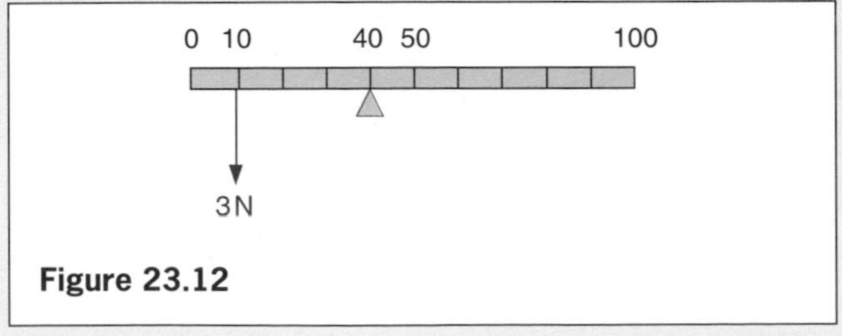

Figure 23.12

4 a A uniform metre rule is balanced at the 30 cm mark when a load of 0.80 N is hung at the zero mark, Figure 23.13a.
(i) At what point on the rule is the centre of gravity of the rule?
(ii) Show with an arrow drawn on a copy of the diagram the weight of the rule acting through the centre of gravity.
(iii) Calculate the weight of the rule.

b Figure 23.13b shows the jib of a building site crane. What is the purpose of the concrete block at one end of the jib?

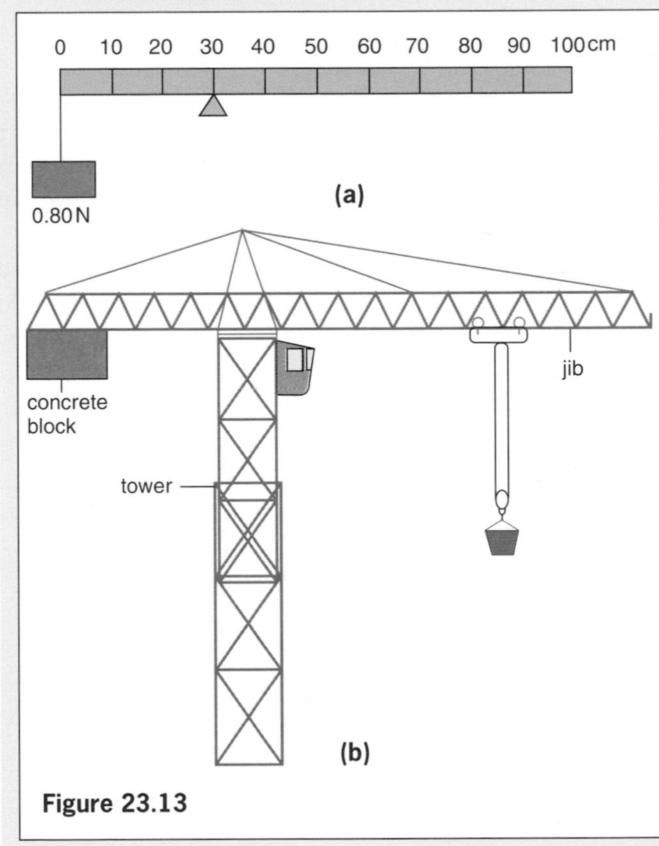

Figure 23.13

Checklist

After studying this chapter you should be able to

- recall that an object behaves as if its whole weight were concentrated at its centre of gravity,
- describe an experiment to find the centre of gravity of an object,
- connect the stability of an object to the position of its centre of gravity.

24 ADDING FORCES

Forces and resultants	***Vectors and scalars***
Examples of addition of forces	***Practical work*** Parallelogram law.

Forces and resultants

Force has both magnitude (size) and direction. It is represented in diagrams by a straight line with an arrow to show its direction of action.

Usually more than one force acts on an object. As a simple example, an object resting on a table is pulled downwards by its weight W and pushed upwards by a force R due to the table supporting it, Figure 24.1. Since the object is at rest, the forces must balance, i.e. $R = W$.

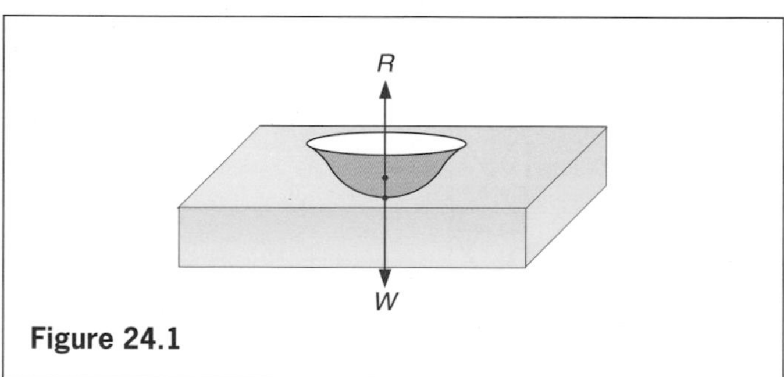

Figure 24.1

In structures like a giant oil platform, Figure 24.2, two or more forces may act at the same point. It is then often useful for the design engineer to know the value of the single force, i.e. the **resultant**, which has exactly the same effect as these forces. If the forces act in the same straight line the resultant is found by simple addition or subtraction as shown in Figure 24.3, but if they do not they are added by the 'parallelogram law'.

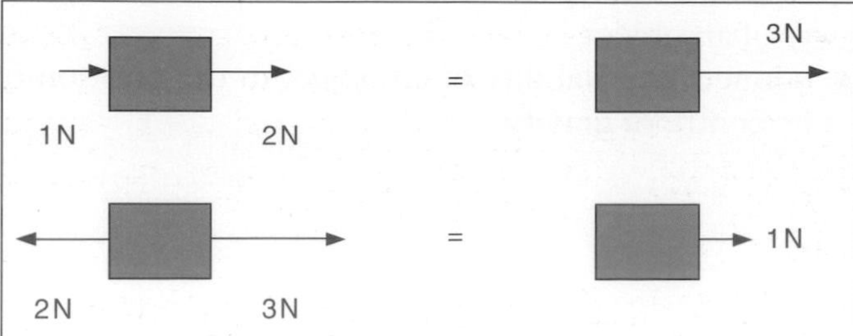

Figure 24.3 The resultant of forces acting in the same straight line is found by addition or subtraction

Figure 24.2 The design of an off-shore oil platform requires an understanding of the combination of many forces

Practical work

Parallelogram law

Arrange the apparatus as in Figure 24.4a with a sheet of paper behind it on a vertical board. We have to find the resultant of forces P and Q.

Read the values of P and Q from the spring balances. Mark on the paper the directions of P, Q and W as shown by the strings. Remove the paper and using a scale of 1 cm to represent 1 N, draw OA, OB and OD to represent the three forces P, Q and W which act at O, Figure 24.4b. (W = weight of the 1 kg mass = 9.8 N, therefore OD = 9.8 cm.)

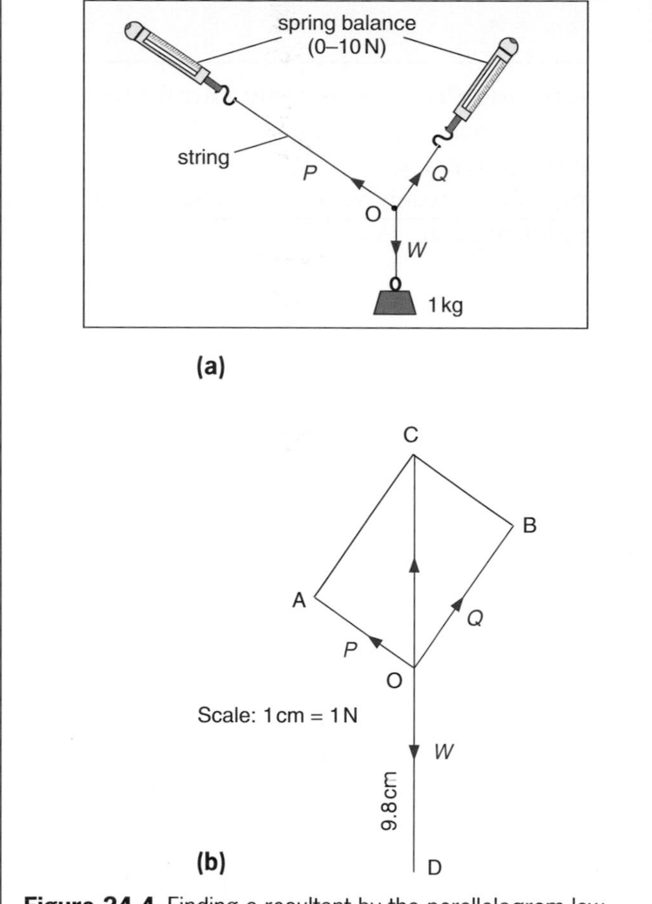

Figure 24.4 Finding a resultant by the parallelogram law

P and *Q* together are balanced by *W* and so their resultant must be a force equal and opposite to *W*.

Complete the parallelogram OACB. Measure the diagonal OC; if it is equal in size (i.e. 9.8 cm) and opposite in direction to *W* then it represents the resultant of *P* and *Q*.

The **parallelogram law** for adding two forces is:

If two forces acting at a point are represented in size and direction by the sides of a parallelogram drawn from the point, their resultant is represented in size and direction by the diagonal of the parallelogram drawn from the point.

Worked example

Find the resultant of two forces of 4.0 N and 5.0 N acting at an angle of 45° to each other.

Using a scale of 1.0 cm = 1.0 N, draw parallelogram ABCD with AB = 5.0 cm, AC = 4.0 N and angle CAB = 45°, Figure 24.5. By the parallelogram law, the diagonal AD represents the resultant in magnitude and direction; it measures 8.3 cm and angle BAD = 21°.

∴ Resultant is a force of 8.3 N acting at an angle of 21° to the force of 5.0 N.

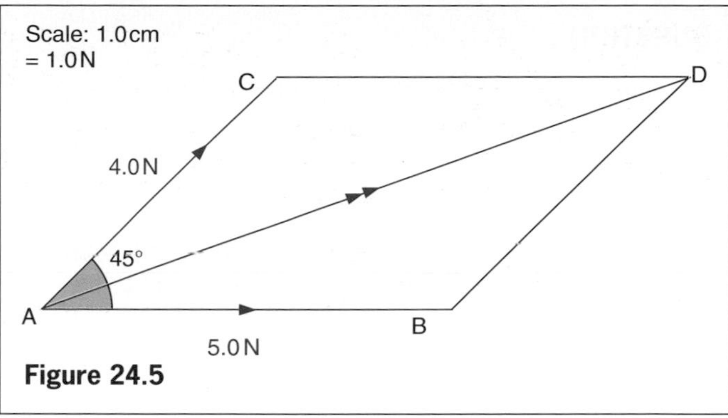

Figure 24.5

Examples of addition of forces

1. Two people carrying a heavy bucket. The weight of the bucket is balanced by the resultant *F* of F_1 and F_2, Figure 24.6a.

2. Two tugs pulling a ship. The resultant of T_1 and T_2 is forwards, Figure 24.6b, and so the ship moves forwards (as long as the resultant is greater than the resistance to motion of the sea and the wind).

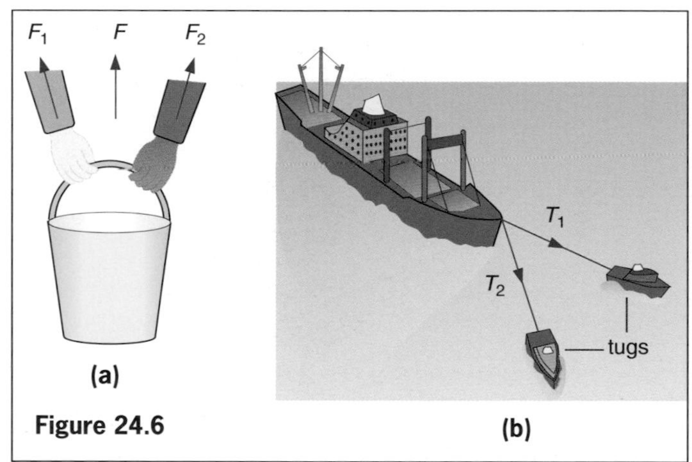

Figure 24.6

Vectors and scalars

A **vector** quantity is one such as force which is described completely only if both its size (magnitude) and direction are stated. It is not enough to say, for example, a force of 10 N, but rather a force of 10 N acting vertically downwards.

A vector can be represented by a straight line whose length represents the magnitude of the quantity and whose direction gives its line of action. An arrow on the line shows which way along the line it acts.

A **scalar** quantity has magnitude only. Mass is a scalar and is completely described when its value is known. Scalars are added by ordinary arithmetic; vectors are added geometrically by the parallelogram law which ensures that their directions as well as their magnitudes are considered.

QUESTIONS

1 Using a scale of 1 cm to represent 10 N find the size and direction of the resultant of forces of 30 N and 40 N acting at **a** right angles to each other, **b** 60° to each other.

2 In Figure 24.7 the lines represent two forces acting at a point O. Which of the single forces could be the resultant of the two? *(E.A.)*

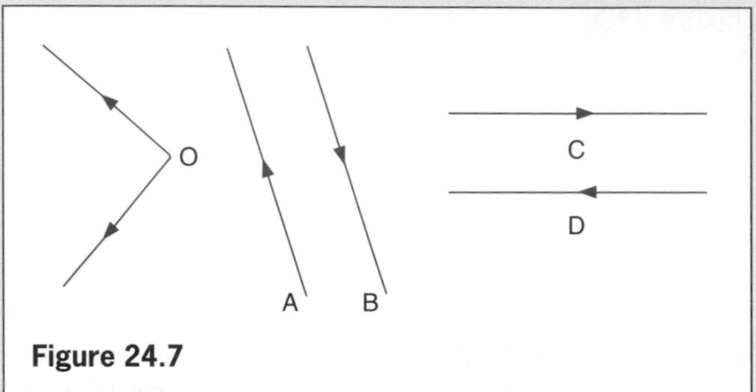

Figure 24.7

3 Jo, Daniel and Helen are pulling a metal ring. Jo pulls with a force of 100 N and Daniel with a force of 140 N at an angle of 70° to Jo. If the ring does not move what force is Helen exerting?

Checklist

After studying this chapter you should be able to

- combine forces graphically to find their resultant using the parallelogram law,
- distinguish between vectors and scalars and give examples of each.

25 ENERGY TRANSFER

Forms of energy	***Energy of food***
Energy transfers	***Combustion of fuels***
Energy measurements Work: the joule. Power.	***Practical work*** Measuring power.
Energy conservation Principle. Efficiency.	

Energy is a theme that pervades all branches of science. It links a wide range of phenomena and enables us to explain them. It exists in different forms and when something happens, it is likely to be due to energy being transferred from one form to another. **Energy transfer** is needed to enable people, computers, machines and other devices to work and for processes and changes to occur. For example, in Figure 25.1, the water skier can only be pulled along by the boat if there is energy transfer in its outboard motor from the burning petrol to its rotating propeller.

Figure 25.1 Energy transfer in action

Forms of energy

(a) Chemical energy

Food and fuels, like oil, gas, coal and wood, are concentrated stores of chemical energy. The energy of food is released by chemical reactions in our bodies, and during the transfer to other forms we are able to do useful jobs. Fuels cause energy transfers when they are burnt in an engine or a boiler.

Batteries are compact sources of chemical energy, which in use is transferred to electrical energy.

(b) Potential energy (p.e.)

This is the energy a body has because of its position or condition. A body above the Earth's surface, like the water in a mountain reservoir, has p.e. in the form of **gravitational energy**.

Wound-up springs and stretched rubber bands have p.e. because of their 'strained' condition.

(c) Kinetic energy (k.e.)

Any moving body has k.e. and the faster it moves the more k.e. it has. As a hammer strikes a nail in a piece of wood, there is a transfer of energy from the k.e. of the moving hammer to other forms of energy as the nail is driven in.

(d) Electrical energy

This is produced by energy transfers at power stations and in batteries. It is the commonest form of energy used in homes and industry because of the ease of transmission and transfer to other forms.

(e) Heat energy

This is also called **thermal** or **internal energy** and is often the ultimate fate of other forms of energy. It is frequently transferred by conduction, convection or radiation, as we will see later.

(f) Other forms

These include light, sound and nuclear energy.

Energy transfers

(a) Demonstration

The apparatus in Figure 25.2 (overleaf) can be used to show a battery changing **chemical energy** to **electrical energy** which becomes **k.e.** in the electric motor. The motor raises a weight, giving it **p.e.** If the changeover switch is joined to the lamp and the weight allowed to fall, the motor acts as a generator in which there is an energy transfer from **k.e.** to **electrical energy**. When this is supplied to the lamp, it produces a transfer to **heat** and **light energy**.

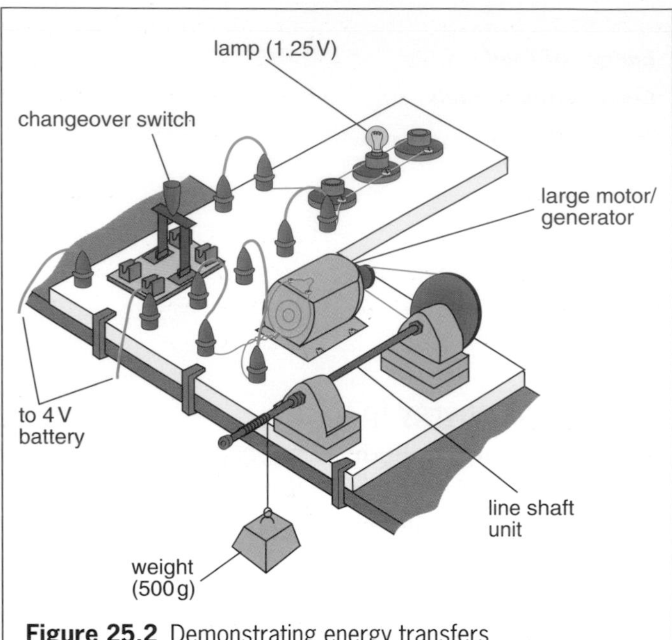

Figure 25.2 Demonstrating energy transfers

(b) Other examples

Study the energy transfers shown in Figures 25.3a to d. Some devices have been invented to cause particular energy transfers. For example, a **microphone** changes sound energy into electrical energy, a **loudspeaker** does the reverse. Belts, chains or gears are used to transfer energy between moving parts, e.g. in a bicycle.

Energy measurements

(a) Work

In science the word **work** has a different meaning from its everyday one. **Work is done when a force moves.** No work is done in the scientific sense by someone standing still holding a heavy pile of books: an upward force is exerted, but no motion results.

If a building worker carries ten bricks up to the first floor of a building he does more work than if he carries only one brick because he has to exert a larger force. Even more work is required if he carries the ten bricks to the second floor. The amount of work done depends on the size of the force applied and the distance it moves. We therefore measure work by

$$\text{work} = \text{force} \times \text{distance moved in direction of force} \quad (1)$$

The unit of work is the **joule** (J) and is **the work done when a force of 1 newton** (N) **moves through 1 metre** (m). For example, if you have to pull with a force of 50 N to move a crate steadily 3 m in the direction of the force, Figure 25.4a, the work done is 50 N × 3 m = 150 N m = 150 J. That is

$$\text{joules} = \text{newtons} \times \text{metres}$$

If you lift a mass of 3 kg vertically through 2 m, Figure 25.4b, you have to exert a vertically upward force equal to the weight of the body, i.e. 30 N (approximately) and the work done is 30 N × 2 m = 60 N m = 60 J.

Note that we must always take the distance in the direction in which the force acts.

(b) Measuring energy transfers

In an energy transfer work is done. **The work done is a measure of the amount of energy trans-**

Figure 25.3 Study these energy transfers

(a) p.e. to k.e.

(b) electrical energy to heat and light energy

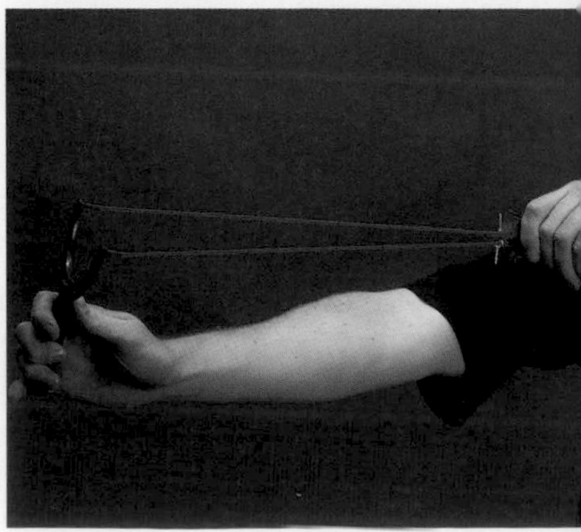

(c) chemical energy (from muscles of hand) to p.e. (strain energy of catapult)

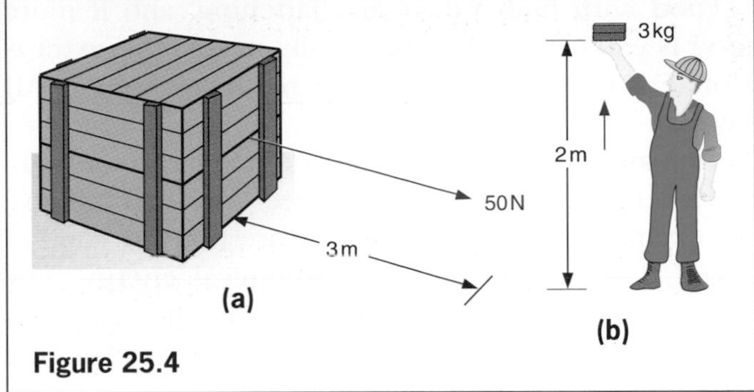

Figure 25.4

ferred. For example, if you have to exert an upward force of 10 N to raise a stone steadily through a vertical distance of 1.5 m, the work done is 15 J. This is also the amount of chemical energy transferred from your muscles to p.e. of the stone. All forms of energy, as well as work, are measured in joules.

(c) Power

The more powerful a car is the faster it can climb a hill, i.e. the faster it does work. The power of a device is the work it does per second, i.e. the rate at which it does work. This is the same as **the rate at which it transfers energy from one form to another**.

$$\text{power} = \frac{\text{work done}}{\text{time taken}} = \frac{\text{energy transfer}}{\text{time taken}} \qquad (2)$$

The unit of power is the **watt** (W) and is **a rate of working of 1 joule per second**, i.e. 1 W = 1 J/s. Larger units are the kilowatt (kW) and the megawatt (MW).

$$1\ \text{kW} = 1000\ \text{W} = 10^3\ \text{W}$$
$$1\ \text{MW} = 1\,000\,000\ \text{W} = 10^6\ \text{W}$$

If a machine does 500 J of work in 10 s its power is 500 J/10 s = 50 J/s = 50 W. A small car develops a maximum power of about 25 kW.

(d) p.e. of water to k.e. of turbine to electrical energy from generator

PRACTICAL WORK

Measuring power

(a) Your own power

Get someone with a stopwatch to time you running up a flight of stairs, the longer the better. Find your weight (in newtons). Calculate the total vertical height (in metres) you have climbed by measuring the height of one stair and counting the number of stairs.

The work you do (in joules) in lifting your weight to the top of the stairs is (your weight) × (vertical height of stairs). Calculate your power (in watts) from equation (2). About 0.5 kW is good. (1 horsepower = 0.75 kW)

(b) Electric motor

This experiment is described in Chapter 54.

Energy conservation

(a) Principle of conservation of energy

This is one of the basic laws of physics and is stated as follows.

Energy cannot be destroyed, it is always conserved.

However, energy is continually being transferred from one form to another. Some forms, such as electrical and chemical energy, are more easily transferred than others, such as heat for which it is hard to arrange a useful transfer. Ultimately all energy transfers result in the surroundings being heated (e.g. as a result of doing work against friction) and the energy is wasted, i.e. spread out and increasingly more difficult to use. For example, when a brick falls its p.e. becomes k.e.; as it hits the ground, its temperature rises and heat and sound are produced. If it seems in a transfer that some energy has disappeared, the 'lost' energy is often converted into non-useful heat. This appears to be the fate of all energy in the Universe and is one reason why new sources of useful energy have to be developed (Chapter 46).

(b) Efficiency of energy transfers

The efficiency of a device is the percentage of the energy supplied to it which is usefully transferred. It is calculated from the expression:

$$\text{efficiency} = \frac{\text{useful energy transferred by device}}{\text{total energy supplied to device}} \times 100\%$$

Table 25.1 lists the efficiencies of some devices and the energy transfers involved.

Table 25.1

Device	% Efficiency	Energy transfer
Large electric motor	90	Electrical to k.e.
Large electric generator	90	K.e. to electrical
Domestic gas boiler	75	Chemical to heat
Compact fluorescent lamp	50	Electrical to light
Steam turbine	45	Heat to k.e.
Car engine	25	Chemical to k.e.
Filament lamp	10	Electrical to light

Energy of food

When food is eaten it reacts with the oxygen we breathe into our lungs and is slowly 'burnt'. As a result chemical energy stored in food becomes thermal energy to warm the body and mechanical energy for muscular movement.

The **calorific value** (or energy value) of a substance is the amount of energy released when 1 kg is completely oxidized.

Calorific value is measured in J/kg. Dietitians sometimes use kilocalories instead of joules. The calorie was the previously used unit of thermal energy and equals 4.2 J. The calorific values of some foods are given in Figure 25.5 in megajoules per kilogram (MJ/kg).

Figure 25.5 Calorific values of some foods in MJ/kg

Food with high values are 'fattening' and if more food is eaten than the body really needs, the extra is stored as fat. The average adult requires about 12 MJ per day.

Our muscles change chemical energy into mechanical energy when we exert a force – to lift a weight, for example. Unfortunately, they are not too good at doing this; of every 100 J of chemical energy they use, they can convert only 25 J into mechanical energy – that is, they are only 25% efficient at changing chemical energy into mechanical energy. The other 75 J becomes thermal energy, much of which the body gets rid of by sweating.

Combustion of fuels

Fuels can be solids like wood and coal, liquids like fuel oil and paraffin, or gases like methane and butane.

Some fuels are better than others for certain jobs. For example, fuels for cooking or keeping us warm should, as well as being cheap, have a high **heating value**. This means that every gram of fuel should produce a large amount of heat energy when burnt. A fuel for a space rocket (e.g. liquid hydrogen) must also burn very quickly so that the gases created expand rapidly and leave the rocket at high speed.

The heating values of some fuels are given in Table 25.2 in kilojoules per gram.

Table 25.2 Heating values of fuels (kJ/g)

Solids	Value	Liquids	Value	Gases	Value
Wood	17	Fuel oil	45	Methane	55
Coal	25–33	Paraffin	48	Butane	50

The thick dark liquid called **petroleum** or **crude oil** is the source of most liquid and gaseous fuels. It is obtained from underground deposits at oil wells in many parts of the world. Natural gas (methane) is often found with it. In an oil refinery different fuels are obtained from petroleum, including fuel oil for industry, diesel oil for lorries, paraffin (kerosene) for jet engines, petrol for cars, as well as butane (bottled gas).

QUESTIONS

1 Name the energy transfers which occur when
  a an electric bell rings,
  b someone speaks into a microphone,
  c a ball is thrown upwards,
  d there is a picture on a television screen,
  e a flashlamp is operating.

2 Name the forms of energy represented by the letters A, B, C and D in the following statement.

In a coal-fired power station, the (A) energy of coal becomes (B) energy which changes water into steam. The steam drives a turbine which drives a generator. A generator transfers (C) energy into (D) energy.

3 How much work is done when a mass of 3 kg (weighing 30 N) is lifted vertically through 6 m?

4 A hiker climbs a hill 300 m high. If she weighs 50 kg calculate the work she does in lifting her body to the top of the hill.

5 In loading a lorry a man lifts boxes each of weight 100 N through a height of 1.5 m.

a How much work does he do in lifting one box?

b How much energy is transferred when one box is lifted?

c If he lifts 4 boxes per minute at what power is he working?

6 A boy whose weight is 600 N runs up a flight of stairs 10 m high in 12 s. What is his average power?

7 **a** When the energy input to a gas-fired power station is 1000 MJ, the electrical energy output is 300 MJ. What is the efficiency of the power station in changing the energy in gas into electrical energy?

b What form does the 700 MJ of 'lost' energy take?

c What is the fate of the 'lost' energy?

Checklist

After studying this chapter you should be able to

- recall the different forms of energy,
- describe energy transfers in given examples,
- use the relation **work done = force × distance moved** to calculate energy transfer,
- define the unit of work,
- recall that power is the rate of energy transfer, give its unit and solve problems,
- describe an experiment to measure your own power,
- state the principle of conservation of energy,
- recall the meaning of **efficiency** of a device,
- recall the meaning of **calorific value** of food,
- explain what is meant by a fuel having a **high heating value**.

26 MACHINES

Force and distance multipliers
Mechanical advantage, velocity ratio.
Efficiency of a machine
Pulleys
Other simple machines
Wheel and axle.
Inclined plane.
Friction
Practical work
Efficiency of a pulley system.

A machine is any device which enables a force (the **effort**) acting at one point to overcome another force (the **load**) acting at some other point. A lever (Chapter 22) is a simple machine, as are pulleys, gears, screws, etc. They are used to build more complicated machines like the crane in Figure 26.3.

Force and distance multipliers

In Figure 26.1 a lever lifts a load of 100 N through 0.50 m when an effort is applied at the other end. The effort can be found from the principle of moments by taking moments about the pivot O as the effort *just begins* to raise the load.

clockwise moment = anticlockwise moment

$$\text{effort} \times 2\text{ m} = \text{load} \times 1\text{ m} = 100\text{ N} \times 1\text{ m}$$

$$\therefore \quad \text{effort} = 100\text{ N m}/2\text{ m} = 50\text{ N}$$

The lever has enabled an effort (E) to raise a load (L) twice as large, i.e. it is a **force multiplier**, but E has had to move twice as far as L. The lever has a **mechanical advantage** (MA) of 2 and a **velocity ratio** (VR) of 2 where

$$\text{MA} = \frac{L}{E} \quad \text{and} \quad \text{VR} = \frac{\text{distance moved by } E}{\text{distance moved by } L}$$

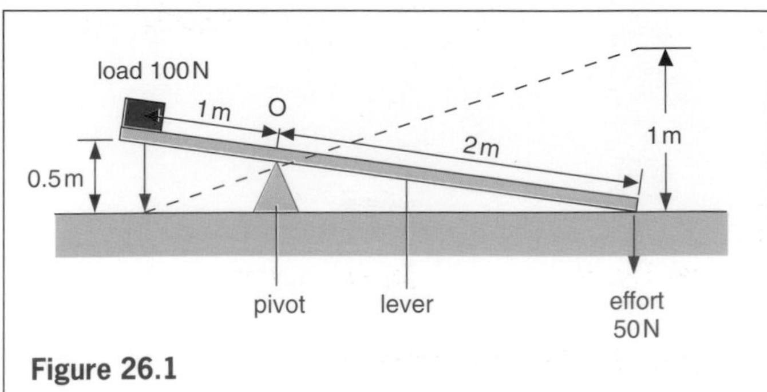

Figure 26.1

The forearm (Figure 22.4c) is a lever which is a **distance multiplier** because the load (which is smaller than the effort) moves farther than the effort applied by the biceps, i.e. its VR (and its MA) is less than 1.

Efficiency of a machine

Machines make work easier and transfer energy from one place to another. No machine is perfect and in practice more work is done by the effort on a machine than by the machine on the load. Work measures energy transfer (or change) and so we can also say that the energy input to a machine is greater than its energy output. Some energy is always wasted to overcome friction and move parts of the machine itself. In Figure 26.1 there is friction at the fulcrum (pivot); the effort will therefore be more than 50 N. As before, we define

$$\text{efficiency} = \frac{\text{energy output}}{\text{energy input}} = \frac{\text{work done on load}}{\text{work done by effort}}$$

This is expressed as a percentage and is always less than 100%.

There is a useful relation between MA, VR and efficiency. From the above equation we can show that

$$\text{percentage efficiency} = \frac{\text{MA}}{\text{VR}} \times 100\%$$

Pulleys

(a) Single fixed pulley, Figure 26.2a

This enables us to **lift** a load L more conveniently by applying a **downward** effort E. E need be only slightly greater than L and if friction in the pulley bearings is negligible then $E = L$ and MA = 1. What is the VR?

(b) Single moving pulley, Figure 26.2b

If the effort applied to the free end of the rope is E the total upward force on the pulley is $2E$ since two parts of the rope support it. A load $L = 2E$ can therefore be raised if the pulley and rope are frictionless and weightless. That is, MA = 2 (but less in practice).

To raise the load by 1 m requires each side of the rope to shorten by 1 m. The free end has therefore to take up 2 m of slack and so VR = 2.

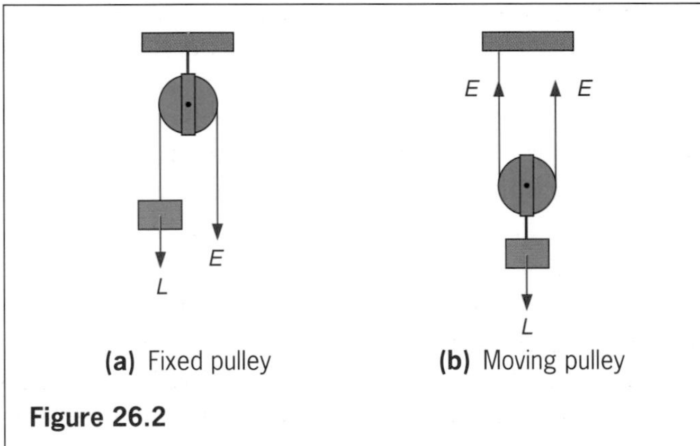

(a) Fixed pulley **(b)** Moving pulley

Figure 26.2

(c) Block and tackle

This type of pulley system is used in cranes (Figure 26.3) and in lifts. It consists of two 'blocks' each with one or more pulleys. In the arrangement of Figure 26.4a the pulleys in the blocks are shown one above the other for clarity; in practice they are side by side on the same axle, Figure 26.4b. The rope passes round each pulley in turn.

The total upward force on the lower block is $4E$ since it is supported by four parts of the rope and a load $L = 4E$ can be raised. Hence MA = 4 if the pulleys are frictionless and weightless. Using the same reasoning as in **(b)** above we see that that VR = 4, i.e. the number of times the rope passes from one block to the other.

PRACTICAL WORK

Efficiency of a pulley system

Set up the system of Figure 26.4b or a similar one. Starting with 50 g in the load pan, add weights to the effort pan until the load just rises steadily. Record the

Figure 26.3 A crane using a 'block and tackle' pulley system

load and effort in a table (100 g has a weight of 1 N) and repeat for greater loads.

The VR can be obtained as explained before. Work out the MA and the efficiency for each pair of readings of load and effort.

Load N	Effort N	MA	Efficiency = $\frac{\text{MA}}{\text{VR}} \times 100\%$

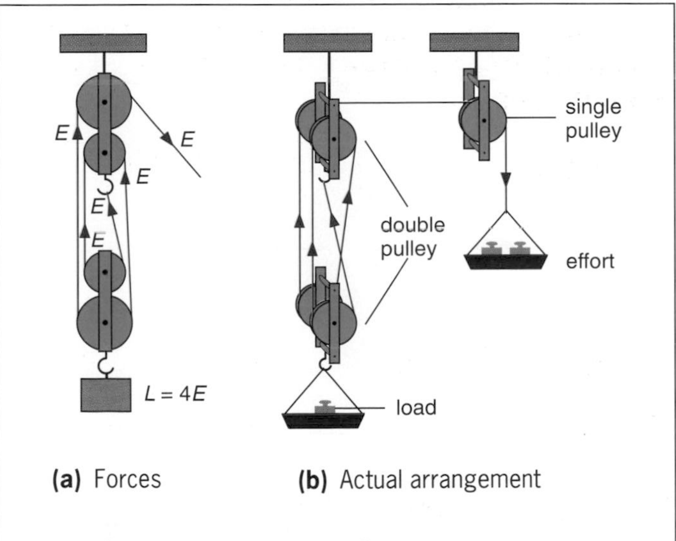

(a) Forces **(b)** Actual arrangement

Figure 26.4 Block and tackle pulleys

Notes:

1 The lower pulley block and the load pan are also raised by the effort but are not included as part of the load. They become less important as the load increases and the MA and the efficiency both increase for this reason. The VR is constant for a particular system.

2 The efficiency is less than 100% because the system is not frictionless and the moving parts are not weightless.

Other simple machines

(a) Wheel and axle

A screwdriver and the steering wheel of a car, Figures 26.5a, b, use the wheel and axle principle. This principle is shown in Figure 26.6; the effort is applied to a rope wound round the wheel and the load is raised by another rope wound oppositely on the axle. A smaller effort (moving a greater distance) is required than without the wheel.

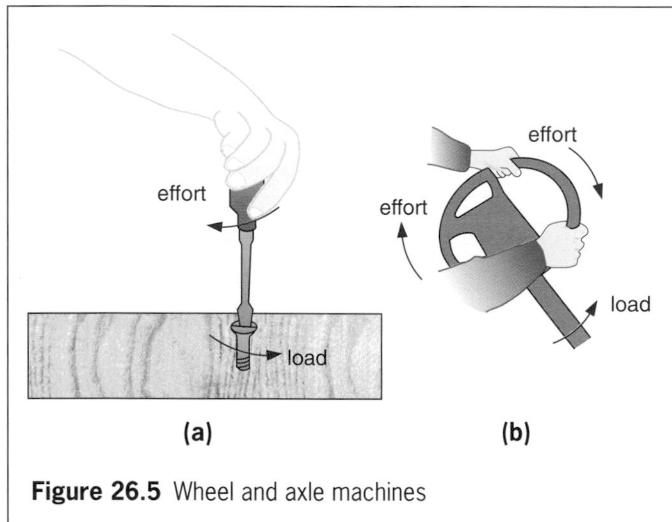

Figure 26.5 Wheel and axle machines

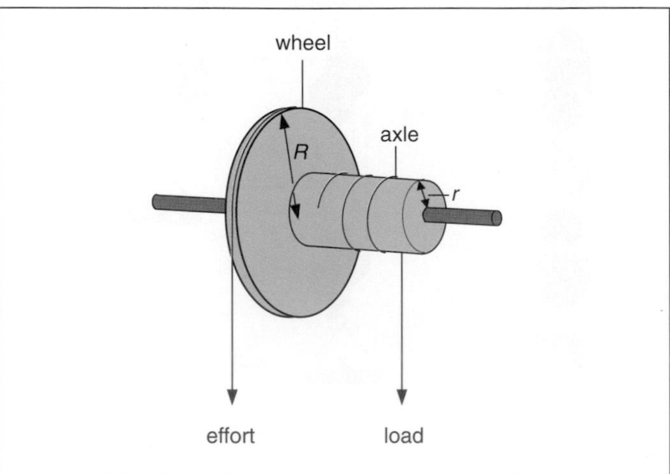

Figure 26.6 The wheel and axle principle: wheels make turning easier

(b) Inclined plane

It is easier to push a load up a plank than to lift it vertically. In Figure 26.7, to raise the load L through a **vertical** height h, the smaller effort E moves a greater distance d equal to the length of the incline.

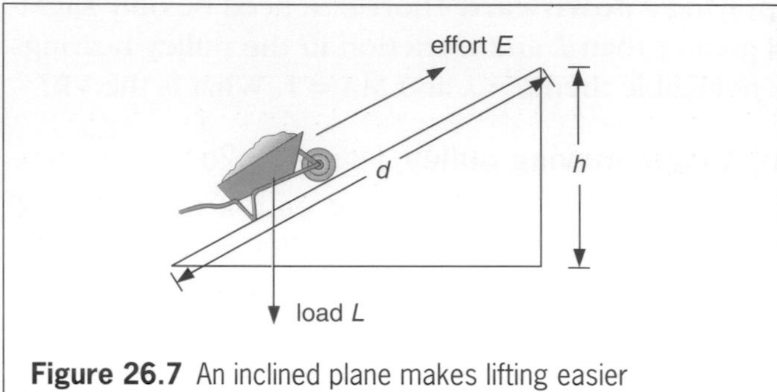

Figure 26.7 An inclined plane makes lifting easier

Friction

Friction is the force that opposes one surface moving, or trying to move, over another. It can be a help or a hindrance. We could not walk if it did not exist between the soles of our shoes and the ground. Our feet would slip backwards, as they tend to if we walk on ice. On the other hand, engineers try to reduce friction to a minimum in the moving parts of machinery by using lubricating oils and ball-bearings.

When a gradually increasing force P is applied through a spring balance to a block on a table, Figure 26.8, it does not move at first. This is because an equally increasing but opposing frictional force F acts where the block and table touch. At any instant P and F are equal and opposite.

If P is increased further the block eventually moves; as it does so F has its maximum value, called **starting** or **static** friction. When it is moving at a steady speed the balance reading is slightly less than that for starting friction. **Sliding** or **dynamic** friction is therefore less than starting or static friction.

A weight on the block increases the force pressing the surfaces together and increases friction.

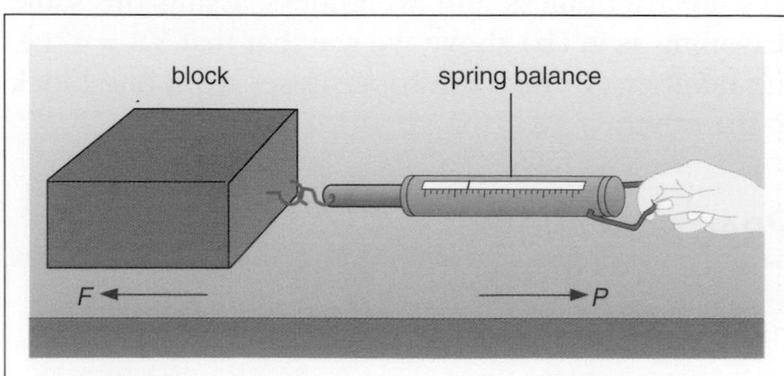

Figure 26.8 Friction opposes motion between surfaces in contact

When work is done against friction, the temperatures of the bodies in contact rise (as you can test by rubbing your hands together); mechanical energy is being changed into heat energy.

Worked example

Lorry A accelerates from rest. Lorry B travels at constant velocity. Explain fully why each lorry must continue to use energy from its fuel during its motion.

(M.E.G. 1986)

Lorry A and Lorry B do work against resistive forces. Lorry A also gains kinetic energy.

QUESTIONS

1 A load of 500 N is raised 0.20 m by a machine in which an effort of 150 N moves 1.0 m. What is

a the work done on the load,

b the work done by the effort,

c the efficiency?

2 An effort of 250 N raises a load of 1000 N through 5 m in a pulley system. If the effort moves 30 m, what is

a the work done in raising the load,

b the work done by the effort,

c the efficiency?

3 For each pulley system shown in Figures 26.9a, b, c, what is (i) the MA, (ii) the VR, and (iii) the efficiency?

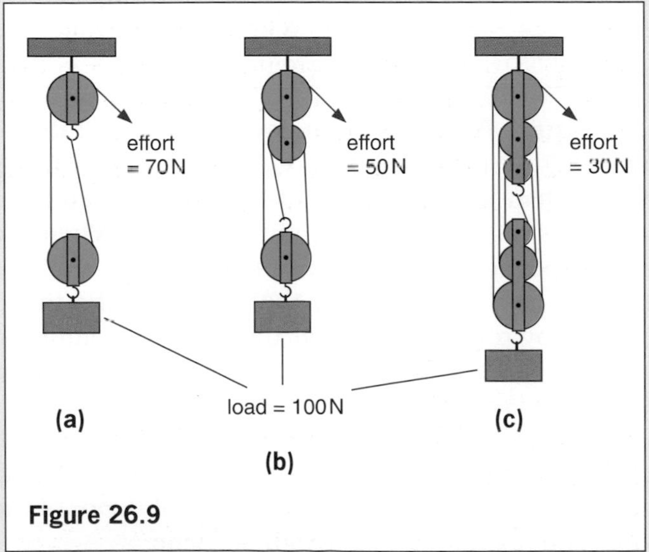

Figure 26.9

4 Is a bicycle a 'distance multiplier' or a 'force multiplier'?

Checklist

After studying this chapter you should be able to

- recognize a machine as a force multiplier or a distance multiplier,
- define the terms **mechanical advantage** (MA), **velocity ratio** (VR) and **efficiency** of a machine and solve problems using them,
- describe an experiment to measure the efficiency of a pulley system,
- recall the use and advantage of a wheel and axle, and an inclined plane,
- recall that friction opposes motion between surfaces in contact,
- recall that heat energy is produced when work is done against friction.

27 PRESSURE IN LIQUIDS

Pressure

To make sense of some effects in which a force acts on a body we have to consider not only the force but also the area on which it acts. For example, wearing skis prevents you sinking into soft snow because your weight is spread over a greater area. We say the **pressure** is less.

Pressure is the force (or thrust) **acting on unit area** (e.g. 1 m^2) and is calculated from

$$\text{pressure} = \frac{\text{force}}{\text{area}}$$

The pressure exerted on the floor by the same box (a) standing on end, (b) lying flat, is shown in Figure 27.1. The unit of pressure is the **pascal** (Pa); it equals 1 newton per square metre (N/m^2) and is quite a small pressure. An apple in your hand exerts about 1000 Pa.

The greater the area over which a force acts the lesser the pressure. This is why a tractor with wide wheels can move over soft ground. The pressure is large when the area is small and accounts for a nail being given a sharp point. Walnuts can be broken in the hand by squeezing two together but not one. Why?

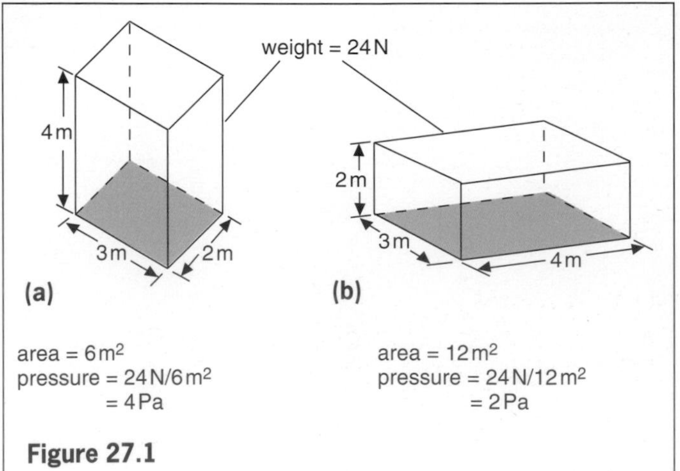

Figure 27.1

Liquid pressure

1. Pressure in a liquid increases with depth because the farther down you go the greater the weight of liquid above. In Figure 27.2a water spurts out fastest and furthest from the lowest hole.

2. Pressure at one depth acts equally in all directions. The can of water in Figure 27.2b has similar holes all round it at the same level. Water comes out as fast and as far from each hole. Hence the pressure exerted by the water at this depth is the same in all directions.

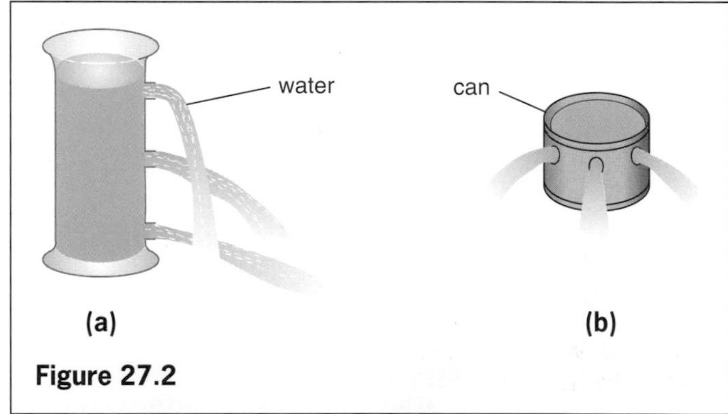

Figure 27.2

3. A liquid finds its own level. In the U-tube of Figure 27.3a the liquid pressure at the foot of P is greater than at the foot of Q because the left-hand column is higher than the right-hand one. When the clip is opened the liquid flows from P to Q until the pressure and the levels are the same, i.e. the liquid finds its own level. Although the weight of liquid in Q is now greater than in P, it acts over a greater area since tube Q is wider.

In Figure 27.3b the liquid is at the same level in each tube and confirms that the pressure at the foot of a liquid column depends only on the **vertical** depth of the liquid and not on the tube width or shape.

4. Pressure depends on the density of the liquid. The denser the liquid, the greater is the pressure at any given depth.

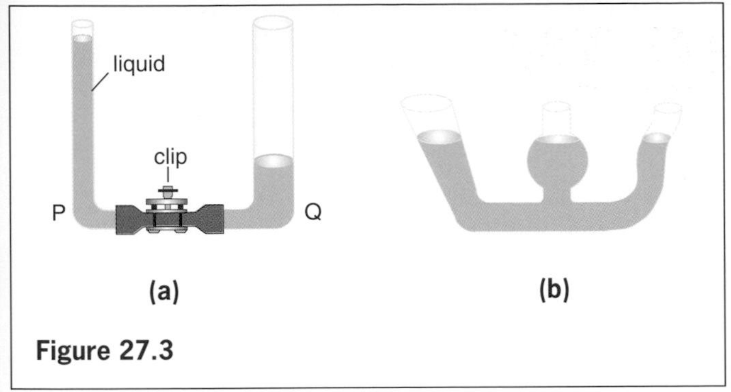

Figure 27.3

Water supply system

A town's water supply often comes from a reservoir on high ground. Water flows from it through pipes to any tap or storage tank that is below the level of water in the reservoir, Figure 27.4. The lower the place supplied the greater is the water pressure at it. In very tall buildings it may be necessary first to pump the water to a large tank in the roof.

Reservoirs for water supply or for hydroelectric power stations are often made in mountainous regions by building a dam at one end of a valley. The dam must be thicker at the bottom than at the top due to the large water pressure at the bottom.

Figure 27.4 Water supply system. Why is the pump needed in the high-rise building?

Hydraulic machines

Liquids are almost incompressible (i.e. their volume cannot be reduced by squeezing) and they 'pass on' any pressure applied to them. Use is made of these facts in hydraulic machines. Figure 27.5 shows the principle on which they work. Suppose a downward force of 1 N acts on a piston A of area 1/100 m^2. The pressure transmitted through the liquid is

$$\text{pressure} = \frac{\text{force}}{\text{area}} = \frac{1\ \text{N}}{1/100\ \text{m}^2} = 100\ \text{N/m}^2$$

$$= 100\ \text{Pa}$$

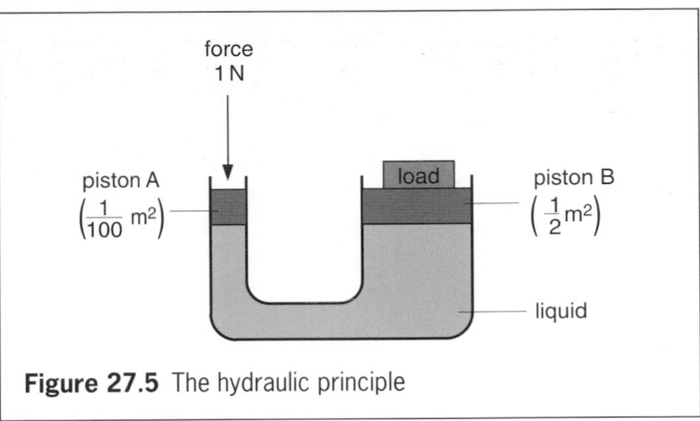

Figure 27.5 The hydraulic principle

This pressure acts on piston B of area $\frac{1}{2}$ m^2. The total upwards force or thrust on B is given by

$$\text{force} = \text{pressure} \times \text{area} = 100\ \text{N/m}^2 \times \tfrac{1}{2}\ \text{m}^2 = 50\ \text{N}$$

A force of 1 N thus produces a force of 50 N. Hydraulic machines are force multipliers.

A **hydraulic jack**, Figure 27.6, has a platform on top of piston B and is used in garages to lift cars. Both valves open only to the right and they allow B to be raised a long way when A moves up and down repeatedly. **Hydraulic fork-lift trucks** and similar machines such as **loaders**, Figure 27.7, work in the same way. In a **hydraulic press** there is a fixed plate above B, and sheets of steel placed between B and the plate can be forged. **Hydraulic car brakes** are shown in Figure 27.8. When the brake pedal is

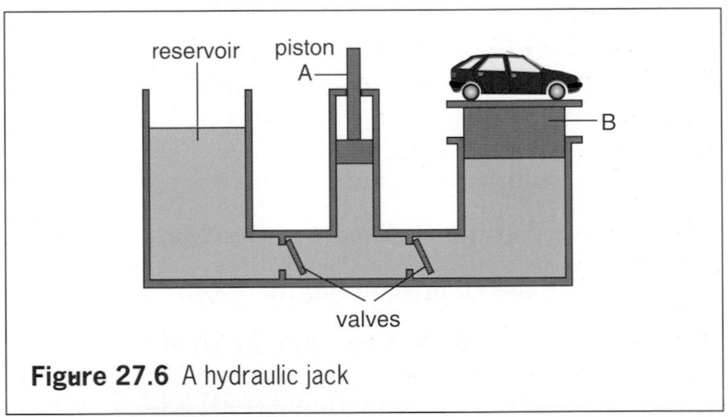

Figure 27.6 A hydraulic jack

Figure 27.7 A hydraulic machine in action

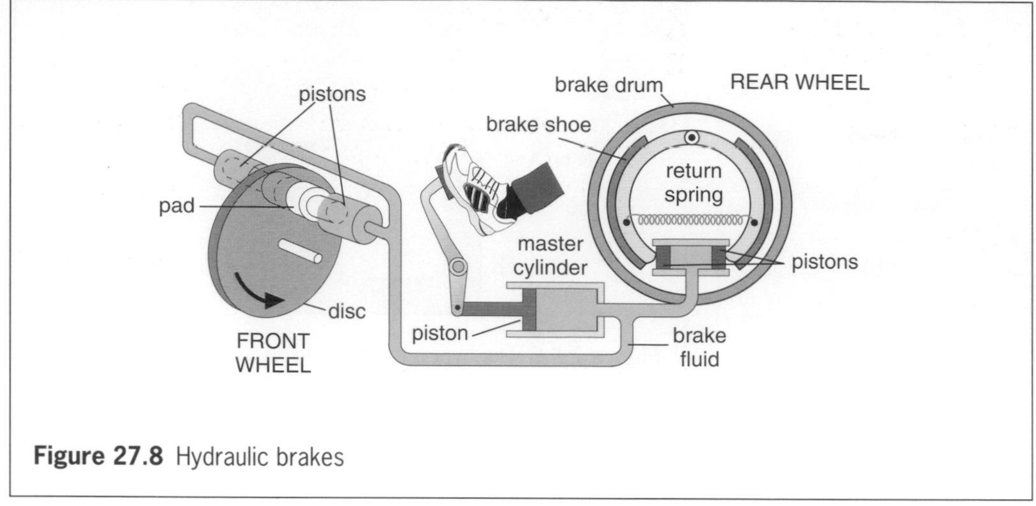

Figure 27.8 Hydraulic brakes

pushed the piston in the master cylinder exerts a force on the brake fluid and the resulting pressure is transmitted equally to eight other pistons (four are shown). These force the brake shoes or pads against the wheels and stop the car.

Expression for liquid pressure

In designing a dam an engineer has to calculate the pressure at various depths below the water surface.

An expression for the pressure at a depth h in a liquid of density d can be found by considering a horizontal area A, Figure 27.9. The force acting vertically downwards on A equals the weight of a liquid column of height h and cross-sectional area A above it. Then

$$\text{volume of liquid column} = hA$$

Since mass = volume × density we can say

$$\text{mass of liquid column} = hAd$$

Taking a mass of 1 kg to have weight 10 N,

$$\text{weight of liquid column} = 10\,hAd$$

$$\therefore \quad \text{force on area } A = 10\,hAd$$

$$\therefore \quad \text{pressure} = \text{force/area} = 10\,hAd/A = 10\,hd$$

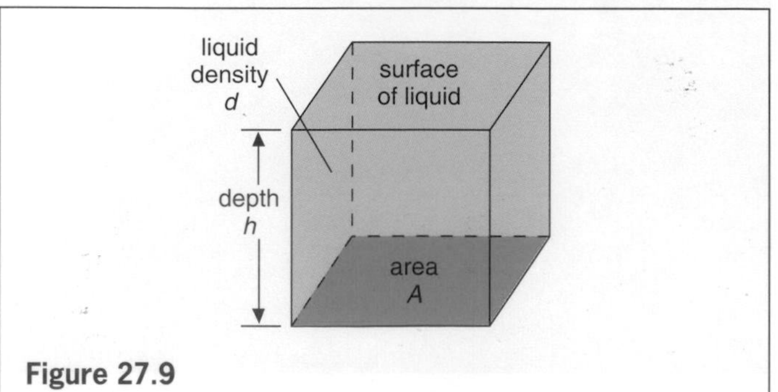

Figure 27.9

This pressure acts equally in all directions at depth h and depends only on h and d. Its value will be in Pa if h is in m and d in kg/m^3.

QUESTIONS

1 A woman in stiletto heels is more likely to damage a wooden floor than an elephant is. Why?

2 a What is the pressure on a surface when a force of 50 N acts on an area of (i) 2.0 m^2, (ii) 100 m^2, (iii) 0.50 m^2?

b A pressure of 10 Pa acts on an area of 3.0 m^2. What is the force acting on the area?

3 A block of concrete weighs 900 N and its base is a square of side 2.0 m. What pressure does the block exert on the ground? *(E.A.)*

4 a What is the volume of the block in Figure 27.10?

b What is the mass of the block if its density is 2000 kg/m^3?

c What is the weight of the block? (Assume a mass of 1 kg has weight 10 N.)

d What pressure is exerted on the ground by the block?

e If the shaded side of the block is on the ground, what effect, if any, will this have on (i) the force exerted by the block on the ground, (ii) the pressure exerted by the block on the ground?

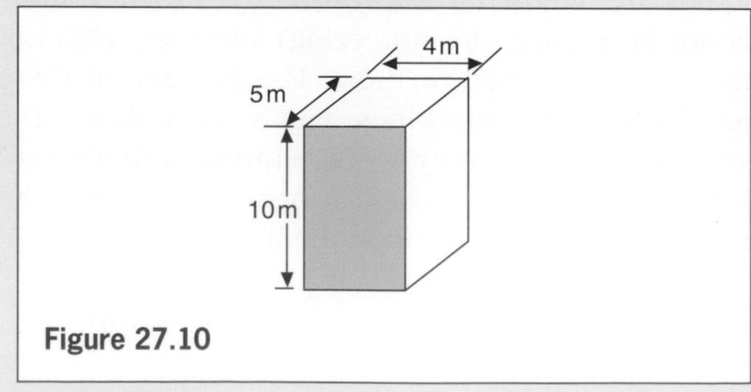

Figure 27.10

5 In a hydraulic press a force of 20 N is applied to a piston of area 0.20 m^2. The area of the other piston is 2.0 m^2. What is

a the pressure transmitted through the liquid,

b the force on the other piston?

6 **a** Why must a liquid and not a gas be used as the 'fluid' in a hydraulic machine?

b On what other important property of a liquid do hydraulic machines depend?

7 What is the pressure 100 m below the surface of sea water of density 1150 kg/m^3?

8 The pressure in a water pipe in the ground floor of a building is 4×10^5 Pa but three floors up it is only 2×10^5 Pa. What is the height between the ground floor and the third floor? (The water in the pipe may be assumed to be stationary; density of water = 1×10^3 kg/m^3.) *(N.I.)*

Checklist

After studying this chapter you should be able to

- define **pressure** and recall its unit,
- connect the pressure in a fluid with its depth and density,
- recall that pressure is transmitted through a fluid and use it to explain the hydraulic jack and hydraulic car brakes,
- use pressure = 10 *hd* to solve problems.

28 ATMOSPHERIC PRESSURE

The air forming the Earth's atmosphere stretches upwards a long way. Air has weight and in a normal room weighs about the same as you do, e.g. 500 N. Owing to its weight the atmosphere exerts a large pressure at sea-level, about 100 000 N/m^2 = 10^5 Pa = 100 kPa. This pressure acts equally in all directions.

We do not normally feel atmospheric pressure because the pressure inside our bodies is almost the same as that outside. A similar balance exists with external objects, unless air is removed from one place. The effect of atmospheric pressure is then noticeable. A space from which all the air has been withdrawn is a **vacuum**.

Air pressure demonstrations

(a) Collapsing can

If air is removed from a can, Figure 28.1a, by a vacuum pump, the can collapses because the air pressure inside becomes less than that outside, Figure 28.1b.

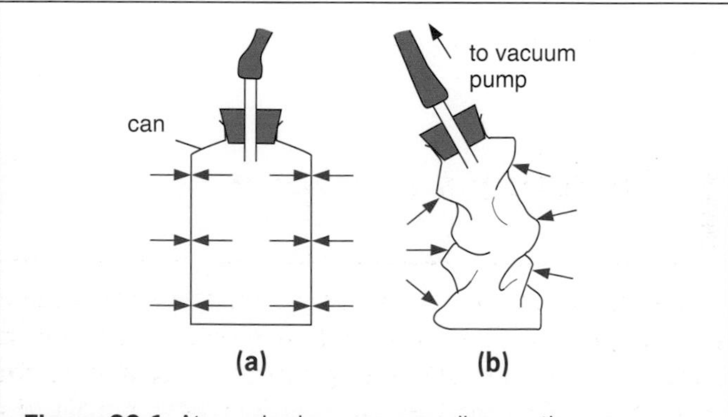

Figure 28.1 Atmospheric pressure collapses the evacuated can

(b) Magdeburg hemispheres

The vacuum pump was invented by von Guericke, the Mayor of Magdeburg. In about 1650 he used it to remove the air from two large hollow metal hemispheres, fitted together to give an airtight sphere. So good was his pump that it took two teams, each of eight horses, to separate the hemispheres. A similar experiment can be done by two people with small brass hemispheres, Figure 28.2.

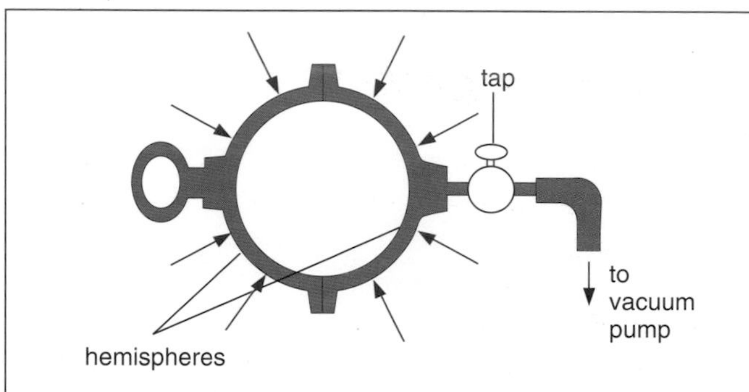

Figure 28.2 Atmospheric pressure holds the hemispheres fast

Using air pressure

(a) Drinking straw, Figure 28.3

When you suck, your lungs expand and air passes into them from the straw. Atmospheric pressure pushing down on the surface of the liquid in the bottle is now greater than the pressure of the air in the straw and so forces the liquid up to your mouth.

Figure 28.3 Atmospheric pressure forces liquid up the straw

(b) Rubber sucker, Figure 28.4

When the sucker is moistened and pressed on a smooth flat surface, the air inside is pushed out. Atmospheric pressure then holds it firmly against the surface. Suckers are used in the home as holders for towels and on the underside of non-slip bath mats, for attaching car licences to windscreens and in industry for lifting metal sheets.

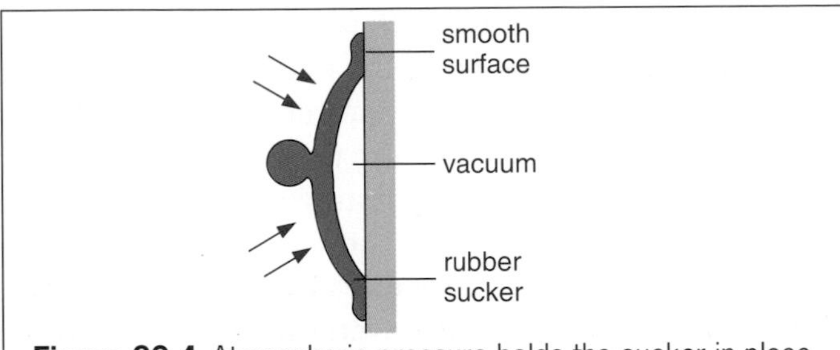

Figure 28.4 Atmospheric pressure holds the sucker in place

Pressure gauges

These measure the pressure exerted by a fluid, i.e. by a liquid or a gas.

(a) Bourdon gauge

This works like the toy in Figure 28.5. The harder you blow into the paper tube, the more it uncurls. In a Bourdon gauge, Figure 28.6, when a fluid pressure is applied, the curved metal tube tries to straighten out and rotates a pointer over a scale. Car oil-pressure gauges and the gauges on gas cylinders are of this type.

Figure 28.5 The harder you blow, the greater the pressure and the more it uncurls

(b) U-tube manometer

In Figure 28.7a each surface of the liquid is acted on equally by atmospheric pressure and the levels are the same. If one side is connected to, for example, the gas supply, Figure 28.7b, the gas exerts a pressure on surface A and level B rises until

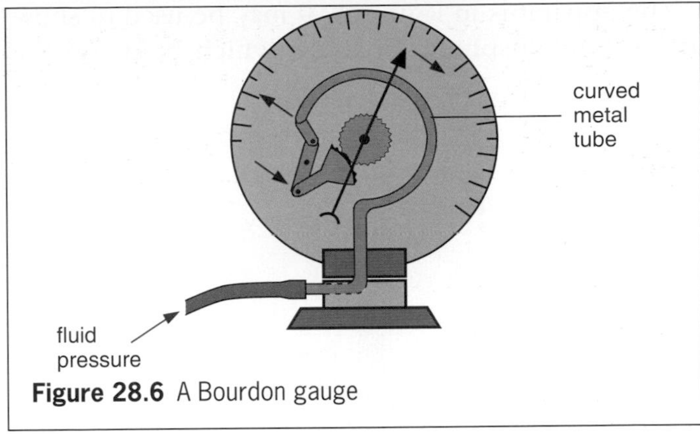

Figure 28.6 A Bourdon gauge

pressure of gas = atmospheric pressure
+ pressure due to liquid column BC

The pressure of the liquid column BC therefore equals the amount by which the gas pressure *exceeds* atmospheric pressure. It equals 10 hd (in Pa) where h is the vertical height of BC (in m) and d is the density of the liquid (in kg/m^3). The height h is called the **head of liquid** and sometimes, instead of stating a pressure in Pa, we say that it is so many cm of water (or mercury for higher pressures).

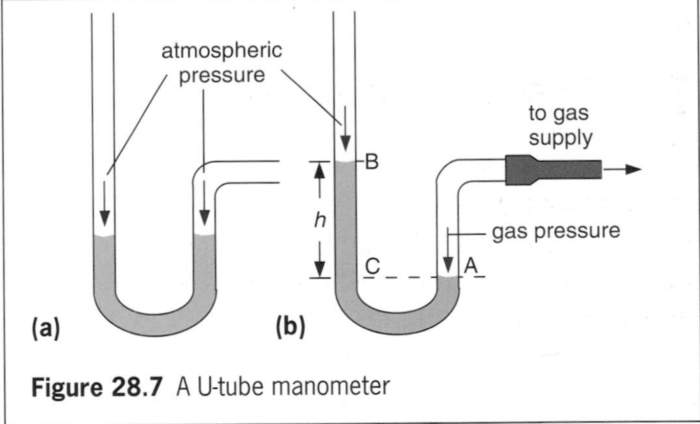

Figure 28.7 A U-tube manometer

Mercury barometer

A barometer is a manometer which measures atmospheric pressure. A simple barometer is shown in Figure 28.8 (overleaf). The pressure at X due to the weight of the column of mercury XY equals the atmospheric pressure on the surface of the mercury in the bowl. XY measures the atmospheric pressure in mm of mercury (mmHg).

The *vertical* height of the column is unchanged if the tube is tilted. Would it be different with a wider tube? Why? The space above the mercury in the tube is a vacuum (except for a little mercury vapour). How could this be tested?

The apparatus in Figure 28.9 may be used to show that it is atmospheric pressure which holds up the column.

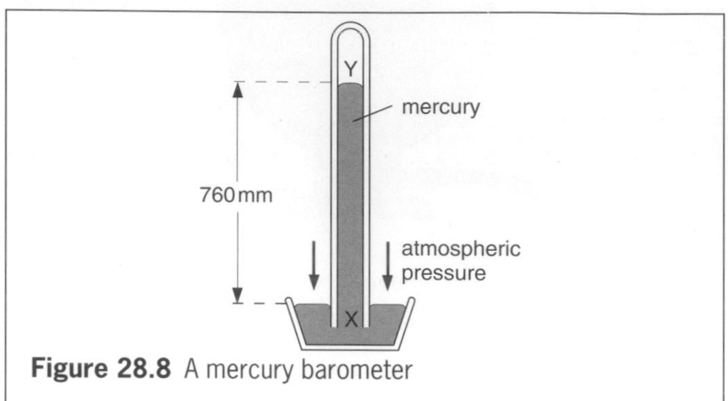

Figure 28.8 A mercury barometer

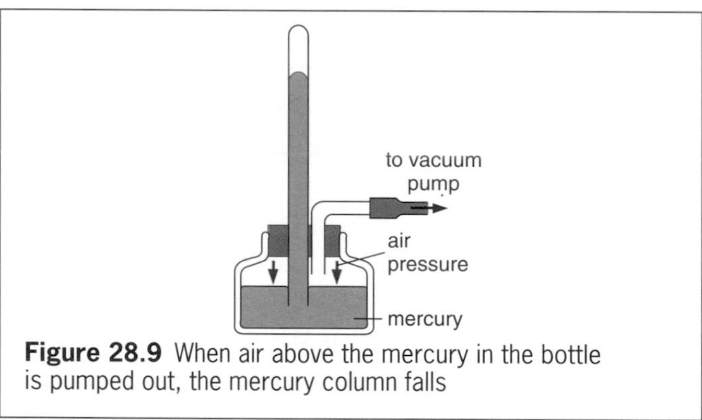

Figure 28.9 When air above the mercury in the bottle is pumped out, the mercury column falls

Aneroid barometer

An aneroid (no liquid) barometer consists of a partially evacuated, thin metal box with corrugated sides to increase its strength, Figure 28.10. The box is prevented from collapsing by a strong spring. If the atmospheric pressure increases, the box caves in slightly, if it decreases the spring pulls it out. A system of levers magnifies this movement and causes a chain to move a pointer over a scale.

Aneroid barometers are used in weather forecasting, high pressure being associated with fine weather. They are also used as altimeters to measure the height of an aircraft since atmospheric pressure decreases with height above sea-level.

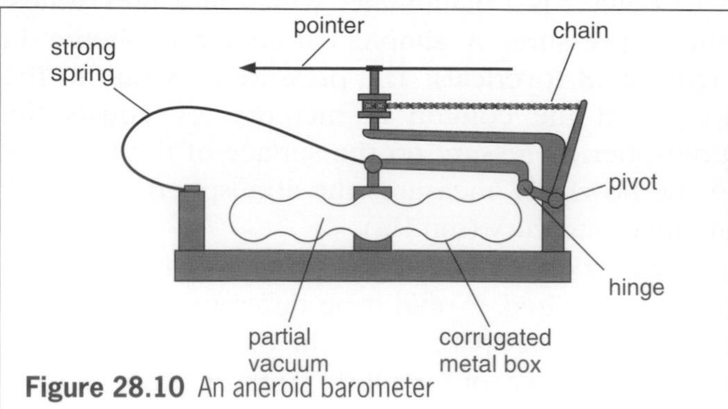

Figure 28.10 An aneroid barometer

Pressure, aviation and diving

Our bodies are designed to work at normal atmospheric pressure. At high altitudes breathing is difficult and modern aircraft have pressurized cabins in which the air pressure is increased sufficiently above that outside to safeguard the crew and passengers.

Ear 'popping' occurs when there are pressure changes, e.g. at aircraft take-off. It is due to the pressure difference between the air in the middle part of the ear and that in the outer ear. The eardrum becomes distorted. Swallowing helps to equalize the pressures.

If astronauts are not in pressurized space capsules, e.g. when on a space walk, they must rely on space suits for supplying a suitable atmosphere, Figure 28.11. Otherwise blood and water in the body would boil and possibly explode.

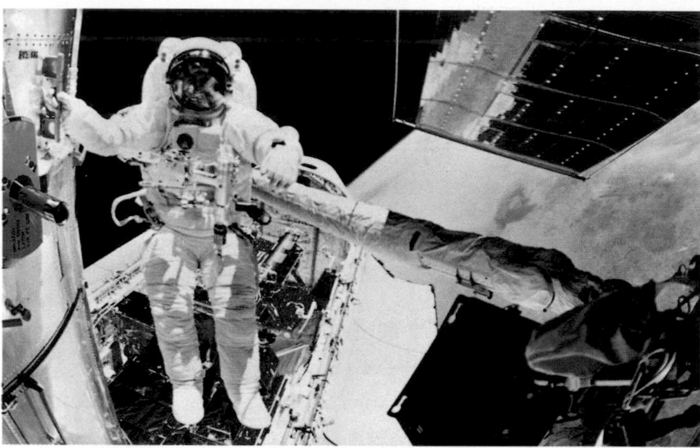

Figure 28.11 Astronaut in a pressurized suit on a space walk to repair the Hubble Telescope

Divers meet high pressures and different dangers. If they breathed air, extra nitrogen might dissolve in the blood and on returning to the surface they would suffer the painful and sometimes fatal condition called 'bends'. This is caused by the dissolved nitrogen forming bubbles in the blood. To prevent this they breathe a mixture of oxygen and helium, Figure 28.12. If divers have been working in deep waters (e.g. 500 m) they are brought to the surface in a diving bell and kept in a decompression chamber for up to several days while the pressure is slowly reduced.

Figure 28.12 Diver equipped with breathing apparatus

QUESTIONS

1 a Figures 28.13a and b show an open-ended and a closed tube manometer connected (at different times) to the same gas cylinder. The mercury barometer reads 760 mm. Assuming no loss in pressure from the gas cylinder, calculate (i) the pressure of gas in the cylinder, (ii) the height of the mercury in Figure 28.13b (i.e. x).

b Draw a diagram of a manometer similar to that in (i) Figure 28.13a, (ii) Figure 28.13b, when it is connected to a gas supply at less than atmospheric pressure (760 mm of mercury). *(W.M.)*

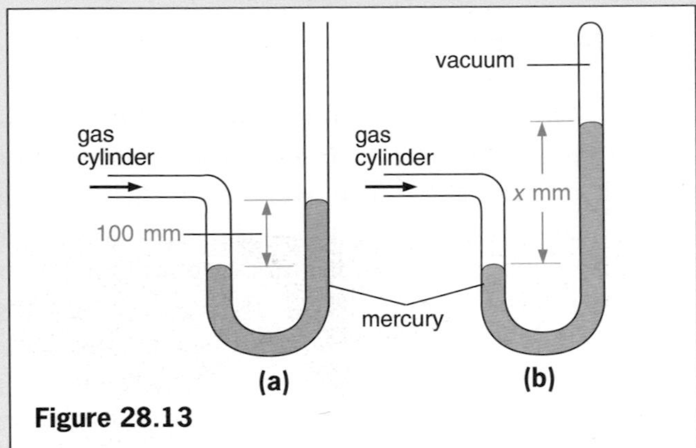

Figure 28.13

2 Figure 28.14 shows a simple barometer.

a What is the region A?

b What keeps the mercury in the tube?

c What is the value of the atmospheric pressure being shown by the barometer?

d What would happen to this reading if the barometer were taken up a high mountain? Give a reason. *(E.M.)*

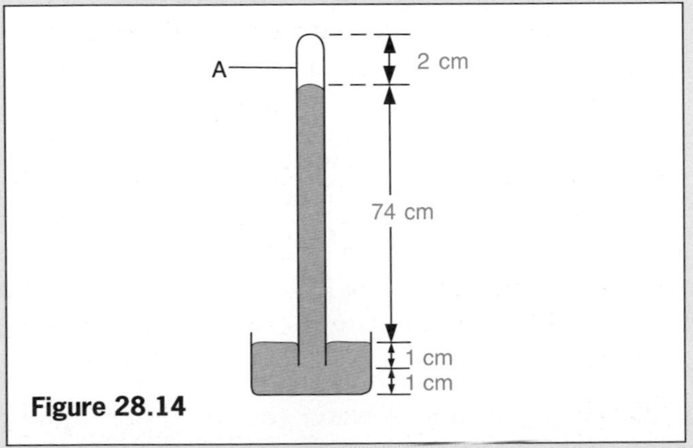

Figure 28.14

3 What would be the height of a water barometer if atmospheric pressure is 1.0×10^5 Pa and the density of water is 1.0×10^3 kg/m^3? (**Hint**. Use $p = 10\ hd$.)

Checklist

After studying this chapter you should be able to

- describe demonstrations of the effects of air pressure,
- describe uses of air pressure,
- describe how a Bourdon gauge or a U-tube manometer may be used to measure gas pressure,
- explain how a simple mercury barometer works,
- explain how an aneroid barometer works and is used as a weather glass and as an altimeter,
- explain certain effects in aviation and diving in terms of pressure, e.g. ear 'popping', the 'bends'.

29 FLOATING, SINKING AND FLYING

Archimedes' principle
Floating
Ships, submarines, balloons

A ship gets support from the water and floats because its average density is less than that of the water (see Chapter 18). Any object in a liquid, whether floating or submerged, is acted on by an upward force or **upthrust**. This makes it seem to weigh less than normal.

Archimedes' principle

In Figure 29.1a the block hanging from the spring balance weighs 10 N in air. When it is completely immersed in water the reading becomes 6 N, Figure 29.1b. The loss in weight of the block is 10 − 6 = 4 N: the upthrust of the water on it is therefore 4 N.

If a can like that in Figure 29.1c is used, full of water to the level of the spout, the water displaced (the overflow) can be collected and weighed. Its weight here is 4 N, the same as the upthrust. (Its volume is exactly equal to the volume of the block.) Experiments with other liquids and also with gases lead to the general conclusion called **Archimedes' principle**.

When a body is wholly or partly submerged in a fluid the upthrust equals the weight of fluid displaced (i.e. pushed aside).

A fluid means either a liquid or a gas. The case of gases will be dealt with later.

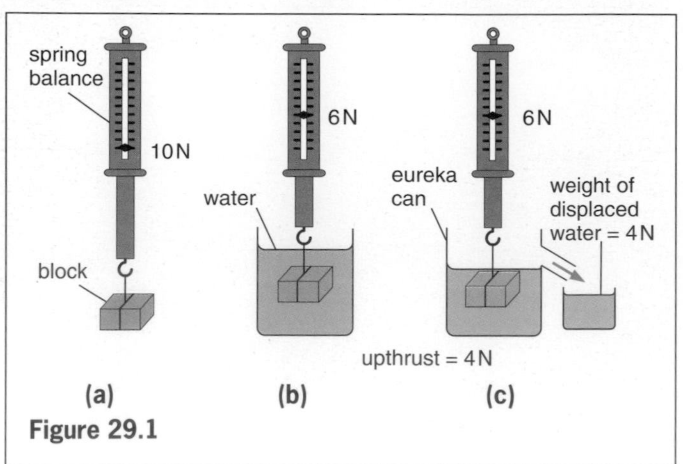

Figure 29.1

Floating

A stone held below the surface of water sinks when released, a cork rises. The weight of the stone is greater than the upthrust on it (i.e. the weight of water displaced) and there is a net or resultant downward force on it, Figure 29.2a. If the cork has the same volume as the stone, it will displace the same weight (and volume) of water. The upthrust on it is therefore the same as for the stone but it is greater than the weight of the cork. The resultant upward force on the cork makes it rise, Figure 29.2b.

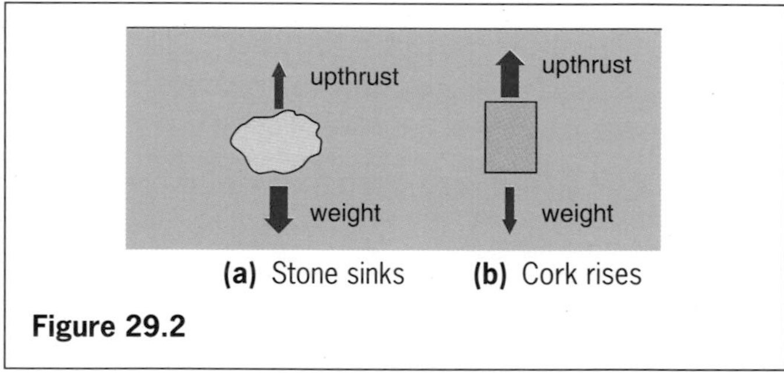

(a) Stone sinks **(b)** Cork rises

Figure 29.2

When a body floats in water the upthrust equals the weight of the body. The net force on the body is zero. This is a case of the **principle of flotation**.

A floating body displaces its own weight of fluid.

For example, a floating block of wood of weight 10 N displaces an amount of water (or any other liquid in which it floats) having weight 10 N (i.e. mass 1 kg).

Ships, submarines, balloons

(a) Ships

A floating ship displaces a weight of water equal to its own weight including that of the cargo. The load lines (called the Plimsoll mark) on the side of a ship show the levels to which it can legally be loaded under different conditions, Figure 29.3. Why is it allowed to take a greater load in summer than in winter?

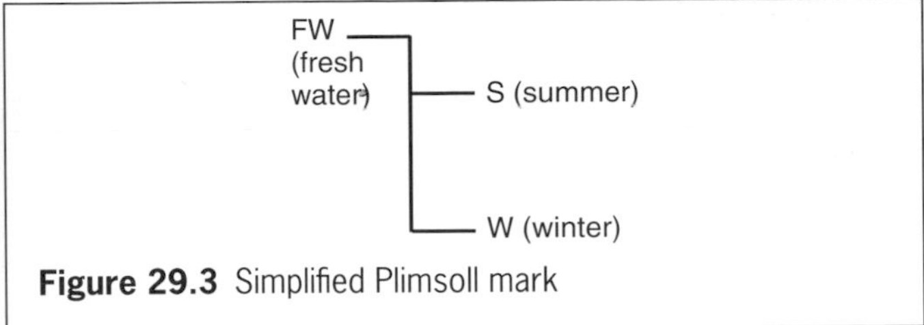

Figure 29.3 Simplified Plimsoll mark

(b) Submarines

A submarine sinks by taking water into its buoyancy tanks. Once submerged, the upthrust is unchanged but the weight of the submarine increases with the inflow of water and it sinks faster. To surface, compressed air is used to blow the water out of the tanks.

In the 'Cartesian diver', Figure 29.4, pressure on the cork forces more water into the bulb. The diver's weight increases and it sinks. Decreasing the pressure causes it to rise.

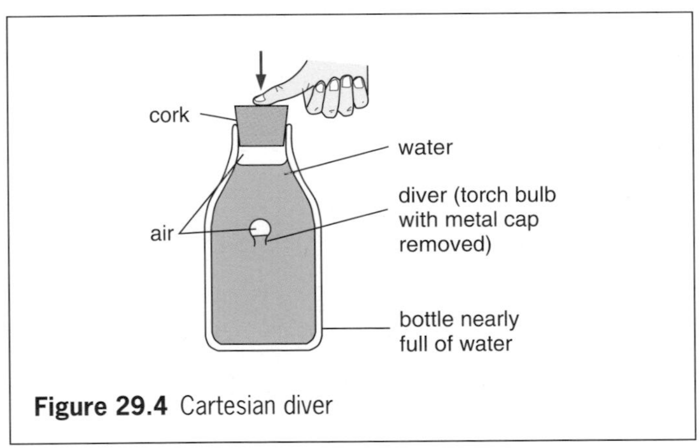

Figure 29.4 Cartesian diver

(c) Balloons

A balloon filled with hydrogen or hot air weighs less than the weight of air it displaces. The upthrust is therefore greater than its weight and the resultant upward force on the balloon causes it to rise.

Meteorological balloons, Figure 29.5 carrying scientific instruments called **radiosondes** are sent into the upper atmosphere. A small radio transmitter sends signals back to Earth which contain information about the temperature, pressure and humidity. They are tracked by radar to give data on wind direction and speed.

(d) Airships

An airship such as that in Figure 29.6 has a plastic gas bag filled with non-flammable helium (a gas less dense than air). It is powered by two car engines which drive swivelling propellers that provide vertical thrust for take-off and landing and horizontal thrust for forward motion. It can climb at about 700 metres a minute, cruise sedately at 80 km per hour (50 mph) and travel 1600 km (1000 miles) on just 550 litres (120 gallons) of fuel.

Airships are used for aerial surveys, photography and advertising, as well as for cargo and passenger carrying.

Figure 29.5 A meteorological balloon

Figure 29.6 An airship

QUESTIONS

1 A metal block is weighed **a** in air, **b** half-submerged in water, **c** fully submerged in water, **d** fully submerged in a strong salt solution.

The readings obtained, though not necessarily in the correct order, were 5 N, 8 N, 10 N and 6 N. Which reading was obtained for each weighing?

2 A block of density 2400 kg/m^3 has a volume of 0.20 m^3. What is **a** its mass, **b** its weight and **c** its apparent weight when completely immersed in a liquid of density 800 kg/m^3?

3 A hot air balloon and its basket together weigh 3000 N and it contains hot air weighing 17 000 N. If it displaces cold air of weight 25 000 N what is the maximum load it can lift?

4 A block of wood of volume 50 cm^3 and density 0.60 g/cm^3 floats on water. What is **a** the mass of the block, **b** the mass of water displaced, **c** the volume immersed in the water?
(Density of water = 1.0 g/cm^3)

5 A swimmer dives off a raft in a pool. Does the raft rise or sink in the water? What happens to the water level in the pool? Give reasons for your answer.

Checklist

After studying this chapter you should be able to

- state Archimedes' principle and use it to solve problems,
- state the principle of flotation and use it to solve problems,
- explain the behaviour of ships, submarines, balloons and airships.

30 FLUID FLOW

There are many occasions in everyday life when fluids, i.e. liquids or gases, have to flow through pipes or when objects such as aircraft have to move through the air. A knowledge and understanding of fluid flow is necessary to ensure that these processes occur efficiently and it is an important aspect of engineering.

Viscosity

Viscosity is a kind of internal frictional force in fluids which opposes motion when they flow. Syrup and motor oil pour slowly and are more **viscous** than water. Gases are less viscous than liquids.

(a) Explanation

All fluids 'stick' to a solid surface due to adhesion so that when they flow their speed must gradually decrease to zero as the wall of the pipe or container is approached. (The existence of a layer at rest is shown by the fact that while large particles of dust can be blown off a shelf, small particles remain which can be removed afterwards with the finger.)

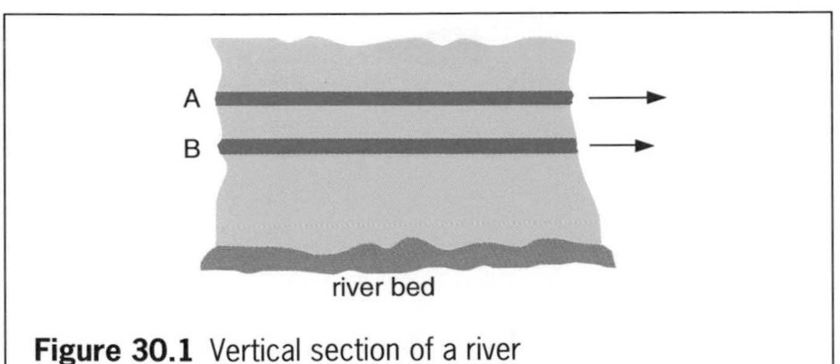

Figure 30.1 Vertical section of a river

The layer of water in contact with the bottom of a river must be at rest (or the river bed would be eroded rapidly). The speeds of higher layers increase the nearer the layer is to the surface. Therefore in Figure 30.1, molecules in layer A are moving faster than those in B below it. Molecules in A may cross to B and vice versa. As a result, A will slow down slightly and B will speed up slightly so creating a dragging force, i.e. friction, between the layers as they slide over each other. This hinders the flow of water, i.e. causes viscosity.

It follows also that water in the centre of a river flows faster than water nearer the banks; the layers in contact with the bank are at rest.

(b) Factors affecting viscosity

In general the viscosity of a liquid decreases rapidly as the temperature increases. This is due to the molecules being further apart as the temperature rises and so it is easier for them to move past each other. 'Viscostatic' motor oils, however, have about the same viscosity whether cold or hot.

Many pure liquids (e.g. water) and gases have the same viscosity whatever their flow speed; they are called **Newtonian fluids**. This is not so for **thixotropic** fluids such as non-drip paints whose viscosity decreases as the flow speed is increased (in the case of paints by being brushed).

(c) Lubricating oils

Viscosity is an essential property of a lubricating oil if it is to keep apart two solid surfaces moving over each other. On the other hand too high a viscosity causes unnecessary resistance to motion.

Fluid flow in a pipe

When a fluid flows *slowly* along a pipe, the flow is said to be **steady** and lines, called **streamlines**, are drawn to represent it. In Figure 30.2a (overleaf) they are parallel to the walls of the pipe.

If the flow is *very fast* and exceeds a certain **critical speed**, the flow becomes **turbulent** and the fluid is churned up. The streamlines are no longer straight and parallel, and **eddies** like those in Figure 30.2b are formed. The resistance to flow increases as a result.

Steady and turbulent flow can be viewed by injecting a thin stream of coloured dye from a jet at the entrance to a horizontal glass tube of water whose rate of flow can be adjusted.

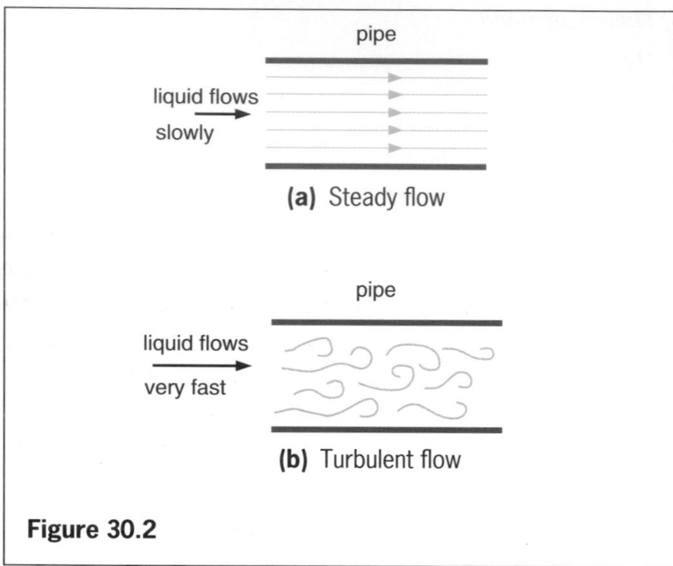

Figure 30.2

PRACTICAL WORK

Investigating fluid flow in tubes

(a) The rate of flow of water through tubes of different lengths and diameters can be investigated using apparatus like that in Figure 30.3. The level of water in the tank never changes so the pressure causing the flow remains constant.

(b) The viscosity of different liquids can be compared with the apparatus in Figure 30.4.

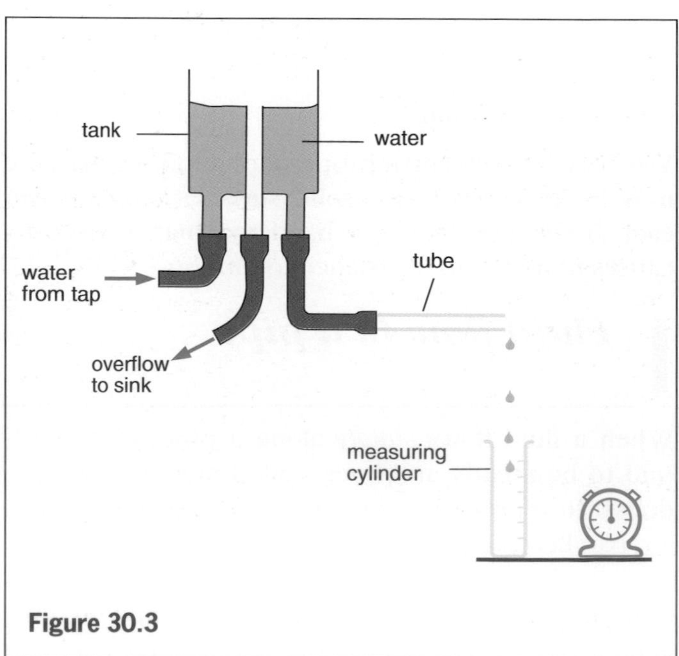

Figure 30.3

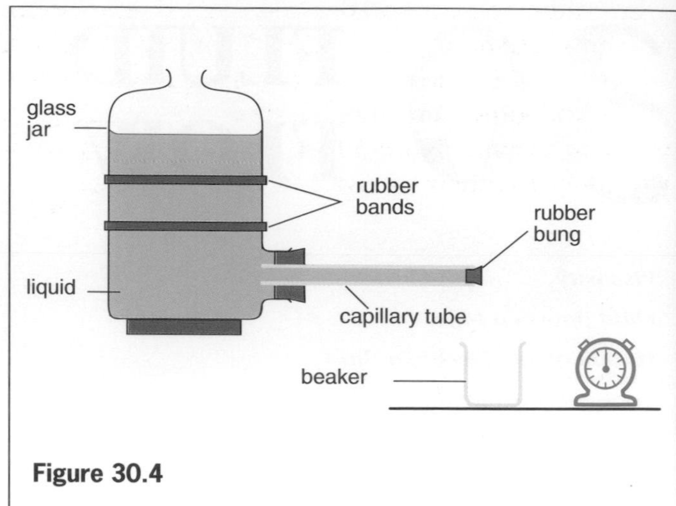

Figure 30.4

Motion of an object in a fluid

The behaviour of a fluid when an object is moving in it is similar to what occurs when a fluid flows in a pipe.

If the object, e.g. a small sphere, moves slowly, **streamlines** like those in Figure 30.5a can be drawn to show how the apparent motion of the fluid would appear to someone on the moving sphere. They represent **steady** flow.

If the speed of the sphere increases, a **critical speed** is reached when the flow breaks up and eddies are formed behind the sphere as in Figure 30.5b. The flow becomes **turbulent** and the viscous drag on the sphere increases sharply.

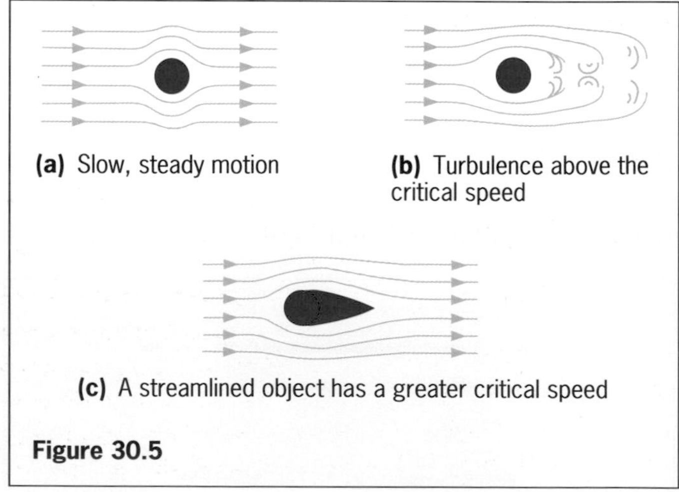

(a) Slow, steady motion

(b) Turbulence above the critical speed

(c) A streamlined object has a greater critical speed

Figure 30.5

The critical speed can be raised by changing the shape of the object, so reducing the drag and causing steady flow to replace turbulent flow. This is called **streamlining** the object and Figure 30.5c shows how it is done for a sphere. The pointed tail can be regarded as filling the region where eddies occur in turbulent motion, thus ensuring that the streamlines merge again behind the sphere.

Streamlining is especially important in the design of high-speed aircraft and other fast-moving vehicles, Figure 30.6a. In nature, streamlining enables fish, dolphins and other creatures to move faster through water or the air, Figure 30.6b, sometimes a matter of importance to their survival.

Figure 30.6a Formula One racing cars have a streamlined design

Figure 30.6b Dolphins have streamlined bodies to assist their movement in water

PRACTICAL WORK

Objects falling in a liquid

The time of fall of small ball-bearings dropped into a viscous liquid such as glycerine, engine oil or clear syrup can be studied using the apparatus of Figure 30.7. The glass jar should be as wide and as tall as possible. To reduce the risk of air bubbles sticking to the ball-bearings, they should be dipped in some of the liquid and thereby coated with it, before dropping. The investigations that can be carried out include:

(i) measurement of the speed of a ball-bearing over a distance at different depths,
(ii) the effect of using balls of different diameters, and
(iii) a comparison of the viscosities of different liquids.

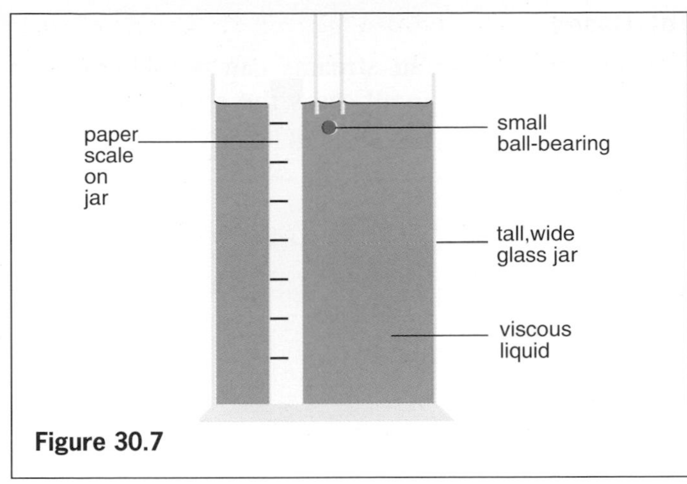

Figure 30.7

At the start the weight W of the ball causing it to fall vertically downwards under gravity is greater than the upwards viscous force V opposing its motion. As it speeds up, V increases until eventually V equals W. The forces acting on the ball then balance and it moves with constant speed (see Chapter 34), called the **terminal speed**.

Bernoulli's principle

As we have seen (Chapter 27), the pressure is the same at all points on the same level in a fluid at rest; this is not so when the fluid is moving.

(a) Liquids

When a liquid flows steadily through a uniform tube the pressure falls steadily, as shown by the decreasing height of liquid in the three vertical tubes (manometers, Chapter 28) in Figure 30.8a. In Figure 30.8b the pressure falls in the narrow part B but rises again in the wider part C. Since the same volume of liquid passes through B in a certain time as enters A, the liquid must be moving faster in B than in A. Therefore a decrease of pressure occurs when the speed of the liquid increases. Conversely an increase of pressure accompanies a fall in speed. This effect, called **Bernoulli's principle**, is stated as follows.

> When the speed of a fluid in smooth flow increases, the pressure in the fluid decreases and vice versa.

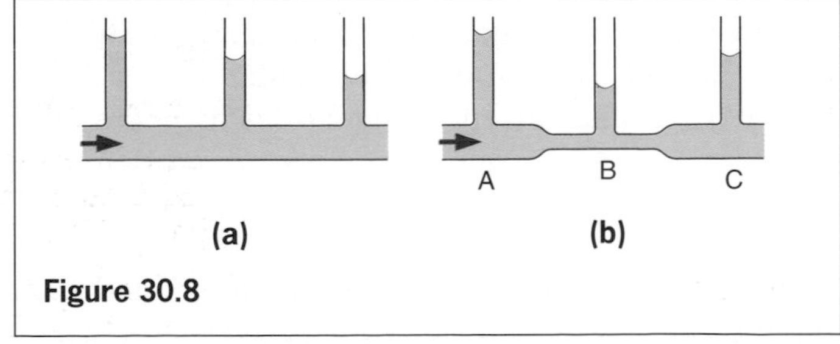

Figure 30.8

(b) Gases

Bernoulli effects in air streams can be shown as in Figures 30.9a, b, c. In all cases the reason is that the pressure falls in the fast-moving air stream.

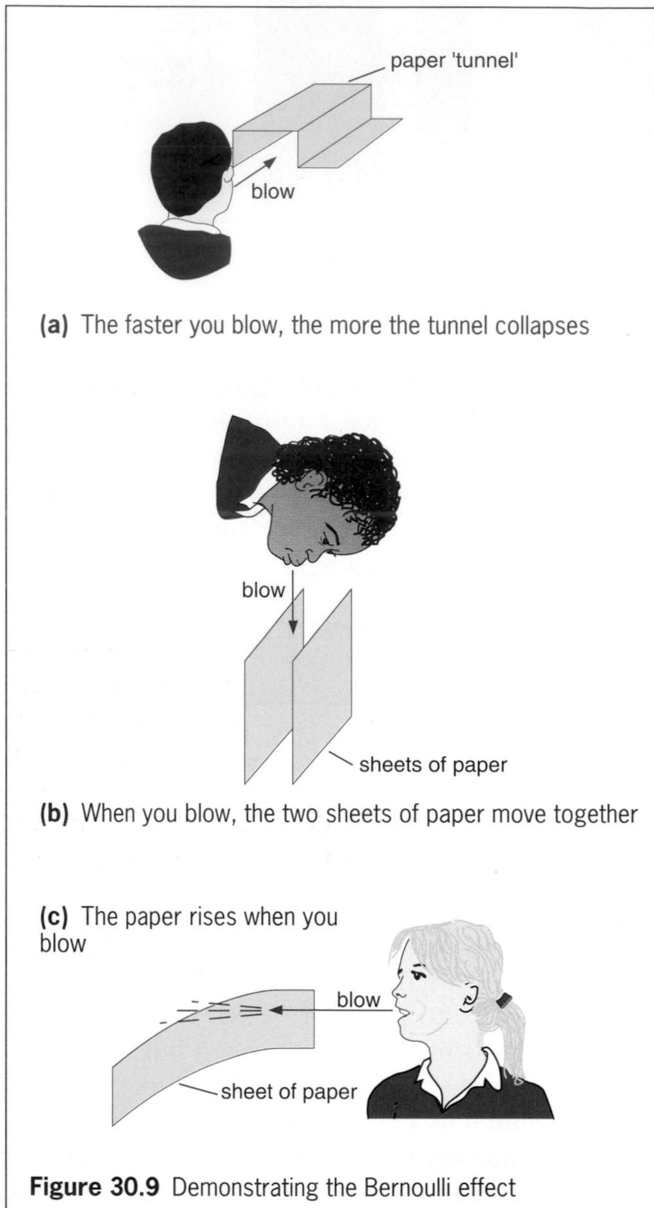

Figure 30.9 Demonstrating the Bernoulli effect

Applications of Bernoulli

(a) Jets and nozzles

A slow stream of water from a tap can be changed into a fast jet by narrowing the outlet with a finger. The more it is narrowed, the greater is the increase of speed and so the greater is the pressure drop. Several devices with jets or nozzles use this effect. Figures 30.10a, b show the action of a Bunsen burner and a paint spray. It is also used in carburettors and filter pumps.

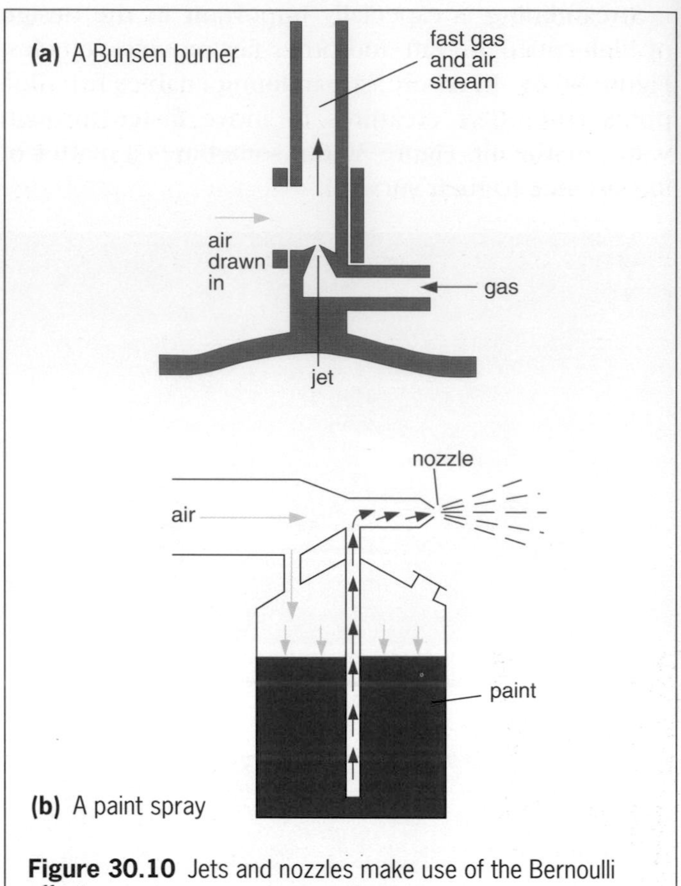

Figure 30.10 Jets and nozzles make use of the Bernoulli effect

(b) Spinning ball

If a tennis ball is 'cut' or a golf ball 'sliced', it spins as it travels through the air and experiences a sideways force which causes it to curve in flight. This is due to air being dragged round by the spinning ball, thereby increasing the air flow on one side and decreasing it on the other. A pressure difference is created, Figure 30.11. The swing of a spinning cricket ball is complicated by its raised seam.

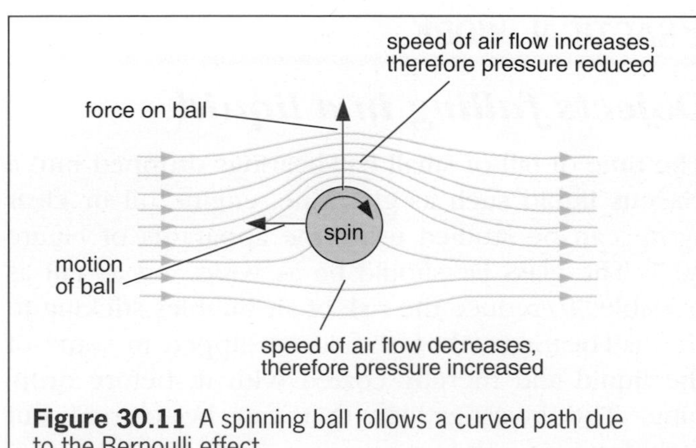

Figure 30.11 A spinning ball follows a curved path due to the Bernoulli effect

(c) Aerofoil

This is a device which is shaped so that when a fluid flows over it, a force is produced at right angles to the direction of flow. Examples of aerofoils are aircraft wings, propellers and turbine blades.

An aircraft wing has a curved upper surface and a flat under surface so that air passing over the top of the wing has to travel further and so faster than over the bottom. This is represented by the streamlines being closer together above than below the aerofoil, Figure 30.12. By Bernoulli's principle it follows that the pressure underneath is increased and that above reduced, and the resultant upward force on the wing provides the 'lift' for the aircraft.

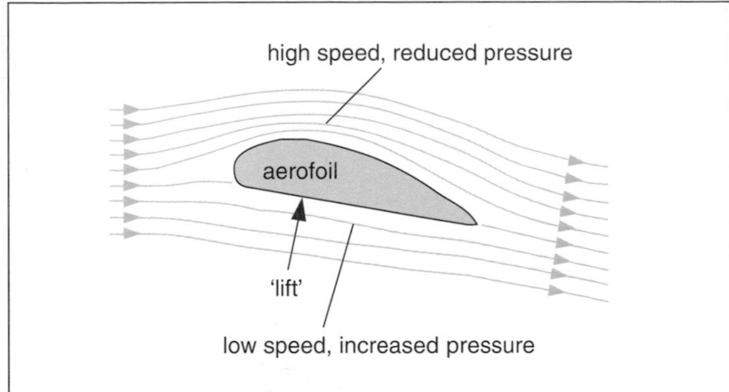

Figure 30.12 When fluid flows over an aerofoil a force is produced at right angles to the flow

The lift increases when the angle between the wing and the air flow, called the 'angle of attack', increases by raising the nose of the aircraft. At a certain angle (about 16°) the flow separates from the upper surface, i.e. the air stops flowing smoothly over the top of the wing. Lift is lost almost completely, the flow downstream becomes very turbulent, drag increases sharply, the aircraft stalls and starts to dive. The pilot loses control until lift can be obtained again.

The sail of a yacht 'tacking into the wind', Figure 30.13a, is another example of an aerofoil. The air flow over the sail produces a pressure increase on the windward side and a decrease on the leeward side. The resultant force is at right angles to the sail, Figure 30.13b.

Figure 30.13a A yacht 'tacking into the wind'

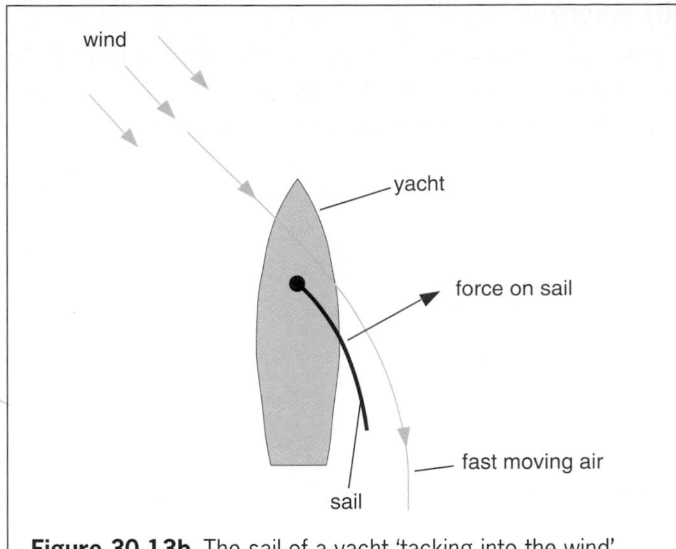

Figure 30.13b The sail of a yacht 'tacking into the wind' acts as an aerofoil

Theory of flight

In straight, level flight an aeroplane has four forces acting on it, Figure 30.14. Its **weight** acts downwards and is balanced by the upward force of **lift**, produced by the wings as explained in the previous section. The engines (jet or propeller) provide the forward **thrust** which acts against the **drag** due to air resistance.

There are three basic controls in an aircraft.

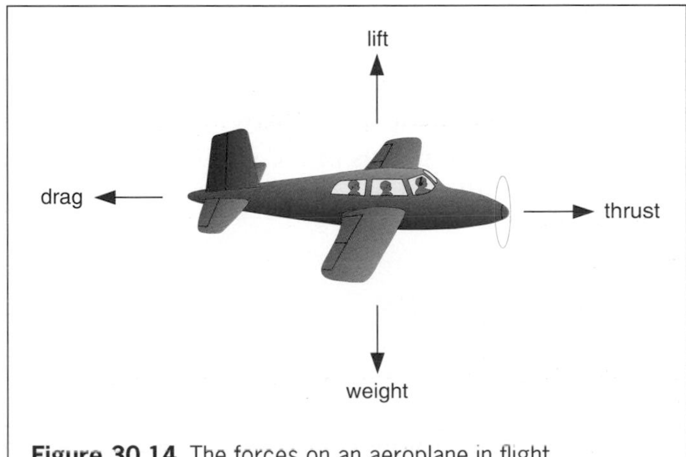

Figure 30.14 The forces on an aeroplane in flight

(a) Elevators

These are hinged flaps on the tail plane, Figure 30.15 (overleaf), and control the up-and-down movement of the nose of the aeroplane, known as **pitch**. They are operated by moving the control column to and fro. Raising the elevators as in (a) directs the nose upwards and the aeroplane climbs; lowering them as in (b) lowers the nose and the aeroplane starts to dive.

(b) Ailerons

These are also hinged flaps, at the rear edge of the wings, and enable the aircraft to **bank**, i.e. tilt with one wing lower than the other, when one aileron is raised and the other lowered, Figure 30.16a. They are operated by sideways movements of the control column.

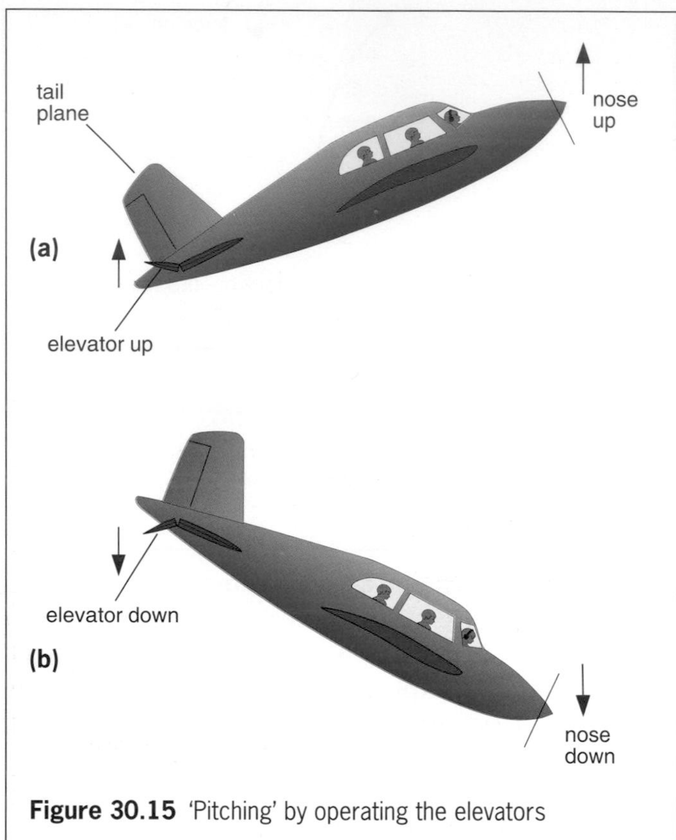

Figure 30.15 'Pitching' by operating the elevators

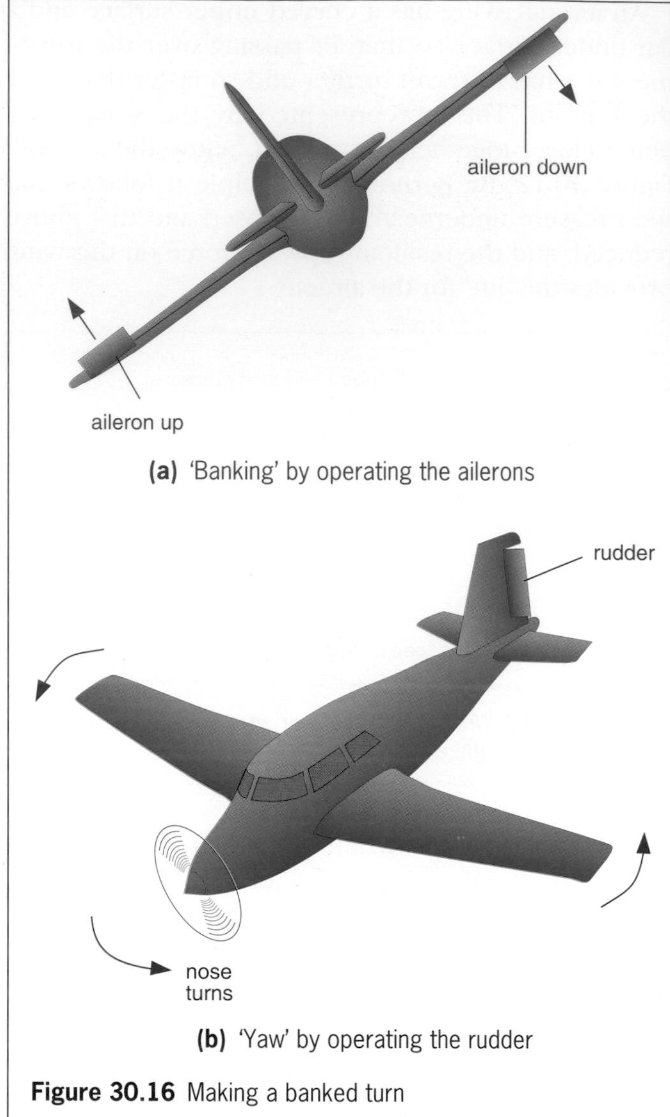

Figure 30.16 Making a banked turn

(c) Rudder

This is on the tail plane and turns the nose of the plane from side to side, called **yaw**, Figure 30.16b. It is operated by foot pedals in the flight deck. By using the ailerons at the same time, the plane can make a banked turn.

Helicopters

The rotor of a helicopter has two or more wing-shaped blades (aerofoils) which generate lift when they rotate. As with fixed-wing aircraft, the amount of lift increases if the angle of attack (i.e. the pitch) of the rotor blades increases, Figure 30.17. It is by changing this that the pilot makes the helicopter ascend, hover or descend. For these manoeuvres, the lifting force acts vertically upwards at right angles to the plane of rotation of the rotor.

To fly forwards, the whole rotor is tilted forwards, altering the direction of the lifting force so that it is inclined forwards slightly from the vertical, Figure 30.18a. This causes forward motion. To go sideways or backwards, the pilot tilts the rotor in the required direction.

The rotation of the rotor tends to make the helicopter turn in the opposite direction (see Newton's third law of motion, Chapter 34). This is counteracted by the tail rotor which rotates in a vertical plane, Figure 30.18b.

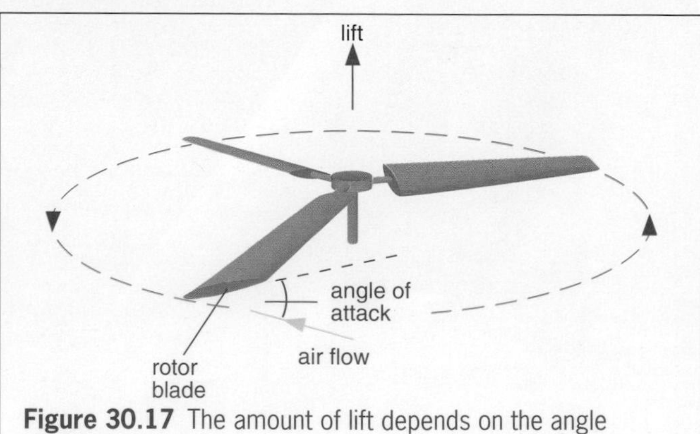

Figure 30.17 The amount of lift depends on the angle of attack

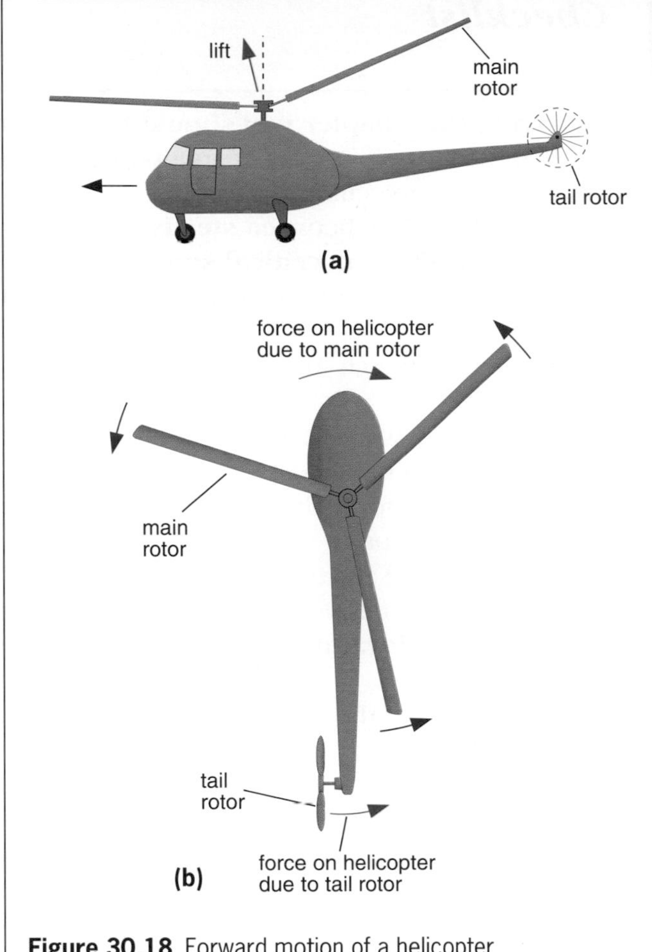

Figure 30.18 Forward motion of a helicopter

QUESTIONS

1 In a model yacht race on a river five identical yachts A, B, C, D, E are released at the same time from different points along the starting line, as in Figure 30.19.

a Why should C win?

b As they approach the finishing line they all slow down. Why?

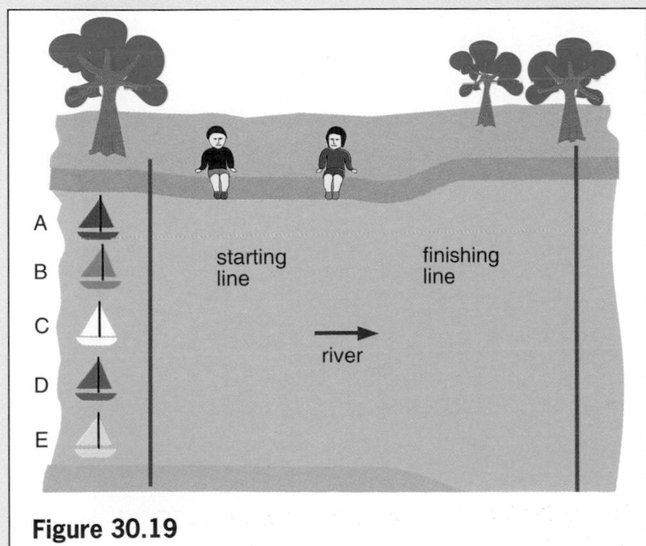

Figure 30.19

2 How would you expect the rate of flow of a liquid through a pipe to depend on (i) the width of the tube, (ii) the length of the tube and (iii) the viscosity of the liquid?

3 Explain the following tricks which you could also try.

a If you blow hard over the top of a small glass containing a table tennis ball, it jumps out of the glass, Figure 30.20a.

b A table tennis ball can be supported in the stream of air coming from the nozzle of a hairdryer, Figure 30.20b.

c If you suddenly blow really hard just above a small coin near a saucer, Figure 30.20c, it may jump into the saucer.

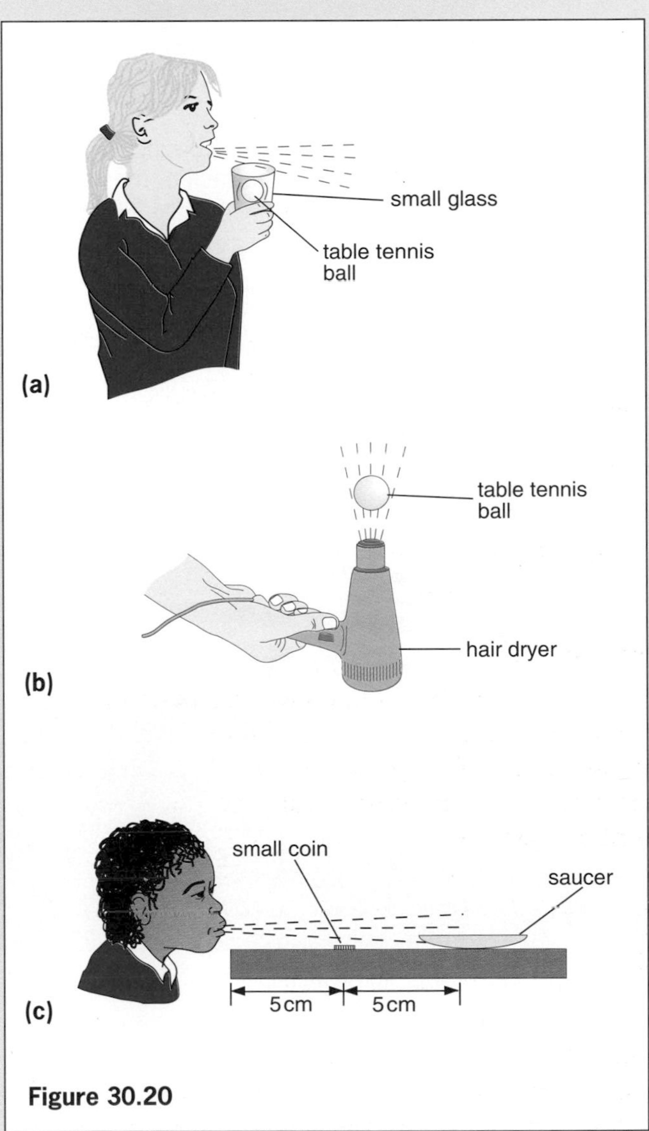

Figure 30.20

4 a Why does streamlining improve the efficiency of a car?

b Wind tunnel tests are carried out on models of new cars, in which the air moves past the car rather than the car moving through the air. The air flow pattern is revealed by fine smoke trails. For the car in Figure 30.21 eddies occur at the foot of the windscreen and the flow becomes unsteady behind the car. Make a sketch to show how you would improve the streamlining.

Figure 30.21

5 a Name the manoeuvres that a fixed-wing aircraft performs using (i) the angle of attack of the wings, (ii) the elevators, (iii) the ailerons and (iv) the rudder.

b How does a helicopter change its flight direction?

c Most of a hydrofoil is out of the water when it moves, Figure 30.22a. By referring to Figure 30.22b showing the streamlines for the flow of water over the aerofoils, explain how the lift occurs.

Figure 30.22a A hydrofoil in motion

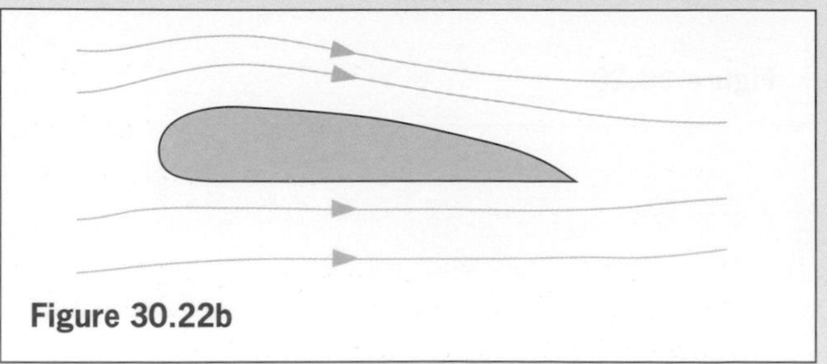

Figure 30.22b

Checklist

After studying this chapter you should be able to

- recall that viscosity is a kind of frictional force in fluids and give a molecular explanation,
- explain the difference between **steady** and **turbulent** flow, using the term **critical speed**,
- describe experiments to investigate the flow of liquids in tubes,
- explain using a sketch the purpose of **streamlining** an object,
- describe an experiment to investigate the time of fall of small objects in liquids,
- state Bernoulli's principle and describe some effects and applications,
- explain the action of an aerofoil,
- describe the manoeuvres of an aeroplane with reference to the **elevators**, the **ailerons** and the **rudder**,
- explain the action of the rotors of a helicopter.

ADDITIONAL QUESTIONS

The first questions under each topic heading are 'basic' questions intended for all students; the following questions marked with a stripe are 'higher' questions for those seeking grades E to A*.

Moments and levers

1 The diagram shows a nail being removed from a piece of wood by a claw hammer.

a (i) What physical principle or law is involved in the withdrawal of the nail? (ii) The nail begins to move when a force of 100 N is applied at the end of the handle. What is the frictional resistance exerted on the nail by the wood?

b State *one* more example of a lever of the same type as the claw hammer. (*N.W.*)

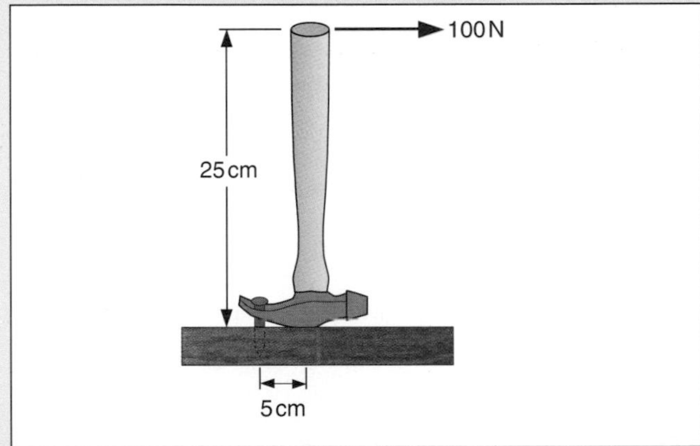

2 The uniform plank shown below weighs 200 N, rests on two trestles A and B, and supports a boy of weight 500 N in the position shown. *P* and *Q* are the reaction forces at A and B.

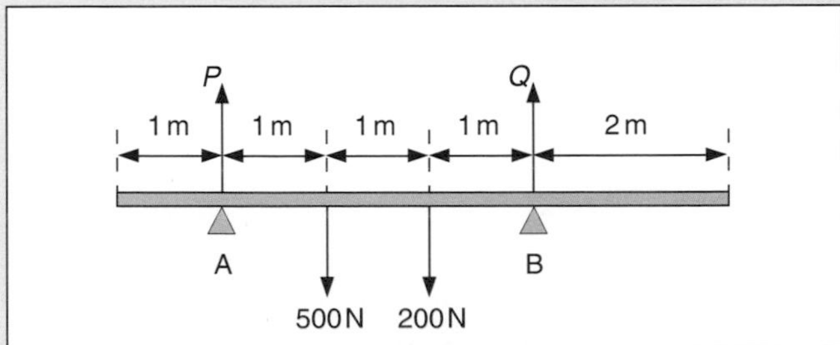

a Write down the moment of each force about A.

b Use the principle of moments to find *Q*.

c What is the total upward force $P + Q$?

d What is the value of *P*?

Adding forces

3 The resultant of the two forces shown in the diagram makes an angle of 30° with the force of 40 N. Determine, graphically or otherwise, the values of the resultant and of the force *P*. (*L.*)

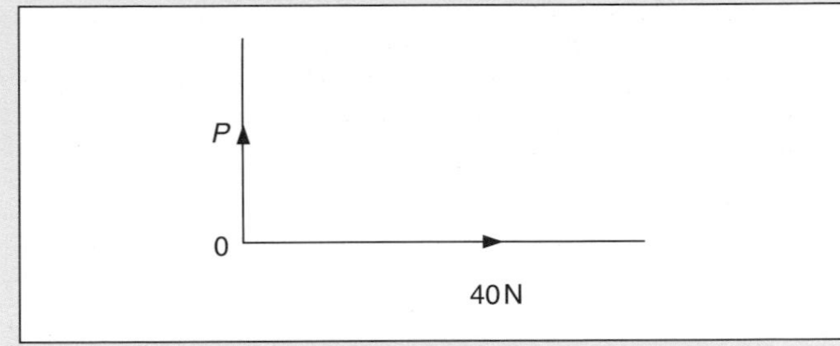

4 A string attached to two hooks A, B, supports a weight *W* attached at a point C (see diagram below). CA makes an angle with the vertical of 30° and CB an angle of 45°. If the tension in BC is 100 N, find, by drawing or calculation, the tension in AC and the weight *W*. (*O. and C.*)

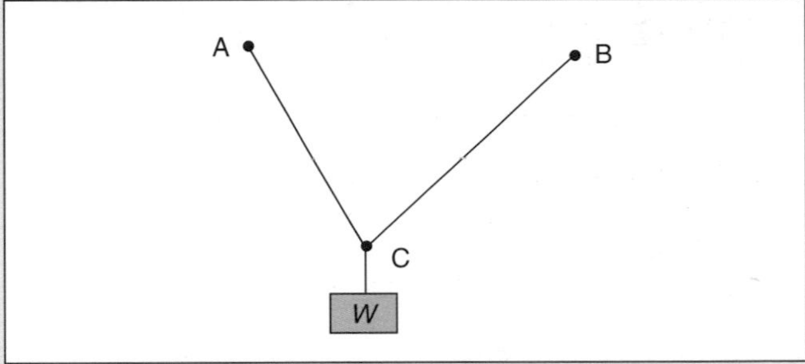

Energy transfer; machines

5 State what energy transfers occur in **a** a hairdryer, **b** a refrigerator, **c** an audio system.

6 A boy is lifting sandbags from the floor onto a shelf 1.6 m high. The pull of the Earth on each sandbag is 10 N. The boy lifts 80 sandbags in 100 s.

a How much useful work is done in lifting the sandbags?

b What is the total p.e. when all the bags are on the shelf?

c What was the boy's useful power output?

d The actual power output was greater than answer **c**. Suggest two possible reasons for this difference.

One sandbag fell from the shelf.

e What is its p.e. when it is halfway to the ground?

f What is its k.e. when it is halfway to the ground?

g What happens to its k.e. when it hits the ground? Give two possible answers.

The diagram (overleaf) shows the boy's arm when he is measuring the pull of the Earth on a sandbag. He holds the balance still to take a reading.

h How much work is he doing as he checks the reading?

i Give a rough estimate (from the diagram) of the tension in the boy's biceps as he holds the balance. Explain how you arrived at this value. *(E.M.)*

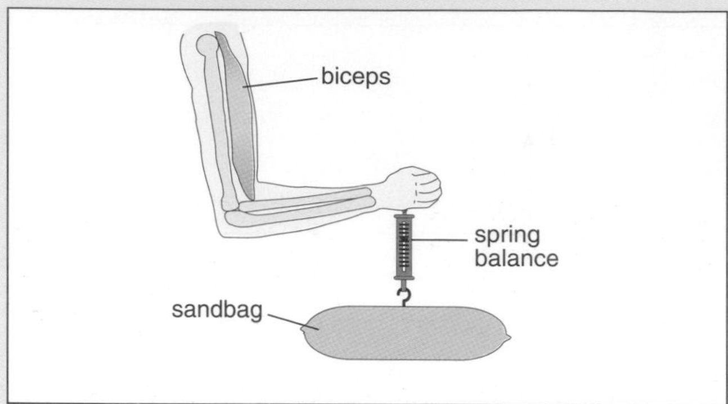

7 An escalator carries 60 people of average mass 70 kg to a height of 5 m in one minute. Find the power needed to do this.

8 **a** A load of 40 N is raised 0.50 m by a pulley system when the effort of 10 N moves 2.5 m. Calculate the efficiency.

b Using a pulley system operating at an efficiency of 75% a man lifts a load of weight 1800 N using an effort of 400 N. Calculate
(i) the work done by the effort when it moves through 6 m,
(ii) the distance the load moves when the effort moves 6 m.

Pressure

9 **a** A mercury barometer is shown in the diagram. Copy it and mark the mercury levels when the tube is tilted from position A to positions B and C.

b How would the mercury height be affected if air got into the tube in position A?

c Why is mercury suitable as a liquid for use in a barometer?

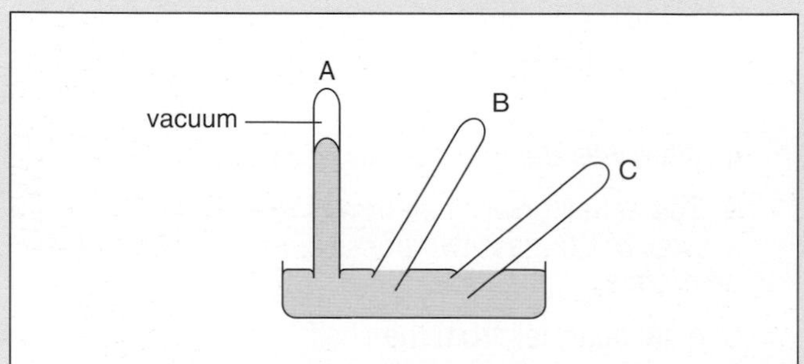

10 What is meant by the term **pressure**? Explain the fact that when someone uses her thumb to push a drawing pin into a block of wood, the pressure on the wood is greater than the pressure on the thumb.

Calculate the pressure on the thumb when the force exerted is 20 N and the top of the pin is square, the length of each side being 10 mm. *(C.)*

Floating, sinking and flying

11 A body of mass 2 kg is suspended from a spring balance, calibrated in newtons, which reads 17 N when the body is completely submerged in water. The intensity of the Earth's gravitational field is 10 N/kg (i.e. the acceleration due to gravity is 10 m/s^2).

a What is the upthrust (in newtons) of the water on the body?

b What is the mass (in kg) of water displaced by the body?

c If the density of water is 1000 kg/m^3, what is the volume (in m^3) of water displaced? *(J.M.B./A.L./N.W. 16+ part qn.)*

12 A rectangular block of wood, mass 0.32 kg, measuring 10 cm $\times$ 10 cm $\times$ 5 cm, floats in a liquid of density 1600 kg/m^3 with its large face horizontal. What is the height in cm of the upper face of the block above the surface of the liquid?

A 0.4 **B** 2.0 **C** 2.5 **D** 3.0 **E** 4.8 *(N.I.)*

Fluid flow

13 Give molecular explanations of the following.

a The viscosity of a fluid affects how it flows through a pipe.

b The viscosity of some fluids decreases if the temperature increases.

MOTION AND ENERGY

31 VELOCITY AND ACCELERATION

Speed

If a car travels 300 km from Liverpool to London in 5 hours, its **average speed** is 300 km/5 h = 60 km/h. The speedometer would certainly not read 60 km/h for the whole journey but might vary considerably from this value. That is why we state the average speed. If a car could travel at a constant speed of 60 km/h for 5 hours, the distance covered would still be 300 km. It is *always* true that

$$\text{average speed} = \frac{\text{distance moved}}{\text{time taken}}$$

To find the actual speed at any instant we would need to know the distance moved in a very short interval of time. This can be done by multiflash photography. In Figure 31.1 the golfer is photographed whilst a flashing lamp illuminates him 100 times a second. The speed of the club-head as it hits the ball is about 190 km/h.

Figure 31.1 Multiflash photograph of a golf swing

Velocity

Speed is the distance travelled in unit time, **velocity is the distance travelled in unit time in a stated direction**. If two cars travel due north at 20 m/s, they have the same speed of 20 m/s and the same velocity of 20 m/s *due north*. If one travels north and the other south, their speeds are the same but not their velocities since their directions of motion are different. Speed is a scalar and velocity a vector quantity (see Chapter 24).

$$\text{velocity} = \frac{\text{distance moved in a stated direction}}{\text{time taken}}$$

The velocity of a body is **uniform** or **constant** if it moves with a steady speed in a straight line. It is not uniform if it moves in a curved path. Why?

The units of speed and velocity are the same, e.g. km/h, m/s, and

60 km/h = 60 000 m/3600 s = 17 m/s

Distance moved in a stated direction is called the **displacement**. It is a vector, unlike distance which is a scalar. Velocity may also be defined as

$$\text{velocity} = \frac{\text{displacement}}{\text{time taken}}$$

Acceleration

When the velocity of a body changes we say it **accelerates**. If a car starting from rest and moving due north has velocity 2 m/s after 1 second, its velocity has increased by 2 m/s in 1 s and its acceleration is 2 m/s per second due north. We write this as 2 m/s^2. **Acceleration is the change of velocity in unit time.**

$$\text{acceleration} = \frac{\text{change of velocity}}{\text{time taken for change}}$$

For a steady increase of velocity from 20 m/s to 50 m/s in 5 s

$$\text{acceleration} = \frac{(50 - 20)\ \text{m/s}}{5\ \text{s}} = 6\ \text{m/s}^2$$

Acceleration is also a vector and both its magnitude and direction should be stated. However, at present we will consider only motion in a straight line and so the magnitude of the velocity will equal the speed, and the magnitude of the acceleration will equal the change of speed in unit time.

The speeds of a car accelerating on a straight road are shown below.

Time (s)	0	1	2	3	4	5	6
Speed (m/s)	0	5	10	15	20	25	30

The speed increases by 5 m/s every second and the acceleration of 5 m/s^2 is said to be **uniform**.

An acceleration is positive if the velocity increases and negative if it decreases. A negative acceleration is also called a **deceleration** or **retardation**.

Tickertape timer: tape charts

A tickertape timer enables us to measure speeds and hence accelerations. One type, Figure 31.2, has a steel strip which vibrates 50 times a second and a carbon paper disc which makes dots at 1/50 s intervals on the paper tape being pulled through it. 1/50 s is called a 'tick'.

The distance between successive dots equals the average speed of whatever is pulling the tape in, say, cm per 1/50 s, i.e., cm per tick. The 'tentick' ($\frac{1}{5}$ s) is also used as a unit of time. Since ticks and tenticks are small we drop the 'average' and just refer to the 'speed'.

Tape charts are made by sticking successive strips of tape, usually tentick lengths, side by side. That in Figure 31.3a represents a body moving with **uniform speed** since equal distances have been moved in each tentick interval.

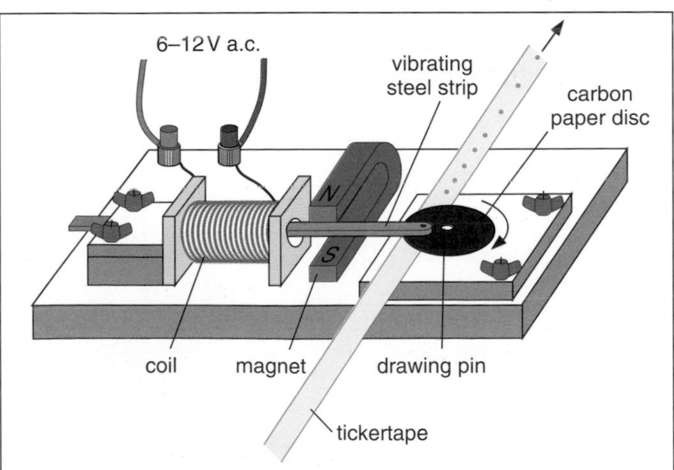

Figure 31.2 Tickertape timer using a carbon paper disc

The chart in Figure 31.3b is for **uniform acceleration**: the 'steps' are of equal size showing that the speed increased by the same amount in every tentick ($\frac{1}{5}$ s). The acceleration (average) can be found from the chart as follows.

The speed during the *first* tentick is 2 cm/$\frac{1}{5}$ s or 10 cm/s. During the *sixth* tentick it is 12 cm/$\frac{1}{5}$ s or 60 cm/s. And so during this interval of 5 tenticks, i.e. 1 second, the change of speed is (60 − 10) cm/s = 50 cm/s.

$$\text{acceleration} = \frac{\text{change of speed}}{\text{time taken}}$$

$$= \frac{50\ \text{cm/s}}{1\ \text{s}}$$

$$= 50\ \text{cm/s}^2$$

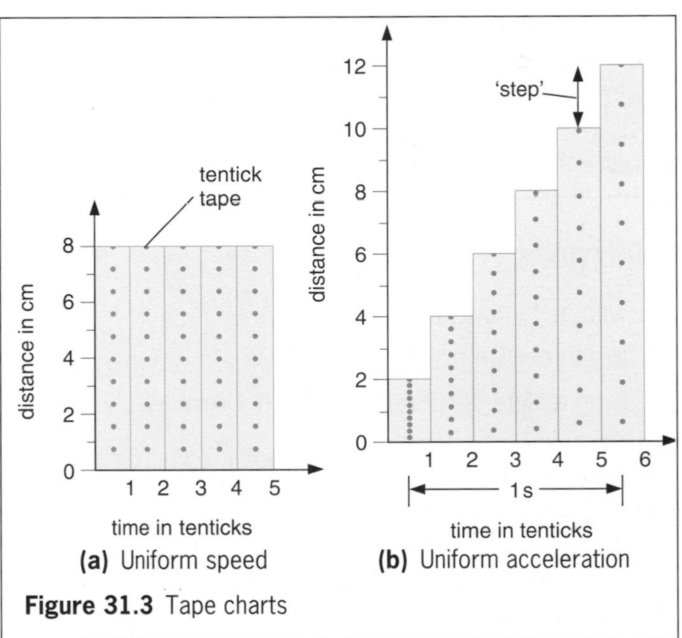

(a) Uniform speed **(b)** Uniform acceleration

Figure 31.3 Tape charts

PRACTICAL WORK

Analysing motion

(a) Your own motion

Pull a 2 m length of tape through a tickertape timer as you walk away from it quickly, then slowly, then speeding up again and finally stopping, Figure 31.4a.

Cut the tape into tentick lengths and make a tape chart. Write labels on it to show where you speeded up, slowed down, etc.

(b) Trolley on a sloping runway

Attach a length of tape to a trolley and release it at the top of a runway, Figure 31.4b. The dots will be very crowded at the start – ignore them; but beyond cut the tape into tentick lengths.

Make a tape chart. Is the acceleration uniform? What is its average value?

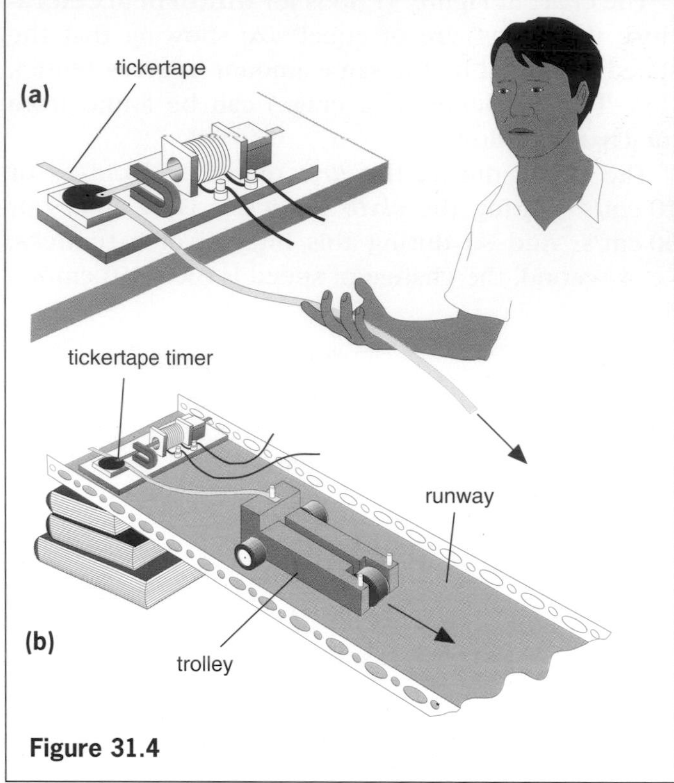

Figure 31.4

QUESTIONS

1 What is the average speed of

a a car which travels 400 m in 20 s,

b an athlete who runs 1500 m in 4 minutes?

2 A train increases its speed **steadily** from 10 m/s to 20 m/s in 1 minute.

a What is its average speed during this time in m/s?

b How far does it travel while increasing its speed?

3 A motor cyclist starts from rest and reaches a speed of 6 m/s after travelling with uniform acceleration for 3 s. What is his acceleration?

4 The tape in Figure 31.5 was pulled through a timer by a trolley travelling down a runway. It was marked off in tentick lengths.

a What can you say about the trolley's motion?

b Find its acceleration in cm/s^2.

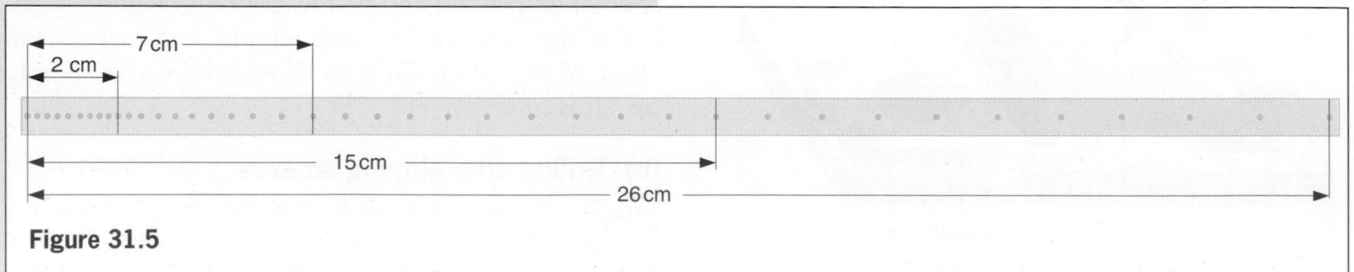

Figure 31.5

5 An aircraft travelling at 600 km/h accelerates steadily at 10 km/h per second. Taking the speed of sound as 1100 km/h at the aircraft's altitude, how long will it take to reach the 'sound barrier'?

6 A vehicle moving with a uniform acceleration of 2 m/s^2 has a velocity of 4 m/s at a certain time. What will its velocity be **a** 1 s later, **b** 5 s later?

7 If a bus travelling at 20 m/s is subject to a steady deceleration of 5 m/s^2, how long will it take to come to rest?

8 Each strip in the tape chart of Figure 31.6 is for a time interval of 1 tentick.

a If the timer makes 50 dots per second, what time intervals are represented by OA and AB?

b What is the acceleration between O and A in (i) cm/tentick^2, (ii) cm/s per tentick, (iii) cm/s^2?

c What is the acceleration between A and B?

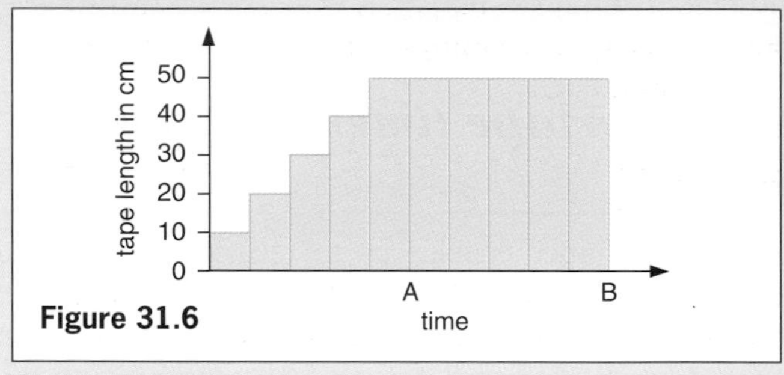

Figure 31.6

Checklist

After studying this chapter you should be able to

- explain the meaning of the terms **speed**, **velocity**, **acceleration**, **displacement**,
- describe how speed and acceleration may be measured using timers and tape charts.

32 GRAPHS AND EQUATIONS

Velocity–time graphs

Distance–time graphs

Equations for uniform acceleration

Velocity–time graphs

If the velocity of a body is plotted against the time, the graph obtained is a velocity-time graph. It provides a way of solving motion problems. Tape charts are crude velocity-time graphs which show the velocity changing in jumps rather than smoothly, as occurs in practice.

The area under a velocity-time graph measures the distance travelled.

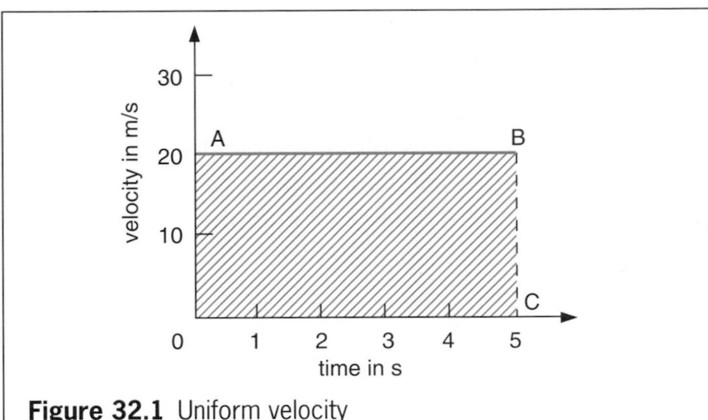

Figure 32.1 Uniform velocity

In Figure 32.1, AB is the velocity-time graph for a body moving with a **uniform velocity** of 20 m/s. Since distance = average velocity × time, after 5 s it will have moved 20 m/s × 5 s = 100 m. This is the shaded area under the graph, i.e. rectangle OABC.

In Figure 32.2, PQ is the velocity-time graph for a body moving with **uniform acceleration**. At the start of the timing the velocity is 20 m/s but it increases steadily to 40 m/s after 5 s. If the distance covered equals the area under PQ, i.e. the shaded area OPQS, then

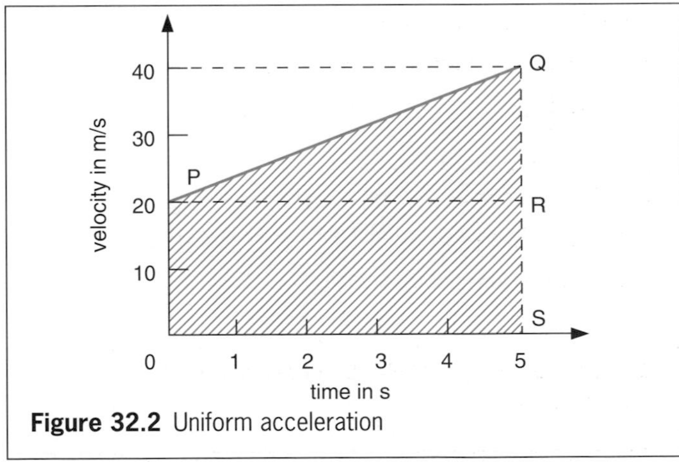

Figure 32.2 Uniform acceleration

distance = area of rectangle OPRS + area of triangle PQR

$= \text{OP} \times \text{OS} + \frac{1}{2} \times \text{PR} \times \text{QR}$
(area of a triangle = $\frac{1}{2}$ base × height)

$= 20\ \text{m/s} \times 5\ \text{s} + \frac{1}{2} \times 5\ \text{s} \times 20\ \text{m/s}$

$= 100\ \text{m} + 50\ \text{m} = 150\ \text{m}$

Notes:

1. When calculating the area from the graph the unit of time must be the same on both axes.

2. This rule for finding distances travelled is true even if the acceleration is not uniform.

The slope or gradient of a velocity-time graph represents the acceleration of the body.

In Figure 32.1 the slope of AB is zero, as is the acceleration. In Figure 32.2 the slope of PQ is QR/PR = 20/5 = 4: the acceleration is 4 m/s^2.

Distance–time graphs

A body travelling with uniform velocity covers equal distances in equal times. Its distance-time graph is a straight line, like OL in Figure 32.3 (overleaf) for a velocity of 10 m/s. The slope of the graph = LM/OM = 40 m/ 4 s = 10 m/s, which is the value of the velocity. The following statement is true in general.

The slope or gradient of a distance-time graph represents the velocity of the body.

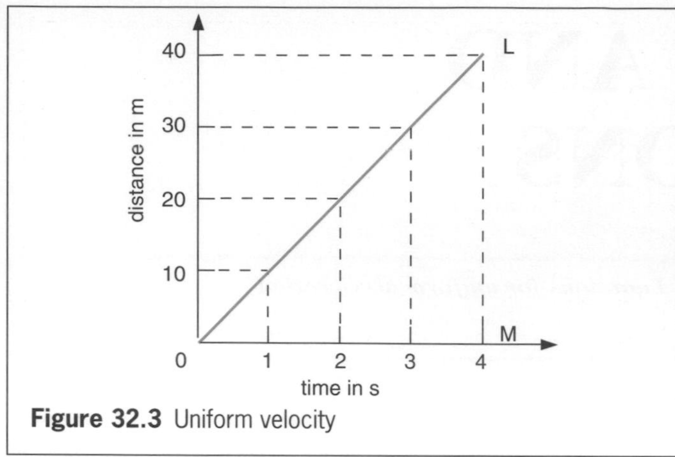

Figure 32.3 Uniform velocity

When the velocity of the body is changing, the slope of the distance–time graph varies, Figure 32.4, and at any point equals the slope of the tangent. For example, the slope of the tangent at T is AB/BC = 40 m/2 s = 20 m/s. The velocity at the instant corresponding to T is therefore 20 m/s.

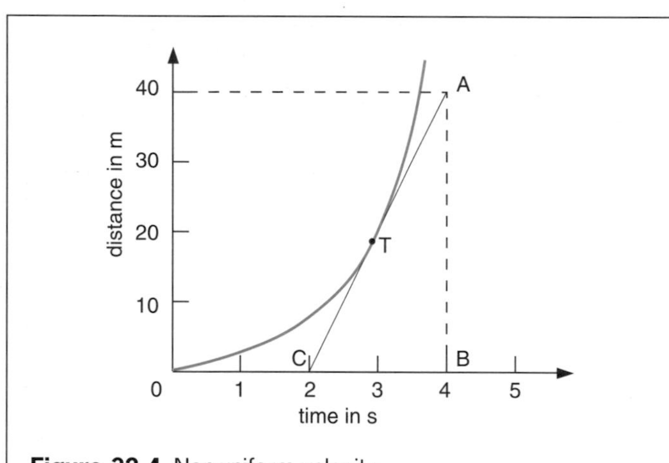

Figure 32.4 Non-uniform velocity

Equations for uniform acceleration

Problems on bodies moving with **uniform acceleration** can often be solved quickly using the **equations of motion**.

First equation

If a body is moving with uniform acceleration a and its velocity increases from u to v in time t, then

$$a = \frac{\text{change of velocity}}{\text{time taken}} = \frac{v - u}{t}$$

$$\therefore \quad at = v - u$$

or $$v = u + at \qquad (1)$$

Note that the initial velocity u and the final velocity v refer to the start and the finish of the *timing* and do not necessarily mean the start and finish of the motion.

Second equation

The velocity of a body moving with uniform acceleration increases steadily. Its average velocity therefore equals half the sum of its initial and final velocities, that is,

$$\text{average velocity} = \frac{u + v}{2}$$

If s is the distance moved in time t, then since average velocity = distance/time = s/t,

$$\frac{s}{t} = \frac{u + v}{2} \qquad (2)$$

Third equation

From equation (1), $v = u + at$
From equation (2),

$$\text{average velocity} = \frac{u + u + at}{2} = \frac{2u + at}{2}$$

$$= u + \tfrac{1}{2}at$$

$$\therefore \quad \frac{s}{t} = u + \tfrac{1}{2}at$$

and

$$s = ut + \tfrac{1}{2}at^2 \qquad (3)$$

Fourth equation

This is obtained by eliminating t from equations (1) and (3). We have

$$v = u + at$$

$$\therefore \quad v^2 = u^2 + 2uat + a^2t^2$$

$$= u^2 + 2a(ut + \tfrac{1}{2}at^2)$$

But $$s = ut + \tfrac{1}{2}at^2$$

$$\therefore \quad v^2 = u^2 + 2as \qquad (4)$$

If we know any *three* of u, v, a, s and t, the others can be found from the equations.

Worked example

A cyclist starts from rest and accelerates at 1 m/s^2 for 20 seconds. He then travels at a constant speed for 1 minute and finally decelerates at 2 m/s^2 until he stops. Find his maximum speed in km/h and the total distance covered in metres.

First stage

$$u = 0 \quad a = 1\ \text{m/s}^2 \quad t = 20\ \text{s}$$

We have $v = u + at = 0 + 1\ \text{m/s}^2 \times 20\ \text{s}$

$$= 20\ \text{m/s}$$

$$= \frac{20}{1000} \times 60 \times 60 = 72\ \text{km/h}$$

The distance s moved in the first stage is given by

$$s = ut + \tfrac{1}{2}at^2 = 0 \times 20\ \text{s} + \tfrac{1}{2} \times 1\ \text{m/s}^2 \times 20^2\ \text{s}^2$$

$$= \tfrac{1}{2} \times 1\frac{\text{m}}{\text{s}^2} \times 400\ \text{s}^2 = 200\ \text{m}$$

Second stage

$$u = 20\ \text{m/s (constant)} \quad t = 60\ \text{s}$$

$$\text{distance moved} = \text{speed} \times \text{time} = 20\ \text{m/s} \times 60\ \text{s}$$

$$= 1200\ \text{m}$$

Third stage

$$u = 20\ \text{m/s} \quad v = 0 \quad a = -2\ \text{m/s}^2\ \text{(a deceleration)}$$

We have $v^2 = u^2 + 2as$

$$\therefore \quad s = \frac{v^2 - u^2}{2a} = \frac{0 - (20)^2\ \text{m}^2/\text{s}^2}{2 \times (-2)\ \text{m/s}^2} = \frac{-400\ \text{m}^2/\text{s}^2}{-4\ \text{m/s}^2}$$

$$= 100\ \text{m}$$

Answers

Maximum speed = 72 km/h

Total distance moved = 200 m + 1200 m + 100 m = 1500 m

QUESTIONS

1 The distance–time graph for a girl on a cycle ride is shown in Figure 32.5.

a How far did she travel?

b How long did she take?

c What was her average speed in km/h?

d How many stops did she make?

e How long did she stop for altogether?

f What was her average speed *excluding* stops?

g How can you tell from the shape of the graph when she travelled fastest? Over which stage did this happen?

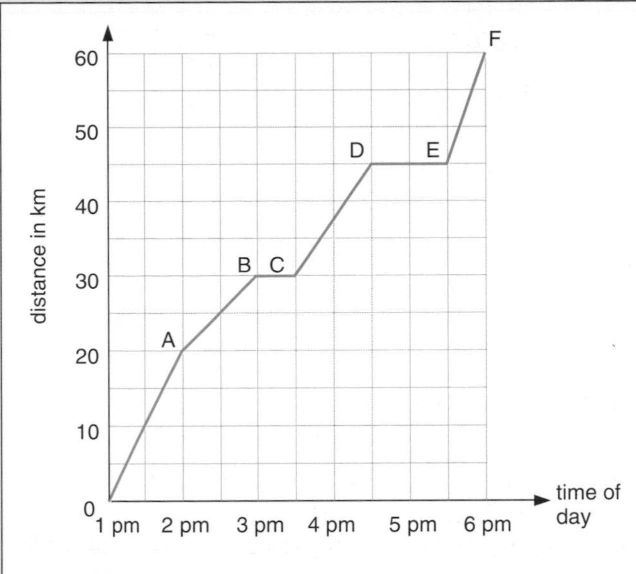

Figure 32.5

2 Figure 32.6a shows the distance–time graph for a moving object and Figure 32.6b shows the speed–time graph for another moving object.

Describe the motion, if any, of the objects in the regions **a** UV, **b** VW, **c** XY, **d** YZ. *(E.A.)*

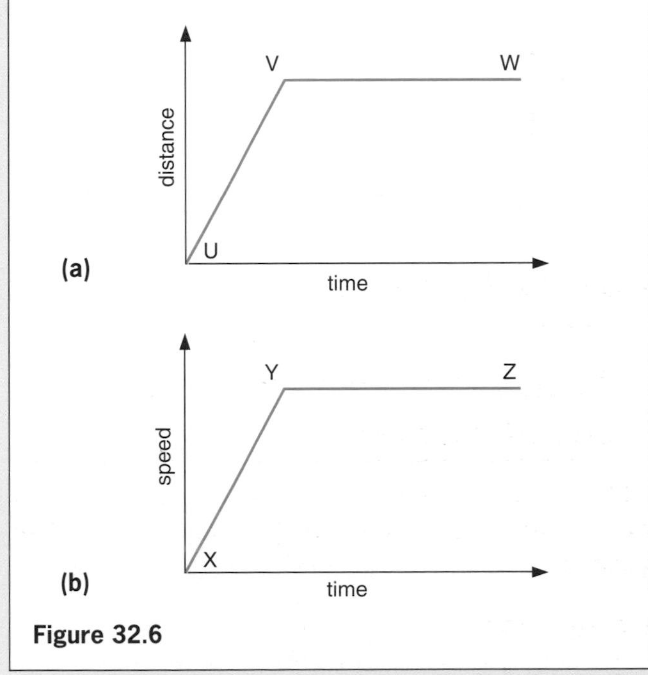

Figure 32.6

3 The graph in Figure 32.7 (overleaf) represents the distance travelled by a car plotted against time.

a How far has the car travelled at the end of 5 seconds?

b What is the speed of the car during the first 5 seconds?

c What has happened to the car after A?

d Draw a graph showing the speed of the car plotted against time during the first 5 seconds.

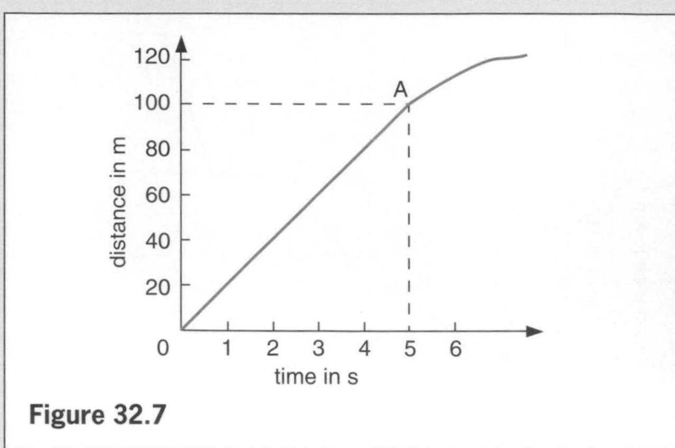

Figure 32.7

4 Figure 32.8 shows an incomplete velocity–time graph for a boy running a distance of 100 m.

a What is his acceleration during the first 4 seconds?

b How far does the boy travel during (i) the first 4 seconds, (ii) the next 9 seconds?

c Copy and complete the graph showing clearly at what time he has covered the distance of 100 m. Assume his speed remains constant at the value shown by the horizontal portion of the graph.

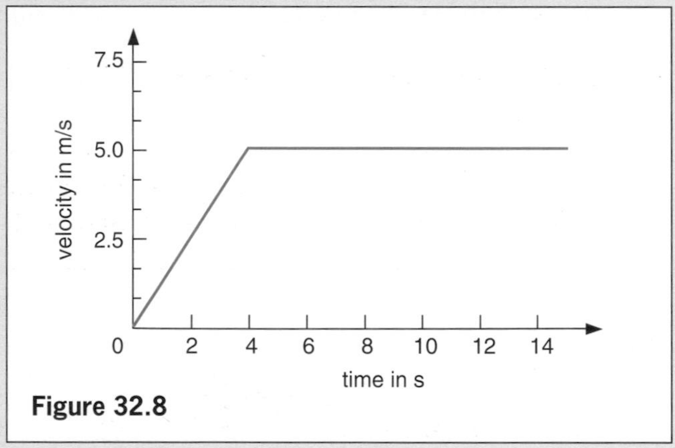

Figure 32.8

5 A body starts from rest and reaches a speed of 5 m/s after travelling with uniform acceleration in a straight line for 2 s. Calculate the acceleration of the body. *(E.A.)*

6 A body starts from rest and moves with a uniform acceleration of 2 m/s^2 in a straight line.

a What is its velocity after 5 s?

b How far has it travelled in this time?

c After how long will the body be 100 m from its starting point? *(J.M.B./A.L./N.W.16+)*

7 A car accelerates from 4 m/s to 20 m/s in 8 s. How far does it travel in this time?

8 A motor cyclist travelling at 12 m/s decelerates at 3 m/s^2.

a How long does he take to come to rest?

b How far does he travel in coming to rest?

Checklist

After studying this chapter you should be able to

- draw, interpret and use velocity–time and distance–time graphs to solve problems,
- recall and use the four equations of motion for solving problems.

33 FALLING BODIES

Acceleration due to gravity	***Projectiles***
Measuring g	***Practical work*** Motion of a falling body.
Distance–time graphs	

In air, a coin falls faster than a small piece of paper. In a vacuum they fall at the same rate as may be shown with the apparatus of Figure 33.1. The difference in air is due to **air resistance** having a greater effect on light bodies than on heavy bodies. The air resistance to a light body is big when compared with the body's weight. With a dense piece of metal the resistance is negligible at low speeds.

There is a story, untrue we now think, that in the sixteenth century the Italian scientist Galileo dropped a small iron ball and a large cannon ball ten times heavier from the top of the Leaning Tower of Pisa, Figure 33.2. And we are told, to the surprise of onlookers who expected the cannon ball to arrive first, that they reached the ground almost simultaneously. You will learn more about air resistance in the next chapter.

PRACTICAL WORK

Motion of a falling body

Arrange things as in Figure 33.3 and investigate the motion of a 100 g mass falling from a height of about 2 m.

Construct a tape chart using *one tick* lengths. Choose as dot '0' the first one you can distinguish clearly. What does the tape chart tell you about the motion of the falling mass?

Acceleration due to gravity

All bodies falling freely under the force of gravity do so with **uniform acceleration** if air resistance is

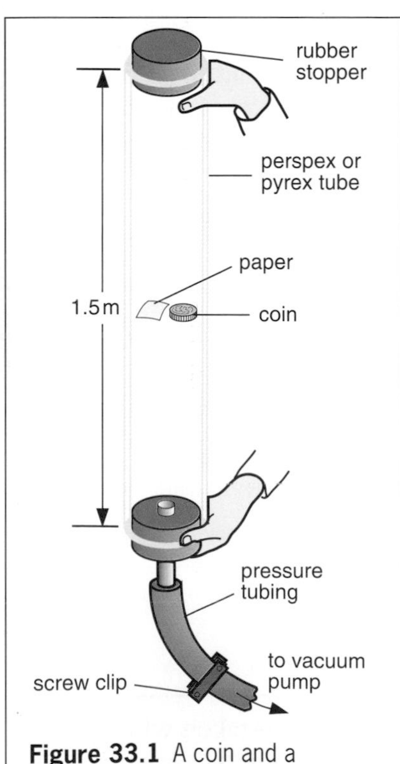

Figure 33.1 A coin and a piece of paper fall at the same rate in a vacuum

Figure 33.2 The Leaning Tower of Pisa, where Galileo is said to have experimented with falling objects

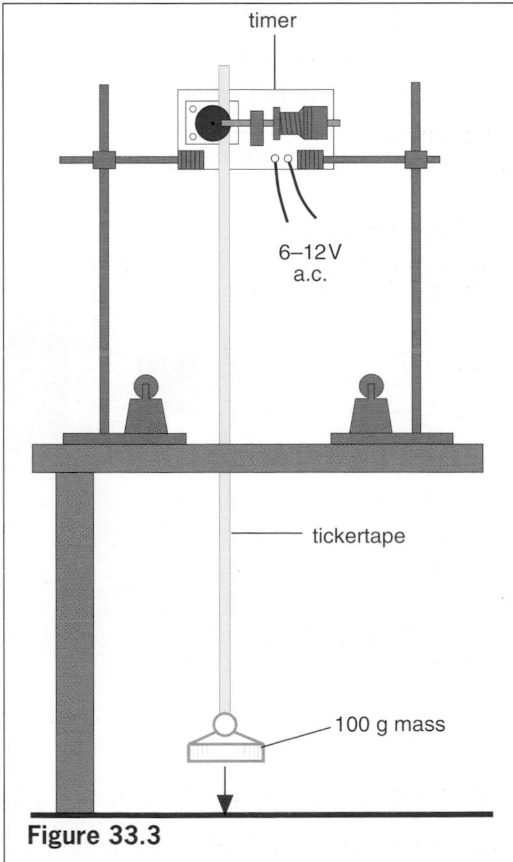

Figure 33.3

negligible (i.e. the 'steps' in the previous tape chart should all be equal).

This acceleration, called the **acceleration due to gravity**, is denoted by the italic letter g. Its value varies over the Earth. In Britain it is about 9.8 m/s^2 or near enough 10 m/s^2. The velocity of a free-falling body therefore increases by 10 m/s every second. A ball shot straight upwards with a velocity of 30 m/s decelerates by 10 m/s every second and reaches its highest point after 3 s.

In calculations using the equations of motion, g replaces a. It is given a positive sign for falling bodies (i.e. $a = +10$ m/s^2) and a negative sign for rising bodies since they are decelerating (i.e. $a = -10$ m/s^2).

Measuring g

Using the arrangement in Figure 33.4 the time for a steel ball-bearing to fall a known distance is measured by an electric stopclock.

When the two-way switch is changed to the 'down' position, the electromagnet releases the ball and simultaneously the clock starts. At the end of its fall the ball opens the 'trap-door' on the impact switch and the clock stops.

The result is found from the third equation of motion $s = ut + \frac{1}{2}at^2$, where s is the distance fallen (in m), t is the time taken (in s), $u = 0$ (the ball starts from rest) and $a = g$ (in m/s^2). Hence

$$s = \tfrac{1}{2}gt^2 \quad \text{or} \quad g = 2s/t^2$$

Air resistance is negligible for a dense object such as a steel ball-bearing falling a short distance.

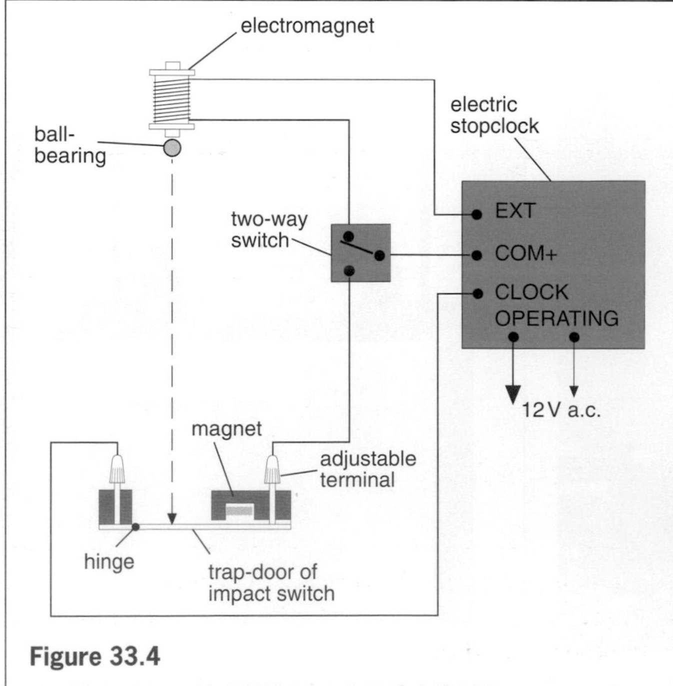

Figure 33.4

Worked example

A ball is projected vertically upwards with an initial velocity of 30 m/s. Find **a** its maximum height, **b** the time taken to return to its starting point. Neglect air resistance and take $g = 10$ m/s^2.

a We have $u = 30$ m/s, $a = -10$ m/s^2 (a deceleration) and $v = 0$ since the ball is momentarily at rest at its highest point. Substituting in $v^2 = u^2 + 2as$,

$$0 = 30^2\ \text{m}^2/\text{s}^2 + 2(-10\ \text{m/s}^2)s$$

or

$$-900\ \text{m}^2/\text{s}^2 = -s \times 20\ \text{m/s}^2$$

$\therefore$

$$s = \frac{-900\ \text{m}^2/\text{s}^2}{-20\ \text{m/s}^2} = 45\ \text{m}$$

b If t is the time to reach the highest point, we have, from $v = u + at$

$$0 = 30\ \text{m/s} + (-10\ \text{m/s}^2)t$$

or

$$-30\ \text{m/s} = -t \times 10\ \text{m/s}^2$$

$\therefore$

$$t = \frac{-30\ \text{m/s}}{-10\ \text{m/s}^2} = 3\ \text{s}$$

The downward trip takes exactly the same time as the upward one and so the answer is 6 s.

Distance–time graphs

For a body falling freely from rest we have

$$s = \tfrac{1}{2}gt^2$$

A graph of s against t is shown in Figure 33.5a and for s against t^2 in Figure 33.5b. The latter is a straight line through the origin since $s \propto t^2$ (g being constant at one place).

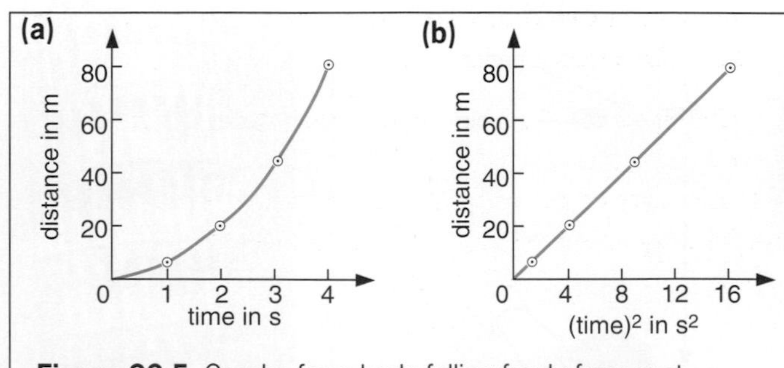

Figure 33.5 Graphs for a body falling freely from rest

Projectiles

The photograph in Figure 33.6 was taken whilst a lamp emitted regular flashes of light. One ball was **dropped** from rest and the other, a 'projectile', was **thrown sideways** at the same time. Their vertical accelerations (due to gravity) are equal, showing that a projectile falls like a body which is dropped from rest. Its horizontal velocity does not affect its vertical motion.

The horizontal and vertical motions of a body are independent and can be treated separately.

For example if a ball is thrown horizontally from the top of a cliff and takes 3 s to reach the beach below we can calculate the height of the cliff by considering the vertical motion only. We have $u = 0$ (since the ball has no vertical velocity initially), $a = g = +10\ \mathrm{m/s^2}$ and $t = 3$ s. The height s of the cliff is given by

$$s = ut + \tfrac{1}{2}at^2$$

$$= 0 \times 3\ \mathrm{s} + \tfrac{1}{2}(+10\ \mathrm{m/s^2})3^2\ \mathrm{s^2} = 45\ \mathrm{m}$$

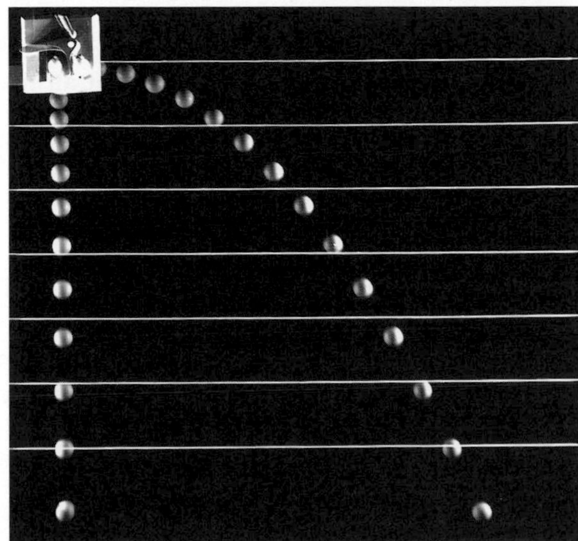

Figure 33.6 Comparing free fall and projectile motion using multiflash photography

Projectiles such as cricket balls and explosive shells are projected from near ground level and at an angle. The horizontal distance they travel, i.e. their **range**, depends on

(i) the **speed** of projection – the greater this is, the greater the range, and
(ii) the **angle** of projection – it can be shown that, neglecting air resistance, the range is a maximum when the angle is 45°, Figure 33.7.

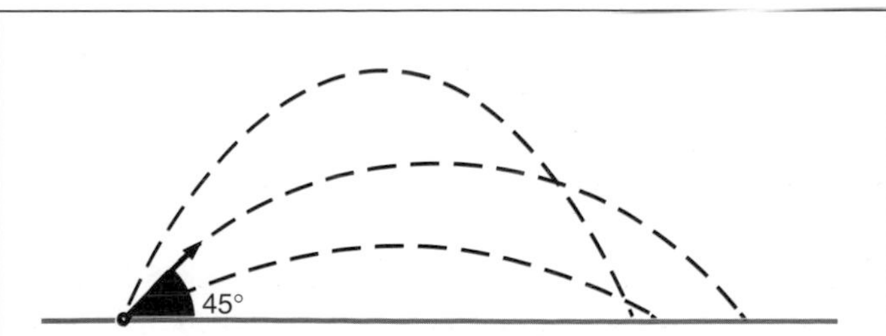

Figure 33.7 The range is greatest for an angle of projection of 45°

QUESTIONS

1 A stone falls from rest from the top of a high tower. Ignoring air resistance and taking $g = 10\ \mathrm{m/s^2}$,

a what is its velocity after (i) 1 s, (ii) 2 s, (iii) 3 s, (iv) 5 s?

b how far has it fallen after (i) 1 s, (ii) 2 s, (iii) 3 s, (iv) 5 s?

2 A body falls from a cliff 125 m high. Find the time of fall. How far did the body fall in half this time? (Take the acceleration due to gravity as $10\ \mathrm{m/s^2}$.) *(E.A.)*

3 A mass is projected upwards with a velocity of 10 m/s. If the acceleration due to gravity is $10\ \mathrm{m/s^2}$, what is the maximum height reached in metres?

A 1 **B** 5 **C** 10 **D** 20 **E** 100 *(N.I.)*

4 An object is dropped from a helicopter at a height of 45 m above the ground. If the helicopter is at rest, how long does the object take to reach the ground and what is its velocity on arrival? ($g = 10\ \mathrm{m/s^2}$)

5 An object is released from an aircraft travelling horizontally with a constant velocity of 200 m/s at a height of 500 m. Ignoring air resistance and taking $g = 10\ \mathrm{m/s^2}$ find

a how long it takes the object to reach the ground,

b the horizontal distance covered by the object between leaving the aircraft and reaching the ground.

6 A gun pointing vertically upwards is fired from an open car B moving with uniform velocity. When the bullet returns to the level of the gun, car B has travelled to C and another car A has reached the position occupied by B when the gun was fired, Figure 33.8. Are the occupants of A or B in danger?

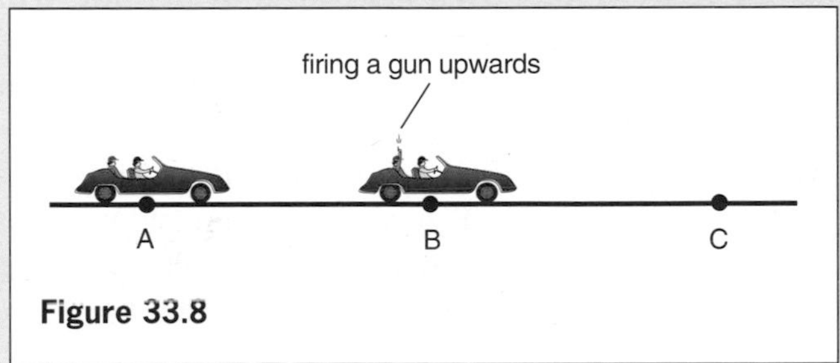

Figure 33.8

Checklist

After studying this chapter you should be able to

- describe the behaviour of falling objects and solve problems on them,
- describe an experiment to find the acceleration due to gravity.

34 NEWTON'S LAWS OF MOTION

First law

Friction and air resistance cause a car to come to rest when the engine is switched off. If these forces were absent we believe that a body, once set in motion, would go on moving forever with a constant speed in a straight line. That is, force is not needed to keep a body moving with uniform velocity so long as no opposing forces act on it.

This idea was proposed by Galileo and is summed up in Newton's first law:

> A body stays at rest, or if moving it continues to move with uniform velocity, unless an external force makes it behave differently.

It seems that the question we should ask about a moving body is not 'what keeps it moving' but 'what changes or stops its motion'.

The smaller the external forces opposing a moving body, the smaller is the force needed to keep it moving with uniform velocity. A hovercraft, which floats on a cushion of air, Figure 34.1, can skim across water or land with little frictional opposition, so that relatively little power is needed to maintain motion.

Figure 34.1 Friction is much reduced for a hovercraft

Mass and inertia

The first law is another way of saying that all matter has a built-in opposition to being moved if it is at rest or, if it is moving, to having its motion changed. This property of matter is called **inertia** (from the Latin word for laziness).

Its effect is evident on the occupants of a car which stops suddenly; they lurch forwards in an attempt to continue moving, and this is why seat belts are needed. The reluctance of a stationary object to move can be shown by placing a large coin on a piece of card on your finger, Figure 34.2. If the card is flicked *sharply* the coin stays where it is while the card flies off.

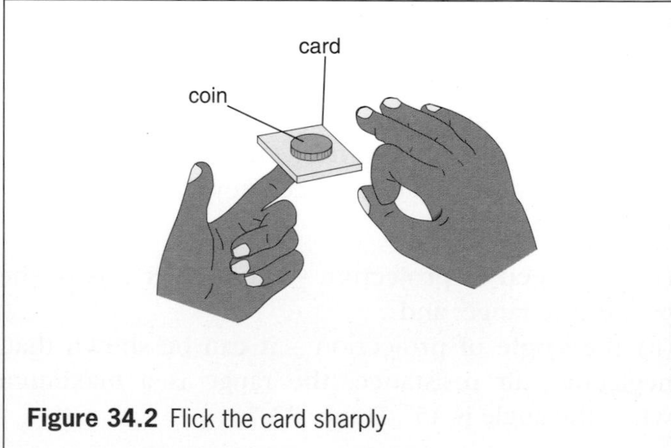

Figure 34.2 Flick the card sharply

The larger the mass of a body the greater is its inertia, i.e. the more difficult it is to move it when at rest and to stop it when in motion. Because of this we consider that **the mass of a body measures its inertia**. This is a better definition of mass than the one given earlier (Chapter 17) in which it was stated to be the 'amount of matter' in a body.

Effect of force and mass on acceleration

The apparatus consists of a trolley to which a force is applied by a stretched length of elastic, Figure 34.3. The velocity of the trolley is found from a tickertape timer.

First compensate the runway for friction by raising one end until the trolley runs down with uniform velocity when given a push. The dots on the tape should be equally spaced. There is now no resultant force on the trolley and any acceleration produced later will be due only to the force caused by the stretched elastic.

(a) Force and acceleration (mass constant)

Fix one end of a short length of elastic to the rod at the back of the trolley and stretch it until the other end is level with the front of the trolley. Before attaching a tape, practise pulling the trolley down the runway, keeping the same stretch on the elastic. After a few trials you should be able to produce a steady accelerating force. Now do it with a tape.

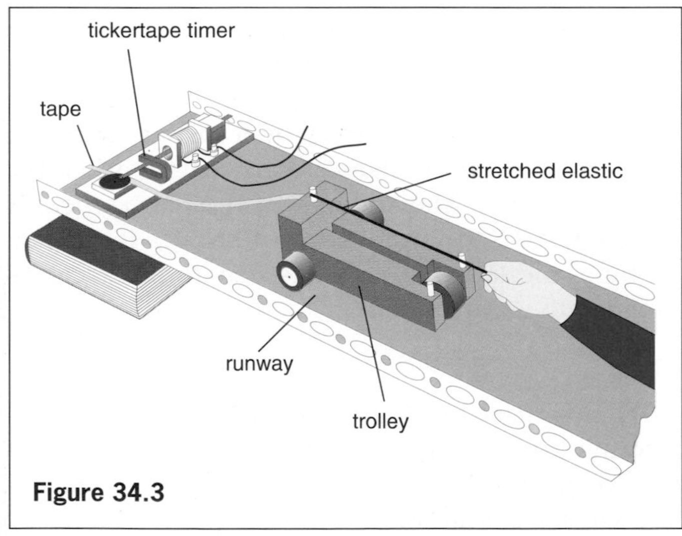

Figure 34.3

Repeat with fresh tapes using first two and then three *identical* pieces of elastic, stretched side by side by the same amount, to give two and three units of force.

Make a tape chart for each force and use it to find the acceleration produced in cm/tentick^2 (see Chapter 31). Ignore the start of the tape (where the dots are too close) and the end (where the force may not be steady). Does a steady force cause a steady acceleration? Put the results in a table. Do they suggest any relationship between a and F?

Force (F) (no. of pieces of elastic)	1	2	3
Acceleration (a) (cm/tentick^2)			

(b) Mass and acceleration (force constant)

Do the experiment as in **(a)** using two pieces of elastic (i.e. constant F) to accelerate first one trolley, then two (stacked one above the other) and finally three. Check the friction compensation of the runway each time.

Find the accelerations from the tape charts and tabulate the results. Do they suggest any relationship between a and m?

Mass (m) (no. of trolleys)	1	2	3
Acceleration (a) (cm/tentick^2)			

Second law

The previous experiment should show roughly that the acceleration a is

(i) directly proportional to the applied force F for a fixed mass, i.e. $a \propto F$, and

(ii) inversely proportional to the mass m for a fixed force, i.e $a \propto 1/m$.

Combining the results into one equation, we get

$$a \propto F/m \quad \text{or} \quad F \propto ma$$

Therefore $\quad F = kma$

where k is the constant of proportionality.

One newton is defined as the force which gives a mass of 1 kg an acceleration of 1 m/s^2, i.e. 1 N = 1 kg m/s^2

Hence if $m = 1$ kg and $a = 1$ m/s^2 then $F = 1$ N. Substituting in $F = kma$ we get $k = 1$ and so we can write

$$F = ma$$

This is Newton's second law of motion. When using it two points should be noted. First, F is the resultant (or unbalanced) force causing the acceleration a. Second, F must be in newtons, m in kilograms and a in metres per second squared, otherwise k is not 1. The law shows that a will be largest when F is large and m small.

You should now appreciate that when the forces acting on a body do not balance, i.e. there is a net (resultant) force, they cause a **change** of motion, i.e. the body accelerates or decelerates. However, if they balance, there is no change of motion but there may be a change of shape. In that case internal forces in the body (i.e. forces between neighbouring atoms) balance the external forces.

Worked examples

1 A block of mass 2 kg is pushed along a table with a constant velocity by a force of 5 N. When the push is increased to 9 N what is **a** the resultant force, **b** the acceleration?

When the block moves with constant velocity the forces acting on it are balanced. The force of friction opposing its motion must therefore be 5 N.

a When the push is increased to 9 N the resultant (unbalanced) force F on the block is $(9-5)\text{ N} = 4\text{ N}$ (since the frictional force is still 5 N).

b The acceleration a is obtained from $F = ma$ where $F = 4\text{ N}$ and $m = 2\text{ kg}$.

$$\therefore \quad a = F/m = 4\text{ N}/2\text{ kg} = \frac{4\text{ kg m/s}^2}{2\text{ kg}} = 2\text{ m/s}^2$$

2 A car of mass 1200 kg travelling at 72 km/h is brought to rest in 4 s. Find the **a** average deceleration, **b** average braking force, **c** distance moved during the deceleration.

a The deceleration is found from $v = u + at$ where $v = 0$.

$$u = 72\text{ km/h} = \frac{72 \times 1000}{60 \times 60} = 20\text{ m/s}$$

since 1 km = 1000 m and 1 hour = 3600 s

and $\qquad t = 4\text{ s}$

Hence $\qquad 0 = 20\text{ m/s} + a \times 4\text{ s}$

or $\qquad -20\text{ m/s} = a \times 4\text{ s}$

$$\therefore \quad a = -20\frac{\text{m}}{\text{s}} \times \frac{1}{4\text{ s}} = -5\text{ m/s}^2$$

b The average braking force F is given by $F = ma$ where $m = 1200\text{ kg}$ and $a = -5\text{ m/s}^2$. Therefore

$$F = 1200\text{ kg} \times (-5)\text{ m/s}^2 = -6000\text{ kg m/s}^2$$
$$= 6000\text{ N}$$

c To find the distance moved s we use $s = ut + \frac{1}{2}at^2$

$$\therefore \quad s = 20\text{ m/s} \times 4\text{ s} + \tfrac{1}{2} \times (-5)\text{ m/s}^2 \times (4^2)\text{ s}^2$$
$$= 80\text{ m} - 40\text{ m} = 40\text{ m}$$

Weight and gravity

The weight W of a body is the force of gravity acting on it which gives it an acceleration g when it is falling freely near the Earth's surface. If the body has mass m, then W can be calculated from $F = ma$ if we put $F = W$ and $a = g$ to give

$$W = mg$$

Taking $g = 10\text{ m/s}^2$ and $m = 1\text{ kg}$, this gives $W = 10\text{ N}$, i.e. a body of mass 1 kg has weight 10 N. Similarly a body of mass 2 kg has weight 20 N and so on. While the mass of a body is always the same, its weight varies depending on the value of g. On the Moon the acceleration due to gravity is only about 1.6 m/s^2, and so a mass of 1 kg has a weight of just 1.6 N there.

The weight of a body is directly proportional to its mass, which explains why g is the same for all bodies. The greater the mass of a body, the greater is the force of gravity on it but it does not accelerate faster when falling because of its greater inertia (i.e. its greater resistance to acceleration).

Gravitational field

The force of gravity acts through space and can cause a body, not in contact with the Earth, to fall to the ground. It is an invisible, action-at-a-distance force. We try to 'explain' its existence by saying that the Earth is surrounded by a **gravitational field** which exerts a force on any body in the field. Later, magnetic and electric fields will be considered.

> The strength of a gravitational field is defined as the force acting on unit mass in the field.

Measurement shows that on the Earth's surface a mass of 1 kg experiences a force of 9.8 N, i.e. its weight is 9.8 N. The strength of the Earth's field is therefore 9.8 N/kg (near enough 10 N/kg). It is denoted by g, the letter also used to denote the acceleration due to gravity. Hence

$$g = 9.8\text{ N/kg} = 9.8\text{ m/s}^2$$

We now have two ways of regarding g. When considering bodies **falling freely** we can think of it as an acceleration of 9.8 m/s^2, but when a body of known mass is **at rest** and we wish to know the force of gravity (in N) acting on it we think of g as the Earth's gravitational field strength of 9.8 N/kg.

Third law

> If a body A exerts a force on body B, then body B exerts an equal but opposite force on body A.

The law states that forces never occur singly but always in pairs as a result of the action between two bodies. For example, when you step forwards from rest your foot pushes backwards on the Earth and the Earth exerts an equal and opposite force forward on you. Two bodies and two forces are involved. The small force you exert on the large mass of the Earth gives no noticeable acceleration to the Earth but the equal force it exerts on your very much smaller mass causes you to accelerate.

Note that the equal and opposite forces **do not act on the same body**; if they did, there could never be any resultant forces and acceleration would be impossible.

An appreciation of the third law and the effect of friction is desirable when stepping from a rowing boat, Figure 34.4. You push backwards on the boat and, although the boat pushes you forwards with an equal force, it is itself now moving backwards (because friction with the water is slight), and this reduces your forwards motion by the same amount – and you might fall in!

Figure 34.4 The boat moves backwards when you step forwards!

Air resistance: terminal velocity

When an object falls in air, the air resistance (fluid friction) opposing its motion **increases as its speed rises**, so reducing its acceleration. Eventually, air resistance acting upwards equals the weight of the object acting downwards. The resultant force on the object is then zero since the two opposing forces balance. The object falls at a constant velocity, called its **terminal velocity**, whose value depends on the size, shape and weight of the object.

A small dense object, e.g. a steel ball-bearing, has a high terminal velocity and falls a considerable distance with a constant acceleration of 9.8 m/s^2 before air resistance equals its weight. A light object, e.g. a raindrop, or one with a large surface area, e.g. a parachute, has a low terminal velocity and only accelerates over a comparatively short distance before air resistance equals its weight. A sky diver, Figure 34.5, has a terminal velocity of more than 50 m/s (100 m.p.h.).

Figure 34.5 Synchronized sky divers

Objects falling in liquids behave similarly (Chapter 30).

Explanation of Bernoulli's principle

In Figure 30.8b (p. 113) the liquid speeds up going from the wide part A of the tube to the narrower part B, i.e. it is accelerated. Therefore, since $F = ma$, the force at A, and so also the pressure at A, must be greater than the force and pressure at B. Between B and C the liquid slows down due to the pressure at C being greater than that at B.

QUESTIONS

1 Which one of the diagrams in Figure 34.6 shows the arrangement of forces which gives the block M the greatest acceleration?

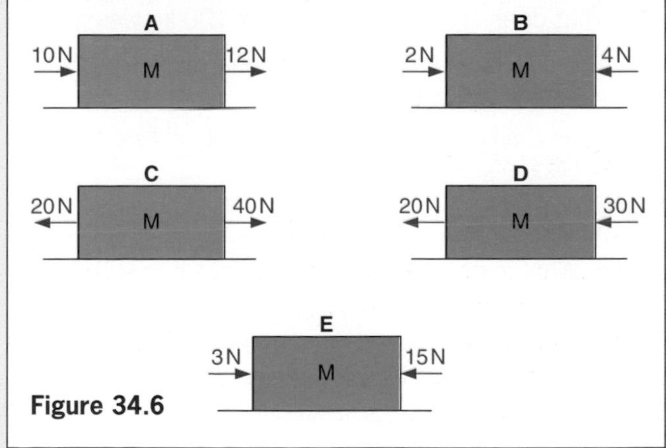

Figure 34.6

2 In Figure 34.7 if *P* is a force of 20 N and the object moves with **constant velocity** what is the value of the opposing force *F*?

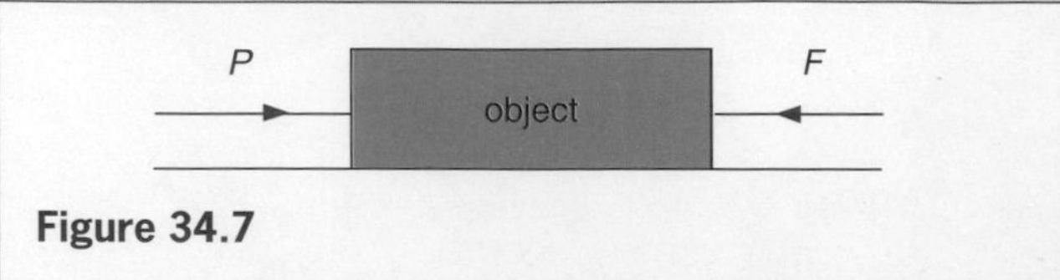

Figure 34.7

3 **a** What resultant force produces an acceleration of 5 m/s^2 in a car of mass 1000 kg?

b What acceleration is produced in a mass of 2 kg by a resultant force of 30 N?

4 A car of mass 500 kg accelerates steadily from rest to 40 m/s in 20 s.

a What is its acceleration?

b What resultant force produces this acceleration?

c The actual force will be greater. Why?

d Explain *two* advantages of keeping the mass of a car as small as possible.

5 A block of mass 500 g is pulled from rest on a horizontal frictionless bench by a steady force *F* and travels 8 m in 2 s. Find **a** the acceleration, **b** the value of *F*.

6 Starting from rest on a level road a girl can reach a speed of 5 m/s in 10 s on her bicycle. Find **a** the acceleration, **b** the average speed during the 10 s, **c** the distance she travels in 10 s.

Eventually, even though she is still pedalling as fast as she can, she stops accelerating and her speed reaches a maximum value. Explain in terms of the forces acting why this happens.

7 A trailer of mass 1000 kg is towed by means of a rope attached to a car moving at a steady speed along a level road. The tension in the rope is 400 N. Why is it not zero?

The car starts to accelerate steadily. If the tension in the rope is now 1650 N, with what acceleration is the trailer moving? *(O. and C.)*

8 What does an astronaut of mass 100 kg weigh

a on Earth where the gravitational field strength is 10 N/kg,

b on the Moon where the gravitational field strength is 1.6 N/kg?

9 A rocket has a mass of 500 kg.

a What is its weight on Earth where g = 10 N/kg?

b At lift-off the rocket engine exerts an upward force of 25 000 N. What is the resultant force on the rocket? What is its initial acceleration?

10 Figure 34.8 shows the forces acting on a raindrop which is falling to the ground.

a (i) *A* is the force which causes the raindrop to fall. What is this force called? (ii) *B* is the total force opposing the motion of the drop. State *one* possible cause of this force.

b What happens to the raindrop when force *A* = force *B*?

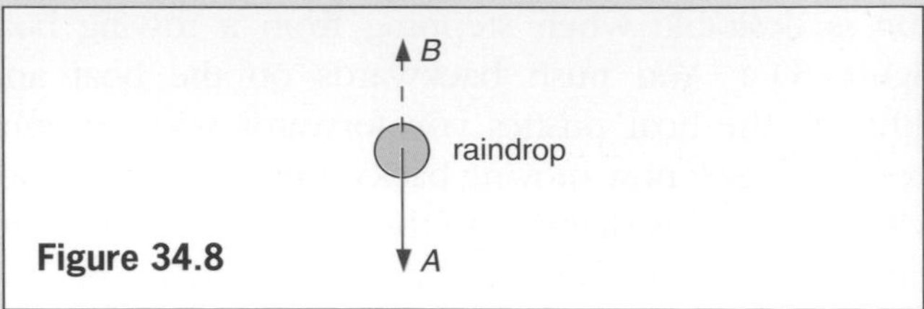

Figure 34.8

11 Explain the following using *F* = *ma*.

a A racing car has a powerful engine and is made of strong but lightweight material.

b A car with a small engine can still accelerate rapidly.

Checklist

After studying this chapter you should be able to

- describe an experiment to investigate the relationship between force, mass and acceleration,
- define the unit of force,
- state Newton's three laws of motion and use them to solve problems,
- define the strength of the Earth's gravitational field,
- describe the motion of an object falling in air.

35 MOMENTUM

Conservation of momentum	*Sport and momentum*
Explosions	*Discovery of the neutron*
Rockets and jets	*Practical work* Collisions and momentum. Explosions and momentum.
Force and momentum	

Momentum is a useful quantity to consider when bodies are involved in collisions and explosions. It is defined as the **mass of the body multiplied by its velocity** and is measured in kilogram metre per second (kg m/s) or newton second (N s).

momentum = mass × velocity

A 2 kg mass moving at 10 m/s has momentum 20 kg m/s, the same as the momentum of a 5 kg mass moving at 4 m/s.

PRACTICAL WORK

Collisions and momentum

Friction-compensate a runway as before (Chapter 34). Place one trolley at rest halfway down the runway and another at the top with a length of tickertape from it passing through a timer, Figure 35.1. Each trolley should have a strip of Velcro fitted so that it 'sticks' to the other on collision.

Give the top trolley a good push. It will move forwards with uniform velocity and should hit the second trolley so that they travel on as one.

From the tape find the velocity of the moving trolley before the collision and the common velocity of both trolleys after the collision (in cm/tentick say).

Repeat the experiment with another trolley stacked on top of the one to be pushed so that two are moving before the collision and three after.

Copy and complete the tables of results.

Before collision (m_2 at rest)

Mass m_1 (no. of trolleys)	**Velocity** v (cm/tentick)	**Momentum** m_1v
1 2		

After collision (m_1 and m_2 together)

Mass $m_1 + m_2$ (no. of trolleys)	**Velocity** v_1 (cm/tentick)	**Momentum** $(m_1 + m_2)v_1$
2 3		

Do the results suggest any connection between the momentum before the collision and after it?

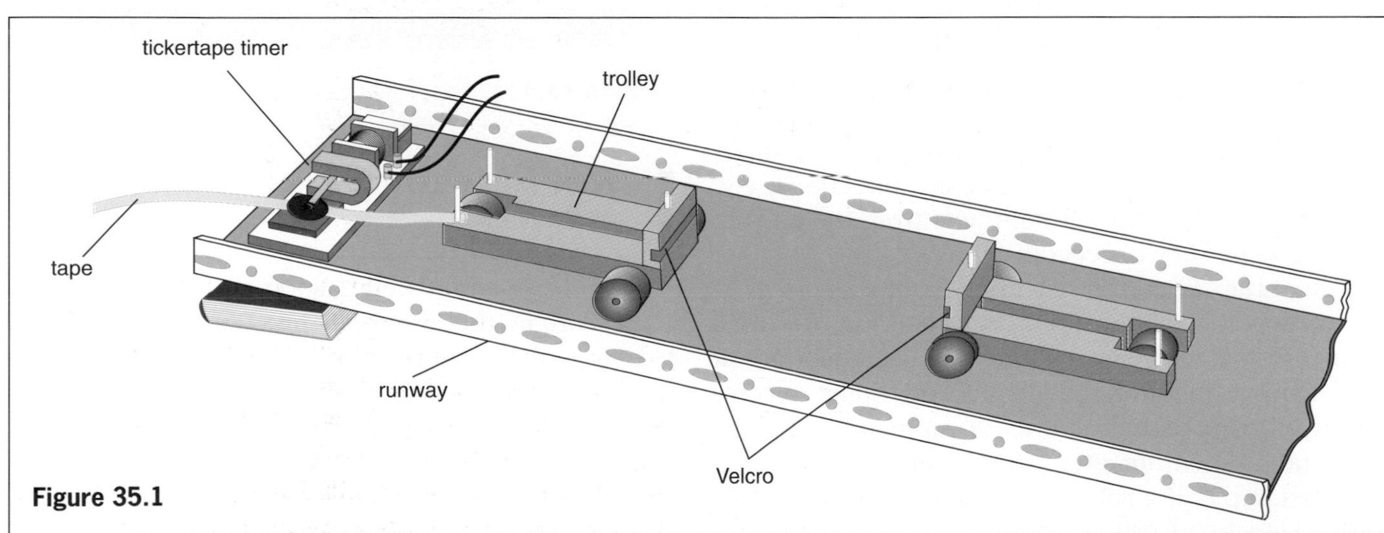

Figure 35.1

Conservation of momentum

When two or more bodies act on one another, as in a collision, the total momentum of the bodies remains constant, provided no external forces act (e.g. friction).

This statement is called the **principle of conservation of momentum**. Experiments like those above show that it is true for all types of collisions.

As an example, suppose a truck of mass 60 kg moving with velocity 3 m/s collides and couples with a stationary truck of mass 30 kg, Figure 35.2a. The two move off together with the same velocity v which we can find as follows, Figure 35.2b.

Total momentum before is

$$(60\ \text{kg} \times 3\ \text{m/s} + 30\ \text{kg} \times 0\ \text{m/s}) = 180\ \text{kg m/s}$$

Total momentum after is

$$(60\ \text{kg} + 30\ \text{kg})v = 90\ \text{kg} \times v$$

Since momentum is not lost

$$90\ \text{kg} \times v = 180\ \text{kg m/s} \quad \text{or} \quad v = 2\ \text{m/s}$$

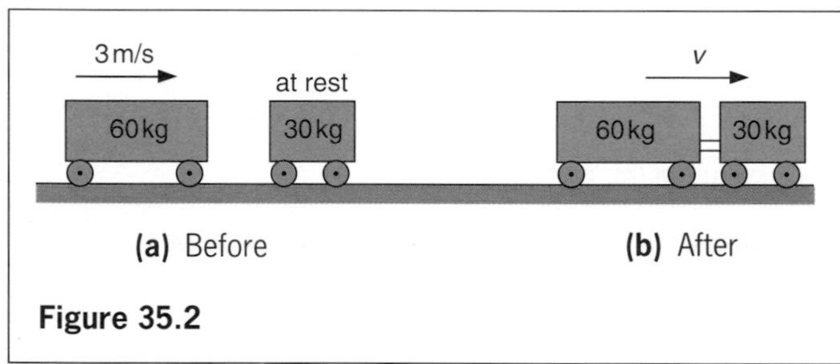

Figure 35.2

Explosions

Momentum, like velocity, is a vector since it has both magnitude and direction. Vectors cannot be added by ordinary addition unless they act in the same direction. If they act in exactly opposite directions, e.g. east and west, the smaller subtracts from the greater, or if the same they cancel out.

Momentum is conserved in an explosion such as occurs when a rifle is fired. Before firing, the total momentum is zero since both rifle and bullet are at rest. During the firing the rifle and bullet receive **equal** but **opposite** amounts of momentum so that the **total** momentum after firing is zero.

For example, if a rifle fires a bullet of mass 0.01 kg with a velocity of 300 m/s then

$$\text{forward momentum of bullet} = 0.01\ \text{kg} \times 300\ \text{m/s} = 3\ \text{kg m/s}$$

$$\therefore \quad \text{backward momentum of rifle} = 3\ \text{kg m/s}$$

If the rifle has mass m, it recoils (kicks back) with a velocity v such that

$$mv = 3\ \text{kg m/s}$$

Taking $m = 6$ kg gives $v = 3/6$ m/s $= 0.5$ m/s.

PRACTICAL WORK

Explosions and momentum

The principle of conservation of momentum can be tested experimentally for 'explosions' with the apparatus in Figure 35.3 arranged as shown. One tape from each trolley goes to a tickertape timer.

Tap one of the buffer rods to release the spring inside. The trolleys fly apart. Work out the velocities v_1 and v_2 of each from the tickertapes.

Since the trolleys are initially at rest

$$\text{total momentum before explosion} = 0$$

If the trolleys have masses m_1 and m_2 then

$$\text{total momentum after explosion} = m_1v_1 - m_2v_2$$

If the principle holds for explosions, $m_1v_1 - m_2v_2$ should be zero. (For trolleys of equal mass it simplifies matters to take '1 trolley' as the unit of mass (as before), then $m_1 = m_2 = 1$.)

Repeat the experiment with another trolley stacked on top of one of the trolleys so that, for example, $m_1 = 1$ and $m_2 = 2$.

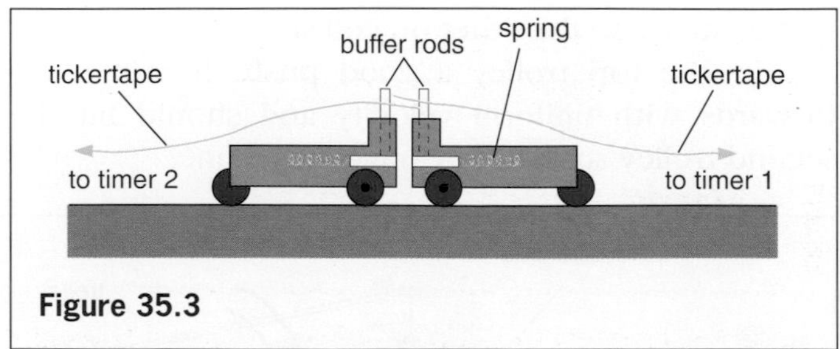

Figure 35.3

Rockets and jets

If you release an inflated balloon with its neck open, it flies off in the opposite direction to that of the escaping air. In Figure 35.4 the air has momentum to the left and the balloon moves to the right with equal momentum.

This is the principle of rockets and jet engines. In both, a high-velocity stream of hot gas is produced by burning fuel and leaves the exhaust with large

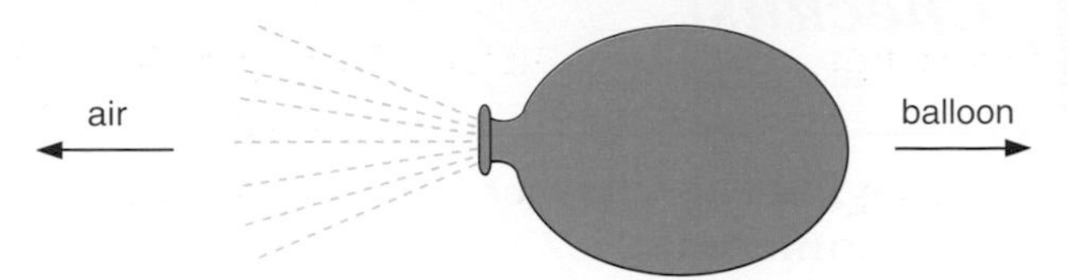

Figure 35.4 A deflating balloon demonstrates the principle of a rocket or a jet engine

momentum. The rocket or jet engine itself acquires an equal forward momentum. Space rockets carry their own oxygen supply; jet engines use the surrounding air.

Force and momentum

If a steady force F acting on a body of mass m increases its velocity from u to v in time t, the acceleration a is given by

$$a = (v - u)/t \quad \text{(from } v = u + at\text{)}$$

Substituting for a in $F = ma$,

$$F = \frac{m(v-u)}{t} = \frac{mv - mu}{t}$$

$$\therefore \qquad Ft = mv - mu$$

mv is the final momentum, mu the initial momentum and Ft is called the **impulse**. Therefore

$$\text{force} = \frac{\text{change of momentum}}{\text{time}} = \text{rate of change of momentum}$$

This is another version of the second law of motion. For some problems it is more useful than $F = ma$.

Sport and momentum

The good cricketer or tennis player 'follows through' with the bat or racket when striking the ball, Figure 35.5a. The force applied then acts for a longer time, the impulse is greater and so also is the gain of momentum (and velocity) of the ball.

When we want to stop a moving ball such as a cricket ball, however, its momentum has to be reduced to zero. An impulse is then required in the form of an opposing force acting for a certain time. While any number of combinations of force and time will give a particular impulse, the 'sting' can be removed from the catch by drawing back the hands as the ball is caught, Figure 35.5b. A smaller average force is then applied for a longer time.

Figure 35.5a Batsman 'following through' after hitting the ball

Figure 35.5b Cricketer drawing back the hands on catching the ball

Discovery of the neutron

The principle of conservation of momentum was responsible for leading to the discovery of the neutron in 1932 by Chadwick. He found that in collisions between alpha particles (Chapter 63) and the element beryllium, the principle only held if it was assumed that a particle was produced which had about the same mass as the proton but no electric charge. It was named the **neutron**.

QUESTIONS

1 What is the momentum in kg m/s of a 10 kg truck travelling at **a** 5 m/s, **b** 20 cm/s, **c** 36 km/h?

2 A ball X mass 1 kg travelling at 2 m/s has a head-on collision with an identical ball Y at rest. X stops and Y moves off. What is Y's velocity?

3 A boy with mass 50 kg running at 5 m/s jumps on to a 20 kg trolley travelling in the same direction at 1.5 m/s. What is their common velocity?

4 A girl of mass 50 kg jumps out of a rowing boat of mass 300 kg on to the bank with a horizontal velocity of 3 m/s. With what velocity does the boat begin to move backwards?

5 A truck of mass 500 kg moving at 4 m/s collides with another truck of mass 1500 kg moving in the same direction at 2 m/s. What is their common velocity just after the collision if they move off together?

6 The velocity of a body of mass 10 kg increases from 4 m/s to 8 m/s when a force acts on it for 2 s.

a What is the momentum before the force acts?

b What is the momentum after the force acts?

c What is the momentum gain per second?

d What is the value of the force?

7 A rocket of mass 10 000 kg uses 5.0 kg of fuel and oxygen to produce exhaust gases ejected at 5000 m/s. Calculate the increase in its velocity.

8 A rocket launched vertically sends out 50 kg of exhaust gases every second with a velocity of 200 m/s.

a What is the upward force on the rocket?

b If the mass of the rocket is 500 kg, what is its initial upward acceleration?

Checklist

After studying this chapter you should be able to

- define **momentum**,
- describe experiments to demonstrate the principle of conservation of momentum,
- state and use the principle of conservation of momentum to solve problems,
- understand the action of rocket and jet engines,
- state the relationship between force and rate of change of momentum and use it to solve problems.

36 KINETIC AND POTENTIAL ENERGY

Energy and its different forms were discussed earlier (Chapter 25). Here we will consider kinetic energy (k.e.) and potential energy (p.e.) in more detail.

Kinetic energy

Kinetic energy is the energy a body has because of its motion.

We can obtain an expression for k.e. Suppose a body of mass m starts from rest and is acted on by a steady force F which gives it a uniform acceleration a. If the velocity of the body is v when it has travelled a distance s, then, using $v^2 = u^2 + 2as$, we get, since $u = 0$,

$$v^2 = 2as \quad \text{or} \quad a = v^2/2s$$

Substituting in $F = ma$,

$$F = m\left(\frac{v^2}{2s}\right) \quad \text{or} \quad Fs = \tfrac{1}{2}mv^2$$

Fs is the work done on the body to give it velocity v and therefore equals its k.e. Hence

$$\text{kinetic energy} = E_k = \tfrac{1}{2}mv^2$$

Figure 36.1 Kinetic energy depends on the square of the velocity

If m is in kg and v in m/s, then k.e. is in J. For example, a cricket ball of mass 0.2 kg (200 g) moving with velocity 20 m/s has k.e. $= \frac{1}{2}mv^2 = \frac{1}{2} \times 0.2\ \text{kg} \times (20)^2\ \text{m}^2/\text{s}^2 = 0.1 \times 400\ \text{kg m/s}^2 \times \text{m} = 40\ \text{N m} = 40\ \text{J}$.

Since k.e. depends on v^2, a vehicle, such as a high-speed train, Figure 36.1, travelling at 200 km/h (125 m.p.h.) has four times the k.e. it has at 100 km/h.

Potential energy

Potential energy is the energy a body has because of its position or condition.

A body above the Earth's surface is considered to have an amount of gravitational p.e. equal to the work that has been done against gravity by the force used to raise it. To lift a body of mass m through a vertical height h at a place where the Earth's gravitational field strength is g, needs a force equal and opposite to the weight mg of the body. Hence

$$\text{work done by force} = \text{force} \times \text{vertical height} = mg \times h$$

$$\therefore \quad \text{potential energy} = E_p = mgh$$

When m is in kg, g in N/kg (or m/s^2) and h in m, the p.e. is in J. For example, if $g = 10$ N/kg, the p.e. gained by a 0.1 kg (100 g) mass raised vertically by 1 m is $0.1\ \text{kg} \times 10\ \text{N/kg} \times 1\ \text{m} = 1\ \text{N m} = 1\ \text{J}$.

Note. Strictly speaking we are concerned with changes in p.e. from that which a body has at the Earth's surface, rather than with actual values. The expression for p.e. is therefore more correctly written

$$\Delta E_p = mgh$$

where Δ (pronounced delta) is the Greek capital D and stands for 'a change in'.

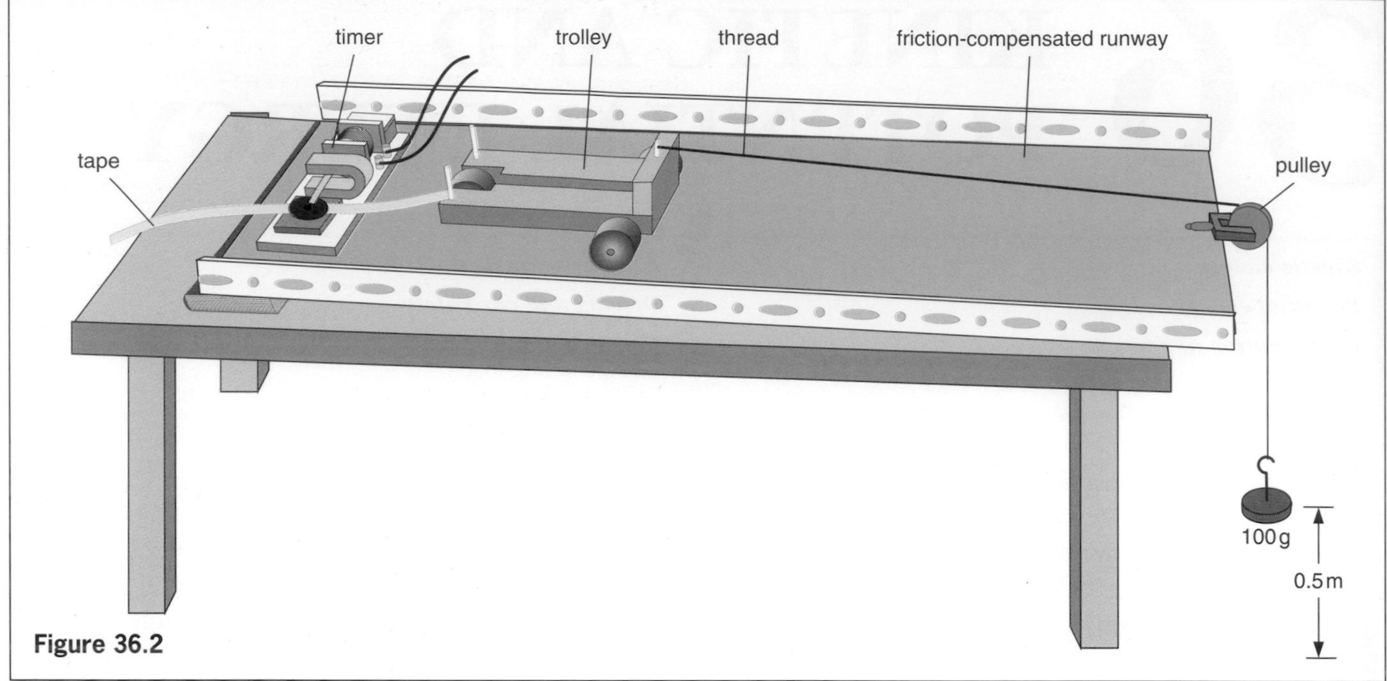

Figure 36.2

PRACTICAL WORK

Change of p.e. to k.e.

Friction-compensate a runway and arrange the apparatus as in Figure 36.2 with the bottom of the 0.1 kg (100 g) mass 0.5 m from the floor.

Start the timer and release the trolley. It will accelerate until the falling mass reaches the floor; after that it moves with *constant* velocity v.

From the tickertape measure v in m/s (50 ticks = 1 s). Find the mass of the trolley in kg. Work out:

k.e. gained by trolley and 0.1 kg mass = J
p.e. lost by 0.1 kg mass = J

Compare and comment on the results.

Conservation of energy

A mass m at height h above the ground has p.e. = mgh, Figure 36.3. When it falls, its velocity increases and it gains k.e. at the expense of its p.e. If it starts from rest and air resistance is negligible, its velocity v on reaching the ground is given by

$$v^2 = u^2 + 2as = 0 + 2gh = 2gh$$

Also, as it reaches the ground, its k.e. is

$$\tfrac{1}{2}mv^2 = \tfrac{1}{2}m \times 2gh = mgh$$

$$\therefore \qquad \text{loss of p.e.} = \text{gain of k.e.}$$

This is an example of the **principle of conservation of energy** which was discussed in Chapter 25.

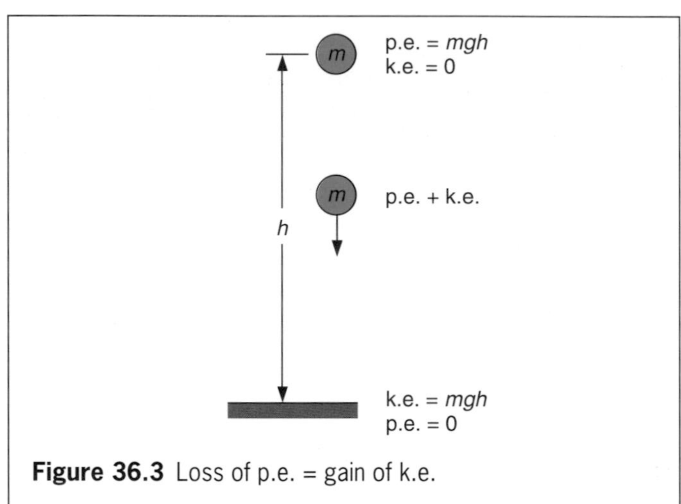

Figure 36.3 Loss of p.e. = gain of k.e.

In the case of a pendulum k.e. and p.e. are interchanged continually. The energy of the bob is all p.e. at the end of the swing and all k.e. as it passes through its central position. In other positions it has both p.e. and k.e., Figure 36.4. Eventually all the energy is changed to heat as a result of overcoming air resistance.

Elastic and inelastic collisions

In all collisions momentum is conserved but there is normally a loss of k.e., usually to heat energy and to a small extent to sound energy. The greater the proportion of k.e. lost, the less **elastic** is the collision, i.e. the more **inelastic** it is. In a perfectly elastic collision k.e. is conserved.

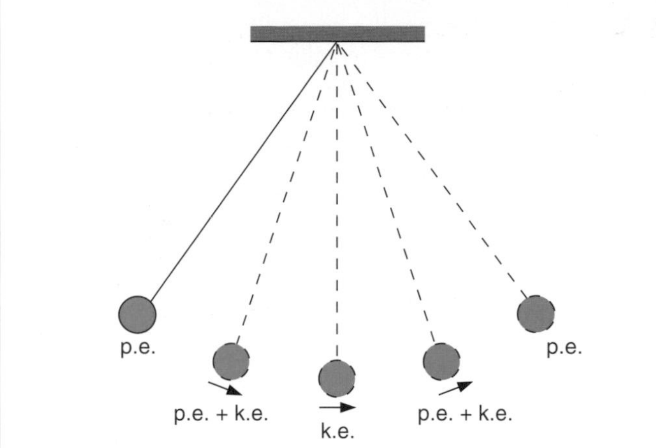

Figure 36.4 Interchange of p.e. and k.e. for a simple pendulum

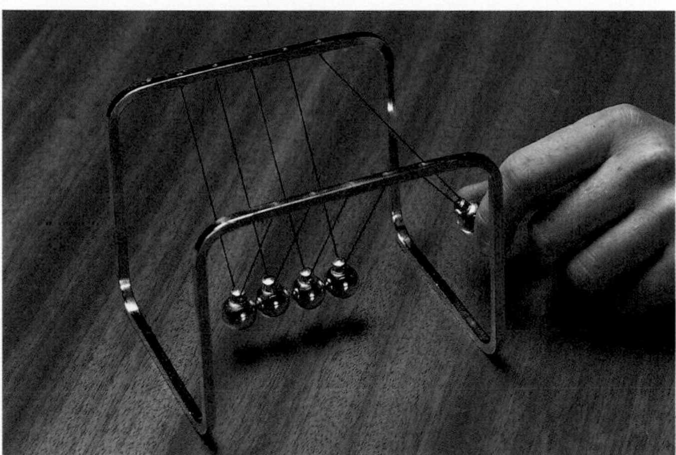

Figure 36.5 Newton's cradle is an instructive toy for studying collisions and conservation

Driving and car safety

(a) Braking distance and speed

In the equation derived earlier

$$Fs = \tfrac{1}{2}mv^2$$

if F is the **steady** braking force applied to a car of mass m moving with speed v, then s and t are the braking distance and time respectively needed to stop it. Since F and m are constant we can say

$$s \propto v^2$$

That is, the **braking distance is directly proportional to the square of the speed**, i.e. if v is doubled, s is quadrupled. The **thinking distance** (i.e. the distance travelled while the driver is reacting before applying the brakes) has to be added to the **braking distance** to obtain the **overall stopping distance** and can make a big difference (see question 12, p. 148). The state of the car, the road surface and the alertness of the driver are also factors.

(b) Car design and safety

When a car stops rapidly in a collision, large forces are produced on the car and its passengers, and their k.e. has to be dissipated.

Crumple zones at the front and rear collapse in such a way that the k.e. is absorbed gradually, Figure 36.6.

Seat belts exert a backwards force (of 10 000 N or so) over about 0.5 m, which is roughly the distance between the front seat occupants and the windscreen. In a car travelling at 15 m/s (34 m.p.h.), the effect felt by anyone *not* using a seat belt is the same as that produced by jumping off a building 12 m high!

Air bags in some cars inflate and protect the driver from injury by the steering wheel.

Head restraints ensure that if the car is hit from behind, the head goes forwards with the body and this prevents damage to the top of the spine.

All these are **secondary** safety devices which aid **survival** in the event of an accident. **Primary** safety factors help to **prevent** accidents and depend on the car's roadholding, brakes, steering, handling and above all on the driver since most accidents are due to driver error.

The chance of being killed in an accident is about *five* times less if seat belts are worn and head restraints are installed.

Figure 36.6 Car in an impact test showing the collapse of the front crumple zone

Worked example

A boulder of mass 4 kg rolls over a cliff and reaches the beach below with a velocity of 20 m/s.

a What is the k.e. of the boulder just before it lands?
b What is its p.e. on the cliff?
c How high is the cliff?

a Mass of boulder $= m = 4$ kg
Velocity of boulder as it lands $= v = 20$ m/s

$\therefore$ k.e. of boulder as it lands $= E_k = \frac{1}{2}mv^2$
$= \frac{1}{2} \times 4\text{ kg} \times (20)^2\text{ m}^2/\text{s}^2$
$= 800\text{ kg m/s}^2 \times \text{m} = 800\text{ N m}$
$= 800\text{ J}$

b Applying the principle of conservation of energy (and neglecting energy lost in overcoming air resistance),

p.e. of boulder on cliff = k.e. as it lands

$$\therefore \quad \Delta E_p = E_k = 800\ \text{J}$$

c If h is the height of the cliff,

$$\Delta E_p = mgh$$

$$\therefore \quad h = \frac{\Delta E_p}{mg} = \frac{800\ \text{J}}{4\ \text{kg} \times 10\ \text{m/s}^2} = \frac{800\ \text{N m}}{40\ \text{kg m/s}^2}$$

$$= \frac{800\ \text{kg m/s}^2 \times \text{m}}{40\ \text{kg m/s}^2} = 20\ \text{m}$$

QUESTIONS

1 Calculate the k.e. of

a a 1 kg trolley travelling at 2 m/s,

b a 2 g (0.002 kg) bullet travelling at 400 m/s,

c a 500 kg car travelling at 72 km/h.

2 **a** What is the velocity of an object of mass 1 kg which has 200 J of k.e.?

b Calculate the p.e. of a 5 kg mass when it is (i) 3 m, (ii) 6 m, above the ground. (g = 10 N/kg)

3 A 100 g steel ball falls from a height of 1.8 m onto a metal plate and rebounds to a height of 1.25 m. Find

a the p.e. of the ball before the fall (g = 10 m/s^2),

b its k.e. as it hits the plate,

c its velocity on hitting the plate,

d its k.e. as it leaves the plate on the rebound,

e its velocity of rebound.

4 A body of mass 5 kg falls from rest and has a k.e. of 1000 J just before touching the ground. Assuming there is no friction and using a value of 10 m/s^2 for the acceleration due to gravity, calculate

a (i) the loss in p.e. during the fall, (ii) the height from which the body has fallen.

b Name an important principle which applies in this situation. *(J.M.B./A.L./N.W.16+)*

5 At what height above the ground must a mass of 5 kg be to have a p.e. equal in value to the k.e. possessed by a mass of 5 kg moving with a velocity of 10 m/s? (Assume g = 10 m/s^2.)

A 1 m **B** 5 m **C** 10 m **D** 50 m **E** 100 m

(J.M.B.)

6 It is estimated that 7×10^6 kg of water pours over the Niagara Falls every second. If the Falls are 50 m high, and if all the energy of the falling water could be harnessed, what power would be available? (g = 10 N/kg)

Checklist

After studying this chapter you should be able to

- define **kinetic energy** (k.e.),
- perform calculations using $E_k = \frac{1}{2}mv^2$,
- define **potential energy** (p.e.),
- calculate changes in p.e. using $\Delta E_p = mgh$,
- apply the principle of conservation of energy to simple mechanical systems, e.g. a swinging pendulum,
- recall the meaning of **elastic** and **inelastic** collisions,
- recall that the braking distance of a vehicle is proportional to the square of the speed,
- describe secondary safety devices in cars.

37 CIRCULAR MOTION

Centripetal force	***Practical work***
Rounding a bend	Investigating circular motion.
Looping the loop	

There are many examples of bodies moving in circular paths – 'chair-o-planes' at a fun fair, clothes in a spin dryer, the planets going round the Sun and the Moon circling the Earth. 'Throwing the hammer' is a sport practised at Highland Games in Scotland, Figure 37.1, in which the hammer is whirled round and round before it is released. When a car turns a corner it may follow an arc of a circle.

Figure 37.1 'Throwing the hammer'

Centripetal force

In Figure 37.2 a ball attached to a string is being whirled round in a horizontal circle. Its direction of motion is constantly changing. At A it is along the tangent at A; shortly afterwards, at B, it is along the tangent at B; and so on.

Velocity has both size and direction; speed has only size. Velocity is speed in a stated direction and if the direction of a moving body changes, even if its speed does not, then its velocity has changed. A change of velocity is an acceleration and so during its whirling motion the ball is accelerating.

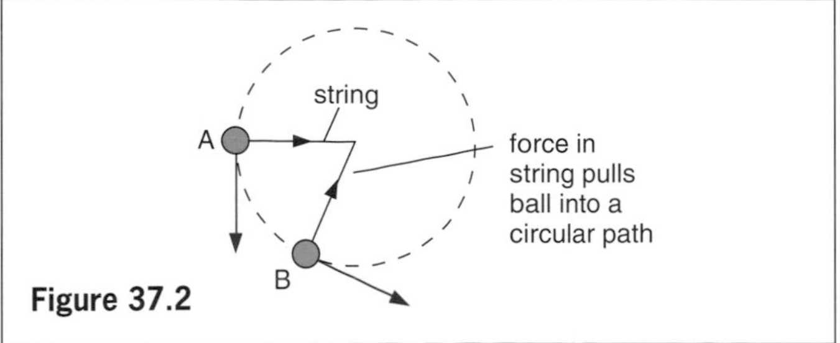

Figure 37.2

It follows from Newton's first law of motion that **if we consider a body moving in a circle to be accelerating** then there must be a force acting on it to cause the acceleration. In the case of the whirling ball it is reasonable to say the force is provided by the string pulling inwards on the ball. Like the acceleration, the force acts towards the centre of the circle and keeps the body at a fixed distance from the centre.

A larger force is needed if

(i) the speed of the ball is increased,
(ii) the radius of the circle is decreased,
(iii) the mass of the ball is increased.

The rate of change of direction, i.e. the acceleration, is increased by (i) and (ii) and, from $F = ma$, if m increases, F must. Should the force be greater than the string can bear, the string breaks and the ball flies off with steady speed in a straight line along the **tangent**, i.e. in the direction of travel when the string broke (as the first law of motion predicts). It is not thrown outwards.

The force which acts **towards the centre** and keeps a body in a circular path is called the **centripetal force** (centre-seeking force). Whenever a body moves in a circle (or circular arc) there must be a centripetal force acting on it. In throwing the hammer it is the pull of the athlete's arms acting on the hammer towards the centre of the whirling path. When a car rounds a bend a frictional force is exerted inwards by the road on the car's tyres.

PRACTICAL WORK

Investigating circular motion

Use the apparatus in Figure 37.3 (overleaf) to investigate the various factors which affect circular motion. Make sure the rubber bung is **tied securely** to the string and that **the area around you is clear of other students**. The paper clip acts as an indicator to aid keeping the radius of the circular motion constant.

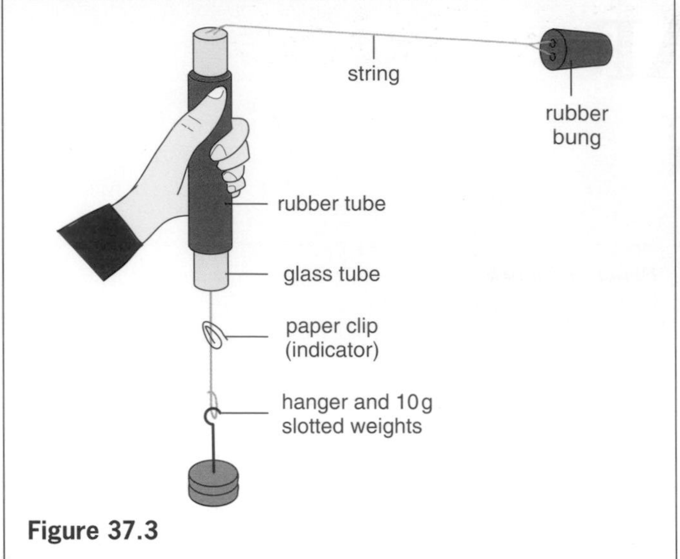

Figure 37.3

Rounding a bend

When a car rounds a bend a frictional force is exerted **inwards** by the road on the car's tyres so providing the centripetal force needed to keep it in its curved path, Figure 37.4a. The successful negotiation of a bend on a flat road therefore depends on the tyres and the road surface being in a condition that enables them to provide a sufficiently large frictional force – otherwise skidding occurs.

Racing cars are fitted with tyres called 'slicks' (which have no tread pattern) for dry tracks and 'treads' for wet tracks. Why?

Safe cornering that does not rely on friction is achieved by 'banking' the road as in Figure 37.4b. The centripetal force is then supplied by the part of the contact force *N* from the road surface on the car which acts horizontally. A bend in a railway track is also banked so that the outer rail is not strained by having to supply the centripetal force, by pushing inwards on the wheel flanges.

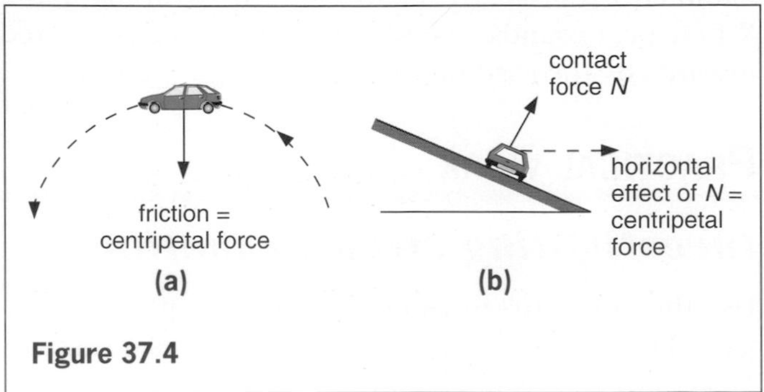

Figure 37.4

Looping the loop

A pilot who is not strapped into his aircraft can loop the loop without falling downwards at the top of the loop. A bucket of water can be swung round in a vertical circle without spilling. Some fairground attractions, Figure 37.5, give similar effects. Can you suggest what provides the centripetal force for each of these three cases (i) at the top of the loop and (ii) at the bottom of the loop?

Figure 37.5
Looping the loop at a fairground

QUESTIONS

1 An apple is whirled round in a horizontal circle on the end of a string which is tied to the stalk. It is whirled faster and faster and at a certain speed the apple is torn from the stalk. Why?

2 A car rounding a bend travels in an arc of a circle.

a What provides the centripetal force?

b Is a larger or a smaller centripetal force required if (i) the car travels faster, (ii) the bend is less curved, (iii) the car has more passengers?

Checklist

After studying this chapter you should be able to

- explain circular motion in terms of an unbalanced centripetal force,
- describe an experiment to investigate the factors affecting circular motion,
- explain how the centripetal force arises for a car rounding a bend.

ADDITIONAL QUESTIONS

The first questions under each topic heading are 'basic' questions intended for all students; the following questions marked with a stripe are 'higher' questions for those seeking grades E to A*.

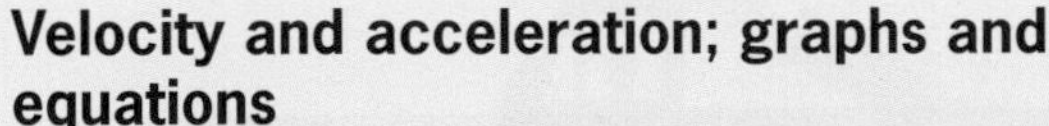

Velocity and acceleration; graphs and equations

1 a A body travelling with uniform velocity covers a distance of 840 m in 1 minute. What is its velocity in m/s?

b If the body begins to accelerate uniformly at 0.2 m/s^2, what will be its velocity after a *further* minute?

c How far will the body have travelled from the instant at which it began to accelerate?

d If the body is then retarded uniformly and comes to rest in a *further* distance of 338 m, what is the value of the retardation? *(J.M.B./A.L./N.W.16+)*

2 The approximate velocity–time graph for a car on a 5-hour journey is shown below. (There is a very quick driver change midway to prevent driving fatigue!)

a State in which of the regions OA, AB, BC, CD, DE the car is (i) accelerating, (ii) decelerating, (iii) travelling with uniform velocity.

b Calculate the value of the acceleration, deceleration or constant velocity in each region.

c What is the distance travelled over each region?

d What is the total distance travelled?

e Calculate the average velocity for the whole journey.

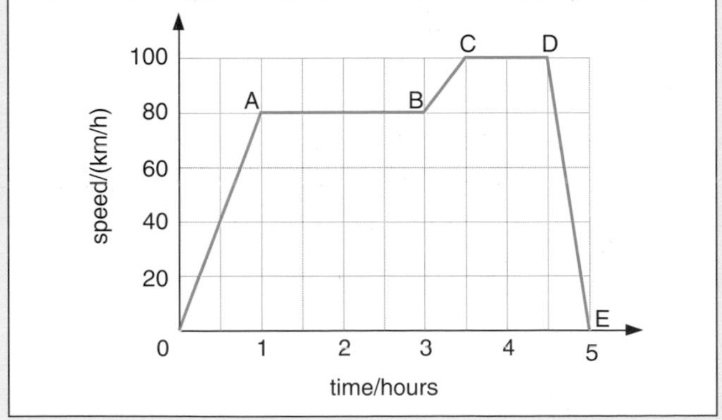

3 The distance–time graph for a motor cyclist riding off from rest follows.

a Describe the motion.

b How far does the motorbike move in 30 seconds?

c Calculate the speed.

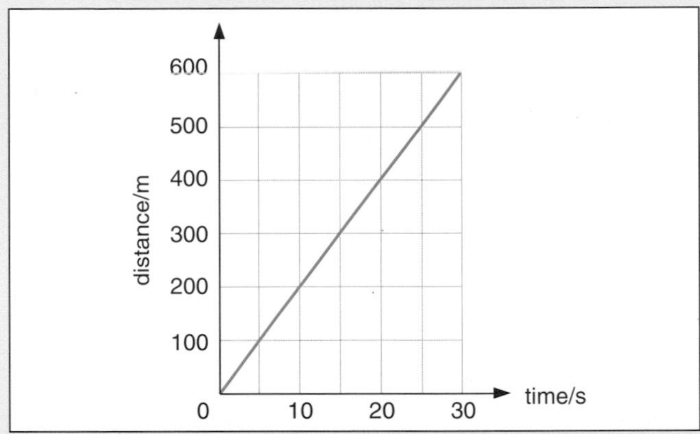

4 The distance from the starting point of a steel ball which starts from rest and rolls with uniform acceleration down an inclined plane is noted at one second intervals. The data are shown in the table.

Time in s	0	1	2	3	4
Distance in cm	0	2	8	18	32

The speed at the end of the fourth second in cm/s is

A 8 **B** 14 **C** 16 **D** 32 **E** 64 *(N.I.)*

5 a Define the terms 'velocity' and 'acceleration'. Choose *one* of these terms and explain what is meant when the quantity is said to be 'uniform'.

b A car runs at a constant speed of 15 m/s for 300 s and then accelerates uniformly to a speed of 25 m/s over a period of 20 s. This speed is maintained for 300 s before the car is brought to rest with uniform deceleration in 30 s.

Draw a velocity–time graph to represent the journey described above.

From the graph find
(i) the acceleration while the velocity changes from 15 m/s to 25 m/s,
(ii) the total distance travelled in the time described,
(iii) the average speed over the time described. *(J.M.B.)*

6 A small pebble is thrown **horizontally** at 20 m/s from the top of a cliff 80 m high. Ignoring air resistance and taking $g = 10\ m/s^2$, calculate

a how long the pebble takes to reach the ground,

b the distance from the foot of the cliff to where the pebble strikes the ground,

c the vertical velocity of the pebble as it hits the ground, and

d the horizontal velocity of the pebble as it hits the ground.

Newton's laws of motion

7 If the forces acting on a moving body cancel each other out (i.e. are in equilibrium) the body will

A move in a straight line at a steady speed
B slow down to a steady slower speed
C speed up to a steady faster speed
D be brought to a state of rest. *(W.M.)*

8 What is meant by **acceleration**? What evidence have you for believing that a moving body will proceed at a constant speed in a straight line indefinitely unless it is acted on by a resultant force?

A cyclist of mass 50 kg riding a cycle of mass 10 kg raises his speed from 2 m/s to 5 m/s by accelerating uniformly for 6 s while travelling due north along a horizontal road. What horizontal force is exerted on the machine by the road, and in which direction? Does the cyclist exert any force on the machine and, if so, in what direction?

How far does he travel in the 6 s? *(O. and C.)*

Momentum; kinetic and potential energy

9 The diagram shows a truck of mass 3 kg moving at 10 m/s along a horizontal track about to collide with another stationary truck of mass 2 kg. After the collision, the trucks link and move together. The speed of the trucks immediately after the collision is

A 1 m/s **B** 2 m/s **C** 6 m/s **D** 6.67 m/s
E 10 m/s *(O.L.E./S.R.16+)*

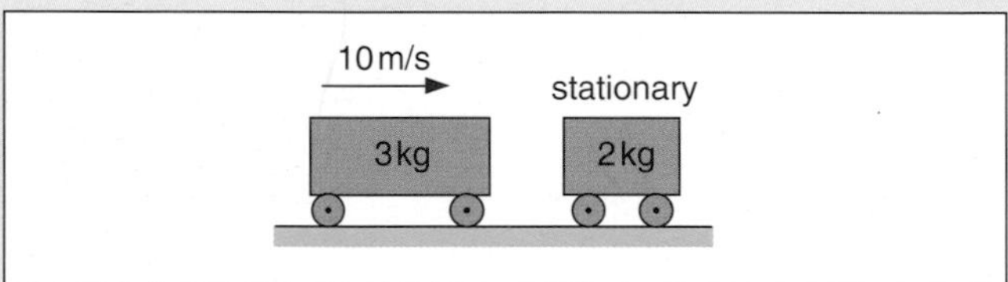

10 The speed of the rocket shown, of mass 20 000 kg, has to be increased by 5.0 m/s. If it burns 10 kg of fuel and oxygen in doing this, what is the speed of the exhaust gases produced?

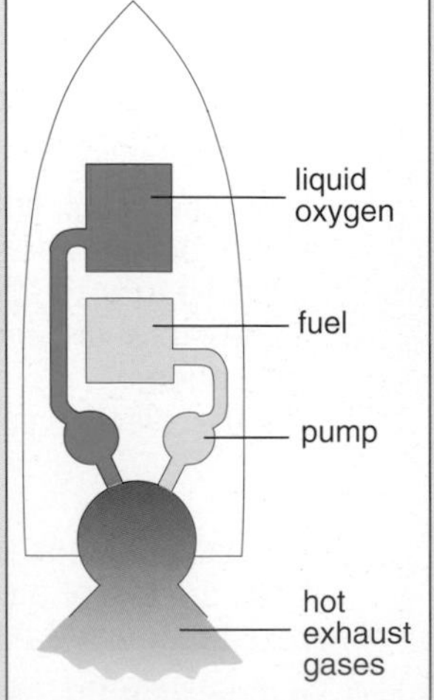

11 A car of mass 1000 kg has a head-on collision in a 50 km/h (30 m.p.h.) zone with a truck of mass 2500 kg in which both vehicles stop dead. If the truck's speed, as detected by police radar, was 30 km/h, what was the car's speed and was the law being broken? (Neglect the momentum of the truck's load and of the passengers after the collision.)

12 The table gives various stopping distances in **feet** for a car travelling at different speeds, assuming a good road surface, a good car and an alert driver. Study the table and using the fact that the **braking distance is directly proportional to the square of the speed**, suggest values for A to J.

m.p.h.	20	30	40	60
Thinking distance	20	30	40	60
Braking distance (dry)	20	45	A	B
Braking distance (wet)	40	90	C	D
Overall stopping distance (dry)	40	E	F	G
Overall stopping distance (wet)	60	H	I	J

13 A body of mass 20 kg, moving with uniform acceleration, has an initial momentum of 200 kg m/s and after 10 s the momentum is 300 kg m/s. What is the acceleration of the body?

A 0.5 m/s^2 **B** 5 m/s^2 **C** 25 m/s^2
D 50 m/s^2 **E** 100 m/s^2 *(J.M.B.)*

14 A small steel ball of mass 80 g is released from rest at a height of 1.25 m above a rigid horizontal metal plate. After the rebound the ball rises vertically to a height of 1.00 m above the plate. Calculate

a the velocity of the ball just before impact,

b the momentum of the ball just before impact,

c the kinetic energy of the ball just before impact,

d the loss of energy on impact. Give reasons for this loss of energy. (g = 10 m/s^2) *(S.)*

HEAT AND ENERGY

38 THERMOMETERS

Liquid-in-glass thermometer	***Other thermometers***
Scale of temperature	Resistance thermometer. Constant-volume gas thermometer. Thermochromic liquids.
Clinical thermometer	
Thermocouple thermometer	***Heat and temperature***

The **temperature** of a body tells us how hot it is and is measured by a thermometer, usually in **degrees Celsius** (°C). The kinetic theory (Chapter 20) regards temperature as a measure of the average k.e. of the molecules of the body. The greater this is, the faster the molecules move and the higher the temperature of the object.

There are different kinds of thermometer, each type being more suitable than another for a certain job. The one in Figure 38.1 is being used to find the temperature of a furnace.

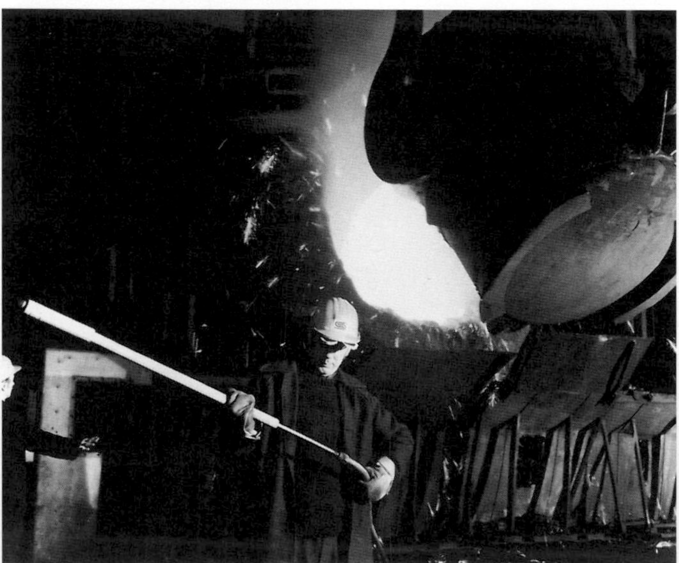

Figure 38.1 Use of a thermocouple thermometer to measure the temperature of a steel furnace

Liquid-in-glass thermometer

In this type the liquid in a glass bulb expands up a capillary tube when the bulb is heated. The liquid must be easily seen, expand (or contract) rapidly and by a large amount over a wide range of temperature. It must not stick to the inside of the tube or the reading will be high when the temperature is falling.

Mercury and coloured alcohol are in common use. Mercury freezes at −39 °C and boils at 357 °C; alcohol freezes at −115 °C and boils at 78 °C and is therefore more suitable for low temperatures.

Scale of temperature

A scale and unit of temperature are obtained by choosing two temperatures, called the **fixed points**, and dividing the range between them into a number of equal divisions or **degrees**.

On the Celsius scale (named after the Swedish scientist who suggested it), **the lower fixed point is the temperature of pure melting ice** and is taken as 0 °C. **The upper fixed point is the temperature of the steam above water boiling at normal atmospheric pressure** (760 mmHg) and is taken as 100 °C.

When the fixed points have been marked on the thermometer, the distance between them is divided into 100 equal degrees, Figure 38.2. The thermometer now has a scale, i.e. it has been calibrated or graduated.

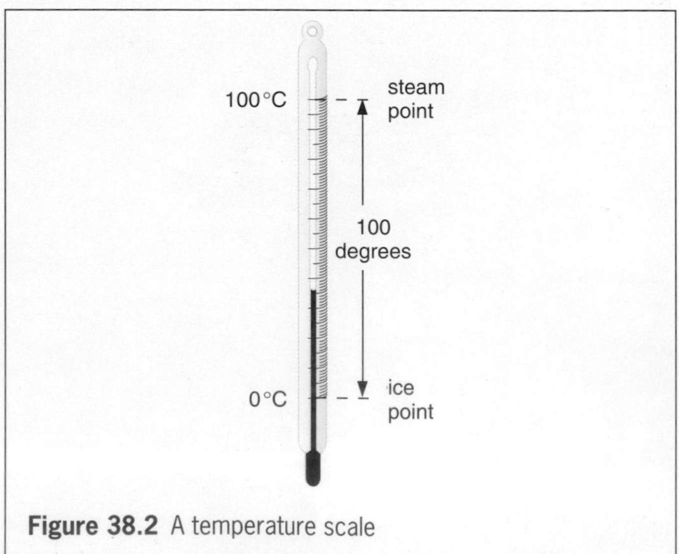

Figure 38.2 A temperature scale

Clinical thermometer

This is a special type of mercury-in-glass thermometer used by doctors and nurses. Its scale only extends over a few degrees on either side of the normal body temperature of 37 °C, Figure 38.3.

The tube has a constriction (i.e. a narrower part) just beyond the bulb. When the thermometer is placed under the tongue the mercury expands, forcing its way past the constriction. When the thermometer is removed (after 1 minute) from the mouth, the mercury in the bulb cools and contracts, breaking the mercury thread at the constriction. The mercury beyond the constriction stays in the tube and shows the body temperature. After use the mercury is returned to the bulb – by a flick of the wrist.

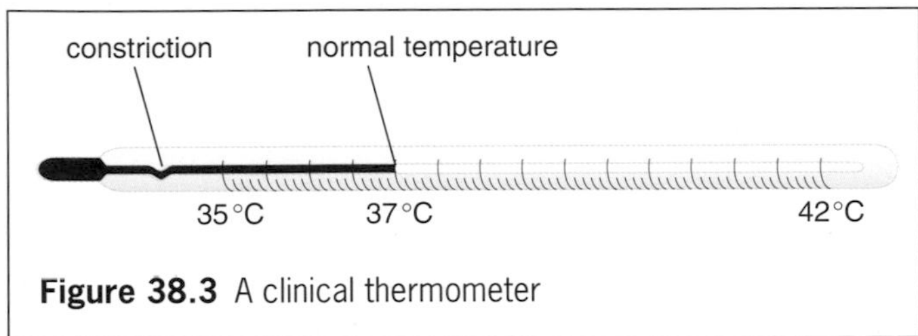

Figure 38.3 A clinical thermometer

Thermocouple thermometer

A thermocouple consists of two wires of different materials, e.g. copper and iron, joined together, Figure 38.4. When one junction is at a higher temperature than the other an electric current flows and produces a deflection on a galvanometer (a sensitive ammeter) which depends on the temperature difference.

Thermocouples are used in industry to measure a wide range of temperatures from −250 °C up to about 1500 °C (Figure 38.1), especially changing ones and those of small objects.

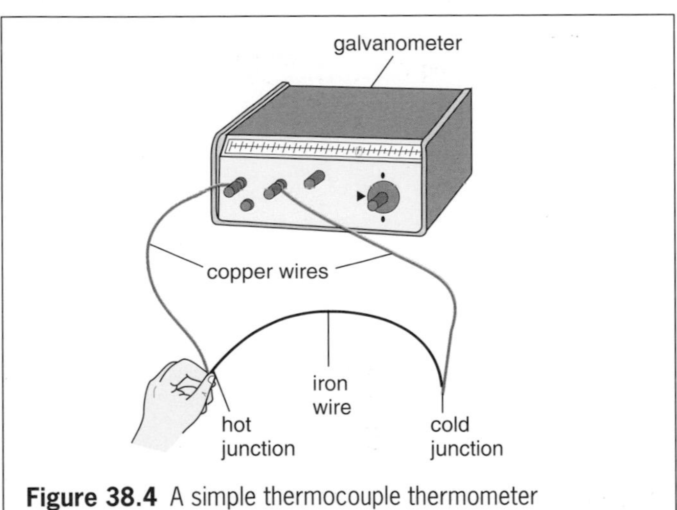

Figure 38.4 A simple thermocouple thermometer

Other thermometers

One type of **resistance thermometer** uses the fact that the electrical resistance (Chapter 51) of a platinum wire increases with temperature according to the equation

$$\frac{\theta}{100} = \frac{R_\theta - R_0}{R_{100} - R_0}$$

θ is the required temperature, R_{100}, R_0 and R_θ are the resistances of the wire at 100 °C, 0 °C and θ. It can measure temperatures accurately in the range −200 °C to 1200 °C but it is bulky and best for steady temperatures. A **thermistor** (Chapter 65) can also be used but over a small range, e.g. −5 °C to 70 °C; its resistance decreases with temperature.

The **constant-volume gas thermometer** uses the change in pressure of a gas to measure temperatures over a wide range. It is an accurate but bulky instrument, basically similar to the apparatus of Figure 40.2.

Thermochromic liquids which change colour with temperature have a limited range around room temperatures.

Heat and temperature

It is important not to confuse the temperature of a body with the heat energy that can be obtained from it. For example, a red-hot spark from a fire is at a higher temperature than the boiling water in a saucepan. In the boiling water the average k.e. of the molecules is lower than in the spark; but since there are many more water molecules, their total energy is greater, and therefore more heat energy can be supplied by the water than the spark.

Heat passes from a body at a higher temperature to one at a lower temperature. This is due to the average k.e. (and speed) of the molecules in the 'hot' body falling as a result of having collisions with molecules of the 'cold' body whose average k.e., and therefore temperature, increases. When the average k.e. of the molecules is the same in both bodies, they are at the same temperature. For example, if the red-hot spark landed in the boiling water, heat would pass from it to the water even though much more heat energy could be obtained from the water.

Heat is also called **thermal** or **internal** energy; it is the energy a body has because of the kinetic energy *and* the potential energy of its molecules. Increasing the temperature of a body increases its heat energy due to the k.e. of its molecules increasing. But as we will see later (Chapter 42), the heat energy of a body can also be increased by increasing the p.e. of its molecules.

QUESTIONS

1 1530 °C 120 °C 55 °C 37 °C 19 °C 0 °C −12 °C −50 °C
From the above list of temperatures choose the most likely value for *each* of the following:

a the melting point of iron,

b the temperature of a room that is comfortably warm,

c the melting point of pure ice at normal pressure,

d the lowest outdoor temperature recorded in London in winter,

e the normal body temperature of a healthy person.

2 In order to make a mercury thermometer which will measure small changes in temperature accurately

A decrease the volume of the mercury bulb
B put the degree markings farther apart
C decrease the diameter of the capillary tube
D put the degree markings closer together
E leave the capillary tube open to the air.

3 a How must a property behave to measure temperature?

b Name three properties that qualify.

c Name a suitable thermometer for measuring
(i) a steady temperature of 1000 °C,
(ii) the changing temperature of a small object, and
(iii) a winter temperature at the North Pole.

4 Why does a clinical thermometer

a have a constriction just above the bulb,

b cover only a narrow range of temperature,

c have a very fine bore? (*E.A.*)

Checklist

After studying this chapter you should be able to

- define the fixed points on the Celsius scale,
- recall the properties of mercury and alcohol as thermometric liquids,
- describe clinical and thermocouple thermometers,
- recall some other types of thermometer and the thermometric properties on which they depend,
- distinguish between heat and temperature and recall that temperature decides the direction of heat flow.

39 EXPANSION OF SOLIDS AND LIQUIDS

In general, when matter is heated it expands and when cooled it contracts. If the changes are resisted large forces are created which are sometimes useful but at other times are a nuisance.

According to the kinetic theory (Chapter 20) the molecules of solids and liquids are in constant vibration. When heated they vibrate faster and force each other a little farther apart. Expansion results, and this is greater for liquids. The linear (length) expansion of solids is small and for the effect to be noticed the solid must be long and/or the temperature change large.

Uses of expansion

(a) Shrink-fitting

In Figure 39.1 the axles have been shrunk by cooling in liquid nitrogen at $-196\,°C$ until the gear wheels can be slipped on to them. On regaining normal temperature the axles expand to give a very tight fit.

Figure 39.1 Shrink-fitting of axles into gear wheels

(b) Riveting metal plates

A white-hot rivet is placed in the rivet hole and its end hammered flat. On cooling it contracts and pulls the plates together, Figure 39.2. Steel plates are riveted in shipbuilding.

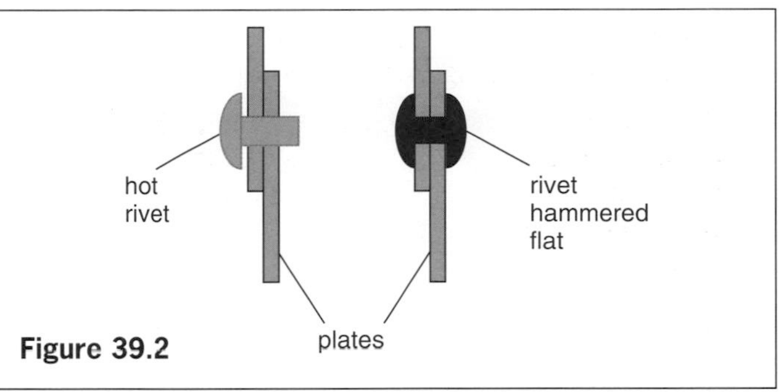

Figure 39.2

Precautions against expansion

(a) Railway lines

Previously gaps were left between the lengths of rail to allow for expansion in summer. They caused a 'clickety-click' sound as the train passed over them.

Figure 39.3 Tapered overlap of rails

Today rails are welded into lengths of about 1 km and are held by concrete 'sleepers' that can withstand the large forces created without buckling. Also, at the joints the ends are tapered and overlap, Figure 39.3. This gives a smoother journey and allows some expansion near the ends of each length of rail.

(b) Bridges

One end is fixed and the other rests on rollers which permit movement.

Bimetallic strip

If equal lengths of two different metals, e.g. copper and iron, are riveted together so that they cannot move separately, they form a bimetallic strip, Figure 39.4a. When heated, copper expands more than iron and to allow this, the strip bends with copper on the outside, Figure 39.4b. If they had expanded equally the strip would have stayed straight.

Bimetallic strips have many uses.

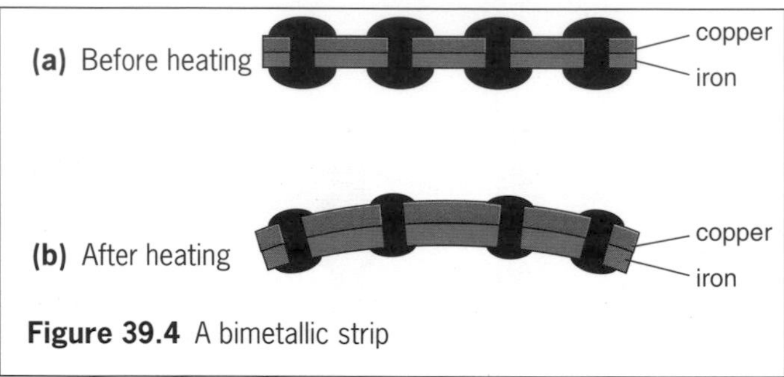

Figure 39.4 A bimetallic strip

(a) Fire alarm

Heat from the fire makes the bimetallic strip bend and complete the electrical circuit, so ringing the alarm bell, Figure 39.5a.

A bimetallic strip is also used in this way to work the flashing direction indicator lamps in a car, being warmed by an electric heating coil wound round it.

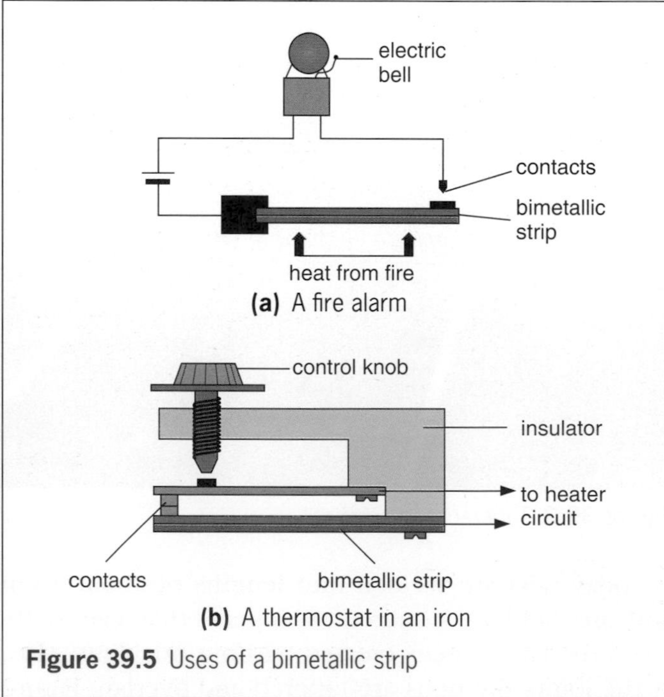

Figure 39.5 Uses of a bimetallic strip

(b) Thermostat

A thermostat keeps the temperature of a room or an appliance constant. The one in Figure 39.5b uses a bimetallic strip in the electrical heating circuit of, for example, an electric iron.

When the iron reaches the required temperature the strip bends down, breaks the circuit at the contacts and switches off the heater. After cooling a little the strip remakes contact and turns the heater on again. A near-steady temperature results.

If the control knob is screwed down, the strip has to bend more to break the heating circuit and this needs a higher temperature.

Linear expansivity

An engineer has to allow for the linear expansion of a bridge when designing it. The expansion can be calculated if (i) the length of the bridge, (ii) the range of temperature it will experience and (iii) the **linear expansivity** of the material to be used are known.

> The linear expansivity of a substance is the increase in length of 1 m for a 1 °C rise in temperature.

The linear expansivity of a material is found by experiment. For steel it is 0.000 012 per °C. This means that 1 m will become 1.000 012 m for a temperature rise of 1 °C. A steel bridge 100 m long will expand by 0.000 012×100 m for each 1 °C rise in temperature. If the maximum temperature **change** expected is 60 °C (e.g. from −15 °C to +45 °C), the expansion will be 0.000 012 °C × 100 m × 60 °C = 0.072 m = 7.2 cm. In general,

> expansion = linear expansivity × original length × temperature rise

Unusual expansion of water

As water is cooled to 4 °C it contracts, as we would expect. However between 4 °C and 0 °C it expands, surprisingly. **Water therefore has a maximum density at 4 °C.**

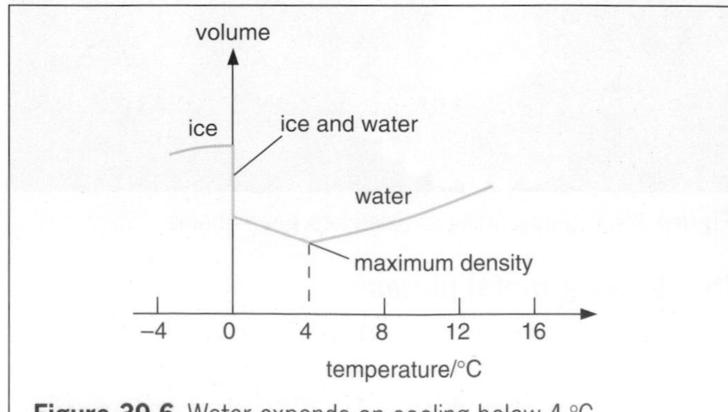

Figure 39.6 Water expands on cooling below 4 °C

At 0 °C, when it freezes, a considerable expansion occurs and every 100 cm^3 of water becomes 109 cm^3 of ice. This accounts for the bursting of water pipes in very cold weather, and for the fact that ice is less dense than water and so floats.

These changes are represented in Figure 39.6.

The unusual expansion of water between 4 °C and 0 °C explains why fish survive in a frozen pond. The water at the top of the pond cools first, contracts and being denser sinks to the bottom. Warmer less dense water rises to the surface to be cooled. When all the water is at 4 °C the circulation stops. If the temperature of the surface water falls below 4 °C, it becomes less dense and *remains at the top*, eventually forming a layer of ice at 0 °C. Temperatures in the pond are then as in Figure 39.7.

The expansion of water between 4 °C and 0 °C is due to the breaking up of the groups which water molecules form above 4 °C. The new arrangement requires a larger volume and more than cancels out the contraction due to the fall in temperature.

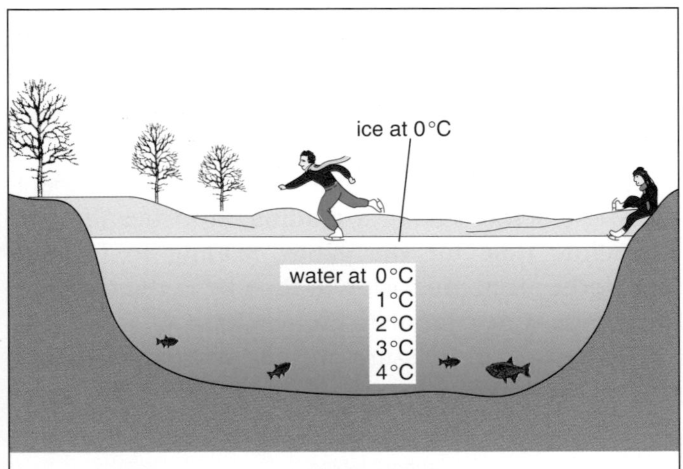

Figure 39.7 Fish can survive in a frozen pond

QUESTIONS

1 Explain why

a the metal lid on a glass jam jar can be unscrewed easily if the jar is inverted for a few seconds with the *lid* in very hot water,

b furniture may creak at night after a warm day,

c concrete roads are laid in sections with pitch between them.

2 A bimetallic strip is made from aluminium and copper. When heated it bends in the direction shown in Figure 39.8.

Which metal expands more for the same rise in temperature?

Draw a diagram to show how the bimetallic strip would appear if it were cooled to below room temperature.

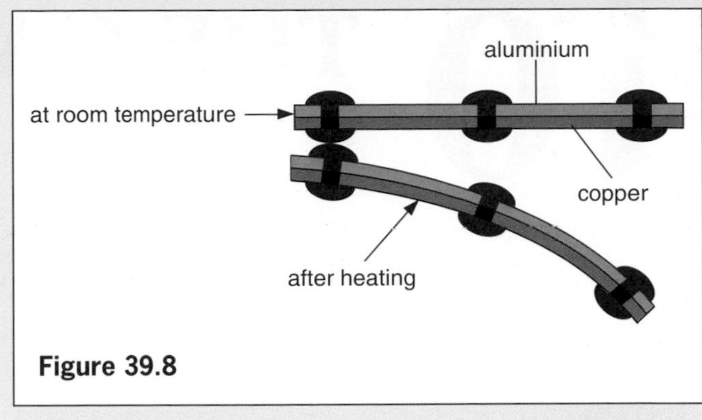

Figure 39.8

3 The linear expansivities of some common substances are given below.

aluminium	0.000 03 per °C	glass	0.000 009 per °C
concrete	0.000 01 per °C	platinum	0.000 009 per °C
copper	0.000 02 per °C	steel	0.000 01 per °C

a Rods of aluminium, copper and steel are the same length at a temperature of 0 °C. Which will be the longest at 300 °C?

b If the aluminium rod is 1.000 m long at 0 °C, what will be its length at 300 °C?

c Why is steel a suitable material for reinforcing concrete?

d Which of the substances listed above would be most suitable for carrying a current of electricity through the walls of a glass vessel?

e How might a steel cylinder liner be fitted tightly inside a cylinder block made of aluminium?

4 Using the values for linear expansivities in question 3 calculate the expansion of

a 100 m of copper pipe heated through 50 °C,

b 50 cm of steel pipe heated through 80 °C,

c 200 m of aluminium pipe heated from 10 °C to 60 °C.

5 When a particular substance at a certain temperature is heated, it expands. When the same substance at the same temperature is cooled, it also expands.

a What is the substance?

b What is the temperature? *(E.M.)*

Checklist

After studying this chapter you should be able to

- describe uses of expansion, including the bimetallic strip,
- describe precautions taken against expansion,
- define linear expansivity,
- recall that water has its maximum density at 4 °C and explain why a pond freezes at the top first.

40 THE GAS LAWS

When a gas is heated, as air is in a jet engine, its **pressure** as well as its **volume** may change. To study the effect of temperature on these two quantities we must keep one fixed while the other is changed.

PRACTICAL WORK

Effect on volume of temperature (pressure constant) – Charles' law

Arrange the apparatus as in Figure 40.1. The index of concentrated sulphuric acid traps the air column to be investigated and also dries it. Adjust the capillary tube so that the bottom of the air column is opposite a convenient mark on the ruler.

Note the length of the air column (to the *lower* end of the index) at different temperatures but before taking a reading, stop heating and stir well to make sure that the air has reached the temperature of the water. Put the results in a table.

Plot a graph of volume (in cm, since the length of the air column is a measure of it) on the *y*-axis and temperature (in °C) on the *x*-axis.

The pressure of (and on) the air column is constant and equals atmospheric pressure plus the pressure of the acid index.

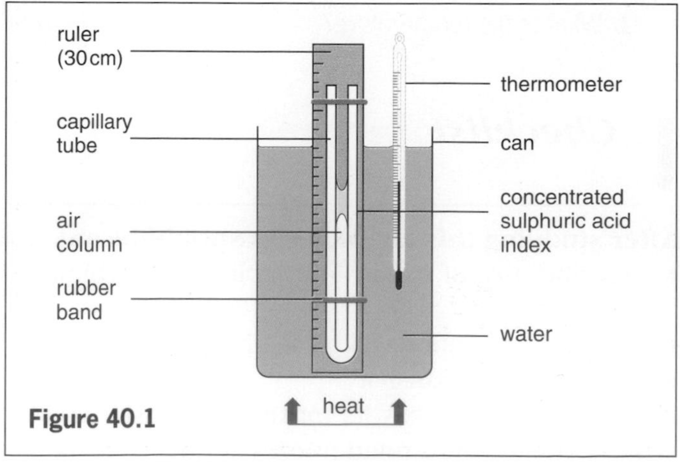

Figure 40.1

PRACTICAL WORK

Effect on pressure of temperature (volume constant) – the Pressure law

The apparatus is shown in Figure 40.2. The rubber tubing from the flask to the pressure gauge should be as short as possible. The flask must be in water almost to the top of its neck and be securely clamped to keep it off the bottom of the can.

Record the pressure over a wide range of temperatures but before taking a reading, stop heating, stir and allow time for the gauge reading to become steady; the air in the flask will then be at the temperature of the water. Tabulate the results.

Plot a graph of pressure on the *y*-axis and temperature on the *x*-axis.

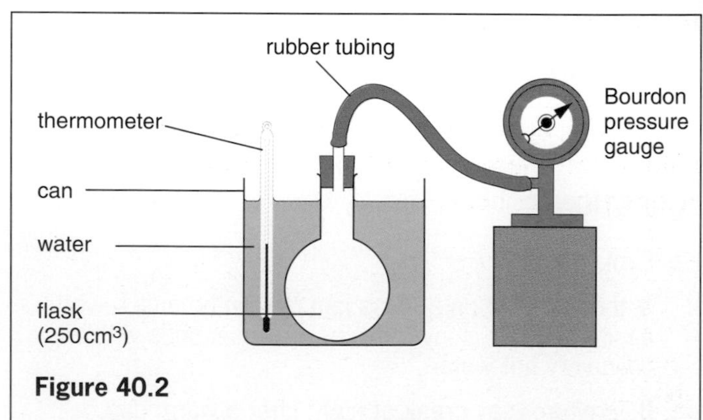

Figure 40.2

Absolute zero

The volume–temperature and pressure–temperature graphs for a gas are straight lines, Figure 40.3. They show that gases expand **uniformly** with temperature as measured on a mercury thermometer, i.e. equal temperature increases cause equal volume or pressure increases.

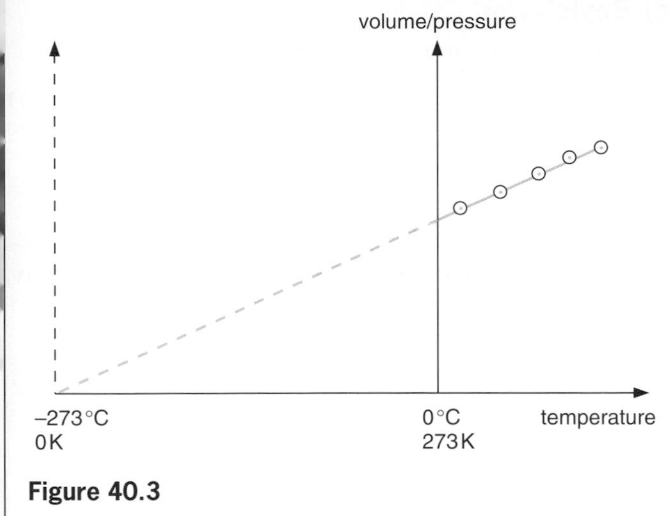

Figure 40.3

The graphs do not pass through the Celsius temperature origin (0 °C). If they are produced backwards they cut the temperature axis at about −273 °C. This temperature is called **absolute zero** because we believe it is the lowest temperature possible. It is the zero of the **absolute** or **Kelvin scale of temperature**.

Degrees on this scale are called **kelvins** and are denoted by K. They are exactly the same size as Celsius degrees. Since −273 °C = 0 K, conversions from °C to K are made by adding 273. For example

$$0\ ^\circ\text{C} = 273\ \text{K}$$
$$15\ ^\circ\text{C} = 273 + 15 = 288\ \text{K}$$
$$100\ ^\circ\text{C} = 273 + 100 = 373\ \text{K}$$

Kelvin or absolute temperatures are represented by the letter T and if θ stands for a Celsius scale temperature then, in general

$$T = 273 + \theta$$

Near absolute zero strange things occur. Liquid helium becomes a 'superfluid'. It cannot be kept in an open vessel because it flows up the inside of the vessel, over the edge and down the outside. Some metals and compounds become 'superconductors' of electricity and a current once started in them flows for ever without a battery. Figure 40.4 shows research equipment that is being used to find materials which are superconductors at very much higher temperatures, e.g. −23 °C.

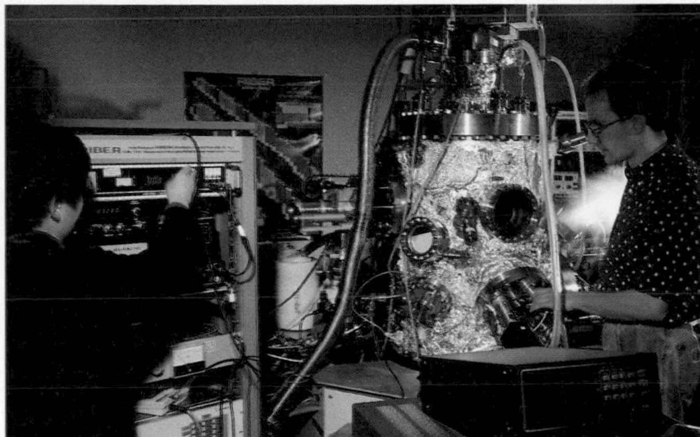
Figure 40.4 Equipment that is being used to make complex composite materials which are superconducting at temperatures far above absolute zero

PRACTICAL WORK

Effect on volume of pressure (temperature constant) – Boyle's law

Changes in the volume of a gas due to pressure changes can be studied using the apparatus in Figure 40.5. The volume V of air trapped in the glass tube is read off on the scale behind. The pressure is altered by pumping air from a foot pump into the space above the oil reservoir. This forces more oil into the glass tube and increases the pressure p on the air in it; p is measured by the Bourdon gauge.

If a graph of pressure against volume is plotted using the results, a curve like that in Figure 40.6a (overleaf) is obtained. Close examination of it shows that if p is doubled, V is halved. That is, **p is inversely proportional to V**. In symbols

$$p \propto \frac{1}{V} \quad \text{or} \quad p = \text{constant} \times \frac{1}{V}$$

$$\therefore \qquad pV = \text{constant}$$

If several pairs of readings p_1V_1, p_2V_2, etc. are taken, then it can be confirmed that $p_1V_1 = p_2V_2 = \text{constant}$. This is **Boyle's law** which is stated as follows.

> The pressure of a fixed mass of gas is inversely proportional to its volume if its temperature is kept constant.

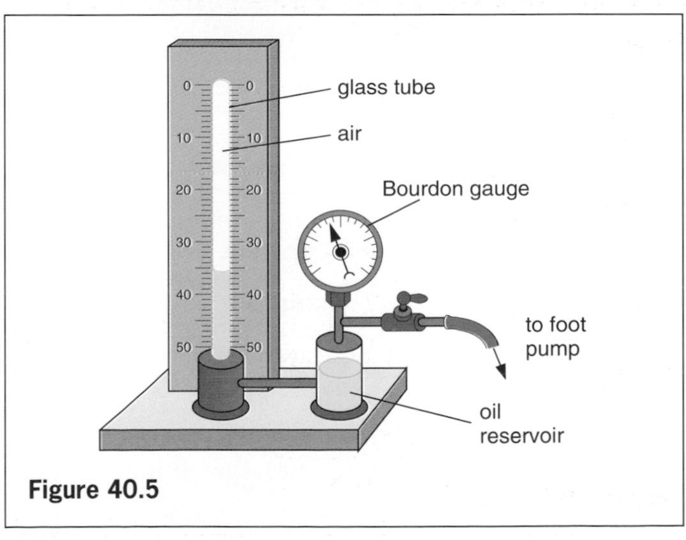

Figure 40.5

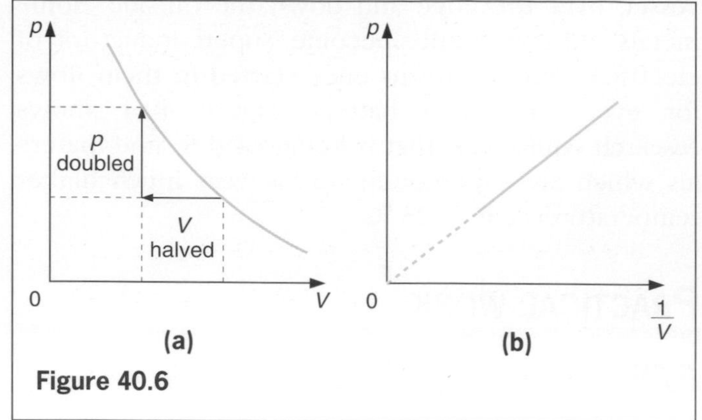

Figure 40.6

Since p is inversely proportional to V, then p is directly proportional to $1/V$. A graph of p against $1/V$ is therefore a straight line through the origin, Figure 40.6b.

The gas laws

Using absolute temperatures the gas laws can be stated in a convenient form for calculations.

(a) Charles' law

In Figure 40.3 the volume–temperature graph does pass through the origin if temperatures are measured on the Kelvin scale. That is, if we take 0 K as the origin. We can then say that the volume V is directly proportional to the absolute temperature T, i.e. doubling T doubles V, etc. Therefore

$$V \propto T \quad \text{or} \quad V = \text{constant} \times T$$

or

$$V/T = \text{constant} \qquad (1)$$

Charles' law may be stated as follows.

> The volume of a fixed mass of gas is directly proportional to its absolute temperature if the pressure is kept constant.

(b) Pressure law

From Figure 40.3 we can say similarly for the pressure p that

$$p \propto T \quad \text{or} \quad p = \text{constant} \times T$$

or

$$p/T = \text{constant} \qquad (2)$$

The Pressure law may be stated as follows.

> The pressure of a fixed mass of gas is directly proportional to its absolute temperature if the volume is kept constant.

(c) Boyle's law

For a fixed mass of gas at constant temperature

$$pV = \text{constant} \qquad (3)$$

These three equations can be combined giving

$$\frac{pV}{T} = \text{constant}$$

For cases in which p, V and T all change from, say, p_1, V_1 and T_1 to p_2, V_2 and T_2, then

$$\frac{p_1V_1}{T_1} = \frac{p_2V_2}{T_2} \qquad (4)$$

Worked example

A cycle pump contains 50 cm^3 of air at 17 °C and a pressure of 1.0 atmosphere. Find the pressure when the air is compressed to 10 cm^3 and its temperature rises to 27 °C.

We have

$p_1 = 1.0$ atmosphere	$p_2 = ?$
$V_1 = 50$ cm^3	$V_2 = 10$ cm^3
$T_1 = 273 + 17 = 290$ K	$T_2 = 273 + 27 = 300$ K

From equation (4) above we get

$$p_2 = p_1 \times \frac{V_1}{V_2} \times \frac{T_2}{T_1}$$

Replacing

$$p_2 = 1 \times \frac{50}{10} \times \frac{300}{290} = 5.2 \text{ atmosphere}$$

Notes:

1. All temperatures must be in K.
2. Any units can be used for p and V so long as they are the same on both sides of the equation.
3. In some calculations the volume of the gas has to be found at s.t.p. (standard temperature and pressure). This is 0 °C and 760 mmHg pressure.

Gases and the kinetic theory

The kinetic theory can explain the behaviour of gases.

(a) Cause of gas pressure

All the molecules in a gas are in rapid motion, with a wide range of speeds, and repeatedly hit the walls of the container in huge numbers per second. The average force and hence the pressure they exert on the walls is constant since pressure is force on unit area.

(b) Boyle's law

If the volume of a fixed mass of gas is halved by halving the volume of the container, Figure 40.7, the number of molecules per cm^3 will be doubled. There will be twice as many collisions per second with the walls, i.e. the pressure is doubled. This is Boyle's law.

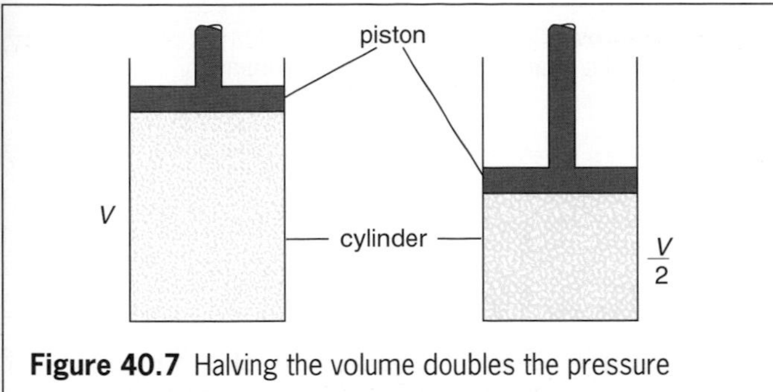

Figure 40.7 Halving the volume doubles the pressure

(c) Temperature

When a gas is heated and its temperature rises, the average speed of its molecules increases. If the volume of the gas is to remain constant, its pressure increases due to more frequent and more violent collisions of the molecules with the walls. If the pressure of the gas is to remain constant, the volume must increase so that the number of collisions does not.

QUESTIONS

1 A gas of volume 2 m^3 at 27 °C is **a** heated to 327 °C, **b** cooled to −123 °C, at constant pressure. What is its new volume in each case?

2 A container holds gas at 0 °C. To what temperature must it be heated for its pressure to double? Assume the container does not expand. *(E.A.)*

3 Figure 40.8a shows air enclosed in a cylinder by an airtight piston. In Figure 40.8b the piston has been pushed in so that the air occupies one-third of the length of the cylinder that it previously occupied. What, if anything, has happened to

a the number of molecules of air,

b the volume of air,

c the pressure of the air? *(E.A.)*

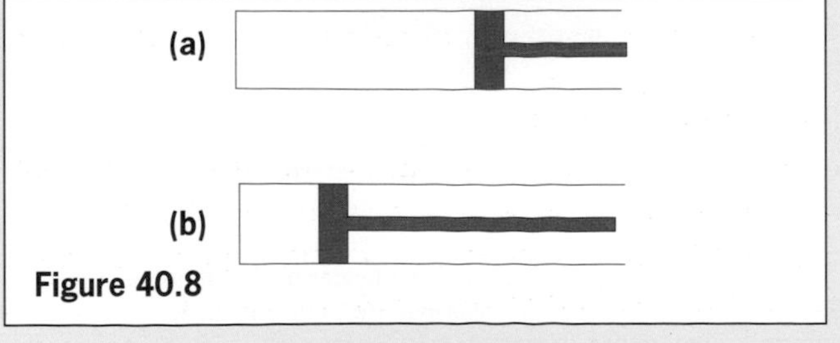

Figure 40.8

4 If a certain quantity of gas has a volume of 30 cm^3 at a pressure of 1×10^5 Pa, what is its volume when the pressure is **a** 2×10^5 Pa, **b** 5×10^5 Pa?

5 A mass of gas occupies a volume of 200 cm^3 at a temperature of 27 °C and a pressure of 1 atmosphere. Calculate the volume when

a the pressure is doubled at constant temperature,

b the absolute temperature is doubled at constant pressure,

c the pressure is $1\frac{1}{2}$ atmospheres and the Celsius temperature is 127 °C. *(E.A.)*

Checklist

After studying this chapter you should be able to

- describe experiments to study the relations between the pressure, volume and temperature of a gas,
- explain the establishment of the Kelvin (absolute) temperature scale from graphs of pressure or volume against temperature and recall the equation connecting the Kelvin and Celsius scales, i.e. $T = 273 + \theta$,
- state Charles' law, the pressure law and Boyle's law and use them to solve problems,
- recall that pV/T = constant and use it to solve problems,
- explain the behaviour of gases using the kinetic theory.

41 SPECIFIC HEAT CAPACITY

The heat equation	***Practical work***
Importance of the high specific heat capacity of water	Finding specific heat capacities: water, aluminium.

If 1 kg of water and 1 kg of paraffin are heated in turn for the same time by the same heater, the temperature rise of the paraffin is about *twice* that of the water. Since the heater gives equal amounts of heat energy to each liquid, it seems that different substances require different amounts of heat to cause the same temperature rise in the same mass, say 1 °C in 1 kg.

The 'thirst' of a substance for heat is measured by its **specific heat capacity** (symbol *c*).

> The specific heat capacity of a substance is the heat required to produce a 1 °C rise in 1 kg.

Heat, like other forms of energy, is measured in joules (J) and the unit of specific heat capacity is the joule per kilogram °C, i.e. J/(kg °C).

In physics the word 'specific' means 'unit mass' is being considered.

1 The heat equation

If a substance has a specific heat capacity of 1000 J/(kg °C) then

1000 J raise the temp. of 1 kg by 1 °C
∴ 2 × 1000 J raise the temp. of 2 kg by 1 °C
∴ 3 × 2 × 1000 J raise the temp. of 2 kg by 3 °C

That is 6000 J will raise the temperature of 2 kg of this substance by 3 °C. We have obtained this answer by multiplying together:

(i) the **mass** in kg,
(ii) the **temperature rise** in °C, and
(iii) the **specific heat capacity** in J/(kg °C).

If the temperature of the substance fell by 3 °C, the heat given out would also be 6000 J. In general, we can write the 'heat equation' as

> heat received or given out
> = mass × temp. change × sp. heat capacity

In symbols

$$Q = m \times \Delta\theta \times c$$

For example, if the temperature of a 5 kg mass of copper of specific heat capacity 400 J/(kg °C) rises from 15 °C to 25 °C, the heat received Q

= 5 kg × (25 − 15) °C × 400 J/(kg °C)
= 5 kg × 10 °C × 400 J/(kg °C) = 20 000 J

PRACTICAL WORK

Finding specific heat capacities

You need to know the power of the 12-volt electric immersion heater to be used. (**Note.** Do not use one with a cracked seal.) A 40-watt heater converts 40 joules of electrical energy into heat energy per second. If the power is not marked on the heater ask about it.[1]

(a) Water

Weigh out 1 kg of water into a container, e.g. an aluminium saucepan. Note the temperature of the water, insert the heater, Figure 41.1, and switch on

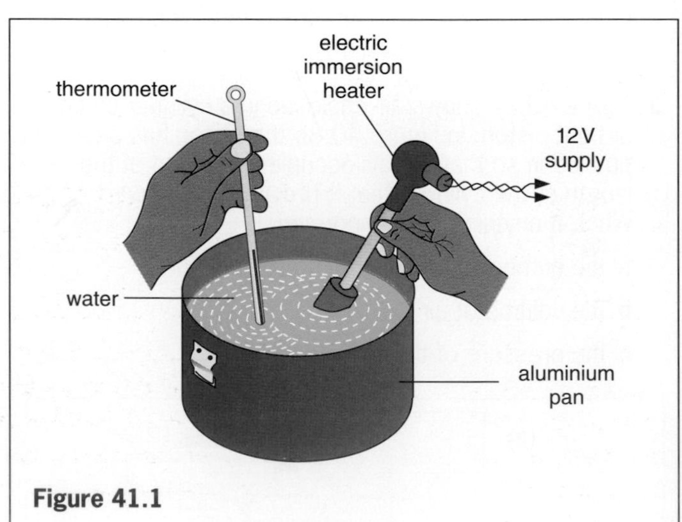

Figure 41.1

[1] The power is found by immersing the heater in water, connecting it to a 12-volt d.c. supply and measuring the current taken (usually 3–4 amperes). Then power in watts = volts × amperes.

the 12 V supply. Stir the water and after 5 minutes switch off, but continue stirring and note the *highest* temperature reached.

Assuming that the heat supplied by the heater equals the heat received by the water, work out the specific heat capacity of water in J/(kg °C), as shown below.

heat received by water (J)
= power of heater (J/s) × time heater on (s)

Rearranging the 'heat equation' we get

$$\text{sp. heat cap. of water} = \frac{\text{heat received by water (J)}}{\text{mass (kg)} \times \text{temp. rise (°C)}}$$

Suggest causes of error in this experiment.

(b) Aluminium

A cylinder weighing 1 kg and having two holes drilled in it is used. Place the immersion heater in the central hole and a thermometer in the other hole, Figure 41.2.

Note the temperature, connect the heater to a 12 V supply and switch it on for 5 minutes. When the temperature stops rising record its highest value.

Calculate the specific heat capacity as before.

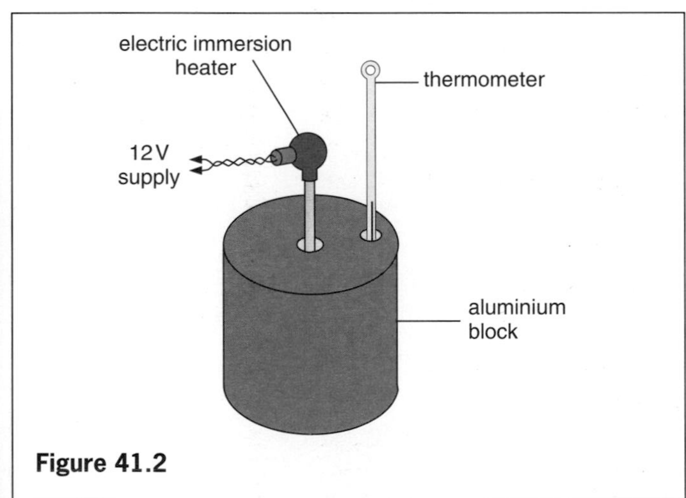

Figure 41.2

Importance of the high specific heat capacity of water

The specific heat capacity of water is 4200 J/(kg °C) and that of soil is about 800 J/(kg °C). As a result, the temperature of the sea rises and falls more slowly than that of the land. A certain mass of water needs five times more heat than the same mass of soil for its temperature to rise by 1 °C. Water also has to give out more heat to fall 1 °C. Since islands are surrounded by water they experience much smaller changes of temperature from summer to winter than large land masses such as Central Asia.

The high specific heat capacity of water (as well as its cheapness and availability) accounts for its use to cool car engines and in the radiators of central heating systems.

Worked examples

1 A tank holding 60 kg of water is heated by a 3 kW electric immersion heater. If the specific heat capacity of water is 4200 J/(kg °C), estimate the time for the temperature to rise from 10 °C to 60 °C.

A 3 kW (3000 W) heater supplies 3000 J of heat energy per second.

Let t = time taken in seconds to raise the temperature of the water by (60 − 10) = 50 °C,

∴ heat supplied to water in time $t = 3000 \times t$ J

From the 'heat equation', we can say

heat received by water
= 60 kg × 4200 J/(kg °C) × 50 °C

Assuming heat supplied = heat received

$$3000 \text{ J/s} \times t = (60 \times 4200 \times 50) \text{ J}$$

$$\therefore \quad t = \frac{(60 \times 4200 \times 50) \text{ J}}{3000 \text{ J/s}} = 4200 \text{ s (70 mins)}$$

2 A piece of aluminium of mass 0.5 kg is heated to 100 °C and then placed in 0.4 kg of water at 10 °C. If the resulting temperature of the mixture is 30 °C, what is the specific heat capacity of aluminium if that of water is 4200 J/(kg °C)?

When two substances at different temperatures are mixed, heat flows from the one at the higher temperature to the one at the lower temperature until both are at the same temperature – the temperature of the mixture. If there is no loss of heat, then in this case

heat given out by aluminium
= heat taken in by water

Using the 'heat equation' and letting c be the sp. heat cap. of aluminium in J/(kg °C), we have

heat given out = 0.5 kg × c × (100 − 30) °C

heat taken in
= 0.4 kg × 4200 J/(kg °C) × (30 − 10) °C

$$\therefore \quad 0.5 \text{ kg} \times c \times 70 \text{ °C} = 0.4 \text{ kg} \times 4200 \text{ J/(kg °C)} \times 20 \text{ °C}$$

$$\therefore \quad c = \frac{(4200 \times 8) \text{ J}}{35 \text{ kg °C}} = 960 \text{ J/(kg °C)}$$

QUESTIONS

1 How much heat is needed to raise the temperature by 10 °C of 5 kg of a substance of specific heat capacity 300 J/(kg °C)?

2 The same quantity of heat was given to different masses of three substances A, B and C. The temperature rise in each case is shown in the table. Calculate the specific heat capacities of A, B and C.

Material	**Mass** (kg)	**Heat given** (J)	**Temp. rise** (°C)
A	1.0	2000	1.0
B	2.0	2000	5.0
C	0.5	2000	4.0

3 1 kg of water is contained in a vessel with a 50 W immersion heater. When the immersion heater is switched on the temperature of the water soon begins to rise at a rate of 1 °C every 2 minutes.

a How much heat is supplied by the immersion heater every 2 minutes?

b From these results what is the approximate specific heat capacity of water? *(O.L.E./S.R.16+)*

4 The jam in a hot 'roly-poly' pudding always seems hotter than the pastry. Why?

Checklist

After studying this chapter you should be able to

- define **specific heat capacity**, *c*,
- solve problems on specific heat capacity using the heat equation $Q = m \times \Delta\theta \times c$,
- describe experiments to measure the specific heat capacity of metals and liquids by electrical heating,
- explain the importance of the high specific heat capacity of water.

42 LATENT HEAT

Latent heat of fusion
Latent heat of vaporization
Latent heat and the kinetic theory
Practical work
Cooling curve of ethanamide.

When a solid is heated, it may melt and **change its state** from solid to liquid. If ice is heated it becomes water. The opposite process, freezing, occurs when a liquid solidifies.

A pure substance melts at a definite temperature, called the **melting point**; it solidifies at the same temperature – sometimes then called the **freezing point**.

PRACTICAL WORK

Cooling curve of ethanamide

Half-fill a test-tube with ethanamide (acetamide) and place it in a beaker of water, Figure 42.1a. Heat the water until all the ethanamide has melted.

Remove the test-tube and arrange it as in Figure 42.1b with a thermometer in the liquid ethanamide. Record the temperature every minute until it has fallen to 70 °C.

Plot a cooling curve of temperature against time. What is the freezing (melting) point of ethanamide?

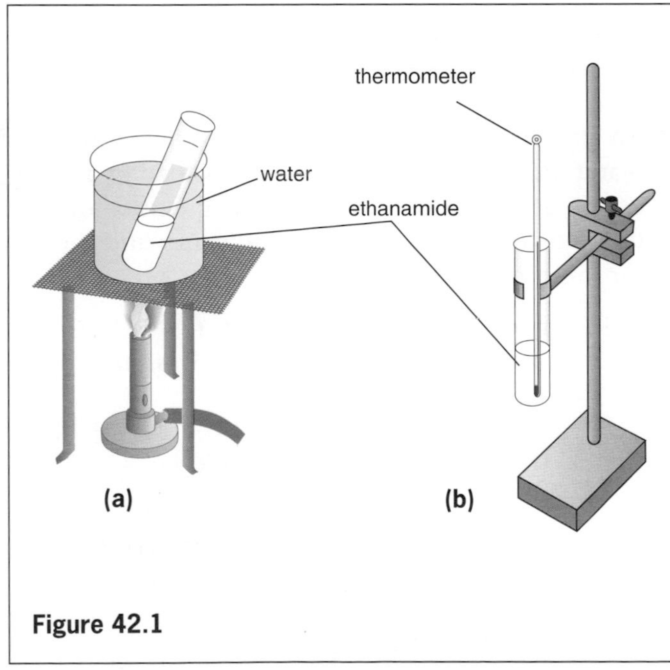

Figure 42.1

Latent heat of fusion

The previous experiment shows that the temperature of liquid ethanamide falls until it starts to solidify (at 82 °C) and remains constant till it has all solidified. The cooling curve in Figure 42.2 is for a pure substance; the flat part AB occurs at the melting point when the substance is solidifying.

During solidification a substance loses heat to its surroundings but its temperature does not fall. Conversely when a solid is melting, the heat supplied does not cause a temperature rise. For example, the temperature of a well-stirred ice–water mixture remains at 0 °C until all the ice is melted.

Heat which is **absorbed** by a solid during melting or **given out** by a liquid during solidification is called **latent heat of fusion**. 'Latent' means hidden and 'fusion' means melting. Latent heat does not cause a temperature change; it seems to disappear.

> The specific latent heat of fusion (l_f) of a substance is the quantity of heat needed to change unit mass from solid to liquid without temperature change.

It is measured in J/kg or J/g. In general, the quantity of heat Q to change a mass m from solid to liquid is given by

$$Q = m \times l_f$$

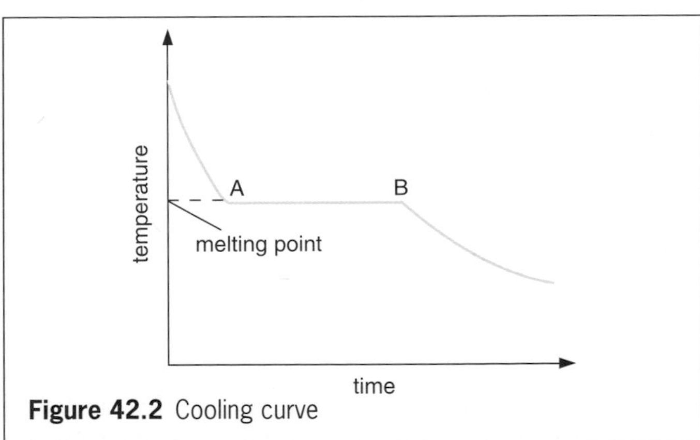

Figure 42.2 Cooling curve

Latent heat of vaporization

Latent heat is also needed to change a liquid into a vapour. The reading of a thermometer in boiling water remains constant at 100 °C even though heat, called **latent heat of vaporization**, is still being absorbed by the water from whatever is heating it. When steam condenses to form water, latent heat is given out.

The specific latent heat of vaporization (l_v) of a substance is the quantity of heat needed to change unit mass from liquid to vapour without change of temperature.

It is measured in J/kg or J/g. In general, the quantity of heat Q to change a mass m from liquid to vapour is given by

$$Q = m \times l_v$$

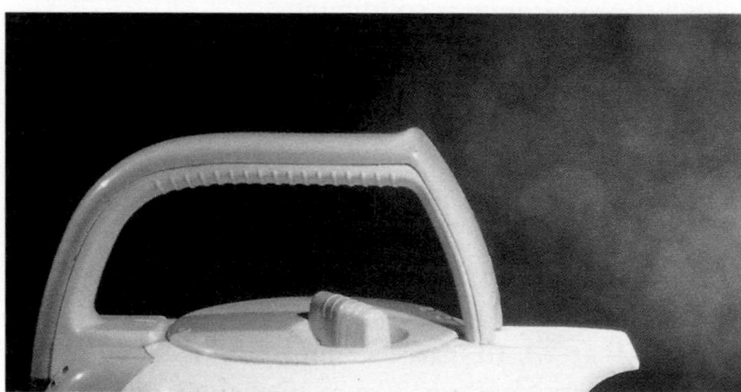

Figure 42.3 Why is a scald from steam often more serious than one from boiling water?

Latent heat and the kinetic theory

(a) Fusion

The kinetic theory explains latent heat of fusion as being the energy which enables the molecules of a solid to overcome the intermolecular forces that hold them in place and when it exceeds a certain value they break free. Their vibratory motion about fixed positions changes to the slightly greater range of movement they have as liquid molecules. Their p.e. increases but not their average k.e. as happens when the heat causes a temperature rise.

(b) Vaporization

If liquid molecules are to overcome the forces holding them together and gain the freedom to move around independently as gas molecules, they need a large amount of energy. They receive this as latent heat of vaporization which, like latent heat of fusion, increases the p.e. of the molecules but not their k.e. It also gives the molecules the energy required to push back the surrounding atmosphere in the large expansion that occurs when a liquid vaporizes.

To change 1 kg of water at 100 °C to steam at 100 °C needs over *five* times as much heat as it does to raise the temperature of 1 kg of water at 0 °C to water at 100 °C (see *Worked example* 1).

Worked examples

The following values are required.

	Water	Ice	Aluminium
Sp. ht. cap. (J/(g °C))	4.2	2.0	0.90
Sp. lat. ht. (J/g)	2300	340	

1 How much heat is needed to change 20 g of ice at 0 °C to steam at 100 °C?

There are three stages in the change.

Heat to change 20 g **ice at 0 °C** to **water at 0 °C**
= mass of ice × sp. lat. ht. of ice
= 20 g × 340 J/g = 6800 J

Heat to change 20 g **water at 0 °C** to **water at 100 °C**
= mass of water × sp. ht. cap. of water × temp. rise
= 20 g × 4.2 J/(g °C) × 100 °C = 8400 J

Heat to change 20 g **water at 100 °C** to **steam at 100 °C**
= mass of water × sp. lat. ht. of steam
= 20 g × 2300 J/g = 46 000 J

∴ Total heat supplied
= 6800 + 8400 + 46 000 = 61 200 J

2 An aluminium can of mass 100 g contains 200 g of water. Both, initially at 15 °C, are placed in a refrigerator at −5.0 °C. Calculate the quantity of heat that has to be removed from the water and the can for their temperatures to fall to −5.0 °C.

Heat lost by **can** in falling from 15 °C to −5.0 °C
= mass of can × sp. ht. cap. of aluminium × temp. fall
= 100 g × 0.90 J/(g °C) × [15 − (−5)] °C
= 100 g × 0.90 J/(g °C) × 20 °C
= 1800 J

Heat lost by **water** in falling from 15 °C to 0 °C
= mass of water × sp. ht. cap. of water × temp. fall
= 200 g × 4.2 J/(g °C) × 15 °C = 12 600 J

Heat lost by **water** at 0 °C freezing to ice at 0 °C
= mass of water × sp. lat. ht. of ice
= 200 g × 340 J/g = 68 000 J

Heat lost by **ice** in falling from 0 °C to −5.0 °C
= mass of ice × sp. ht. cap. of ice × temp. fall
= 200 g × 2.0 J/(g °C) × 5.0 °C = 2000 J

∴ Total heat removed
= 1800 + 12 600 + 68 000 + 2000 = 84 400 J

QUESTIONS

Use values given in *Worked examples*.

1 **a** How much heat will change 10 g of ice at 0 °C to water at 0 °C?

b What quantity of heat must be removed from 20 g of water at 0 °C to change it to ice at 0 °C?

2 **a** How much heat is needed to change 5 g of ice at 0 °C to water at 50 °C?

b If a refrigerator cools 200 g of water from 20 °C to its freezing point in 10 minutes, how much heat is removed per minute from the water?

3 How long will it take a 50 W heater to melt 100 g of ice at 0 °C?

4 Some small aluminium rivets of total mass 170 g and at 100 °C are emptied into a hole in a large block of ice at 0 °C.

a What will be the final temperature of the rivets?

b How much ice will melt?

5 **a** How much heat is needed to change 4 g of water at 100 °C to steam at 100 °C?

b Find the heat given out when 10 g of steam at 100 °C condenses and cools to water at 50 °C.

6 A 3 kW electric kettle is left on for 2 minutes after the water starts to boil. What mass of water is boiled off in this time?

7 **a** Why is ice good for cooling drinks?

b Why do engineers often use superheated steam (steam above 100 °C) to transfer heat?

Checklist

After studying this chapter you should be able to

- describe an experiment to show that during a change of state the temperature stays constant,
- define **specific latent heat of fusion**, l_f,
- define **specific latent heat of vaporization**, l_v,
- explain latent heat using the kinetic theory,
- solve problems on latent heat using $Q = ml$.

43 VAPOURS

Evaporation and boiling
Cooling by evaporation
Vapour pressure
Unsaturated and saturated vapours.
Boiling
Measurement of s.v.p. of water
Liquefaction of gases and vapours

Vapour is produced when a liquid evaporates or boils.

Evaporation and boiling

(a) Evaporation

This occurs at all temperatures at the surface of the liquid. It happens more rapidly when
(i) the **temperature is higher**, since then more molecules in the liquid are moving fast enough to escape from the surface,
(ii) the **surface area** of the liquid is large so giving more molecules a chance to escape because more are near the surface, and
(iii) a **wind** or **draught** is blowing over the surface carrying vapour molecules away from the surface thus stopping them from returning to the liquid and making it easier for more liquid molecules to break free. (Evaporation into a vacuum occurs much more rapidly than into a region where there are air or other molecules.)

(b) Boiling

This occurs for a pure liquid at a definite temperature called its **boiling point** and is accompanied by bubbles that form within the liquid, containing the gaseous or vapour form of the particular substance.

(c) Latent heat of vaporization

Latent heat is needed in both evaporation and boiling and is stored in the vapour, from which it is released when the vapour is cooled or compressed and changes to liquid again.

Cooling by evaporation

In evaporation latent heat is obtained by the liquid from its surroundings, as may be shown by the following demonstration, **done in a fume cupboard**.

(a) Demonstration

Dichloromethane is a **volatile** liquid, i.e. it has a low boiling point and evaporates readily at room temperature, especially when air is blown through it, Figure 43.1. Latent heat is taken first from the liquid itself and then from the water below the can. The water soon freezes causing the block and can to stick together.

Do this in a fume cupboard

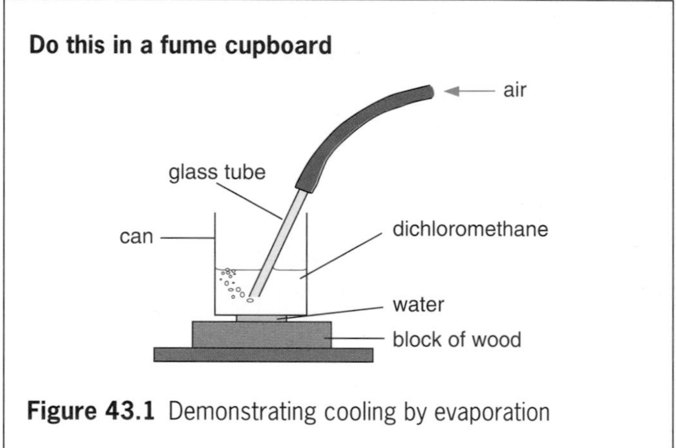

Figure 43.1 Demonstrating cooling by evaporation

(b) Explanation

Evaporation occurs when faster-moving molecules escape from the surface of the liquid. The average speed, and therefore the average k.e., of the molecules left behind decreases, i.e. the temperature of the liquid falls.

(c) Uses

Water evaporates from the skin when we sweat. This is the body's way of using unwanted heat and keeping a constant temperature. After vigorous exercise there is a risk of the body being overcooled, especially in a draught; it is then less able to resist infection.

Ether acts as a local anaesthetic by chilling (as well as cleaning) your arm when you are having an injection. Refrigerators, freezers and air-conditioning systems use cooling by evaporation on a large scale.

Volatile liquids are used in perfumes.

Vapour pressure

(a) Unsaturated and saturated vapours

In a closed vessel containing a liquid and its vapour, the vapour molecules collide with and rebound from the walls of the vessel exerting a pressure on it, called **vapour pressure**. This is in addition to the air pressure caused by the molecules of any air present.

Initially the rate at which molecules leave the liquid surface exceeds that at which they enter it from the vapour, i.e. evaporation occurs more rapidly than condensation and the vapour is said to be **unsaturated**.

Eventually a state of **dynamic equilibrium** is reached when molecules leave and re-enter the liquid at the same rate. The number of vapour molecules in the space above the liquid is then a maximum. The exchange of molecules continues (hence dynamic) but there is a balance between the two streams (hence equilibrium). The vapour is then said to be **saturated**, Figure 43.2. The pressure of a vapour above its liquid in a closed space is called its **saturation vapour pressure** (s.v.p.).

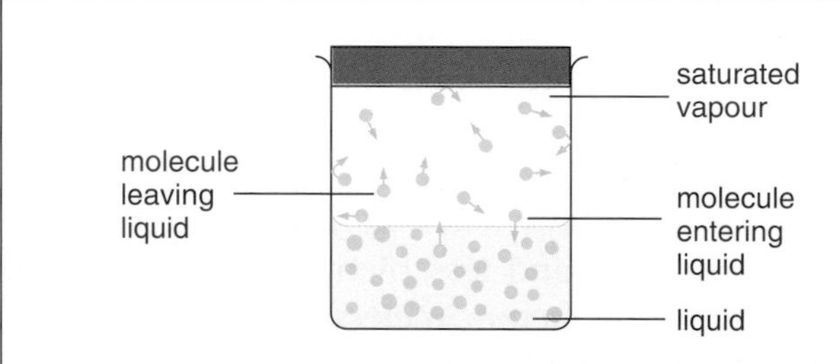

Figure 43.2 Dynamic equilibrium between a saturated vapour and its liquid

(b) Vapours and the gas laws

Vapours obey the three gas laws (Chapter 40) so long as they do not become saturated.

With saturated vapours, the s.v.p. increases with temperature but not according to the Pressure law, as Figure 43.3 shows for water.

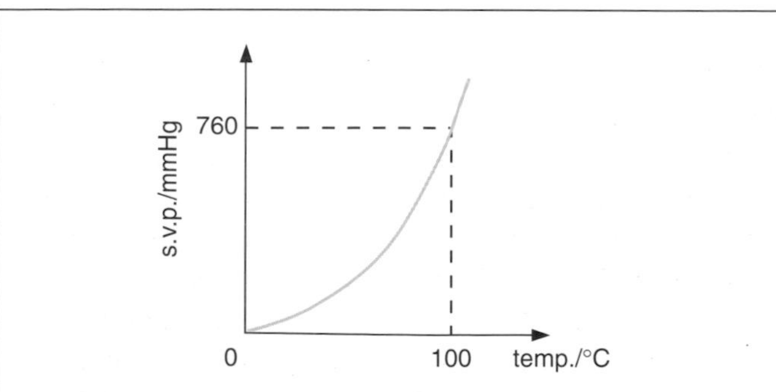

Figure 43.3 Variation of the s.v.p. of water with temperature

The s.v.p. is not affected by changes of volume. If the volume available to the vapour increases, its density momentarily decreases and fewer molecules return to the liquid in a given time than previously. The rate at which molecules leave the surface is still the same. The rate of evaporation therefore exceeds the rate of condensation until the density of the vapour rises to its original value and dynamic equilibrium is restored once more. The s.v.p. then has its initial value but with more vapour molecules creating the same pressure in a larger volume.

Worked example

A sealed vessel contains air and water. The pressures in the vessel at 20 °C and 70 °C are 738 mmHg and 1080 mmHg respectively and there is still some liquid water in it at 70 °C. If the s.v.p. of water at 20 °C is 18 mmHg, find the value at 70 °C.

Apply the Pressure law to the air in the mixture: the water vapour is exerting its s.v.p. at both temperatures since there is always liquid water present.

At 20 °C, p_1 = total pressure − s.v.p. water at 20 °C
$= 738 - 18 = 720$ mmHg
$T_1 = 273 + 20 = 293$ K

At 70 °C, p_2 = total pressure − s.v.p. water at 70 °C
$= 1080 - p$
$T_2 = 273 + 70 = 343$ K

Rearranging $p_1/T_1 = p_2/T_2$ we get

$$p_2 = \frac{p_1 T_2}{T_1} = \frac{720 \times 343}{293} = 843 \text{ mmHg}$$

But $p_2 = 1080 - p$

Hence $p = 1080 - p_2 = 1080 - 843 = 237$ mmHg

∴ s.v.p. water at 70 °C = 237 mmHg

Boiling

When a liquid boils, bubbles of vapour form throughout the liquid and rise to the surface where they burst with a 'gurgling' sound. The pressure inside the bubbles must at least equal the pressure in the surrounding liquid otherwise they would be crushed and collapse. The internal pressure is the s.v.p. at the temperature of the boiling point, since the vapour is in contact with liquid, Figure 43.4 (overleaf). The external pressure is practically equal to atmospheric pressure for a liquid in an open vessel.

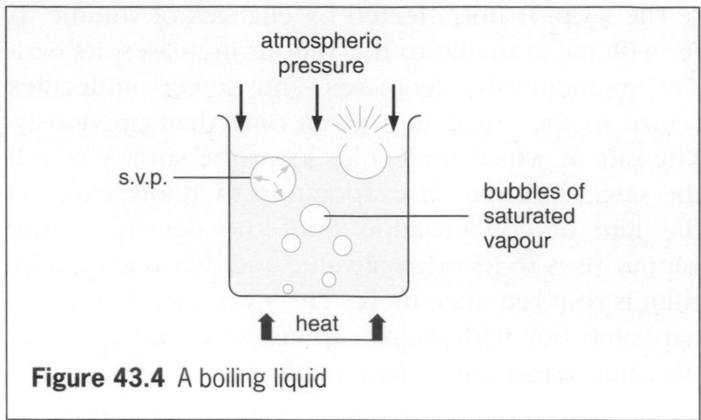

Figure 43.4 A boiling liquid

The boiling point of a liquid is thus the temperature at which its s.v.p. equals the external pressure.

Measurement of s.v.p. of water

The apparatus in Figure 43.5 is suitable for s.v.ps above 50 mmHg (for water from about 50 °C up). The boiling point of the liquid is found at different external pressures; the s.v.p. is then equal to the external pressure at that temperature.

The vacuum pump or compressor enables the pressure to be set to any desired value below or above atmospheric pressure. With cold water circulating through the condenser, the liquid in the flask is heated until it boils. The temperature of the saturated vapour is then given by the thermometer and the difference between the pressure inside the apparatus and atmospheric pressure is shown by the manometer. The reservoir helps to prevent large pressure fluctuations by acting as a buffer.

Liquefaction of gases and vapours

A vapour can be liquefied if it is compressed enough. A gas must be cooled below a certain **critical temperature** T_c before liquefaction by pressure can occur. The critical temperatures for some gases are given in Table 43.1.

Table 43.1

Gas	carbon dioxide	oxygen	air	nitrogen	hydrogen	helium
T_c/K	304	154	132	126	33.3	5.3
T_c/°C	+31	−119	−141	−147	−239.7	−267.7

Low-temperature liquids have many uses. Liquid hydrogen and oxygen are used as the fuel and oxidant respectively in space rockets. Liquid nitrogen is used in industry as a coolant in, for example, shrink- fitting (see Figure 39.1). Certain substances have recently been discovered which behave as superconductors (see Chapter 40) when cooled by liquid nitrogen. Previously this behaviour had only been displayed by some metals and alloys when at the temperature of the more expensive and less plentiful liquid helium. The use of superconductors in electrical power engineering and electronics is being explored.

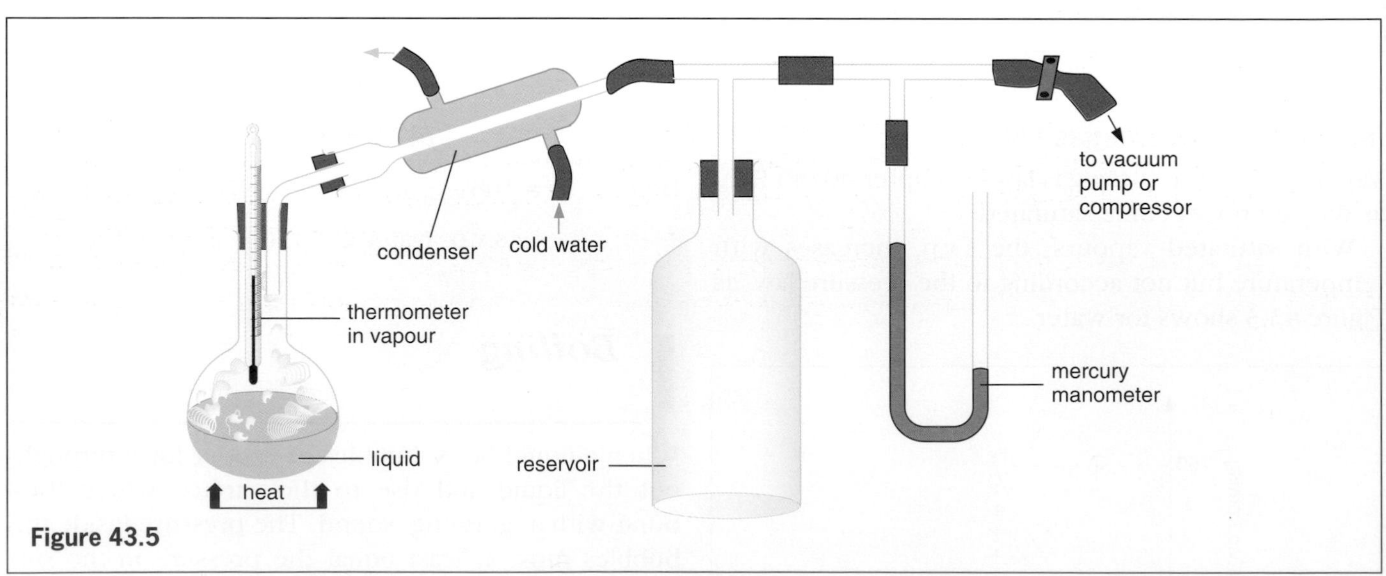

Figure 43.5

QUESTIONS

1 Some water is stored in a bag of a porous material, e.g. canvas, which is hung where it is exposed to a draught of air. Explain why the temperature of the water is lower than that of the air.

2 Explain why a bottle of milk keeps better when it stands in water in a porous pot in a draught.

3 **a** State *two* differences between evaporation and boiling.

b What is the difference between a saturated and an unsaturated vapour?

4 When water in an open beaker is boiling, bubbles appear at the bottom of the beaker and rise quickly. The bubbles grow rapidly as they rise and burst violently at the surface.

a What substance is in these bubbles, and where has it come from?

b State and explain how the pressure in the bubbles compares with atmospheric pressure.

c Why is it necessary to keep supplying energy to boiling water even though its temperature is remaining at 100 °C? *(M.E.G. GCSE 1988)*

5 Are the gas laws obeyed by

a unsaturated vapours,

b saturated vapours?

6 Which of the gases in Table 43.1 could *not* be liquefied by pressure alone if it was at a temperature of 100 K?

7 A vessel contains a mixture of air and unsaturated water vapour at 77 °C and a pressure of 800 mmHg.

a The mixture is cooled to 27 °C and water begins to condense. What is the new pressure of the mixture?

b If it is cooled further to 7 °C, calculate the total pressure now, taking the s.v.p. of water at 27 °C as 26 mmHg and at 7 °C as 8 mmHg.

Checklist

After studying this chapter you should be able to

- distinguish between **evaporation** and **boiling**,
- describe an experiment to show that evaporation causes cooling,
- explain cooling by evaporation using the kinetic theory,
- explain the difference between an **unsaturated** and a **saturated** vapour,
- recall that an unsaturated vapour obeys the gas laws and use these to solve problems,
- recall that the s.v.p. increases with temperature, and that when it reaches the external pressure the liquid boils,
- describe a method of measuring the s.v.p. of water,
- explain the significance of the **critical temperature** in the liquefaction of gases.

44 CONDUCTION AND CONVECTION

Conduction	*Natural convection currents*
Uses of conductors	*Energy losses from buildings*
Conduction and the kinetic theory	*U-value*
Convection in liquids	*Ventilation*
Convection in air	

To keep a building or a house at a comfortable temperature, in winter and summer, requires a knowledge of how heat travels, if it is to be done economically and efficiently.

Conduction

The handle of a metal spoon held in a hot drink soon gets warm. Heat passes along the spoon by **conduction**.

Conduction is the flow of heat through matter from places of higher temperature to places of lower temperature without movement of the matter as a whole.

A simple demonstration of the different conducting powers of various metals is shown in Figure 44.1. A matchstick is fixed to one end of each rod using a little melted wax. The other ends of the rods are heated by a burner. When the temperatures of the far ends reach the melting point of wax, the matches drop off. The match on copper falls first showing it is the best conductor, followed by aluminium, brass and then iron.

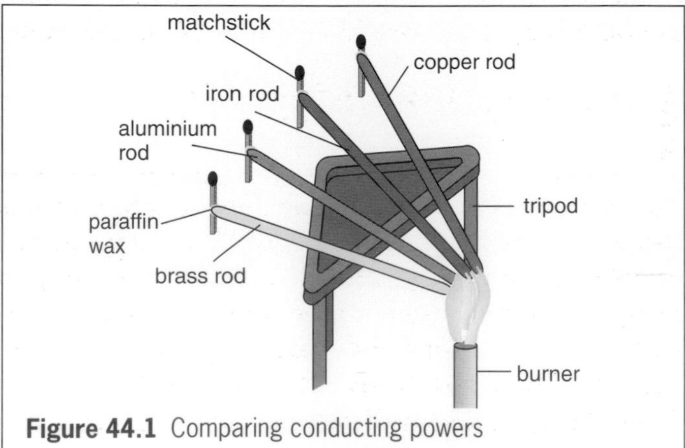

Figure 44.1 Comparing conducting powers

Heat is conducted faster through a rod if it has a large cross-section area, is short and has a large temperature difference between its ends.

Most metals are good conductors of heat; materials such as wood, glass, cork, plastics and fabrics are bad conductors. The arrangement in Figure 44.2 can be used to show the difference between brass and wood. If the rod is passed through a flame several times, the paper over the wood scorches but not over the brass. The brass conducts the heat away from the paper quickly and prevents it reaching the temperature at which it burns. The wood only conducts it away slowly.

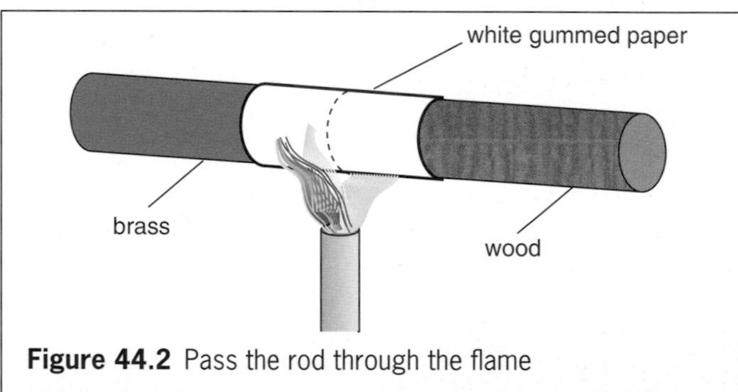

Figure 44.2 Pass the rod through the flame

Metal objects below body temperature *feel* colder than those made of bad conductors because they carry heat away faster from the hand – even if all the objects are at exactly the same temperature.

Liquids and gases also conduct heat but only very slowly. Water is a very poor conductor as shown in Figure 44.3. The water at the top of the tube can be boiled before the ice at the bottom melts.

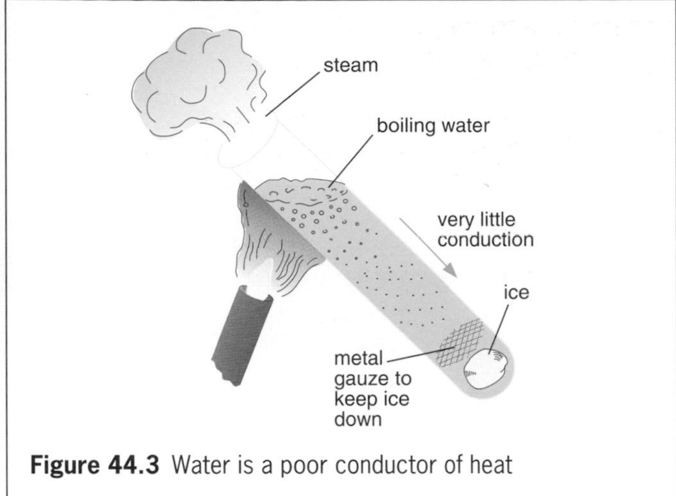

Figure 44.3 Water is a poor conductor of heat

Uses of conductors

(a) Good conductors

These are used whenever heat is required to travel quickly through something. Saucepans, boilers and radiators are made of metals such as aluminium, iron and copper.

(b) Bad conductors (insulators)

The handles of some saucepans are made of wood or plastic. Cork is used for table mats.

Air is one of the worst conductors, i.e. best insulators. This is why houses with cavity walls (i.e. two walls separated by an air space) and double-glazed windows keep warmer in winter and cooler in summer.

Materials which trap air, e.g. wool, felt, fur, feathers, polystyrene, fibreglass, are also very bad conductors. Some of these materials are used as 'lagging' to insulate water pipes, hot water cylinders, ovens, refrigerators and the walls and roofs of houses, Figures 44.4a, b. Others are used to make warm winter clothes.

Figure 44.4a Lagging in a cavity wall provides extra insulation

Figure 44.4b Laying lagging in a house loft

'Wet suits' are worn by divers and water skiers to keep them warm. The suit gets wet and a layer of water gathers between the person's body and the suit. The water is warmed by body heat and stays warm because the suit is made of an insulating fabric, e.g. neoprene, a synthetic rubber.

Conduction and the kinetic theory

Two processes occur in metals. These have a large number of 'free' electrons (Chapter 49) which wander about inside them. When one part of a metal is heated, the electrons there move faster (i.e. their k.e. increases) and farther. As a result they 'jostle' atoms in cooler parts, so passing on their energy and raising the temperature of these parts. This process occurs quickly.

The second process is much slower. The atoms themselves at the hot part make 'colder' neighbouring atoms vibrate more vigorously. This is less important in metals but is the only way conduction occurs in non-metals since these do not have 'free' electrons.

Convection in liquids

Convection is the usual method by which heat travels through fluids, i.e. liquids and gases. It can be shown in water by dropping a few crystals of potassium permanganate down a tube to the bottom of a beaker or flask of water. When the tube is removed and the beaker heated just below the crystals by a *small* flame, Figure 44.5 (overleaf), purple streaks of water rise upwards and fan outwards.

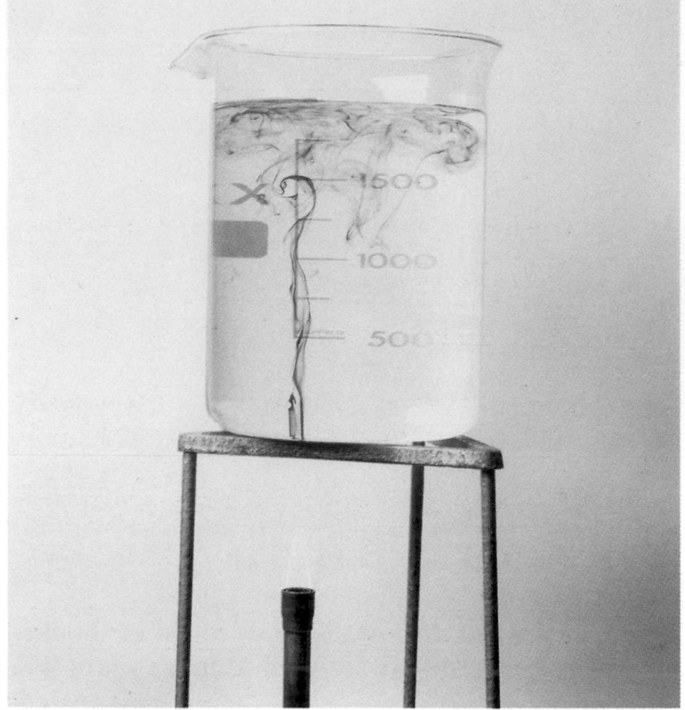

Figure 44.5 Convection currents shown by potassium permanganate in water

Streams of warm moving fluids are called **convection currents**. They arise when a fluid is heated because it expands, becomes less dense and is forced upwards by surrounding cooler, denser fluid which moves under it. We say 'hot water (or hot air) rises'. Warm fluid behaves like a cork released under water: being less dense it bobs up. In convection, however, a fluid floats in a fluid, not a solid in a fluid.

Convection is the flow of heat through a fluid from places of higher temperature to places of lower temperature by movement of the fluid itself.

Convection in air

Black marks often appear on the wall or ceiling above a lamp or a radiator. They are caused by dust being carried upwards in air convection currents produced by the hot lamp or radiator.

A laboratory demonstration of convection currents in air can be given using the apparatus of Figure 44.6. The direction of the convection current created by the candle is made visible by the smoke from the touch paper (made by soaking brown paper in strong potassium nitrate solution and drying it).

Convection currents set up by electric, gas and oil heaters help to warm our homes. Many so-called 'radiators' are really convector heaters.

Where should the input and extraction ducts for cold/hot air be located in a room?

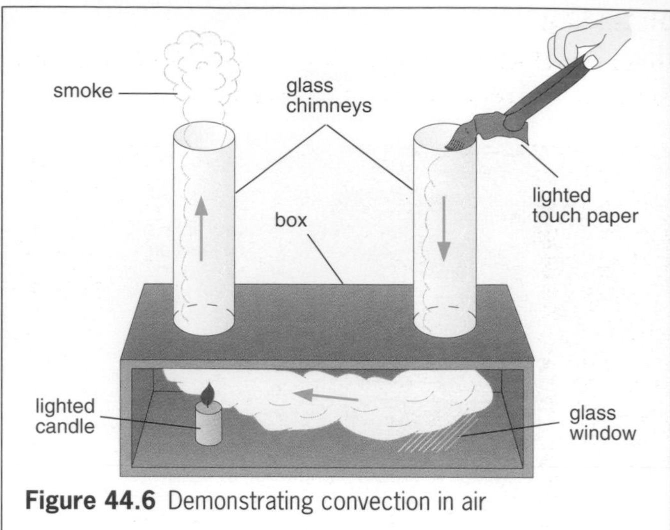

Figure 44.6 Demonstrating convection in air

Natural convection currents

(a) Coastal breezes

During the day the temperature of the land increases more quickly than that of the sea (because the specific heat capacity of the land is much smaller, Chapter 41). The hot air above the land rises and is replaced by colder air from the sea. A breeze from the sea results, Figure 44.7a.

At night the opposite happens. The sea has more heat to lose and cools more slowly. The air above the sea is warmer than that over the land and a breeze blows from the land, Figure 44.7b.

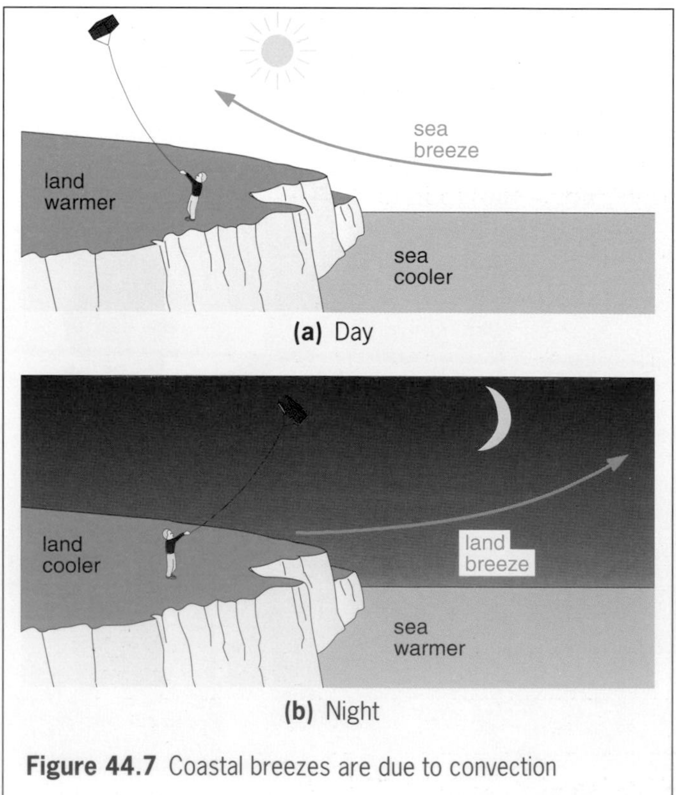

Figure 44.7 Coastal breezes are due to convection

(b) Gliding

Gliders, including 'hang-gliders', Figure 44.8, depend on hot air currents, called **thermals**.

Figure 44.8 By flying from one thermal to another a hang-glider can stay airborne for several hours

Energy losses from buildings

The inside of a building can only be kept at a steady temperature above that outside by heating it at a rate which equals the rate it is losing energy. The loss occurs mainly by conduction through the walls, roof, floors and windows. For a typical house in the UK where no special precautions have been taken, the contribution each of these makes to the total loss is shown in Table 44.1a.

As fuels (and electricity) become more expensive and the burning of fuels becomes of greater environmental concern (Chapter 46), more people are considering it worthwhile to reduce heat losses from their homes. The substantial reduction of this loss which can be achieved, especially by wall and roof insulation, is shown in table (b).

Table 44.1 (a)

Percentage of total energy loss due to				
walls	**roof**	**floors**	**windows**	**draughts**
35	25	15	10	15

(b)

Percentage of each loss saved by				
insulating				
walls	**roof**	**carpets on floors**	**double glazing**	**draught excluders**
65	80	≈30	50	≈60
Percentage of total loss saved = 60				

U-value

This is a term used by heating engineers and is defined as follows.

The *U*-value for a specified heat conductor is the heat energy lost per second through it per square metre when there is a temperature difference of 1 °C between its surfaces.

The rate of heat energy loss through a conductor can therefore be calculated knowing its *U*-value, from

rate of energy loss = *U*-value × surface area × temp. difference

Two *U*-values in joules per second per square metre per °C, i.e. W/(m^2 °C), are given below:

Tiled roof without insulation: 2.2
Tiled roof with 75 mm thick insulation: 0.45

Ventilation

In addition to supplying heat to compensate for the energy losses from a building, a heating system has also to warm the ventilated cold air, needed for comfort, which comes in to replace stale air.

If the rate of heat loss is, say, 6000 J s^{-1}, i.e. 6 kW, and the warming of ventilated air requires 2 kW, then the total power needed to maintain a certain temperature (e.g. 20 °C) in the building is 8 kW. Some of this is supplied by each person's 'body heat', estimated to be roughly equal to a 100 W heater.

QUESTIONS

1 Explain why

a newspaper wrapping keeps hot things hot, e.g. fish and chips, and cold things cold, e.g. ice cream,

b fur coats would keep their owners warmer if they were worn inside out,

c a string vest keeps a person warm even though it is a collection of holes bounded by string.

2 Figure 44.9 illustrates three ways of reducing heat losses from a house.

a As far as you can, explain how each of the three methods reduces heat losses. Draw diagrams where they will help your explanations.

b Why are fibreglass and plastic foam good substances to use?

c Air is one of the worst conductors of heat. What is the point of replacing it by the plastic foam in (ii)?

d A vacuum is an even better heat insulator than air. Suggest one (scientific) reason why the double glazing should not have a vacuum between the sheets of glass.

e The manufacturers of roof lagging suggest that two layers of fibreglass are more effective than one. Describe how you might set up an experiment in the laboratory to test whether this is true.

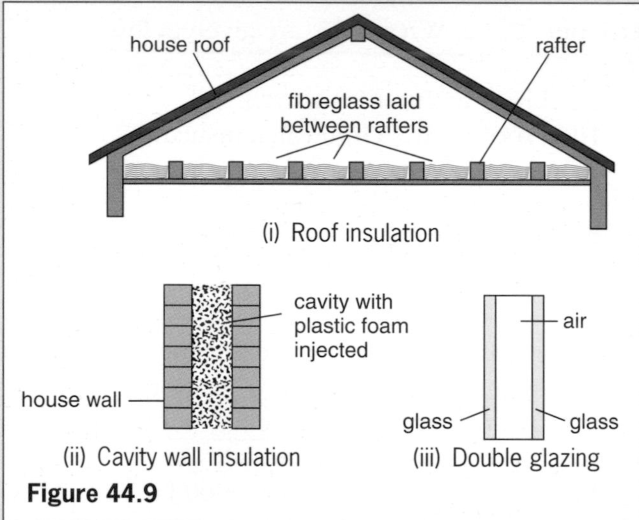

Figure 44.9

3 What is the advantage of placing an electric immersion heater **a** near the top, **b** near the bottom, of a tank of water?

4 a How much heat energy is lost in an hour through a window measuring 2.0 m by 2.5 m when the inside and outside temperatures are 18 °C and −2 °C respectively if the window is
(i) single-glazed, *U*-value 6.0 W/(m^2 °C),
(ii) double-glazed, *U*-value 3.0 W/(m^2 °C)?

b What power of heater is required in case (ii) to maintain this temperature if there are no ventilation losses?

Checklist

After studying this chapter you should be able to

- describe experiments to show the different conducting powers of various substances,
- name good and bad conductors and state uses for each,
- explain conduction using the kinetic theory,
- describe experiments to show convection in fluids (liquids and gases),
- relate convection to phenomena such as land and sea breezes,
- explain the importance of insulating a building,
- recall the definition of ***U*-value** and use it to solve heat loss problems.

45 RADIATION

Good and bad absorbers
Good and bad emitters
Vacuum flask
The greenhouse
Rate of cooling of an object

Radiation is a third way in which heat can travel but whereas conduction and convection both need matter to be present, radiation can occur in a vacuum. It is the way heat reaches us from the Sun.

Radiation has all the properties of electromagnetic waves (Chapter 14), e.g. it travels at the speed of radio waves and gives interference effects. When it falls on an object, it is partly reflected, partly transmitted and partly absorbed; the absorbed part raises the temperature of the object.

Radiation is the flow of heat from one place to another by means of electromagnetic waves.

Radiation is emitted by all bodies above absolute zero and consists mostly of infrared radiation (Chapter 14) but light and ultraviolet are also present if the body is very hot (e.g. the Sun).

Good and bad absorbers

Some surfaces absorb radiation better than others as may be shown using the apparatus in Figure 45.1. The inside surface of one lid is shiny and of the other dull black. The coins are stuck on the outside of each lid with candle wax. If the heater is midway between the lids they each receive the same amount of radiation. After a few minutes the wax on the black lid melts and the coin falls off. The shiny lids stay cool and the wax unmelted.

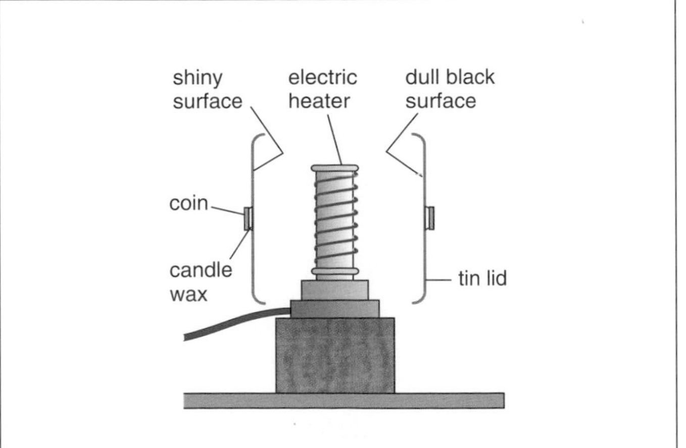

Figure 45.1 Comparing absorbers of radiation

Figure 45.2 Why are buildings in hot countries often painted white?

Dull black surfaces are better absorbers of radiation than white shiny surfaces – the latter are good **reflectors** of radiation. Reflectors on electric fires are made of polished metal because of its good reflecting properties.

Good and bad emitters

Some surfaces also emit radiation better than others when they are hot. If you hold the backs of your hands on either side of a hot copper sheet which has one side polished and the other blackened, Figure 45.3 (overleaf), it will be found that the **dull black surface is a better emitter of radiation than the shiny one**.

The cooling fins on the heat exchangers at the back of a refrigerator are painted black so that they lose heat more quickly. By contrast, saucepans etc. which are polished are poor emitters and keep their heat longer.

In general, surfaces that are good absorbers of radiation are good emitters when hot.

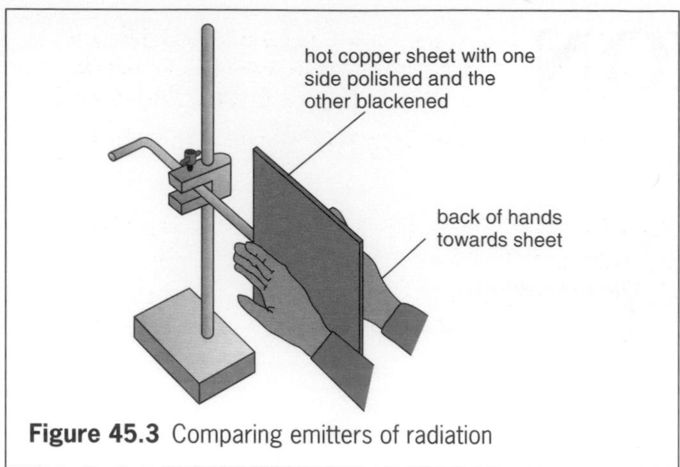

Figure 45.3 Comparing emitters of radiation

Vacuum flask

A vacuum or Thermos flask keeps hot liquids hot or cold liquids cold. It is very difficult for heat to travel into or out of the flask.

Transfer by conduction and convection is minimized by making the flask a double-walled glass vessel with a vacuum between the walls, Figure 45.4. Radiation is reduced by silvering both walls on the vacuum side. Then if, for example, a hot liquid is stored, the small amount of radiation from the hot inside wall is reflected back across the vacuum by the silvering on the outer wall. The slight heat loss which does occur is by conduction up the walls and through the stopper.

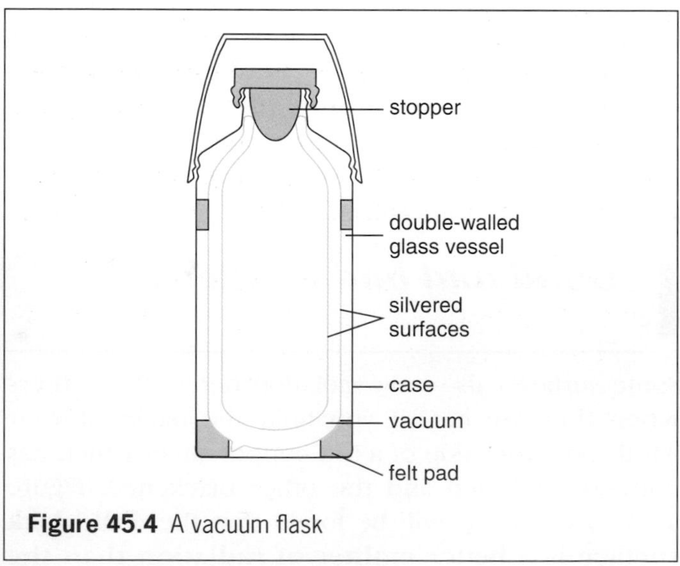

Figure 45.4 A vacuum flask

The greenhouse

The warmth from the Sun is not cut off by a sheet of glass but the warmth from a red-hot fire is. The radiation from very hot bodies like the Sun is mostly in the form of light and short-wavelength infrared. The radiation from less hot objects, e.g. a fire, is largely long-wavelength infrared which, unlike light and short-wavelength infrared, cannot pass through glass.

Light and short-wavelength infrared from the Sun penetrate the glass of a greenhouse and are absorbed by the soil, plants, etc., raising their temperature. These in turn emit infrared but, because of their relatively low temperature, this has a long wavelength and is not transmitted by the glass. The greenhouse thus acts as a 'heat-trap' and its temperature rises.

Carbon dioxide and other gases such as methane in the Earth's atmosphere also act like the glass of a greenhouse in trapping heat and could have serious implications for our climate (Chapter 67).

Rate of cooling of an object

The rate at which an object cools, i.e. at which its temperature falls, can be shown to be proportional to the ratio of its surface area A to its volume V.

For a cube of side l

$$A_1/V_1 = 6 \times l^2/l^3 = 6/l$$

For a cube of side $2\,l$

$$A_2/V_2 = 6 \times 4\,l^2/8\,l^3 = 3/l = \tfrac{1}{2} \times 6/l = \tfrac{1}{2}A_1V_1$$

The larger cube has the smaller A/V ratio and so cools more slowly.

You could investigate this using two aluminium cubes, one having twice the length of side of the other. Each needs holes for a thermometer and an electric heater to raise them to the same starting temperature. Temperature against time graphs for both blocks can then be obtained.

QUESTIONS

1 We feel the heat from a coal fire by

A convection **B** conduction **C** regelation **D** diffusion **E** radiation.

2 a Three beakers are of identical size and shape; one beaker is painted matt black, one is dull white and one is gloss white. The beakers are filled with boiling water. In which beaker will the water cool most quickly? Give a reason.

b State a process, in addition to conduction, convection and radiation, by which heat energy will be lost from the beakers. (E.A.)

3 The door canopy in Figure 45.5 shows in a striking way the difference between white and black surfaces when radiation falls on them. Explain why.

Figure 45.5

4 Figure 45.6 shows an electric heater H placed midway between two flasks A and B. Flask A is shiny on the outside and flask B is blackened on the outside.

a Name the process by which heat travels from the heater to the flasks.

b What happens to the liquid in the tube XY?

c Give a reason for your answer to **b**. *(E.M.)*

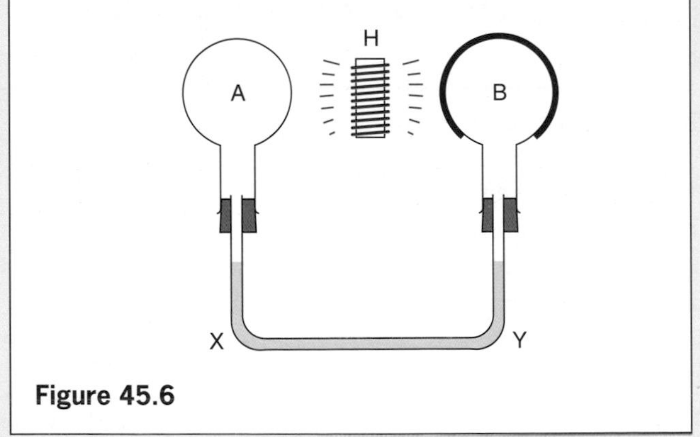

Figure 45.6

5 **a** The Earth has been warmed by the radiation from the Sun for millions of years yet we think its average temperature has remained fairly steady. Why is this?

b Why is frost less likely on a cloudy night than a clear one?

6 A Thermos flask has a silver layer on its thin glass walls to reduce loss of heat by

A convection **B** evaporation **C** conduction **D** radiation. *(W.M.)*

Checklist

After studying this chapter you should be able to

- describe experiments to study factors affecting the absorption and emission of radiation,
- recall that good absorbers are also good emitters,
- explain how a knowledge of heat transfer affects the design of a vacuum flask,
- explain how a greenhouse acts as a 'heat-trap',
- investigate how rate of cooling depends on the ratio of surface area to volume.

46 ENERGY SOURCES

Non-renewable energy sources
Fossil fuels.
Nuclear fuels.

Renewable energy sources
Solar energy.
Wind energy.
Wave energy.
Tidal and hydroelectric energy.
Geothermal energy.
Biomass (vegetable fuels).

Power stations

Economic, environmental and social issues

Energy is needed to heat buildings, to make cars move, to provide artificial light, to make computers work, ... The list is endless. This 'useful' energy needs to be produced in controllable energy transfers (Chapter 25). For example, in power stations a supply of useful energy in the form of electricity is produced. The 'raw materials' for energy production are **energy sources**. These may be **non-renewable** or **renewable**.

Non-renewable energy sources

Once used up these cannot be replaced.

(a) Fossil fuels

These include coal, oil and natural gas formed from the remains of plants and animals which lived millions of years ago and obtained energy originally from the Sun. At present they are our main energy source. Predictions vary as to how long they will last since this depends on what reserves are recoverable and the future demands of a world population expected to increase from about 5000 million in 1990 to 6000 million by the year 2000. Some estimates say oil and gas will run low early in the 21st century but coal should last for 200 years or so.

Burning fossil fuels in power stations and in cars pollutes the atmosphere with harmful gases, i.e. has an effect on the environment which is generally considered to be bad. Acid rain is caused by sulphur dioxide emission, and carbon dioxide emission aggravates the greenhouse effect (Chapters 45 and 67).

(b) Nuclear fuels

Their energy is released in a nuclear reactor (Chapter 64) from uranium, found as an ore in the ground. These also create environmental problems if radiation leakage occurs, and from the need to dispose of dangerous waste materials with long radioactive lives (Chapter 63).

Two advantages of non-renewable fuels are

(i) their high **energy density** (i.e. they are concentrated sources) and the relatively small size of the energy transfer device (e.g. a furnace) which releases their energy, and
(ii) their ready **availability** when energy demand suddenly increases or to meet seasonal changes in demand.

Renewable energy sources

These cannot be exhausted and are non-polluting.

(a) Solar energy

The energy falling on the Earth from the Sun is mostly in the form of light and in an hour equals the total energy used by the world in a year. Unfortunately its low energy density requires large collecting devices and its availability varies. Its greatest potential use is as an energy source for low-temperature water heating. This uses solar panels as the energy transfer devices, Figure 46.1a, which convert light into heat energy. They are common in hot climates to heat swimming pools and to produce domestic hot water at about 70 °C.

Solar energy can also be used to produce high-temperature heating, up to 3000 °C or so, if a large curved mirror (i.e. a solar furnace) focuses the Sun's rays on to a small area. The energy can then be used to turn water to steam for driving the turbine of an electric generator in a power station.

Solar cells, made from semiconducting materials, convert sunlight into electricity directly. Panels of cells connected together supply the electronic equipment in communication and other satellites. They are

Figure 46.1a Solar panels on a house

Figure 46.1b Honda solar-powered car which won the World Solar Challenge race in Darwin, Australia, in 1993

also used for small-scale power generation in remote areas of developing countries where there is no electricity supply. Recent developments have made large-scale generation more cost-effective and there is now a large solar power plant in California. There are many designs for prototype light vehicles run on solar power, Figure 46.1b.

(b) Wind energy

Giant windmills (called wind turbines) with two or three blades each up to 30 m long drive electrical generators. 'Wind farms' of twenty to thirty turbines spaced about 400 m apart, Figure 2b, p. vii, are in an experimental stage. It is predicted that by the year 2025 wind energy could supply up to 10% of the UK's electricity.

There is some environmental objection to wind turbines, since the best sites are often in coastal areas of great natural beauty.

(c) Wave energy

The rise and fall of sea waves has to be transferred by some kind of wave-energy converter into the rotary motion required to drive a generator. It is a difficult problem and the large-scale production of electricity by this means is unlikely in the near future, but small systems are being developed to supply island communities with power.

(d) Tidal and hydroelectric energy

The flow of water from a higher to a lower level from behind a tidal barrage (barrier) or the dam of a hydroelectric scheme is used to drive a water turbine (water wheel) connected to a generator.

One of the largest working tidal schemes is at the estuary of the River Rance in France, Figure 46.2. Feasibility studies have shown that a 10 mile long barrage across the Severn estuary could produce about 7% of today's electrical energy consumption in England and Wales. Such schemes have implications for the environment, including wildlife habitats, and for shipping routes.

In the UK hydroelectric power stations generate about 2% of the electricity supply. Most are located in Scotland and Wales where the average rainfall is higher than in other areas. With good management hydroelectric energy is a reliable energy source, but there are risks connected with the construction of dams, and a variety of problems may result from the impact of a dam on the environment.

Figure 46.2 Tidal barrage across the mouth of the River Rance, France

(e) Geothermal energy

If cold water is pumped down a shaft into hot rocks below the Earth's surface, it comes up another shaft as steam. This can be used to drive a turbine and generate electricity or to heat buildings. The energy heating the rocks is constantly being released by radioactive elements as they decay (Chapter 63).

Geothermal energy is used in some countries for heating and in the USA has operated a small turbogenerator. In the UK there was an experimental project in Cornwall.

(f) Biomass (vegetable fuels)

These include cultivated crops (e.g. oil-seed rape), crop residues (e.g. cereal straw), natural vegetation (e.g. gorse), trees (e.g. spruce) grown for their wood, animal dung and sewage. **Biofuels** such as alcohol (ethanol) and methane gas are obtained from them by fermentation using enzymes or by decomposition by bacterial action in the absence of air, Figure 46.3.

Liquid biofuels can replace petrol and, although they have up to 50% less energy per litre, they are lead- and sulphur-free and so cleaner. **Biogas** is a mix of methane and carbon dioxide with an energy content about two-thirds that of natural gas. It is used for heating and cooking in developing countries.

Figure 46.3 'Digester' (fermentation) tanks producing biofuel

Power stations

The processes involved in the production of electricity at power stations depend on the energy source being used.

(a) Non-renewable sources

These are used in **thermal** power stations to produce heat energy that turns water into steam. The steam drives turbines which in turn drive the generators that produce electrical energy as described in Chapter 60. If fossil fuels are the energy source (usually coal but natural gas is favoured in new stations), the steam is obtained from a boiler. If nuclear fuel is used, i.e. uranium or plutonium, the steam is produced in a heat exchanger as explained in Chapter 64.

The efficiency of thermal power stations is about 30%. They require cooling towers to condense steam from the turbine to water and this is a waste of energy. A block diagram and an energy-transfer diagram for a thermal power station are given in Figure 46.4.

In some recently constructed **gas-fired** power stations, natural gas is burnt in a gas turbine (Chapter 47) linked directly to an electricity generator. The hot exhaust gases from the turbine are not released into the atmosphere but used to produce steam in a boiler. The steam is then used to generate more electricity from a steam turbine driving another generator. The efficiency is claimed to be over 50% without any extra consumption of the fuel. Furthermore, the gas turbines have a near-100% combustion efficiency so very little harmful exhaust gas (i.e. unburnt methane) is produced, and natural gas is almost sulphur-free so the environmental pollution caused is much less than for coal.

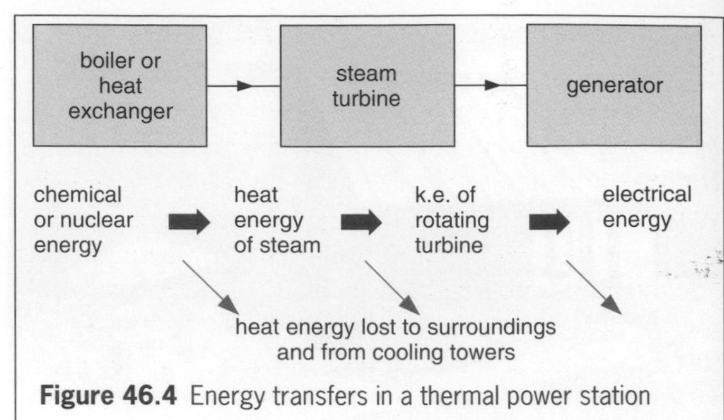

Figure 46.4 Energy transfers in a thermal power station

(b) Renewable sources

In most cases the energy source is used to drive turbines directly, as explained earlier in the cases of hydroelectric, wind, wave, tidal and geothermal schemes.

The block diagram and energy-transfer diagram for a hydroelectric scheme like that in Figure 25.3d are shown in Figure 46.5. The efficiency of a large installation can be as high as 85–90% since many of the causes of loss in thermal power stations (e.g. water cooling towers) are absent. In some cases the generating costs are half those of thermal stations.

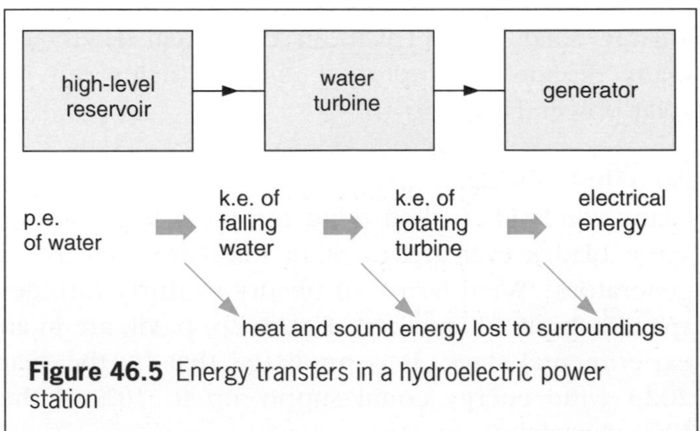

Figure 46.5 Energy transfers in a hydroelectric power station

A feature of some hydroelectric stations is **pumped storage**. Electrical energy cannot be stored, it must be used as it is generated, but the demand varies with the time of day and season. In this system electricity generated at off-peak periods is used to pump water back up from a low-level reservoir to a higher-level one. It is easier to do this than to reduce the output of the generator. At peak times the p.e. of the water in the high-level reservoir is converted back into electrical energy, three-quarters of the electrical energy used to pump the water being produced.

Economic, environmental and social issues

When considering the large-scale generation of electricity, the economic and environmental costs of using various energy sources have to be weighed against the benefits that electricity brings to society as a 'clean', convenient and fairly 'cheap' energy supply.

Environmental problems such as polluting emissions that arise with different energy sources were outlined when each was discussed previously. Apart from people using less energy, how far pollution can be reduced by, for example, installing desulphurization processes in coal-fired power stations, is often a matter of cost.

Although there are no fuel costs with non-polluting renewable sources, the energy is so 'thinly spread' that the capital cost of the initial installation is high. Similarly, despite fuel costs for nuclear power stations being low, the cost of building the stations and of dismantling them at the end of their useful life is high. It has been estimated that to generate a unit of electricity it costs 3.2p in a coal-fired power station, 2.9p in an advanced gas-cooled nuclear reactor and 2.7p in a gas-fired power station.

The reliability of a source has also to be considered as well as how cheaply production can be started up and shut down as demand varies. For example, tides are more predictable than wind or waves.

Renewable sources are still only being used on a small scale globally. The contribution of the main energy sources to the world's consumption at present is given in Table 46.1. The pattern in the UK is similar but France generates nearly three-quarters of its electricity from nuclear plants, for Japan and Taiwan the proportion is one-third, and it is in the developing economies of East Asia where interest in nuclear energy is growing most dramatically. However, the great dependence on fossil fuels worldwide is evident. It is clear the world has an energy problem.

Table 46.1 World use of energy sources

Oil	Coal	Gas	Nuclear	Hydroelectric
40%	28%	23%	7%	2%

Consumption varies from one country to another; North America and Europe are responsible for about two-thirds of the world's energy consumption each year. Table 46.2 shows approximate values for the annual consumption per head of population for different areas.

Table 46.2 Energy consumption per head per year/J $\times 10^9$

N. America	UK	Japan	S. America	China	Africa
330	150	120	40	25	15

The world average is 63×10^9 J per head per year.

QUESTIONS

1 The pie chart in Figure 46.6 shows the percentages of the main energy sources used by a certain country.

a What percentage is supplied by water power?

b Which of the sources is/are renewable?

c What is meant by 'renewable'?

d Name two other renewable sources.

e Why, if energy is always conserved, is it important to develop renewable sources?

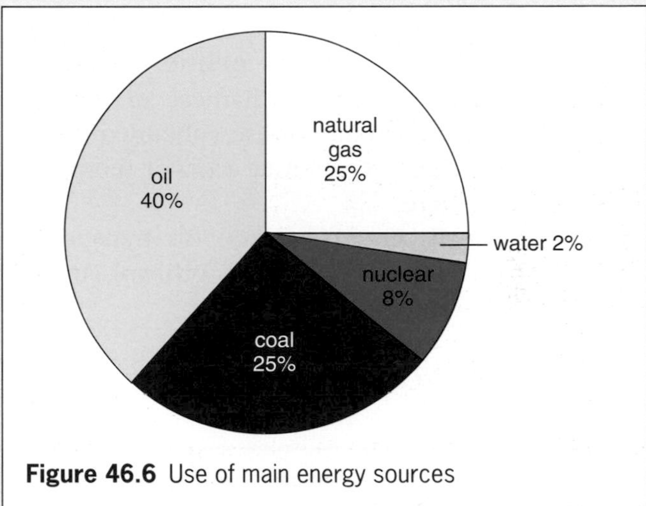

Figure 46.6 Use of main energy sources

2 List *six* properties which you think the ideal energy source should have for generating electricity in a power station.

3 a List *six* social everyday benefits for which electrical energy is responsible.

b Draw up *two* lists of suggestions for saving energy (i) in the home, and (ii) globally.

Checklist

After studying this chapter you should be able to

- distinguish between **renewable** and **non-renewable** energy sources,
- give some advantages and some disadvantages of using non-renewable fuels,
- describe the different ways of harnessing **solar**, **wind**, **wave**, **tidal**, **hydroelectric**, **geothermal** and **biomass** energy,
- describe the energy transfer processes in a thermal and a hydroelectric power station,
- discuss the environmental and economic issues of electricity production and consumption.

47 HEAT ENGINES

Petrol engines
Jet engines (gas turbines)
Steam turbines

A heat engine, also called a **prime mover**, is a machine which extracts mechanical energy (k.e.) from a gas at a high temperature (obtained by burning a fuel) and then expels it at a lower temperature into the atmosphere.

Oil is the main energy source for transport but other alternatives exist including ethanol, methane and electric batteries.

Petrol engines

The action of a 'four-stroke' engine is shown in Figure 47.1.

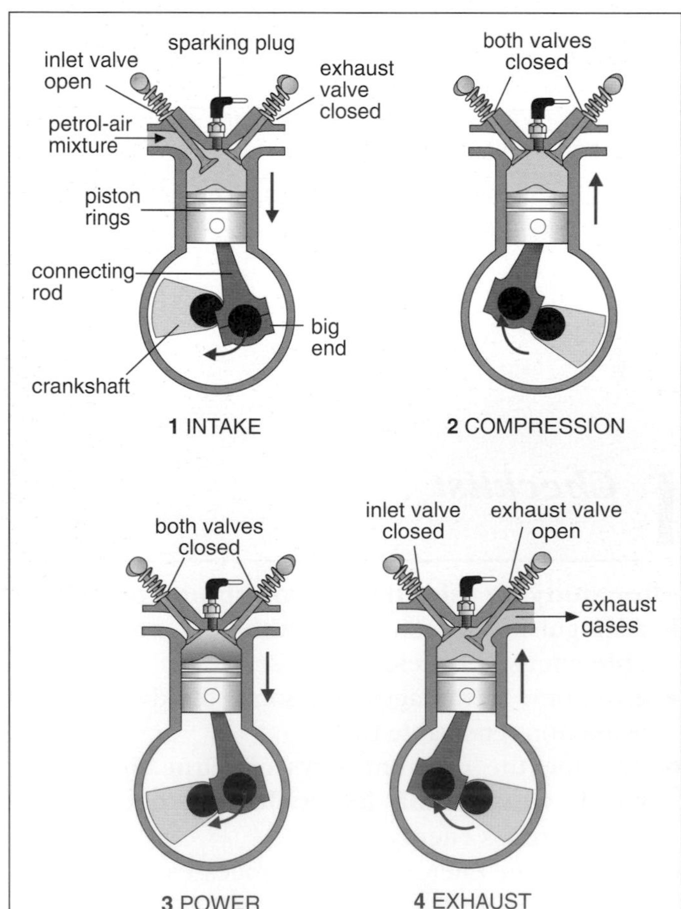

Figure 47.1 A four-stroke petrol engine

1 On the **intake stroke**, the piston moves down (due to the starter motor turning the crankshaft) so reducing the pressure inside the cylinder. The inlet valve opens and the petrol–air mixture from the carburettor is forced into the cylinder by atmospheric pressure.
2 On the **compression stroke**, both valves are closed and the piston moves up, compressing the mixture.
3 On the **power stroke**, a spark jumps across the points of the sparking plug and the mixture burns rapidly, forcing the piston down.
4 On the **exhaust stroke**, the outlet valve opens and the piston rises, pushing the exhaust gases out of the cylinder.

The crankshaft turns a flywheel (a heavy wheel) whose momentum keeps the piston moving between power strokes.

Most cars have at least four cylinders on the same crankshaft, Figure 47.2. Each cylinder 'fires' in turn, giving a power stroke every half revolution of the crankshaft. Smoother running results.

The efficiency (Chapter 25) of petrol engines varies with the load but is around 25%. This means that only 25% of the heat energy supplied becomes kinetic energy; much of the rest is lost with the exhaust gases.

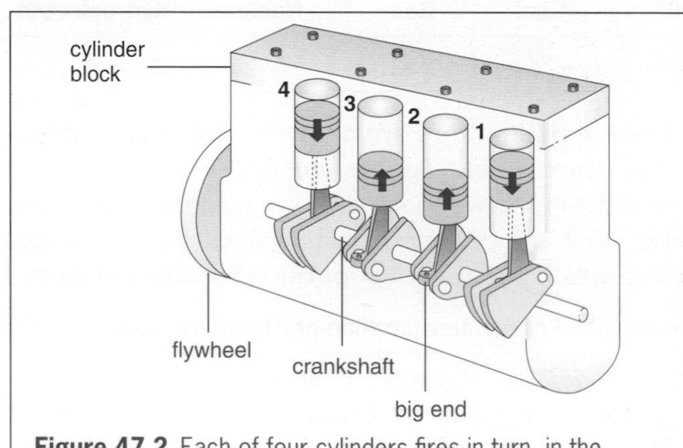

Figure 47.2 Each of four cylinders fires in turn, in the order 1–3–4–2

Jet engines (gas turbines)

There are several kinds of jet engine; Figure 47.3 is a simplified diagram of a turbo-jet. To start the engine, an electric motor sets the compressor rotating. The compressor is like a fan – its blades draw in and compress air at the front of the engine. Compression raises the temperature of the air before it reaches the combustion chamber. Here, fuel (kerosene) is injected and burns to produce a high-speed stream of hot gas which escapes from the rear of the engine, so thrusting it forward (as explained in Chapter 35 when we considered momentum). The exhaust gas also drives a turbine (another fan) which, once the engine is started, turns the compressor, since both are on the same shaft.

Gas turbines are used in gas-fired power stations to drive an electrical generator directly.

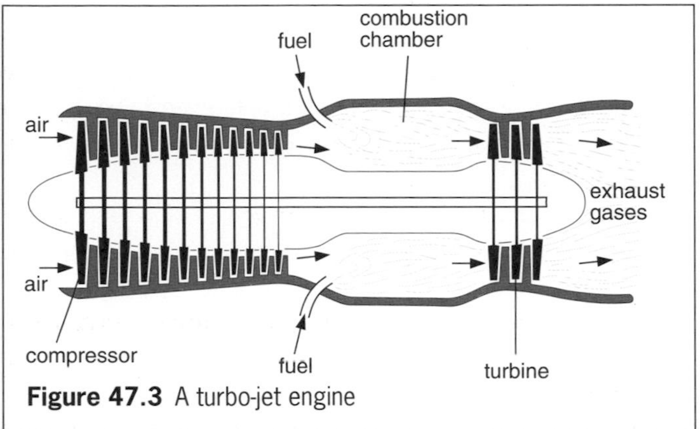

Figure 47.3 A turbo-jet engine

Steam turbines

Steam turbines are used in power stations and nuclear submarines, Figure 47.4. They have efficiencies of about 45%.

The action of a steam turbine resembles that of a water wheel but moving steam not moving water causes the motion. Steam produced in a separate boiler enters the turbine and is directed by the **stator** or **diaphragm** (sets of fixed blades) on to the **rotor** (sets of blades on a shaft that can rotate), Figure 47.5. The rotor revolves and drives whatever is coupled to it, e.g. an electrical generator or a ship's propeller. The steam expands as it passes through the turbine and the size of the blades increases along the turbine to allow for this.

Figure 47.4 Nuclear submarines are powered by a steam turbine, the steam being produced by the heat from a nuclear reactor

Figure 47.5 The rotor of a steam turbine

QUESTIONS

1 Most car engines are four-stroke petrol engines.

a Name the four strokes, in the correct sequence.

b How many times does the crankshaft revolve during the four stokes? *(E.A.)*

2 In a car engine chemical energy is converted into mechanical energy.

a What other forms of energy are produced?

b Some of the mechanical energy is not usefully employed. What happens to it?

c Describe the energy changes involved when the brakes are applied.

3 **a** Explain the actions of (i) a compressor, (ii) a turbine, in a jet engine.

b Draw a labelled diagram of a jet engine, showing the positions of the compressor and the turbine.

c The graphs in Figure 47.6 compare the performances of a propeller-driven aircraft with that of a turbo-jet. Explain what the graphs show.

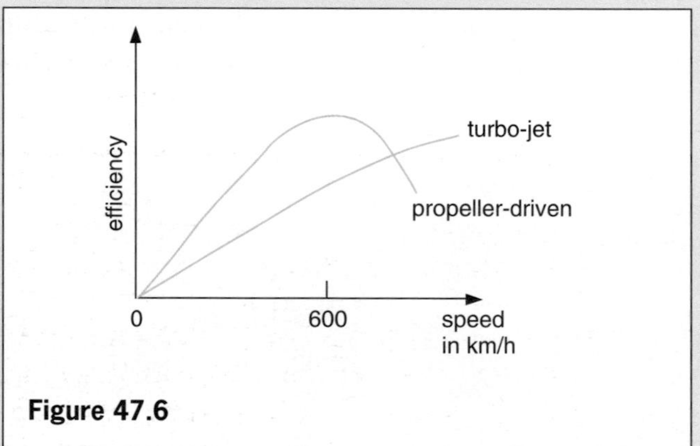

Figure 47.6

Checklist

After studying this chapter you should be able to explain the principles of operation of

- a four-stroke petrol engine,
- a jet engine,
- a steam turbine.

ADDITIONAL QUESTIONS

The first questions under each topic heading are 'basic' questions intended for all students; the following questions marked with a stripe are 'higher' questions for those seeking grades E to A*.

Thermometers

1 a State the advantages and disadvantages of mercury and alcohol as thermometric liquids.

b A clinical thermometer should not be sterilized in boiling water. Why?

Expansion of solids and liquids

2 When a metal bar is heated the increase in length is greater if

1 the bar is long
2 the temperature rise is large
3 the bar has a large diameter.

Which statement(s) is (are) correct?

A 1, 2, 3 **B** 1, 2 **C** 2, 3 **D** 1 **E** 3

3 A bimetallic thermostat for use in an electric iron is shown in the diagram below.

1 It operates by the bimetallic strip bending away from the contact.

2 Metal A has a greater expansivity than metal B.

3 Screwing in the control knob raises the temperature at which the contacts open.

Which statement(s) is (are) correct?

A 1, 2, 3 **B** 1, 2 **C** 2, 3 **D** 1 **E** 3

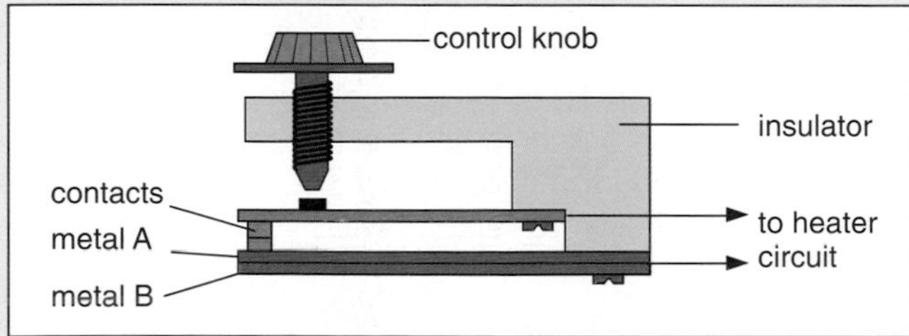

4 Which one of the graphs most nearly shows how the volume of water changes with temperature between −5 °C and +10 °C? *(J.M.B./A.L./N.W.16+)*

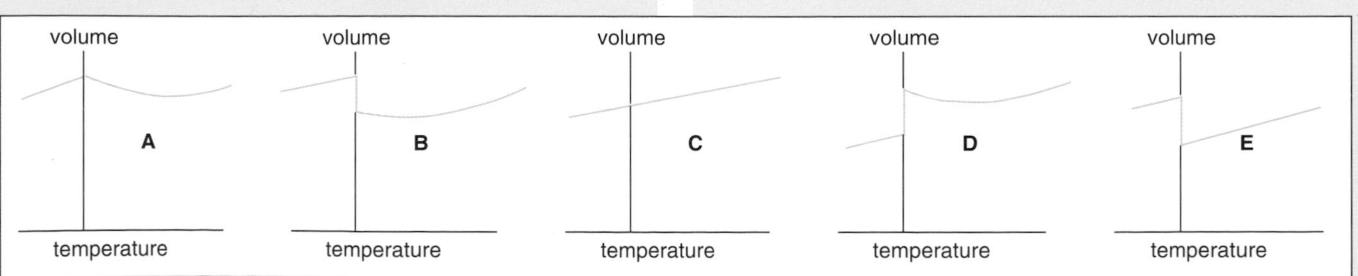

The gas laws

5 The pressure of the air inside a car tyre increases if the car stands for some time in full sunlight. According to kinetic theory this is due to an increase inside the tyre of

A the size of the molecules
B the number of air molecules
C the speed of the air molecules
D the average distance between the air molecules
E the total mass of the air molecules.

(O.L.E./S.R.16+)

6 Describe fully the apparatus to be used, the precautions which should be taken, and the observations to be made to establish the relationship at constant pressure between the volume of a fixed mass of gas and its temperature over the range approximately 0 °C to 100 °C.

Sketch the graph you would expect to obtain and indicate the expected relationship.

A thick-walled steel cylinder used for storing compressed air is fitted with a safety valve which lifts at a pressure of 1.0 Pa. It contains the air at 17 °C and 0.8 Pa. At what temperature will the valve lift? *(L.)*

7 A long capillary tube of uniform bore, and sealed at one end, has dry air trapped in it by a small pellet of mercury. The length of the air column is 145 mm at 17 °C. Find

a the length of the air column at 91 °C,

b the temperature at which the air column is 132 mm long. *(W.)*

Specific heat capacity; latent heat

8 A certain liquid has a specific heat capacity of 3.0 J/(g °C). What mass of the liquid may be heated from 20 °C to 50 °C by 630 J of heat?

A 4.2 g **B** 7.0 g **C** 10.5 g **D** 11.4 g **E** 12.0 g

(J.M.B./A.L./N.W.16+)

9 The diagram shows a graph of temperature against time for a substance which is heated at a constant rate from a low to high temperature. Use the letters to answer the following questions.

a Which part, or parts, of the graph correspond to the substance existing in two states at the same time?

b Over which part is the substance increasing in temperature at the fastest rate?

c Which point of the graph corresponds to the molecules of the substance having the greatest average kinetic energy? *(W.)*

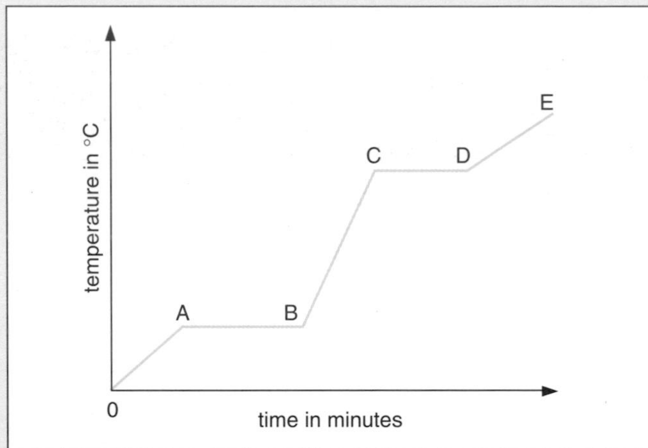

10 An immersion heater rated at 100 W supplies heat for 440 s to 2 kg of paraffin oil. Assuming that the specific heat capacity of paraffin oil is 2.2×10^3 J/(kg °C) and that all the heat from the heater is used to heat the paraffin, the rise in temperature of the paraffin in °C, is

A 1 **B** 10 **C** 20 **D** 40 **E** 100 *(N.I.)*

11 A mass of water was heated in a vessel with an immersion heater of power 40 W. The heater was used to boil the water for 100 s during which time the mass of water decreased by 0.002 kg. Assuming all the heat energy was given to the water, the specific latent heat of vaporization of water is

A $40 \times 100 \times 0.002$ J/kg
B $40/(100 \times 0.002)$ J/kg
C $100/(40 \times 0.002)$ J/kg
D $40 \times 100/0.002$ J/kg
E 40×0.002 J/kg *(J.M.B.)*

Conduction and convection; radiation

12 On a frosty day the metal handlebars of a bicycle feel colder than the rubber grips because

A the rubber is a better absorber of radiation than the metal
B the metal is colder than the rubber
C the rubber has a higher heat capacity than the metal
D the metal is a better conductor of heat than the rubber
E the metal is a better radiator of heat than the rubber. *(O.L.E./S.R.16+)*

13 Explain in terms of heat transfer the function of the following features of a Thermos flask.

a A tight fitting cork or plastic stopper.

b The silver coating on the inside of the outer glass wall and the outside of the inner glass wall.

c The evacuated enclosure between the walls. *(N.I.)*

14 a Describe the differences in the manner in which heat is transmitted by **conduction** and **radiation.**

Describe an experiment which shows that a shiny or white surface is a poorer **absorber** of heat radiation than a dull or black surface. Give a brief explanation of one use made of this fact in everyday life.

b Describe a method of showing that heat radiation is to be found spread over an infrared region in the spectrum of electromagnetic radiation from the Sun.

How does the action of a greenhouse depend on the difference in behaviour between longer and shorter wavelengths of infrared radiation? *(L.)*

15 Calculate the rate of heat energy loss through the four double-brick air-cavity walls of a building measuring 10 m by 10 m along the ground. The walls are 5 m high and the temperatures outside and inside the building are 3 °C and 18 °C respectively. (*U*-value of walls = 2.0 W/m^2 °C)

ELECTRICITY AND MAGNETISM

48 STATIC ELECTRICITY

Figure 48.1 A flash of lightning is nature's most spectacular static electricity effect

A nylon garment often crackles when it is taken off. We say it has become 'charged with static electricity'; the crackles are caused by tiny electric sparks which can be seen in the dark. Pens and combs made of certain plastics become charged when rubbed on the sleeve and can then attract scraps of paper.

Positive and negative charges

When a strip of polythene (white) is rubbed with a cloth it becomes charged. If it is hung up and another rubbed polythene strip is brought near, repulsion occurs, Figure 48.2. Attraction occurs when a rubbed cellulose acetate (clear) strip approaches.

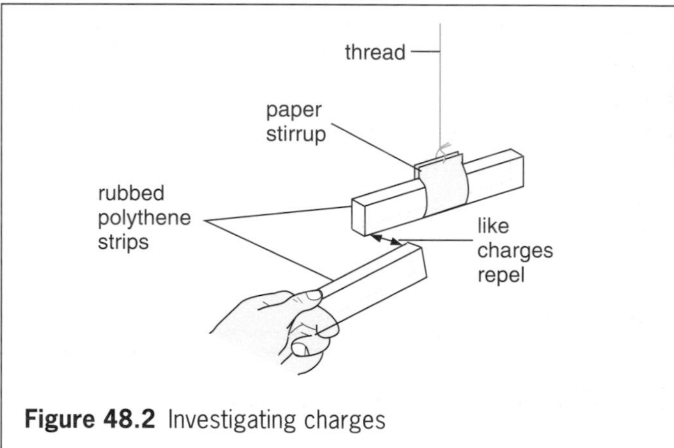

Figure 48.2 Investigating charges

This shows there are two kinds of electric charge. That on cellulose acetate is taken as **positive** (+) and that on polythene is **negative** (−). It also shows that:

> Like charges (+ and + or − and −) repel, unlike charges (+ and −) attract.

The force between electric charges decreases as their separation increases.

Charges, atoms and electrons

There is evidence (Chapter 64) that we can picture an atom as made up of a small central nucleus containing positively charged particles called **protons**, surrounded by an equal number of negatively charged **electrons**. The charges on a proton and an electron are equal and so an atom as a whole is normally electrically neutral, i.e. has no net charge.

Hydrogen is the simplest atom with one proton and one electron, Figure 48.3. A copper atom has 29 protons in the nucleus and 29 surrounding electrons. Every nucleus except hydrogen also contains uncharged particles called **neutrons**.

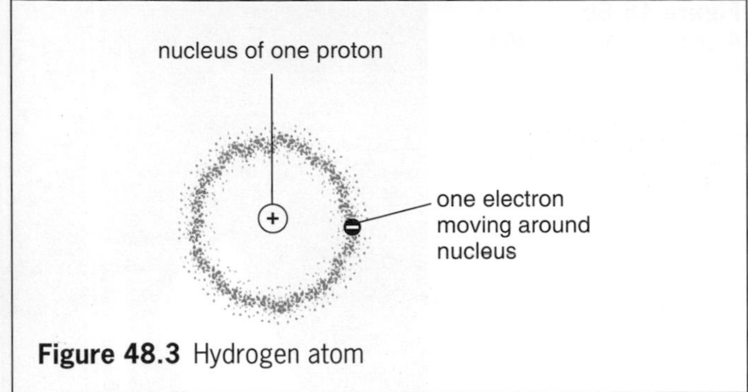

Figure 48.3 Hydrogen atom

The production of charges by rubbing can be explained by supposing that electrons are transferred from one material to the other. For example, when cellulose acetate is rubbed, electrons go from the acetate to the cloth, leaving the acetate short of electrons, i.e. positively charged. The cloth now has more electrons than protons and becomes negatively charged. Note that it is electrons which move, the protons remain in the nucleus.

How does polythene become charged when rubbed?

PRACTICAL WORK

Gold-leaf electroscope

A gold-leaf electroscope consists of a metal cap on a metal rod at the foot of which is a metal plate having a leaf of gold foil attached, Figure 48.4. The rod is held by an insulating plastic plug in a case with glass sides to protect the leaf from draughts.

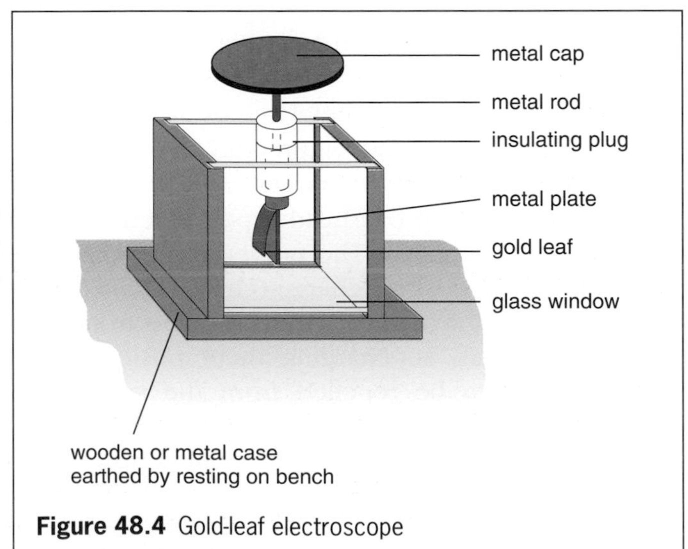

Figure 48.4 Gold-leaf electroscope

(a) Detecting a charge

Bring a charged polythene strip towards the cap: the leaf rises away from the plate. On removing the charged strip the leaf falls again. Repeat with a charged acetate strip.

(b) Charging by contact

Draw a charged polythene strip *firmly across the edge* of the cap. The leaf should rise and stay up when the strip is removed. If it does not, repeat the process but press harder. The electroscope has now become negatively charged by contact with the polythene strip.

(c) Insulators and conductors

Touch the cap of the charged electroscope with different things, e.g. a piece of paper, a wire, your finger, a comb, a cotton handkerchief, a piece of wood, a glass rod, a plastic pen, rubber tubing.

When the leaf falls, charge is passing to or from the ground through you and the material touching the cap. If the fall is rapid the material is a **good conductor**, if slow, it is a poor conductor and if the leaf does not alter the material is a **good insulator**. Record your results.

Electrons, insulators and conductors

In an insulator all electrons are bound firmly to their atoms; in a conductor some electrons can move freely from atom to atom. An insulator can be charged by rubbing because the charge produced cannot move from where the rubbing occurs, i.e. the electric charge is static. A conductor will become charged only if it is held with an insulating handle; otherwise electrons are transferred between the conductor and the ground via our body.

Good insulators are plastics such as polythene, cellulose acetate, Perspex, nylon. All metals and carbon are good conductors. In between are materials that are both poor conductors and (because they conduct to some extent) poor insulators. Examples are wood, paper, cotton, the human body, the Earth. Water conducts and if it were not present in materials like wood and on the surface of, for example, glass, these would be good insulators. Dry air insulates well.

Electrostatic induction

This effect may be shown by bringing a negatively charged polythene strip near to an insulated metal sphere X which is touching a similar sphere Y, Figure 48.5a (overleaf). Electrons in the spheres are repelled to the far side of Y.

If X and Y are separated, with the charged strip still in position, X is left with a positive charge (deficient of electrons) and Y with a negative charge (excess of electrons), Figure 48.5b. The signs of the charges can be tested by removing the charged strip (c) and taking X up to the cap of a positively charged electroscope and Y to a negatively charged one. In both cases the leaf should rise farther.

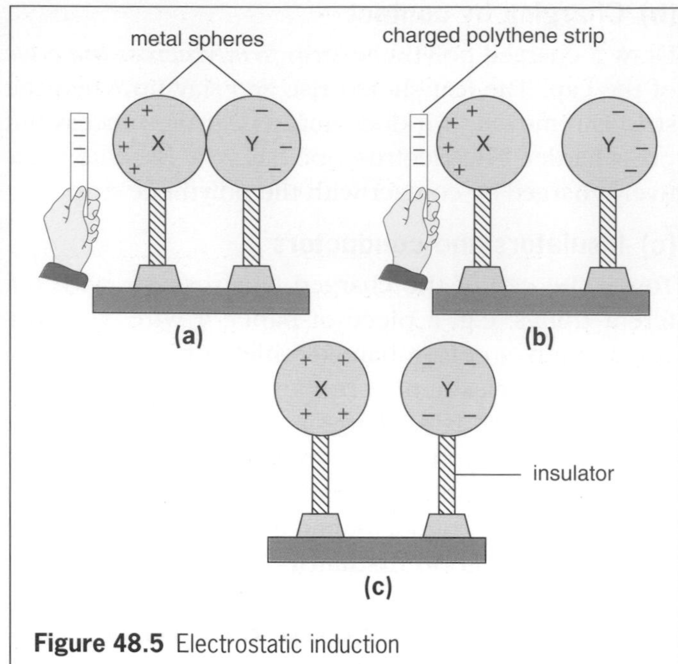

Figure 48.5 Electrostatic induction

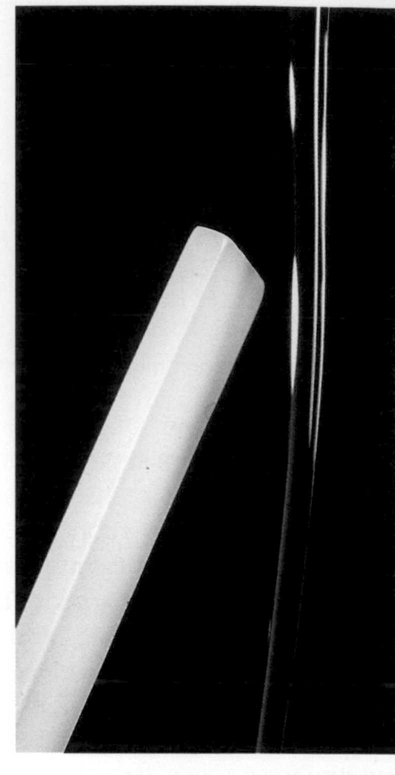

Figure 48.6b A slow stream of water being bent by electrostatic attraction to a charged polythene rod

Attraction of uncharged objects

The attraction of an uncharged object by a charged object near it is due to electrostatic induction.

In Figure 48.6a a small piece of aluminium foil is attracted to a negatively charged polythene rod held just above it. The charge on the rod pushes free electrons to the bottom of the foil (aluminium is a conductor), leaving the top of the foil short of electrons, i.e. with a net positive charge, and the bottom negatively charged. The top of the foil is nearer the rod than the bottom. Hence the force of attraction between the negative charge on the rod and the positive charge on the top of the foil is greater than the force of repulsion between the negative charge on the rod and the negative charge on the bottom of the foil. The foil is therefore pulled to the rod.

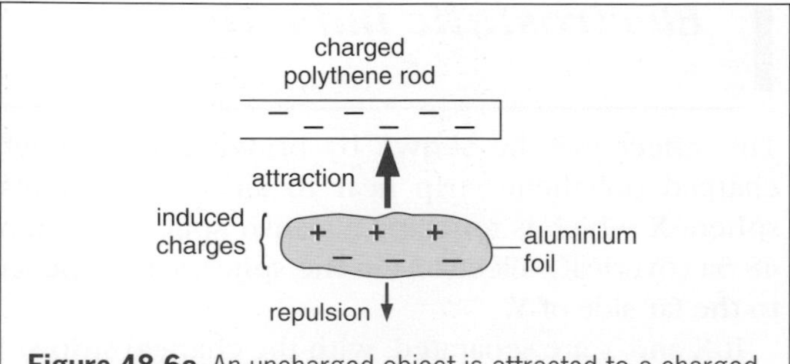

Figure 48.6a An uncharged object is attracted to a charged one

A small scrap of paper, although an insulator, is also attracted by a charged rod. There are no free electrons in the paper but the charged rod pulls the electrons of the atoms in the paper slightly closer (by electrostatic induction) and so distorts the atoms. In the case of a negatively charged polythene rod, the paper behaves as if it had a positively charged top and a negative charge at the bottom.

In Figure 48.6b a slow, uncharged stream of water is attracted by a charged polythene rod.

Dangers of static electricity

(a) Lightning

A tall building is protected by a lightning conductor consisting of a thick copper strip on the outside of the building connecting metal spikes at the top to a metal plate in the ground, Figure 48.7.

Thunderclouds carry charges and a negatively charged one passing overhead repels electrons from the spikes to the Earth. The points of the spikes are left with a large positive charge (charge concentrates on sharp points) which removes electrons from nearby air molecules, so charging them positively and causing them to be repelled from the spike. This effect, called **action at points**, results in an 'electric wind' of positive air molecules streaming upwards to cancel some of the charge on the cloud. If a flash does occur it is less violent and the conductor gives it an easy path to ground.

(b) Refuelling

Sparks from static electricity can be dangerous when flammable vapour is present. For this reason, the tanks in an oil-tanker may be cleaned in an atmosphere of nitrogen – otherwise oxygen in the air could promote a fire.

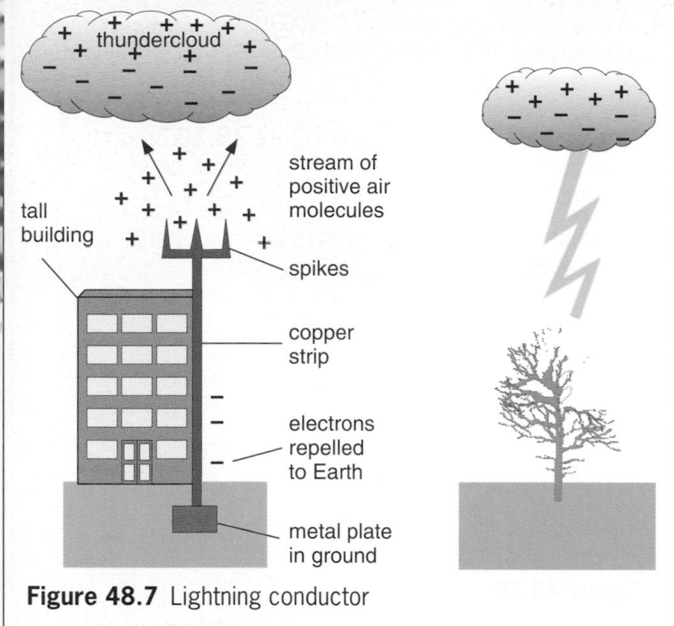

Figure 48.7 Lightning conductor

An aircraft in flight may become charged by 'rubbing' the air. Its tyres are of conducting rubber which lets the charge pass harmlessly to ground on landing, otherwise an explosion could be 'sparked off' when the aircraft refuels. What precautions are taken at petrol pumps to avoid similar charges when a car is refuelled?

(c) Operating theatres

Dust and germs are attracted by charged objects and so it is essential to ensure that equipment and medical personnel are well 'earthed' so that electrons can flow to and from the ground, e.g. no nylon carpets or clothing.

Uses of static electricity

(a) Flue-ash precipitation

An electrostatic precipitator removes dust and ash that goes up the chimneys of coal-burning power stations. It consists of a charged fine wire mesh which gives a similar charge to the rising particles of ash. They are then attracted to plates with an opposite charge. These are tapped from time to time to remove the ash which falls to the bottom of the chimney.

(b) Photocopiers

These contain a charged drum and when the paper to be copied is laid on the glass plate, the light reflected from the white parts of the paper causes the charge to disappear from the corresponding parts of the drum opposite. The charge pattern remaining on the drum corresponds to the dark-coloured printing on the original. Special **toner** powder is then dusted over the drum and sticks to those parts which are still charged. When a sheet of paper passes over the drum, the particles of toner are attracted to it and fused into place by a short burst of heat.

Van de Graaff generator

This produces a continuous supply of charge on a large metal dome when a rubber belt is driven by an electric motor or by hand, Figure 48.8.

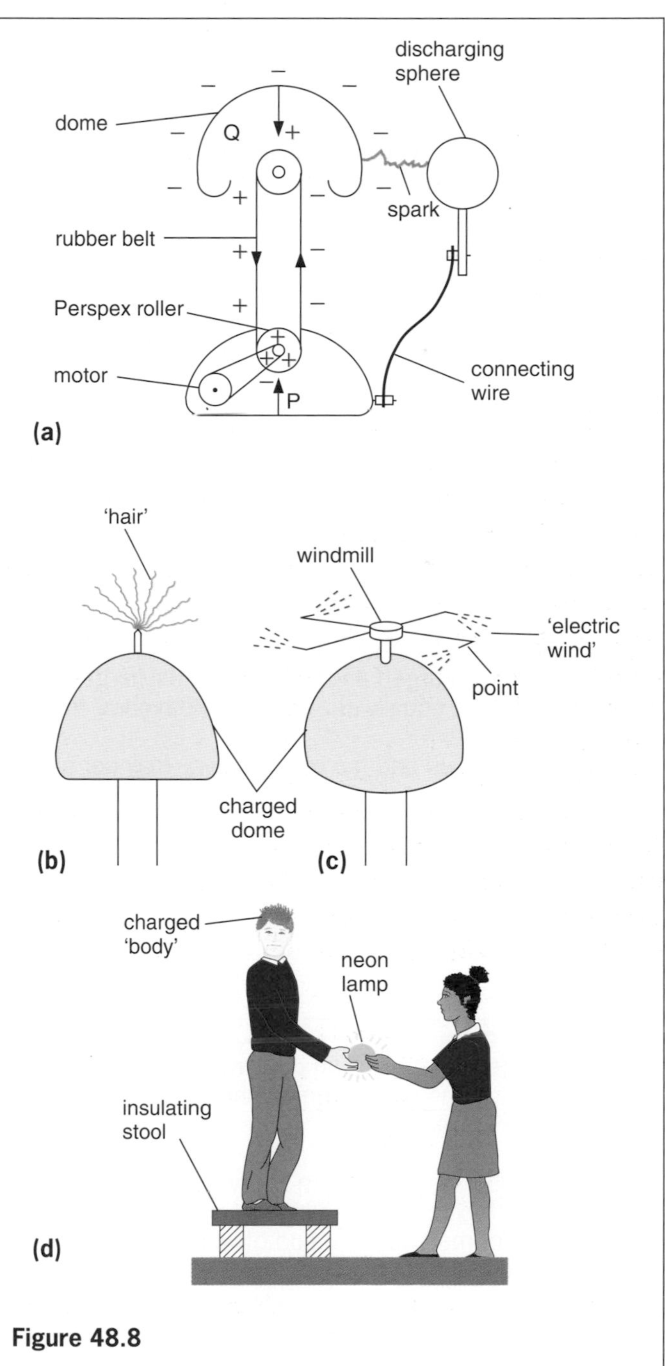

Figure 48.8

(a) Demonstrations

In (a) sparks jump between the dome and the discharging sphere. Electrons flow round a complete path (circuit) from the dome. Can you trace it? In (b) why does the 'hair' stand on end? In (c) the 'windmill' revolves due to the reaction that arises from the 'electric wind' caused by the **action at points** effect, explained above for the lightning conductor. In (d) the 'body' on the insulating stool first gets charged by touching the dome and then lights a neon lamp.

The dome can be discharged harmlessly by bringing your elbow close to it.

(b) Action

Initially a positive charge is produced on the motor-driven Perspex roller due to it rubbing the belt. This induces a negative charge on the 'comb' of metal points P, Figure 48.8a, which are sprayed off by 'action at points' onto the outside of the belt and carried upwards. A positive charge is then induced in the comb of metal points Q and negative charge is repelled to the dome.

QUESTIONS

1 Two light conducting balls, suspended on nylon threads, come to rest with the threads making equal angles with the vertical, Figure 48.9. This shows that:

A the balls are equally and oppositely charged
B the balls are oppositely charged but not necessarily equally charged
C one ball is charged and the other is uncharged
D the balls are equally charged and both carry the same charge
E one is charged and the other may or may not be charged.

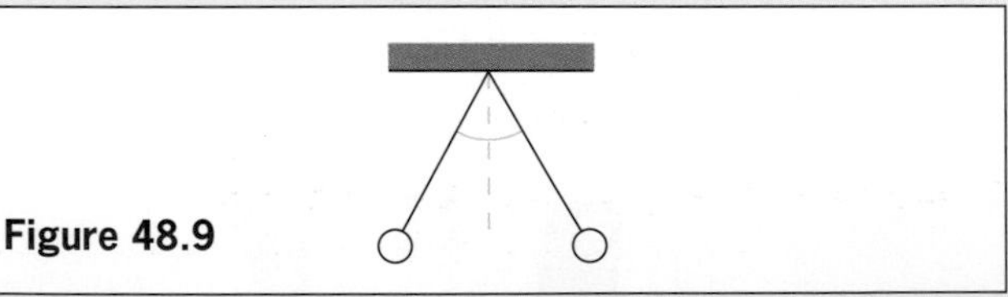

Figure 48.9

2 Explain in terms of electron movement what happens when a polythene rod becomes charged negatively by being rubbed with a cloth.

3 A metal body, insulated from everything else, is given an electrostatic charge. This body is then touched with a rod of material, one end of which is held in the hand. Which of the following statements is/are true?

a If the body retains its charge, the rod is made from a material which is a good insulator.

b If the body loses its charge slowly, the rod is made from a material which is a poor insulator.

c If the body loses its charge rapidly, the rod is made from a material which is a good conductor.

(J.M.B./A.L./N.W.16+)

4 A negatively charged rod is brought close to an uncharged metal sphere which is held on an insulated stand.

a Which of the diagrams in Figure 48.10 best shows the distribution of charge on the sphere when the rod is near?

b What would happen if the sphere was earthed while the negatively charged rod was near? *(E.A.)*

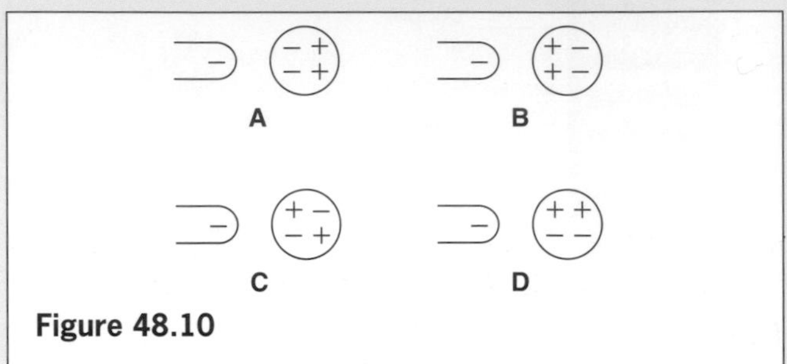

Figure 48.10

Checklist

After studying this chapter you should be able to

- describe how positive and negative charges are produced by rubbing,
- recall that like charges repel and unlike charges attract,
- explain the charging of objects in terms of the motion of negatively charged electrons,
- describe the gold-leaf electroscope, and explain how it can be used to compare electrical conductivities of different materials,
- explain the differences between insulators and conductors,
- describe how a conductor can be charged by induction,
- explain how a charged object can attract uncharged objects,
- give examples of the dangers and the uses of static electricity,
- describe the action of a van de Graaff generator and explain effects observed with its use.

49 ELECTRIC CURRENT

Effects of a current
The ampere and the coulomb
Circuit diagrams
Series and parallel circuits
Direct and alternating current
Practical work
Measuring current.

An electric current consists of moving electric charges. In Figure 49.1 when the van de Graaff is working, the table tennis ball dashes to and fro between the plates and the meter records a small current. As the ball touches each plate it becomes charged and is repelled to the other plate. In this way charge is carried across the gap. This also shows that 'static' charges cause a deflection on a meter just as current electricity produced by a battery.

In a metal, each atom has one or more loosely held electrons that are free to move. When a van de Graaff or a battery is connected across the ends of such a conductor, the free electrons drift slowly along it in the direction from the negative to the positive terminal of a battery. There is then a current of negative charge.

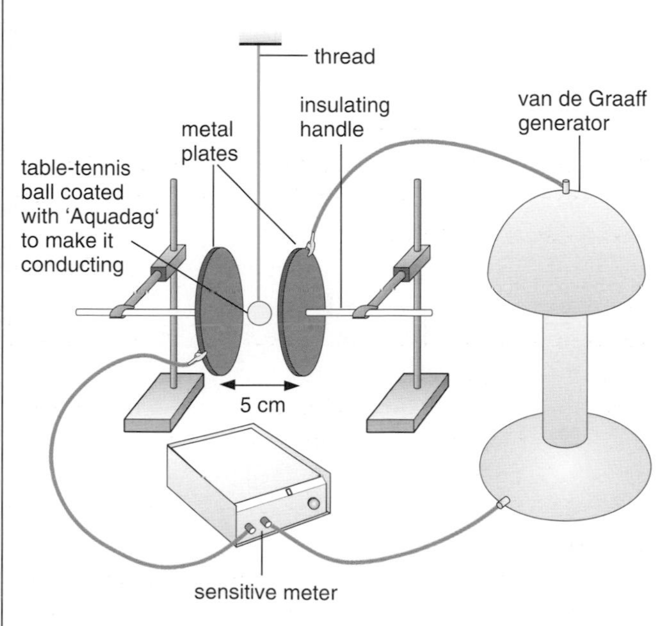

Figure 49.1 Demonstrating that an electric current consists of moving charges

Effects of a current

An electric current has three effects that reveal its existence and which can be shown with the circuit of Figure 49.2.

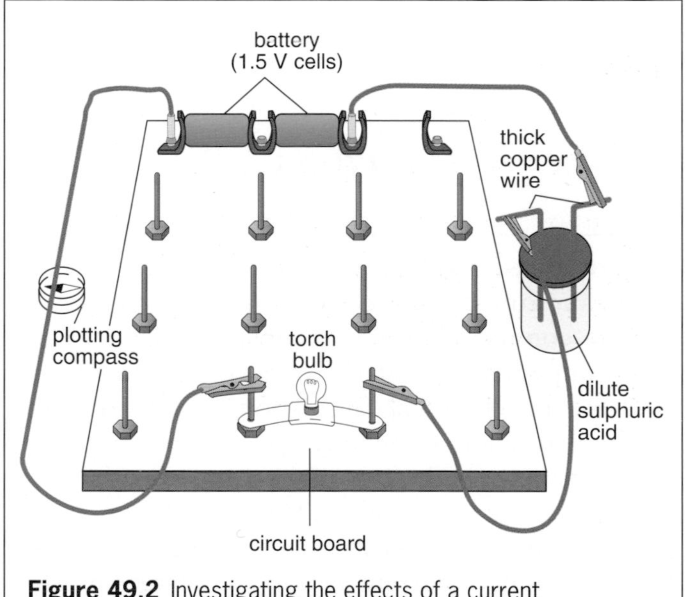

Figure 49.2 Investigating the effects of a current

(a) Heating and lighting

The bulb lights due to a small wire in it (the filament) being made white hot by the current.

(b) Magnetic

The plotting compass is deflected when it is placed in the magnetic field produced round any wire carrying a current.

(b) Chemical

Bubbles of gas are given off at the wires in the acid because of the chemical action of the current.

The ampere and the coulomb

The unit of current is the **ampere** (A) which is defined using the magnetic effect. One milliampere (mA) is one-thousandth of an ampere. Current is measured by an **ammeter**.

The unit of charge, the **coulomb** (C), is defined in terms of the ampere.

One coulomb is the charge passing any point in a circuit when a steady current of 1 ampere flows for 1 second. That is, 1 C = 1 A s.

A charge of 3 C would pass each point in 1 s if the current was 3 A. In 2 s, 3 A × 2 s = 6 A s = 6 C would pass. In general, if a steady current I (amperes) flows for time t (seconds) the charge Q (coulombs) passing any point is given by

$$Q = I \times t$$

This is a useful expression connecting charge and current.

Circuit diagrams

Current must have a complete path (a circuit) of conductors if it is to flow. Wires of copper are used to connect batteries, lamps, etc. in a circuit since copper is a good electrical conductor. If the wires are covered with insulation, e.g. plastic, the ends are bared for connecting up.

The signs or symbols used for various parts of an electrical circuit are shown in Figure 49.3.

Before the electron was discovered scientists agreed to think of current as positive charges moving round a circuit in the direction from positive to negative of a battery. This agreement still stands. Arrows on circuit diagrams show the direction of what we call the **conventional current**, i.e. the direction in which **positive** charges would flow.

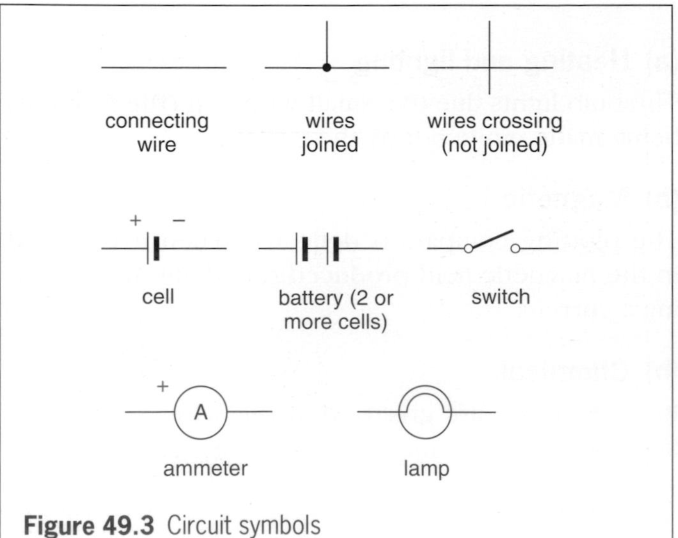

Figure 49.3 Circuit symbols

PRACTICAL WORK

Measuring current

(a) Connect the circuit of Figure 49.4a (on a circuit board if possible) ensuring that the + of the cell (the metal stud) goes to the + of the ammeter (marked red). Note the current.

(b) Connect the circuit of Figure 49.4b. The cells are **in series** (+ of one to − of the other), as are the lamps. Record the current. Measure the current at B, C and D by disconnecting the circuit at each point in turn and inserting the ammeter. What do you find?

(c) Connect the circuit of Figure 49.4c. The lamps are **in parallel**. Read the ammeter. Also measure the currents at P, Q and R. What is your conclusion?

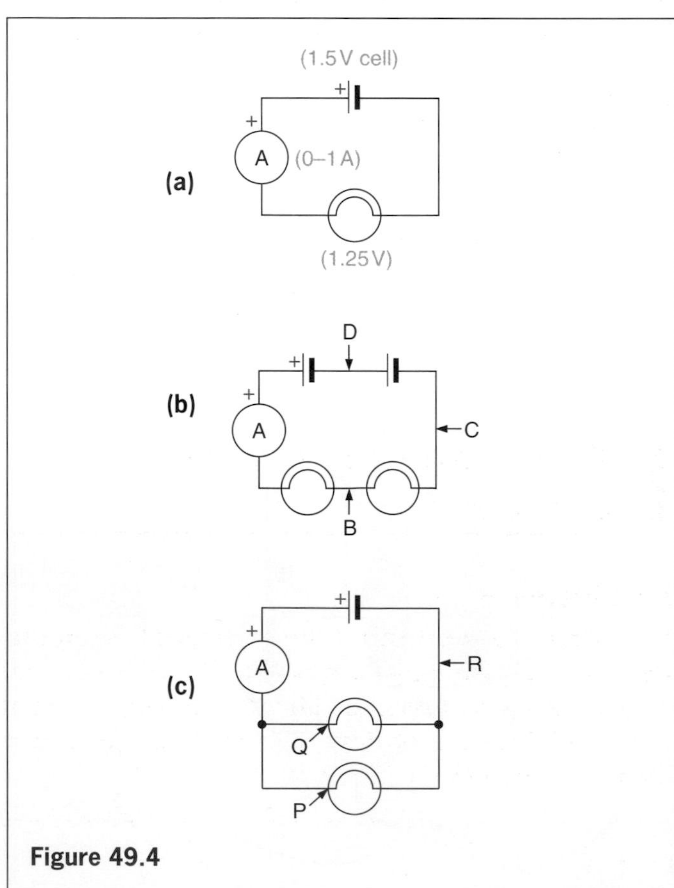

Figure 49.4

Series and parallel circuits

(a) Series

In a series circuit, Figure 49.4b, the different parts follow one after the other and there is just one path for the current to follow. You should have found in the previous experiment that the reading on the ammeter when in the position shown (e.g. 0.2 A) is also obtained at B, C and D. That is, current is not used up.

The current is the same at all points in a series circuit.

(b) Parallel

In a parallel circuit, Figure 49.4c, the lamps are side by side and alternative paths are provided for the current which splits. Some goes through one lamp and the rest through the other. For example, if the ammeter reading was 0.4 A in the position shown, then if the lamps are identical the reading at P would be 0.2 A, as it would be at Q, giving a total of 0.4 A. Whether the current splits equally or not depends on the lamps (as we will see later); it might divide so that 0.3 A goes one way and 0.1 A by the other branch.

The sum of the currents in the branches of a parallel circuit equals the current entering or leaving the parallel section.

Direct and alternating current

(a) Difference

In a **direct current (d.c.)** the electrons flow in one direction only. Graphs for steady and varying d.c. are shown in Figure 49.5. In an **alternating current (a.c.)** the direction of flow reverses regularly, Figure 49.6. The circuit sign for a.c. is —∿— .

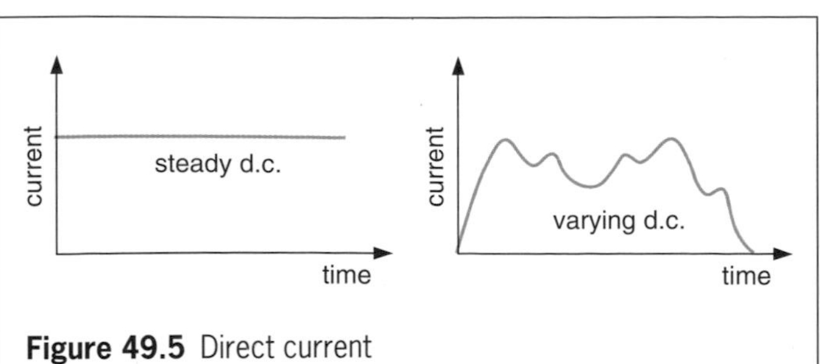

Figure 49.5 Direct current

The pointer of an ammeter for measuring d.c. is deflected one way by d.c.; a.c. makes it move to and fro about the zero if the changes are slow enough, otherwise there is no deflection.

Batteries give d.c.; generators can produce either d.c. or a.c.

(b) Frequency of a.c.

The number of complete alternations or cycles in 1 second is the **frequency** of the a.c. The unit of frequency is the **hertz** (Hz) (formerly the cycle per second, c/s). The frequency of the a.c. in Figure 49.6 is 2 Hz, which means there are two cycles per second, or one cycle lasts $\frac{1}{2} = 0.5$ s. The mains supply in the UK is a.c. of frequency 50 Hz.

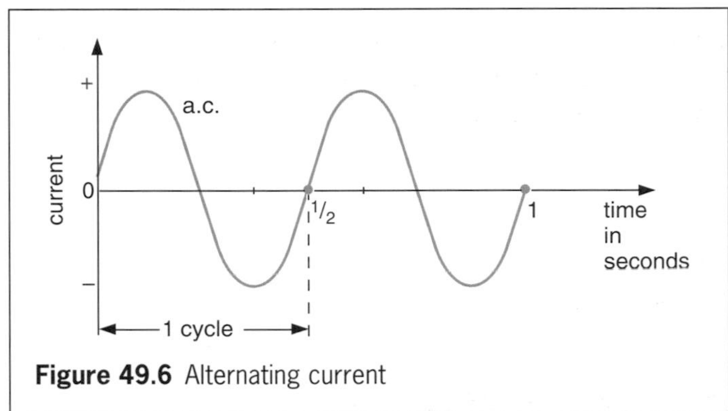

Figure 49.6 Alternating current

QUESTIONS

1 If the current through a floodlamp is 5 A, what charge passes in **a** 1 s, **b** 10 s, **c** 5 minutes?

2 What is the current in a circuit if the charge passing each point is **a** 10 C in 2 s, **b** 20 C in 40 s, **c** 240 C in 2 minutes?

3 **a** Figure 49.7a shows a cell correctly connected to 2 lamps and a meter. Draw a circuit diagram for the same circuit.

b Figure 49.7b shows 3 lamps and a cell. Draw a circuit diagram of the same circuit. *(E.A.)*

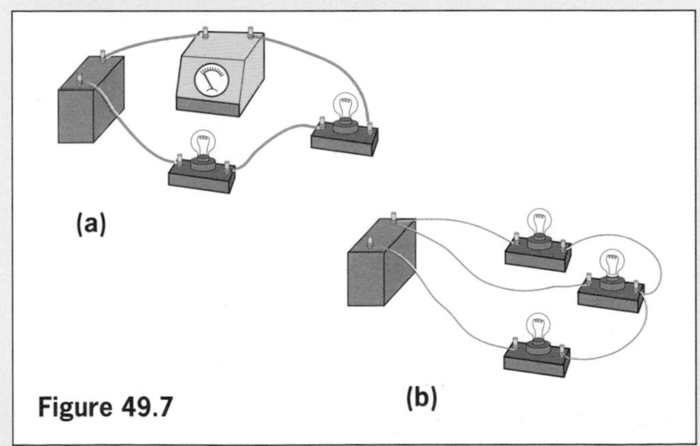

Figure 49.7

4 Study the circuits in Figure 49.8. The switch S is open (there is a break in the circuit at this point). The circuit in which lamp P would not light but lamps Q and R would is:

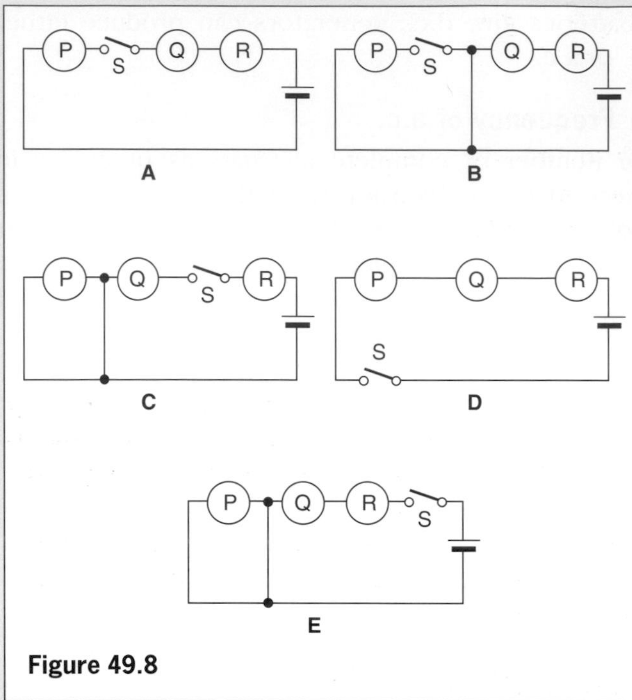

Figure 49.8

5 Using the circuit in Figure 49.9, which of the following statements is correct?

a When S_1 and S_2 are closed A and B are lit.

b With S_1 open and S_2 closed A and B are not lit.

c With S_2 open and S_1 closed A lights and B does not light. *(A.L.)*

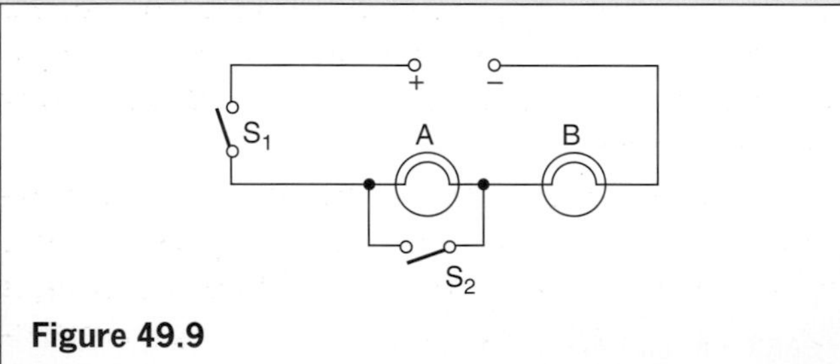

Figure 49.9

6 If the lamps are all the same in Figure 49.10 and if (A_1) reads 0.50 A, what do (A_2), (A_3), (A_4) and (A_5) read?

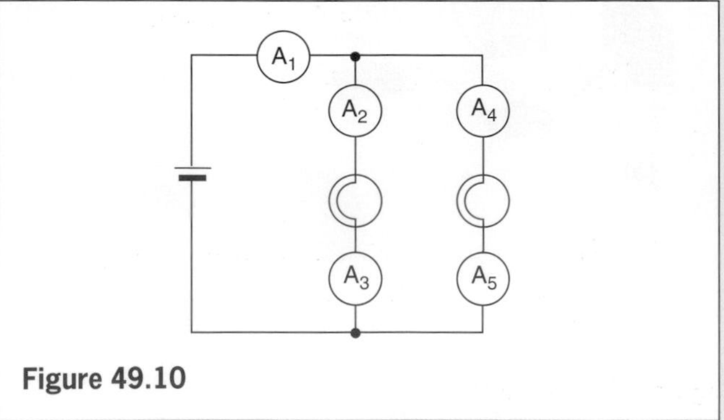

Figure 49.10

Checklist

After studying this chapter you should be able to

- describe a demonstration which shows that an electric current is a flow of charge,
- recall that an electric current in a metal is a flow of negative electrons from the negative to the positive terminal of the battery round a circuit,
- state the three effects of an electric current,
- state the unit of electric current and recall that current is measured by an ammeter,
- define the unit of charge in terms of the unit of current,
- recall the relation $Q = It$ and use it to solve problems,
- use circuit symbols for wires, cells, switches, ammeters and lamps,
- draw and connect simple series and parallel circuits, observing correct polarities for meters,
- recall that the current in a series circuit is the same everywhere in the circuit,
- recall that the sum of the currents in the branches of a parallel circuit equals the current entering or leaving the parallel section,
- distinguish between electron flow and conventional current,
- distinguish between direct current and alternating current,
- recall that frequency of a.c. is the number of cycles per second.

50 POTENTIAL DIFFERENCE

Energy transfers and p.d.	***Voltages round a circuit***
Model of a circuit	***Practical work*** Measuring voltage.
The volt	
Cells and batteries	

A battery transforms chemical energy to electrical energy. Because of the chemical action going on inside it, it builds up a surplus of electrons at one of its terminals (the negative) and creates a shortage at the other (the positive). It is then able to maintain a **flow of electrons**, i.e. an **electric current**, in any circuit connected across its terminals so long as the chemical action lasts.

The battery is said to have a **potential difference (p.d.** for short) at its terminals. Potential difference is measured in **volts** (V) and the term **voltage** is sometimes used instead of p.d. The p.d. of a car battery is 12 V and of the domestic mains supply in the UK 240 V.

Energy transfers and p.d.

In an electric circuit electrical energy is supplied from a source such as a battery and is transferred to other forms of energy by devices in the circuit. A lamp produces heat and light.

If the circuits of Figure 50.1 are connected up, it will be found from the ammeter readings that the current is about the same (0.4 A) in each lamp. However, the mains lamp with a p.d. of 240 V applied to it gives much more light and heat than the car lamp with 12 V across it. In terms of energy, the mains lamp transfers a great deal more electrical energy in a second than the car lamp.

Evidently the p.d. across a device affects the rate at which it transfers electrical energy. This gives us a way of defining the unit of p.d. – the volt.

Figure 50.1 Investigating the effect of p.d. on energy transfer

Model of a circuit

It may help you to understand the definition of the volt, i.e. what a volt is, if you *imagine* that the current in a circuit is formed by 'drops' of electricity, each having a charge of 1 coulomb and carrying equal-sized 'bundles' of electrical energy. In Figure 50.2, Mr Coulomb represents one such 'drop'. As a 'drop' moves around the circuit it gives up all its energy which is changed to other forms of energy. **Note that electrical energy is 'used up', not charge or current.**

In our imaginary representation, Mr Coulomb travels round the circuit and unloads energy as he goes, most of it in the lamp. We think of him receiving a fresh 'bundle' every time he passes through the battery, which suggests he must be travelling very fast. In fact, as we found earlier (Chapter 49), the electrons

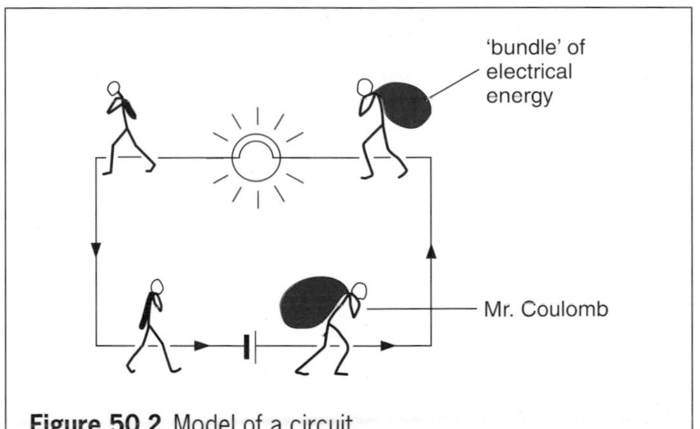

Figure 50.2 Model of a circuit

drift along quite slowly. As soon as the circuit is complete, energy is delivered at once to the lamp, not by electrons directly from the battery but from electrons that were in the connecting wires. The model is helpful but not an exact representation.

The volt

The demonstrations of Figure 50.1 show that the greater the voltage at the terminals of a supply, the larger is the 'bundle' of electrical energy given to each coulomb and the greater is the rate at which light and heat are produced in a lamp.

The p.d. between two points in a circuit is 1 volt if 1 joule of electrical energy is transferred to other forms of energy when 1 coulomb passes from one point to the other.

That is, 1 volt = 1 joule per coulomb (1 V = 1 J/C). If 2 J are given up by each coulomb, the p.d. is 2 V. If 6 J are transferred when 2 C pass, the p.d. is 6 J/2 C = 3 V.

In general if W (joules) is the energy transferred (i.e. the work done) when charge Q (coulombs) passes between two points, the p.d. V (volts) between the points is given by

$$V = W/Q \qquad \text{or} \qquad W = Q \times V$$

If Q is in the form of a steady current I (amperes) flowing for time t (seconds) then $Q = I \times t$ (Chapter 49) and

$$W = I \times t \times V$$

Cells and batteries

A 'battery' consists of two or more **electric cells** (see Chapter 53). Greater voltages are obtained when cells are joined in series, i.e. + of one to − of next. In Figure 50.3a the two 1.5 V cells give a voltage of 3 V at the terminals A, B. Every coulomb in a circuit connected to this battery will have 3 J of electrical energy.

If two 1.5 V cells are connected in parallel, Figure 50.3b, the voltage at terminals P, Q is still 1.5 V but the arrangement behaves like a larger cell and will last longer.

The cells in Figure 50.3c are in opposition and the voltage at X, Y is zero.

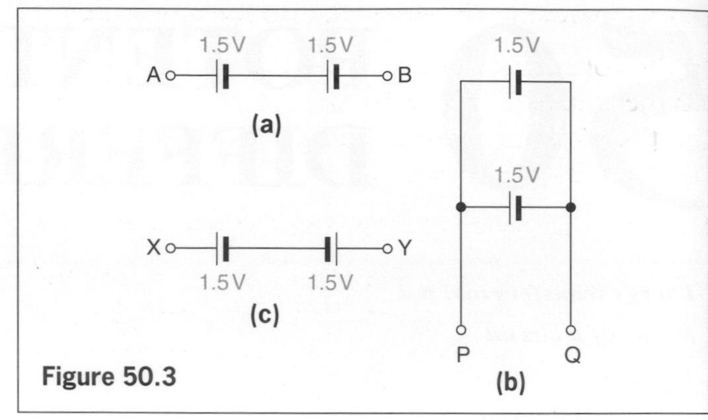

Figure 50.3

PRACTICAL WORK

Measuring voltage

A **voltmeter** is an instrument for measuring voltage or p.d. It looks like an ammeter but has a scale marked in volts. Whereas an ammeter is inserted **in series** in a circuit to measure the current, a voltmeter is connected across that part of the circuit where the voltage is required, i.e. **in parallel**. (We will see later that a voltmeter should have a high resistance and an ammeter a low resistance.)

To prevent damage the + terminal (marked red) must be connected to the point nearest the + of the battery.

(a) Connect the circuit of Figure 50.4a. The voltmeter gives the voltage across the lamp. Read it.

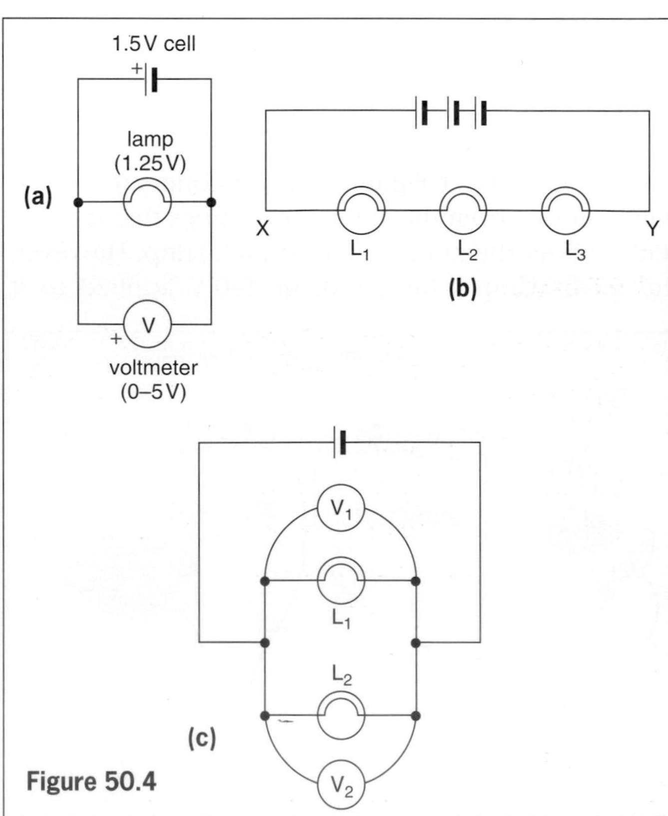

Figure 50.4

(b) Connect the circuit of Figure 50.4b. Measure:

(i) the voltage V between X and Y,
(ii) the voltage V_1 across lamp L_1,
(iii) the voltage V_2 across lamp L_2,
(iv) the voltage V_3 across lamp L_3.

How does the value of V compare with $V_1 + V_2 + V_3$?

(c) Connect the circuit of Figure 50.4c, so that two lamps L_1 and L_2 are in parallel across one 1.5 V cell. Measure the voltages, V_1 and V_2, across each lamp in turn. How do V_1 and V_2 compare?

Voltages round a circuit

(a) Series

In the previous experiment you should have found in the circuit of Figure 50.4b that

$$V = V_1 + V_2 + V_3$$

For example, if $V_1 = 1.4$ V, $V_2 = 1.5$ V and $V_3 = 1.6$ V, then V should be $(1.4 + 1.5 + 1.6) = 4.5$ V.

The voltage at the terminals of a battery equals the sum of the voltages across the devices in the external circuit from one battery terminal to the other.

(b) Parallel

In the circuit of Figure 50.4c $V_1 = V_2$.

The voltages across devices in parallel in a circuit are equal.

QUESTIONS

1 The p.d. across the lamp in Figure 50.5 is 12 V. How many joules of electrical energy are changed into light and heat when

a a charge of 1 C passes through it,

b a charge of 5 C passes through it,

c a current of 2 A flows through it for 10 s?

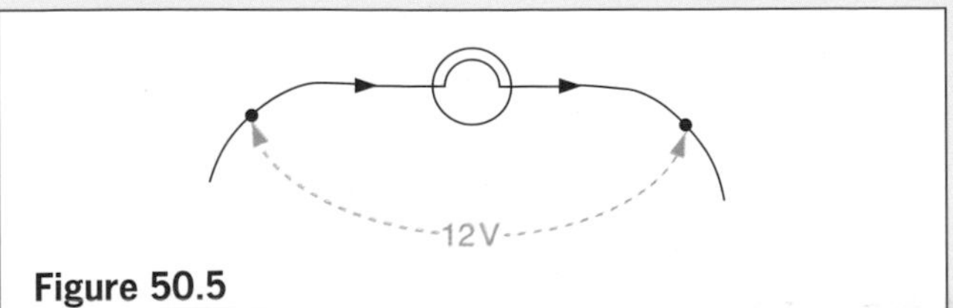

Figure 50.5

2 Three 2 V accumulator cells are connected in series and used as the supply for a circuit.

a What is the p.d. at the terminals of the supply?

b How many joules of electrical energy does 1 C gain on passing through (i) one accumulator, (ii) all three accumulators?

3 A set of Christmas tree lights consists of 20 identical lamps connected in series to a 240 V mains supply. What is the voltage (p.d.) across each lamp?

A 12 V **B** 20 V **C** 240 V **D** 4800 V (E.A.)

4 The symbol for a 1.5 V cell is —|⊢ . Which of the arrangements in Figure 50.6 would produce a battery with a p.d. of 6 V?

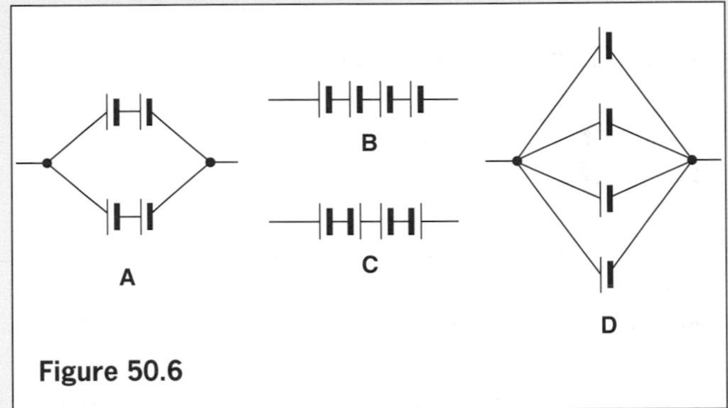

Figure 50.6

5 The lamps and the cells in all the circuits of Figure 50.7 are the same. If the lamp in (a) has its full, normal brightness, what can you say about the brightness of the lamps in (b), (c), (d), (e) and (f)?

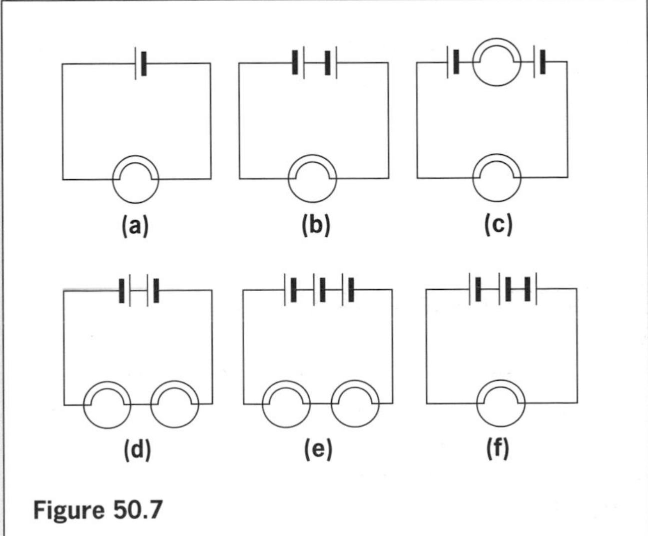

Figure 50.7

6 Three voltmeters (V), (V_1), (V_2) are connected as in Figure 50.8.

a If (V) reads 18 V and (V_1) reads 12 V, what does (V_2) read?

b If the ammeter (A) reads 0.5A, how much electrical energy is changed to heat and light in L_1 in 1 minute?

c Copy Figure 50.8 and mark with a + the positive terminals of the ammeter and voltmeters for correct connection.

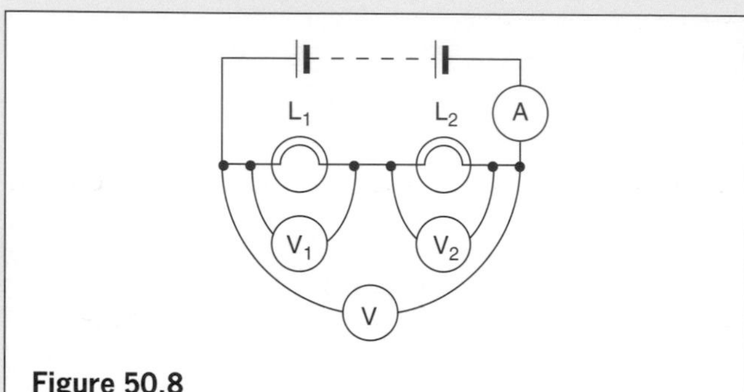

Figure 50.8

7 Three voltmeters are connected as in Figure 50.9. What are the voltmeter readings *x*, *y* and *z* in the table below (which were obtained with three different batteries)?

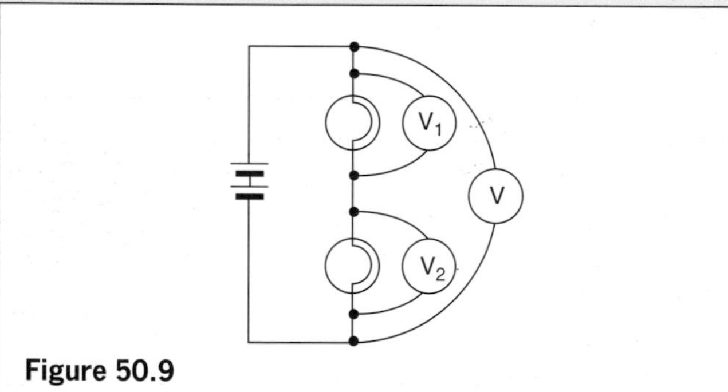

Figure 50.9

V/V	V_1/V	V_2/V
x	12	6
6	4	*y*
12	*z*	4

Checklist

After studying this chapter you should be able to

- describe simple experiments to show the transfer of electrical energy to other forms (e.g. in a lamp),
- recall the definition of the unit of p.d. and that p.d. (also called 'voltage') is measured by a voltmeter,
- demonstrate that the sum of the voltages across any number of components in series equals the voltage across all of those components,
- demonstrate that the voltages across any number of components in parallel are the same,
- work out the voltages of cells connected in series and parallel.

51 RESISTANCE

Electrons move more easily through some conductors than others when a p.d. is applied. The opposition of a conductor to current is called its **resistance**. A good conductor has a low resistance and a poor conductor has a high resistance. The resistance of a wire of a certain material

(i) increases as its length increases,
(ii) increases as its cross-section area decreases,
(iii) depends on the material.

A long thin wire has more resistance than a short thick one of the same material. Silver is the best conductor, but copper, the next best, is cheaper and is used for connecting wire and in electric cables.

The ohm

If the current through a conductor is I when the voltage across it is V, Figure 51.1a, its resistance R is defined by

$$R = \frac{V}{I}$$

This is a reasonable way to measure resistance since the smaller I is for a given V, the greater is R. If V is in volts and I in amperes, R is in **ohms** (Ω: pronounced omega). For example, if $I = 2$ A when $V = 12$ V, then $R = 12\text{ V}/2\text{ A} = 6\ \Omega$.

The ohm is the resistance of a conductor in which the current is 1 ampere when a voltage of 1 volt is applied across it.

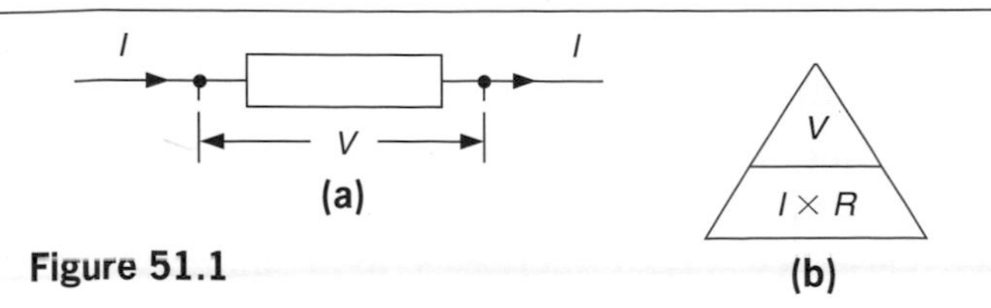

Figure 51.1

Alternatively, if R and I are known, V can be found from

$$V = IR$$

Also, knowing V and R, I can be calculated from

$$I = \frac{V}{R}$$

The triangle in Figure 51.1b is an aid to remembering the three equations. It is used like the 'density triangle' in Chapter 18.

Resistors

Conductors intended to have resistance are called **resistors** (symbol –▭–) and are made either from wires of special alloys or from carbon. Those in radio and television sets have values from a few ohms up to millions of ohms, Figure 51.2a.

Variable resistors are used in electronics (and are then called **potentiometers**) as volume and other controls, Figure 51.2b. Larger current versions are useful in laboratory experiments and consist of a coil of constantan wire (an alloy of 60% copper, 40% nickel) wound on a tube with a sliding contact on a metal bar above the tube, Figure 51.3 (overleaf).

There are two ways of using a variable resistor. It may be used as a **rheostat** for changing the current in a circuit; only one end connection and the sliding contact are then required. In Figure 51.4a moving the

Figure 51.2a Resistor

Figure 51.2b Variable resistor (potentiometer)

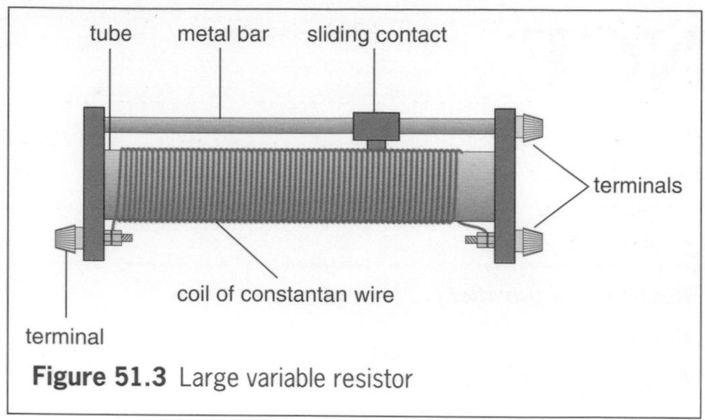

Figure 51.3 Large variable resistor

sliding contact to the left reduces the resistance and increases the current. It can also act as a **potential divider** for changing the p.d. applied to a device, all three connections being used. In Figure 51.4b any fraction from the total p.d. of the battery to zero can be 'tapped off' by moving the sliding contact down.

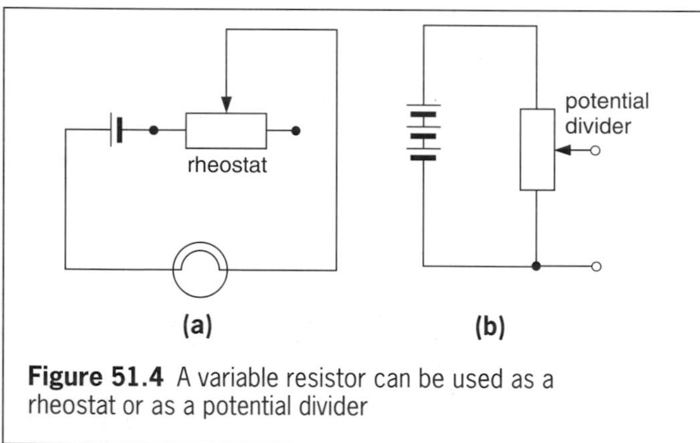

Figure 51.4 A variable resistor can be used as a rheostat or as a potential divider

PRACTICAL WORK

Measuring resistance

The resistance R of a conductor can be found by measuring the current I through it when a p.d. V is applied across it and then using $R = V/I$. This is called the **ammeter–voltmeter method**.

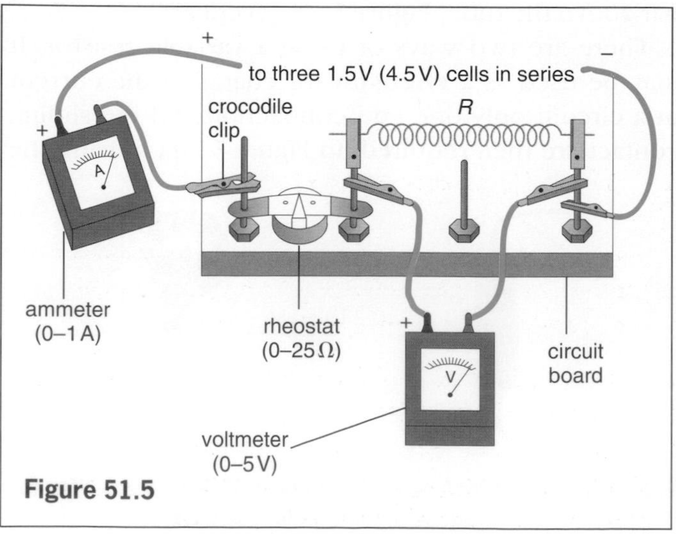

Figure 51.5

Set up the circuit of Figure 51.5 in which the unknown resistance R is 1 metre of SWG 34 constantan wire. Altering the rheostat changes both the p.d. V and the current I. Record in a table, with three columns, five values of I (e.g. 0.10, 0.15, 0.20, 0.25 and 0.30 A) and the corresponding values of V. Work out R for each pair of readings.

Repeat the experiment but instead of the wire use (i) a torch bulb (e.g. 2.5 V, 0.3 A), (ii) a semiconductor diode (e.g. 1 N4001) connected first one way then the other way round, (iii) a thermistor (e.g. TH 7).

I–V graphs: Ohm's law

The results of the previous experiment allow graphs of I against V to be plotted for different conductors.

(a) Metallic conductors

Metals and some alloys give I–V graphs which are a straight line through the origin, Figure 51.6a, so long as their temperature is constant. I is directly proportional to V, i.e. $I \propto V$. Doubling V doubles I, etc. Such conductors obey **Ohm's law**, stated as follows.

> The current through a metallic conductor is directly proportional to the voltage across its ends if the temperature and other conditions are constant.

They are called **ohmic** or **linear** conductors and since $I \propto V$, it follows that V/I = a constant (obtained from the slope of the I–V graph). The resistance of an ohmic conductor therefore does not change when the voltage does.

(b) Semiconductor diodes

The typical I–V graph in Figure 51.6b shows that current passes when the voltage is applied in one direction but is almost zero when it acts in the opposite direction. A diode has a small resistance when connected one way round but a very large resistance when the voltage is reversed. It conducts in one direction only and is a **non-ohmic** conductor. This makes it useful as a **rectifier** for changing alternating current (a.c.) to direct current (d.c.).

(c) Filament lamp

For a filament lamp, e.g. a torch bulb, the I–V graph bends over as V and I increase, Figure 51.6c. That is, the resistance (V/I) increases as I increases and makes the filament hotter.

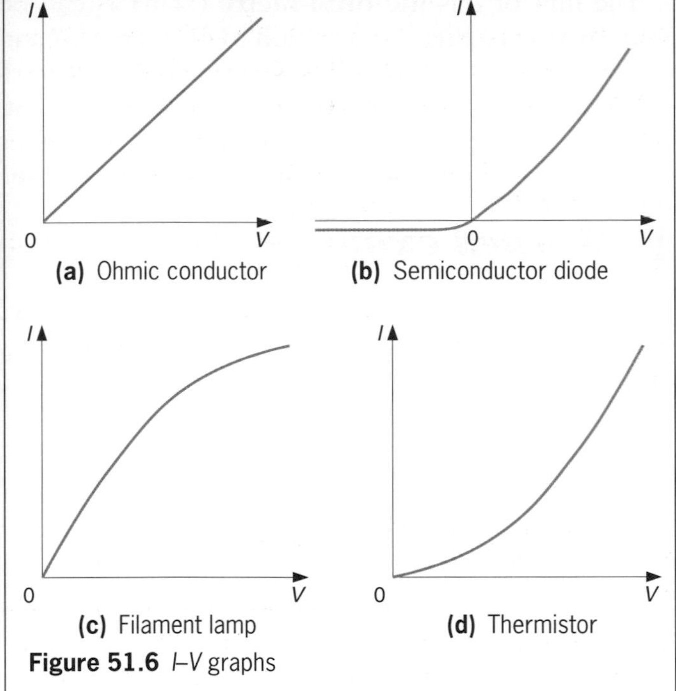

Figure 51.6 I–V graphs

(d) Variation of resistance with temperature

In general, an increase of temperature increases the resistance of metals, as for the filament lamp in Figure 51.6c, but decreases the resistance of semiconductors. The resistance of most **thermistors** decreases if their temperature rises, i.e. their I–V graph bends up, Figure 51.6d.

Resistors in series

The resistors in Figure 51.7 are in series. The **same current** I flows through each and the total voltage V across all three equals the separate voltages across them, i.e.

$$V = V_1 + V_2 + V_3$$

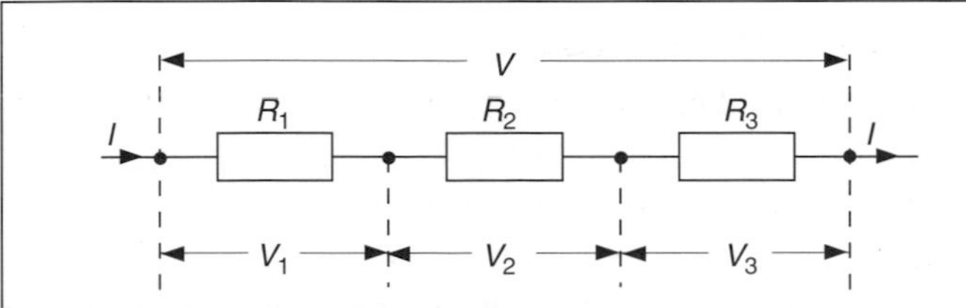

Figure 51.7 Resistors in series

But $V_1 = IR_1$, $V_2 = IR_2$ and $V_3 = IR_3$. Also, if R is the combined resistance, $V = IR$ and so

$$IR = IR_1 + IR_2 + IR_3$$

Dividing both sides by I,

$$R = R_1 + R_2 + R_3$$

Resistors in parallel

The resistors in Figure 51.8 are in parallel. The **voltage** V **between the ends of each is the same** and the total current I equals the sum of the currents in the separate branches, i.e.

$$I = I_1 + I_2 + I_3$$

But $I_1 = V/R_1$, $I_2 = V/R_2$ and $I_3 = V/R_3$. Also, if R is the combined resistance, $V = IR$ and so

$$\frac{V}{R} = \frac{V}{R_1} + \frac{V}{R_2} + \frac{V}{R_3}$$

Dividing both sides by V,

$$\frac{1}{R} = \frac{1}{R_1} + \frac{1}{R_2} + \frac{1}{R_3}$$

For the simpler case of *two* resistors in parallel

$$\frac{1}{R} = \frac{1}{R_1} + \frac{1}{R_2} = \frac{R_2}{R_1R_2} + \frac{R_1}{R_1R_2}$$

$$\therefore \quad \frac{1}{R} = \frac{R_2 + R_1}{R_1R_2}$$

Inverting both sides

$$R = \frac{R_1R_2}{R_1 + R_2} = \frac{\text{product of resistances}}{\text{sum of resistances}}$$

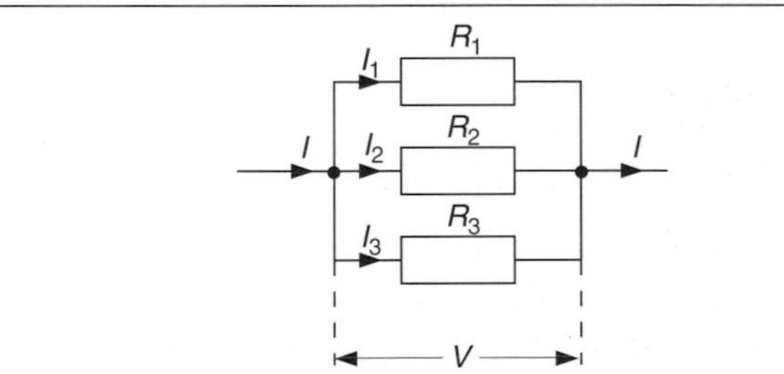

Figure 51.8 Resistors in parallel

Worked example

A p.d. of 24 V from a battery is applied to the network of resistors in Figure 51.9a (overleaf).

a What is the combined resistance of the 6 Ω and 12 Ω resistors in parallel?
b What is the current in the 8 Ω resistor?
c What is the voltage across the parallel network?
d What is the current in the 6 Ω resistor?

a Let R_1 = resistance of 6 Ω and 12 Ω in parallel

$$\therefore \quad \frac{1}{R_1} = \frac{1}{6} + \frac{1}{12} = \frac{2}{12} + \frac{1}{12} = \frac{3}{12}$$

$$\therefore \quad R_1 = \frac{12}{3} = 4\ \Omega$$

b Let R = *total* resistance of circuit = 4 + 8 = 12 Ω. The equivalent circuit is shown in Figure 51.9b and if I is the current in it then since V = 24 V

$$I = \frac{V}{R} = \frac{24\ \text{V}}{12\ \Omega} = 2\ \text{A}$$

∴ current in 8 Ω resistor = 2 A

c Let V_1 = voltage across parallel network

$$\therefore \quad V_1 = I \times R_1 = 2\ \text{A} \times 4\ \Omega = 8\ \text{V}$$

d Let I_1 = current in 6 Ω resistor, then since V_1 = 8 V

$$I_1 = \frac{V_1}{6\ \Omega} = \frac{8\ \text{V}}{6\ \Omega} = \frac{4}{3}\ \text{A}$$

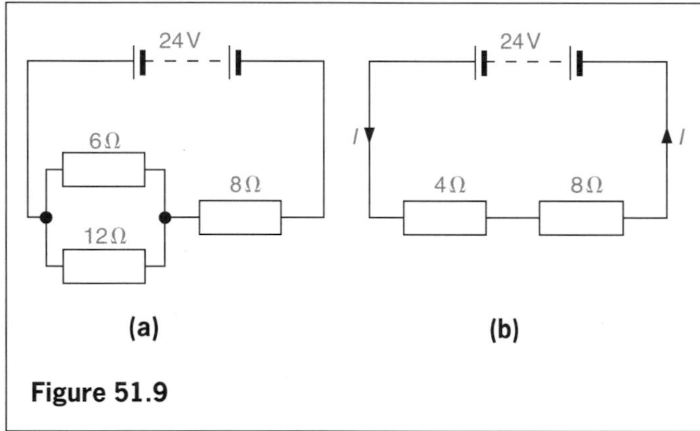

Figure 51.9

Resistivity

Experiments show that the resistance R of a wire of a given material is

(i) directly proportional to its length l, i.e. $R \propto l$,
(ii) inversely proportional to its cross-section area A, i.e. $R \propto l/A$ (doubling A halves R).
Combining these two statements, we get

$$R \propto l/A \qquad \text{or} \qquad R = \rho l/A$$

where ρ is a constant, called the **resistivity** of the material. If we put l = 1 m and A = 1 m^2, then $\rho = R$.

The resistivity of a material is numerically equal to the resistance of a 1 m length of it of cross-section area 1 m^2.

The unit of ρ is the **ohm-metre** (Ω m) as can be seen by rearranging the equation to give $\rho = AR/l$ and inserting units for A, R and l. Knowing ρ for a material, the resistance of any sample of it can be calculated. The resistivities of metals increase at higher temperatures, for most other materials they decrease.

Worked example

Calculate the resistance of a copper wire 1.0 km long and 0.50 mm diameter if the resistivity of copper is 1.7×10^{-8} Ω m.

Converting all units to metres, we get

$$\text{length } l = 1.0\ \text{km} = 1000\ \text{m} = 10^3\ \text{m}$$
$$\text{diameter } d = 0.50\ \text{mm} = 0.50 \times 10^{-3}\ \text{m}$$

If r is the radius of the wire, the cross-section area $A = \pi r^2 = \pi(d/2)^2 = (\pi/4)d^2$

$$\therefore \quad A = \frac{\pi}{4}(0.50 \times 10^{-3})^2\ \text{m}^2 \approx 0.20 \times 10^{-6}\ \text{m}^2$$

$$R = \frac{\rho l}{A} = \frac{(1.7 \times 10^{-8}\ \Omega\ \text{m}) \times (10^3\ \text{m})}{0.20 \times 10^{-6}\ \text{m}^2} = 85\ \Omega$$

QUESTIONS

1 What is the resistance of a lamp when a voltage of 12 V across it causes a current of 4 A?

2 Calculate the p.d. across a 10 Ω resistor carrying a current of 2 A.

3 The p.d. across a 2 Ω resistor is 4 V. What is the current flowing (in ampere)?

A $\frac{1}{2}$ **B** 1 **C** 2 **D** 6 **E** 8 (*J.M.B./A.L./N.W.16+*)

4 Three 5 Ω resistors are connected as shown in Figure 51.10. Their effective total resistance is

A $\frac{3}{5}$ Ω **B** $1\frac{2}{3}$ Ω **C** 5 Ω **D** 15 Ω **E** 125 Ω
(*O.L.E./S.R. 16+*)

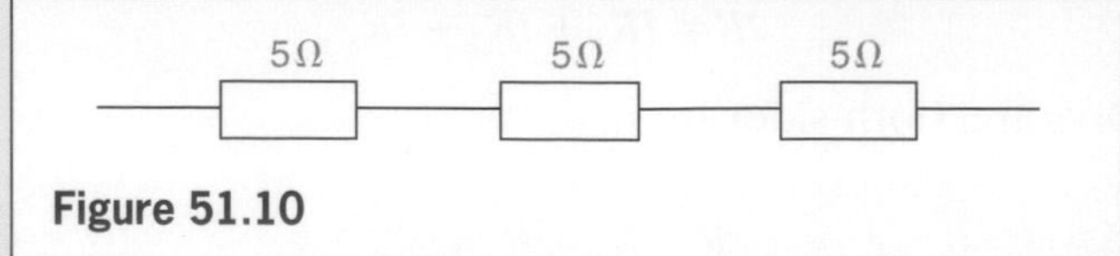

Figure 51.10

5 The resistors R_1, R_2, R_3 and R_4 in Figure 51.11 are all equal in value. What would you expect the voltmeters A, B and C to read, assuming that the connecting wires in the circuit have negligible resistance?

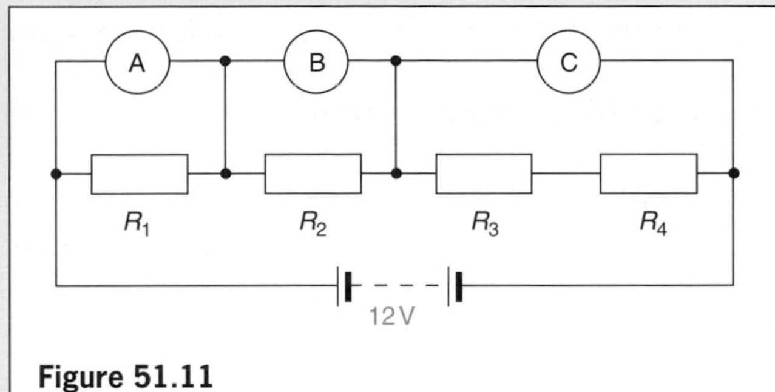

Figure 51.11

6 Calculate the effective resistance between A and B in Figure 51.12. *(E.M.)*

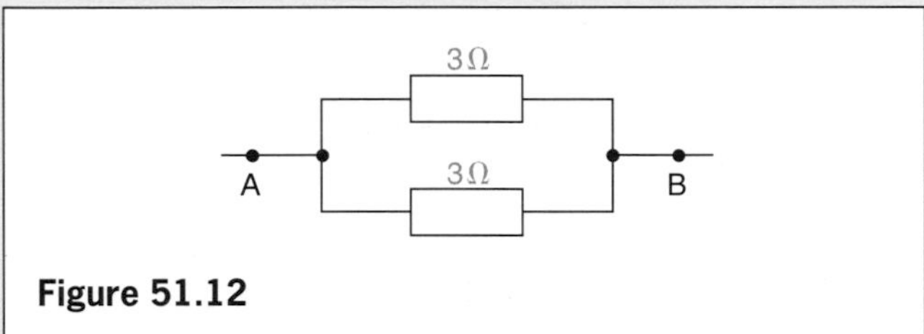

Figure 51.12

7 What is the effective resistance in Figure 51.13 between **a** A and B, **b** C and D?

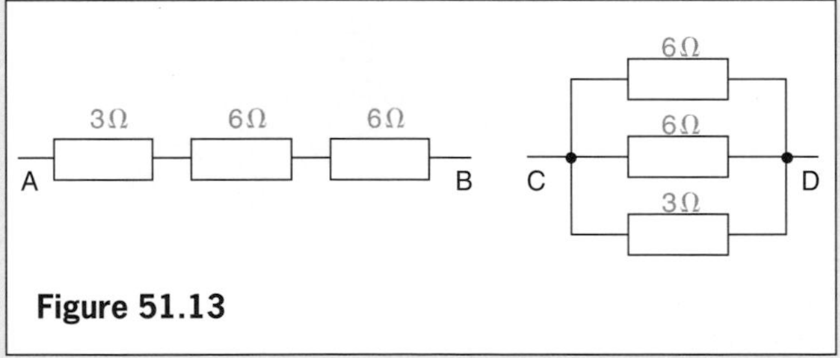

Figure 51.13

8 Figure 51.14 shows three resistors. Their combined resistance in ohms is

A $1\frac{5}{7}$ **B** 14 **C** $1\frac{1}{5}$ **D** $7\frac{1}{2}$ **E** $6\frac{2}{3}$

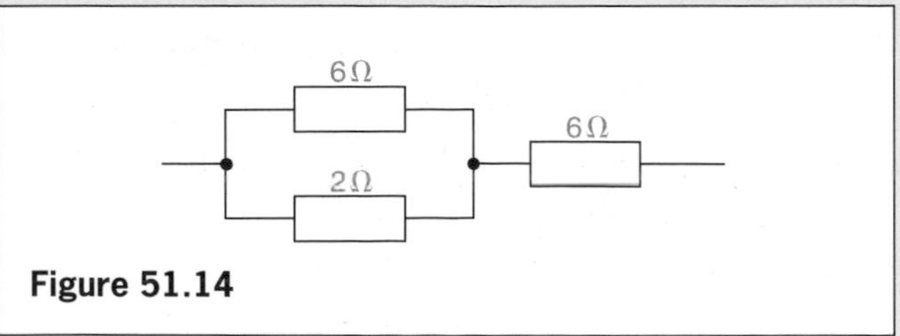

Figure 51.14

Checklist

After studying this chapter you should be able to

- define **resistance** and state the factors on which it depends,
- recall the unit of resistance,
- solve simple problems using $R = V/I$,
- describe experiments using the ammeter–voltmeter method to measure resistance and study the relationship between current and p.d. for (a) metallic conductors, (b) semiconductor diodes, (c) filament lamps, (d) thermistors,
- plot I–V graphs from the results of such experiments and draw appropriate conclusions from them,
- use the formulae for resistors in series and in parallel to solve problems,
- recall the definition of **resistivity** and use it to solve problems.

52 CAPACITORS

Capacitance	*Charging and discharging a capacitor*
Types of capacitor	*Effect of capacitors in d.c. and a.c. circuits*

A capacitor stores electric charge and is useful in many electronic circuits. In its simplest form it consists of two parallel metal plates separated by an insulator, called the **dielectric**, Figure 52.1.

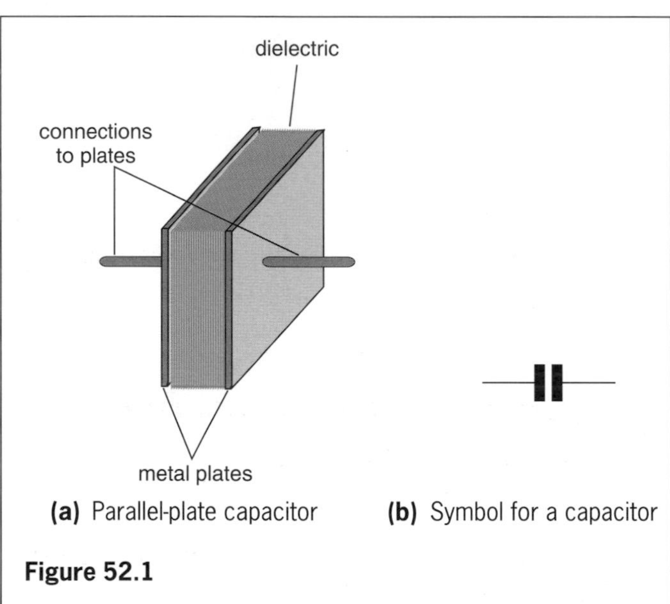

(a) Parallel-plate capacitor **(b)** Symbol for a capacitor

Figure 52.1

Capacitance

The more charge a capacitor can store, the greater is its **capacitance** (C). The capacitance is large when the plates have a large area and are close together. It is measured in **farads** (F) but smaller units such as the **microfarad** (μF) are more convenient.

$$1\ \mu F = 1 \text{ millionth of a farad} = 10^{-6}\ F$$

Types of capacitor

Practical capacitors, with values ranging from about 0.01 μF to 100 000 μF, often consist of two long strips of metal foil separated by long strips of dielectric, rolled up like a 'Swiss roll' as in Figure 52.2. The arrangement allows plates of large area to be close

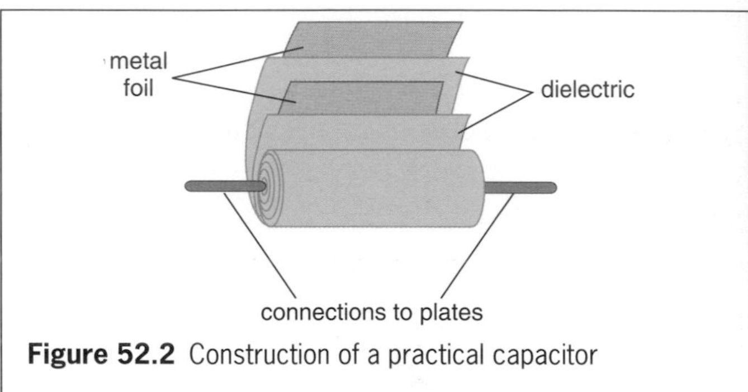

Figure 52.2 Construction of a practical capacitor

together in a small volume. Plastics (e.g. polyesters) are commonly used as the dielectric, with films of metal being deposited on the plastic to act as the plates, Figure 52.3.

Figure 52.3
Polyester capacitor

The **electrolytic** type, Figure 52.4a, has a very thin layer of aluminium oxide as the dielectric between two strips of aluminium foil, giving large capacitances. It is polarized, i.e. it has positive and negative terminals, Figure 52.4b, and these *must* be connected to the + and − respectively of the voltage supply.

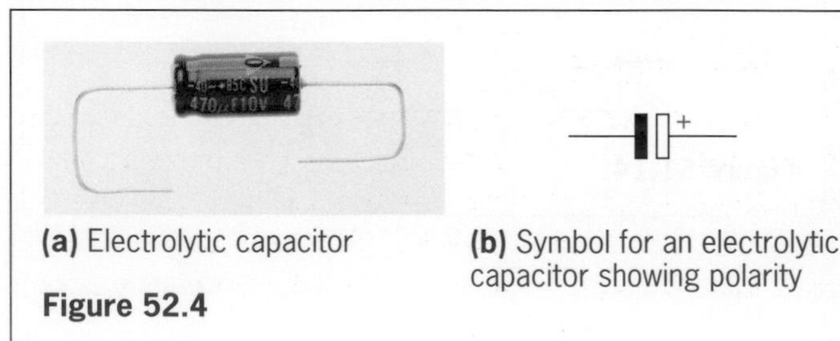

(a) Electrolytic capacitor **(b)** Symbol for an electrolytic capacitor showing polarity

Figure 52.4

Charging and discharging a capacitor

(a) Charging

A capacitor can be charged by connecting a battery across it. In Figure 52.5a, the + of the battery attracts electrons (since they have a negative charge) from plate X and the − of the battery repels electrons to plate Y. A positive charge builds up on plate X (since it loses electrons) and an equal negative charge builds up on Y (since it gains electrons).

During the charging, there is a *brief* flow of electrons round the circuit from X to Y (but not through the dielectric). A momentary current would be detected by a sensitive ammeter. The voltage builds up between X and Y and opposes the battery voltage. Charging stops when these two voltages are equal; the electron flow, i.e. the charging current, is then zero.

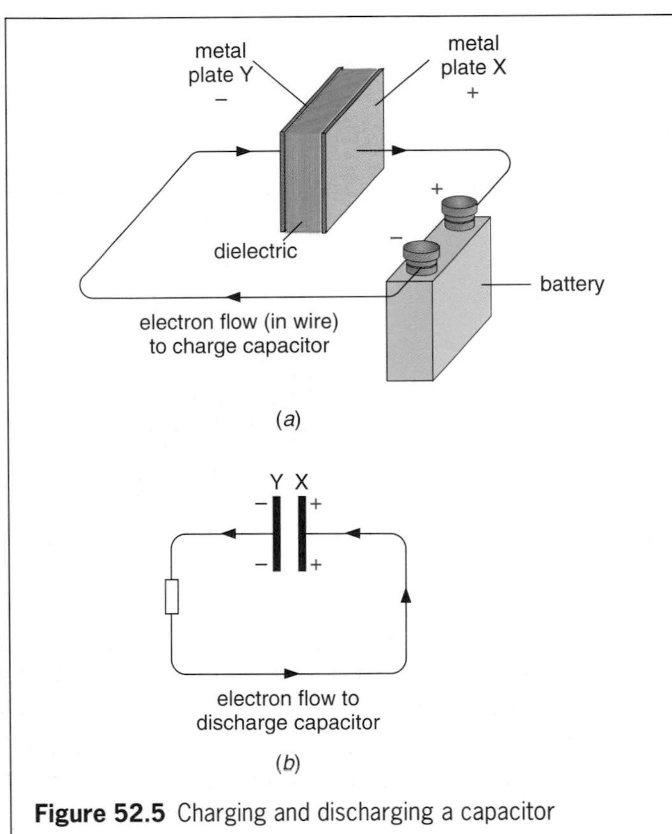

Figure 52.5 Charging and discharging a capacitor

(b) Discharging

When a conductor is connected across a charged capacitor, there is a brief flow of electrons from the negatively charged plate to the positively charged one, i.e. from Y to X in Figure 52.5b. The charge stored by the capacitor falls to zero, as does the voltage across it.

(c) Demonstration

The circuit in Figure 52.6 uses a two-way switch which charges C in position 1 and discharges it in position 2. The larger the values of R and C the longer it takes for the capacitor to charge or discharge. The direction of the deflection of the centre-zero milliammeter reverses for each process. The corresponding changes of capacitor charge (measured by the voltage across it) with time are shown by the graphs in Figures 52.7a, b.

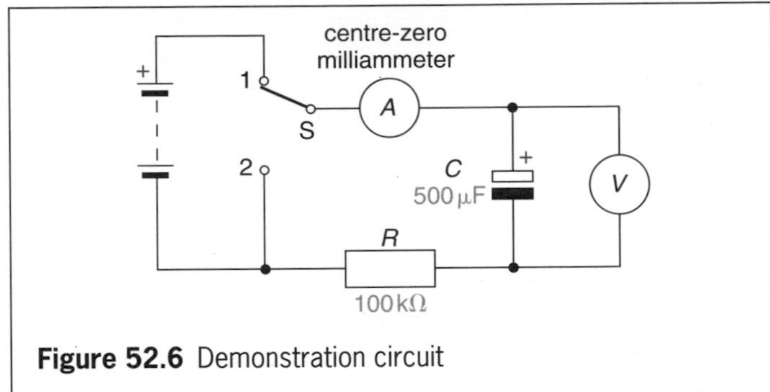

Figure 52.6 Demonstration circuit

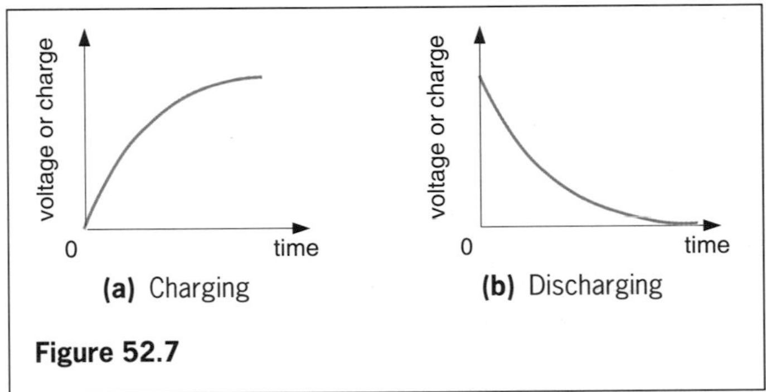

Figure 52.7

Effect of capacitors in d.c. and a.c. circuits

(a) d.c.

In Figure 52.8a the supply is d.c. but the lamp does not light, i.e. a capacitor blocks d.c.

(b) a.c.

In Figure 52.8b the supply is a.c. and the lamp lights, suggesting that a capacitor passes a.c. In fact, no current actually passes *through* the capacitor since its plates are separated by an insulator. But as the a.c. reverses direction, the capacitor charges and discharges causing electrons to flow to and fro rapidly in the wires joining the plates. Thus effectively a.c. flows round the circuit, lighting the lamp.

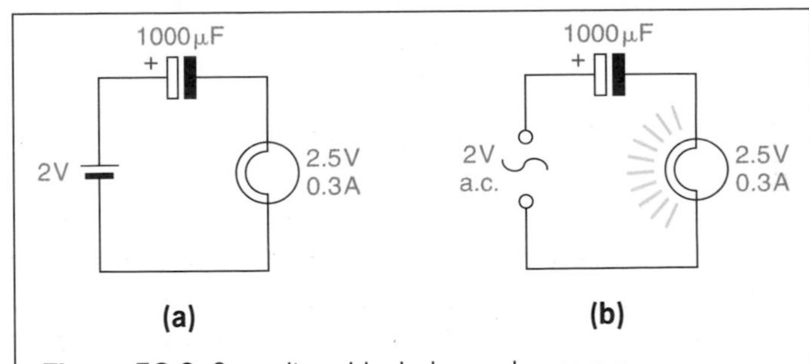

Figure 52.8 Capacitors block d.c. and pass a.c.

QUESTIONS

1 **a** Describe the basic construction of a capacitor.

b What does a capacitor do?

c State two ways of increasing the capacitance of a capacitor.

d Name a unit of capacitance.

2 **a** When a capacitor is being charged, what is the value of the charging current (i) at the start and (ii) at the end of charging?

b Repeat **a** for a capacitor discharging.

3 How does a capacitor behave in a circuit with

a a d.c. supply, **b** an a.c. supply?

Checklist

After studying this chapter you should be able to

- state what a capacitor does and what its capacitance depends on,
- state the unit of capacitance,
- describe the construction of practical capacitors,
- describe in terms of electron motion how a capacitor can be charged and discharged, and sketch graphs of the capacitor voltage with time for charging and discharging through a resistor,
- recall that a capacitor blocks d.c. but passes a.c. and explain why.

53 ELECTROLYSIS AND CELLS

Conduction in liquids
Electrolysis of copper sulphate solution with copper electrodes
Ionic theory
Electroplating
Electric cells
Primary cells
Secondary cells

Conduction in liquids

The apparatus of Figure 53.1 can be used to find which liquids conduct electricity. There is no effect with distilled water; it is not a conductor. However, if a little dilute sulphuric acid or common salt is added to the water the bulb lights up, showing that current is flowing through the solution. Bubbles of gas appear at the copper strips and are evidence of chemical action. Tap water usually contains dissolved salts and allows a small current to flow.

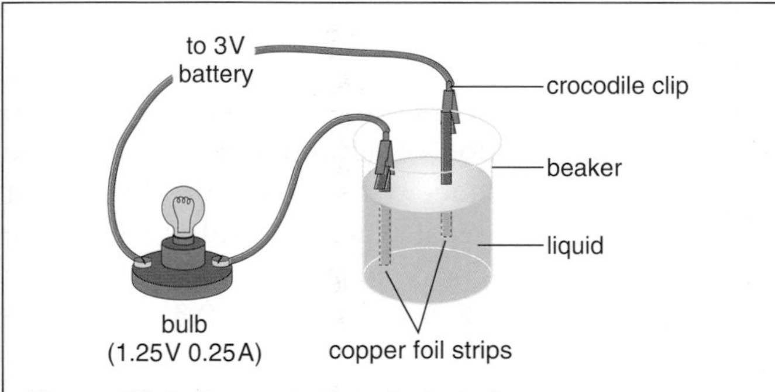

Figure 53.1 Demonstrating electrolysis

The production of chemical action in a liquid by an electric current is called **electrolysis**. The liquid is called an **electrolyte**. Solutions in water of acids, bases and salts are electrolytes. Mercury conducts electricity but no chemical change occurs and so it is not an electrolyte.

The two conductors (wires or plates) where current enters and leaves the liquid are called **electrodes**. The one joined to the positive terminal of the battery at which the current (conventional) enters is the **anode**; the other is the **cathode**.

Electrolysis of copper sulphate solution with copper electrodes

The apparatus and circuit of Figure 53.1 can be used with a strong copper sulphate solution (10 g to 100 cm^3 of water) in the beaker. No bubbles appear at the electrodes but if they are removed after a few minutes the cathode will be seen to be covered with a fresh layer of copper while the anode is dull. It can be shown that the mass of copper lost by the anode (if it is pure) equals that gained by the cathode. Also, the concentration of the copper sulphate solution is unchanged. It seems that copper is transferred from the anode to the cathode.

Ionic theory

Current is considered to be carried in an electrolyte by **ions**. An ion is an atom or group of atoms which has either a positive charge due to losing one or more electrons or a negative charge due to gaining one or more electrons, Figure 53.2a.

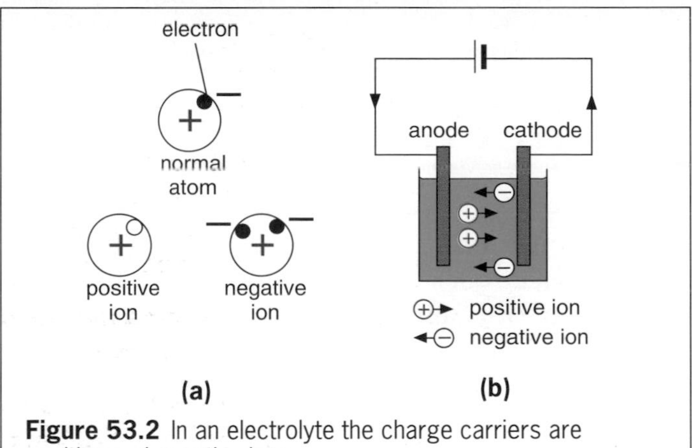

Figure 53.2 In an electrolyte the charge carriers are positive and negative ions

According to the ionic theory, when some chemical compounds are dissolved in water or are melted, they form ions. In electrolysis positive ions are attracted to the negative cathode and negative ions to the positive anode, Figure 53.2b.

Copper sulphate solution contains copper ions and sulphate ions and since copper is deposited on the cathode during electrolysis, the copper ions must be positively charged.

Electroplating

Articles of one metal are often electrically plated with another to improve their appearance, or to prevent rusting if they are made of iron or steel. Cutlery may be made of nickel, plated with silver. Steel is often chromium plated, for example in bicycle parts. The object to be plated is effectively made the cathode, with a salt of the plating metal as electrolyte, e.g. a silver salt for silver plating.

Figure 53.3 Industrial electroplating of tableware

Electric cells

In an electric cell an electrolyte reacts chemically with two electrodes, making one have a positive electric charge (the anode) and the other a negative charge (the cathode). When the two electrodes (or terminals) are connected in a conducting circuit, a current flows and chemical energy is transferred to electrical energy. This is the reverse of what happens in electrolysis.

Two or more cells together are called a battery.

Primary cells

A primary cell is one which is discarded when the chemicals are used up.

(a) Simple cell, Figure 53.4

This has a voltage of about 1.0 V but stops working after a short time due to 'polarization', i.e. the collection of hydrogen gas bubbles on the copper plate. The cell is depolarized by adding potassium dichromate which oxidizes the hydrogen to water. A second defect is 'local action'. This is due to impurities in the zinc and results in the zinc being used up even when current is not supplied. The simple cell is no longer used.

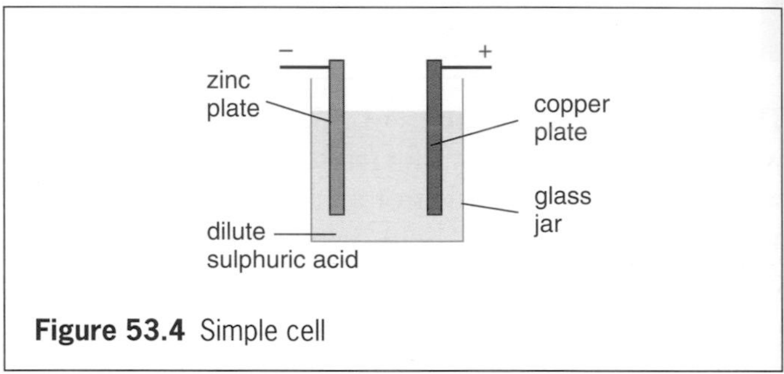

Figure 53.4 Simple cell

(b) Zinc–carbon (Leclanché or dry) cell, Figure 53.5

This has a zinc cathode, a manganese(IV) oxide anode (which also acts as a slow depolarizer) and the electrolyte is a solution of ammonium chloride. The carbon rod is in contact with the anode (but is not involved in the chemical reaction) and is called the 'current collector'. The voltage is 1.5 V. This is the most popular cell for low current (e.g. 0.3 A) or occasional use, e.g. in torches.

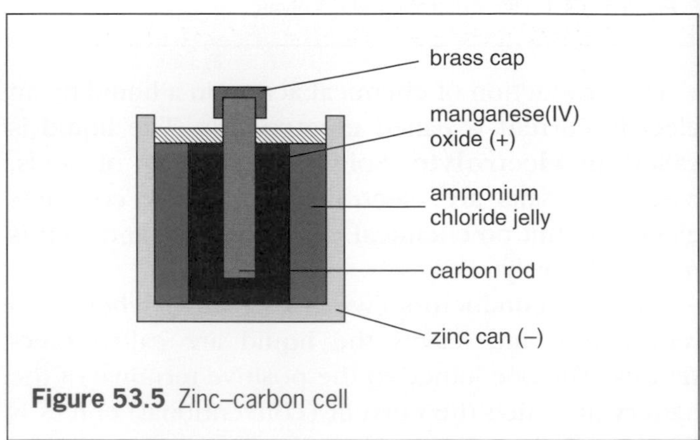

Figure 53.5 Zinc–carbon cell

Secondary cells

Secondary cells or 'accumulators' can be recharged by passing a current through them in the opposite direction to that in which they supply one. They are much more expensive than primary cells.

In the **lead–acid cell**, Figure 53.6, the anode is lead(IV) oxide (brown) and the cathode lead (grey). The electrolyte is dilute sulphuric acid. The voltage is steady at 2.0 V and quite large continuous currents can be supplied. A 12 V car battery consists of six lead-acid cells in series.

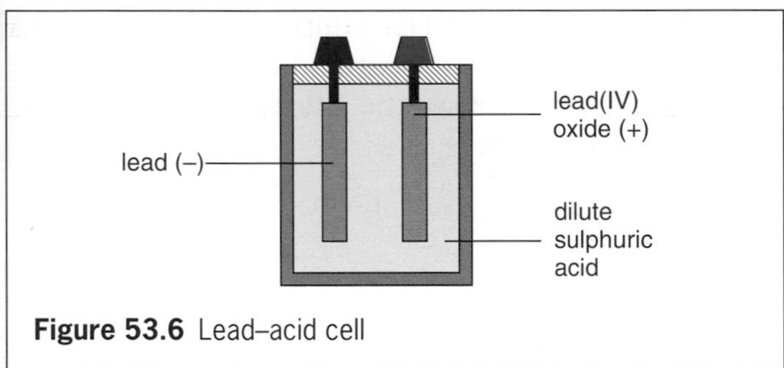

Figure 53.6 Lead–acid cell

QUESTIONS

1 a Name the current-carrying particles in an electrolyte.

b In the electrolysis of copper sulphate solution what would be the effect of increasing (i) the current, (ii) the time for which it passes? (**Hint.** Remember that the current is the charge passing per unit time.)

2 a What is the voltage of (i) a simple cell, (ii) a zinc–carbon cell, (iii) a lead–acid cell?

b What materials are used for the anode and cathode and the electrolyte of each of the cells in **a**?

3 a What is the difference between a primary and a secondary cell?

b Suggest advantages and disadvantages of each type.

Checklist

After studying this chapter you should be able to

- recall that **electrolysis** is the production of chemical action in a liquid by an electric current, that the liquid is called an **electrolyte**, and that the positive electrode is the **anode** and the negative electrode the **cathode**,
- describe a demonstration of the effects of electrolysis with copper electrodes in copper sulphate solution,
- explain electrolysis in terms of ion movement,
- recall that in an electric cell a chemical reaction produces positively and negatively charged electrodes,
- draw and describe the construction of a simple cell, a zinc-carbon cell and a lead-acid cell,
- distinguish between a **primary** cell and a **secondary** cell.

54 ELECTRIC POWER

Power in electric circuits
Electric lighting
Electric heating
Joulemeter
Practical work
Measuring electric power.

Power in electric circuits

In many circuits it is important to know the rate at which electrical energy is transferred into other forms of energy. Earlier (Chapter 25) we said that **energy transfers were measured by the work done** and power was defined by the equation

$$\text{power} = \frac{\text{work done}}{\text{time taken}} = \frac{\text{energy transfer}}{\text{time taken}}$$

In symbols $$P = \frac{W}{t} \qquad (1)$$

where P is in watts (W) if W is in joules (J) and t in seconds (s).

From the definition of p.d. (Chapter 50) we saw that if W is the electrical energy transferred when a steady current I (in amperes) passes for time t (in seconds) through a device (e.g. a lamp) with a p.d. V (in volts) across it, Figure 54.1, then

$$W = ItV \qquad (2)$$

Substituting for W in (1) we get

$$P = \frac{W}{t} = \frac{ItV}{t} = IV$$

Therefore to calculate the power P of an electrical appliance we multiply the current I through it by the p.d. V across it. For example if a lamp on a 240 V supply has a current of 0.25 A through it, its power is

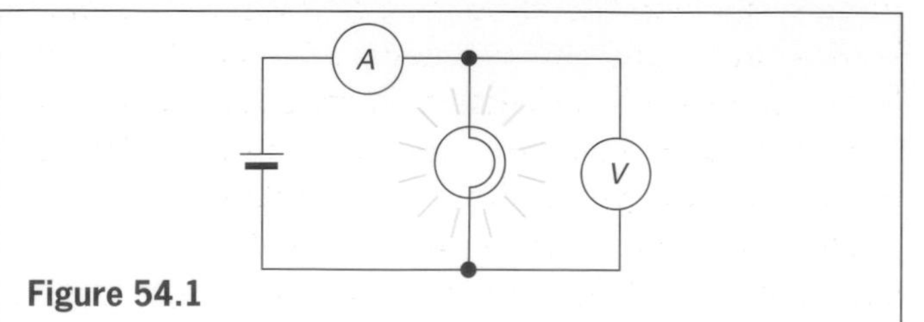

Figure 54.1

240 V × 0.25 A = 60 W. The lamp is transferring 60 J of electrical energy into heat and light each second. Larger units of power are the **kilowatt** (kW) and the **megawatt** (MW) where

$$1\ \text{kW} = 1000\ \text{W} \quad \text{and} \quad 1\ \text{MW} = 1\,000\,000\ \text{W}$$

In units

$$\text{watts} = \text{amperes} \times \text{volts} \qquad (3)$$

It follows from (3) that since

$$\text{volts} = \frac{\text{watts}}{\text{amperes}} \qquad (4)$$

the volt can be defined as a **watt per ampere** and p.d. calculated from (4).

If all the energy is transferred to heat in a resistor of resistance R, then $V = IR$ and the rate of production of heat is given by

$$P = V \times I = IR \times I = I^2R$$

That is, if the current is doubled, four times as much heat is produced per second. Also, $P = V^2/R$.

PRACTICAL WORK

Measuring electric power

(a) Lamp

Connect the circuit of Figure 54.2. Note the ammeter and voltmeter readings and work out the electric power supplied to the bulb in watts.

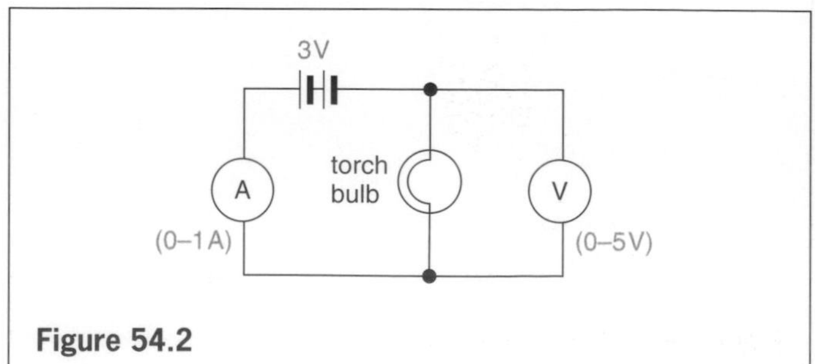

Figure 54.2

(b) Motor

Replace the bulb in Figure 54.2 by a small electric motor. Attach a known mass m (in kg) to the axle of the motor with a length of thin string and find the time t (in s) required to raise the mass through a known height h (in m) at a steady speed. Then the power output P_o (in W) of the motor is given by

$$P_o = \frac{\text{work done in raising mass}}{\text{time taken}} = \frac{mgh}{t}$$

If the ammeter and voltmeter readings I and V are noted while the mass is being raised, the power input P_i (in W) can be found from

$$P_i = IV$$

The efficiency of the motor is given by

$$\text{efficiency} = \frac{P_o}{P_i} \times 100\%$$

Also investigate the effect of a greater mass on (i) the speed, (ii) the power output and (iii) the efficiency of the motor at its rated p.d.

Electric lighting

(a) Filament lamps, Figure 54.3a

The filament is a small coiled coil of tungsten wire, Figure 54.3b, which becomes white hot when current flows through it. The higher the temperature of the filament the greater is the proportion of electrical energy transferred to light and for this reason it is made of tungsten, a metal with a high melting point (3400 °C).

Most lamps are gas-filled and contain nitrogen and argon, not air. This reduces evaporation of the tungsten which would otherwise condense on the bulb and blacken it. The coil is coiled compactly so that it is cooled less by convection currents in the gas.

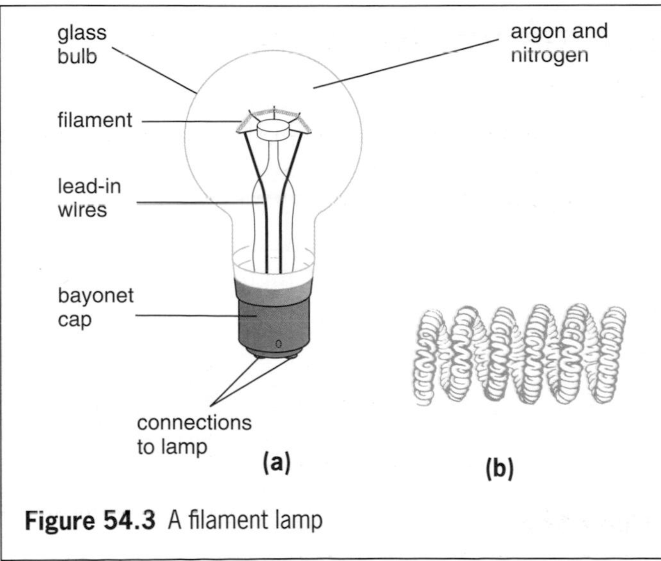

Figure 54.3 A filament lamp

(b) Fluorescent strips

A filament lamp transfers only 10% of the electrical energy supplied to light; the other 90% becomes heat. Fluorescent strip lamps are five times as efficient and may last 3000 hours compared with the 1000-hour life of filament lamps. They cost more to install but running costs are less.

A simplified diagram of a fluorescent strip is shown in Figure 54.4a. When the lamp is switched on, the mercury vapour emits ultraviolet radiation (invisible) which makes the powder on the inside of the tube fluoresce (glow), i.e. light (visible) is emitted. Different powders give different colours.

(c) Compact fluorescent lamps, Figure 54.4b

These energy-saving lamps fit straight into normal light sockets, either bayonet or screw-in. They last up to eight times longer (typically 8000 hours) and use about five times less energy than filament lamps for the *same* light output. For example, a 20 W compact fluorescent is equivalent to a 100 W filament lamp.

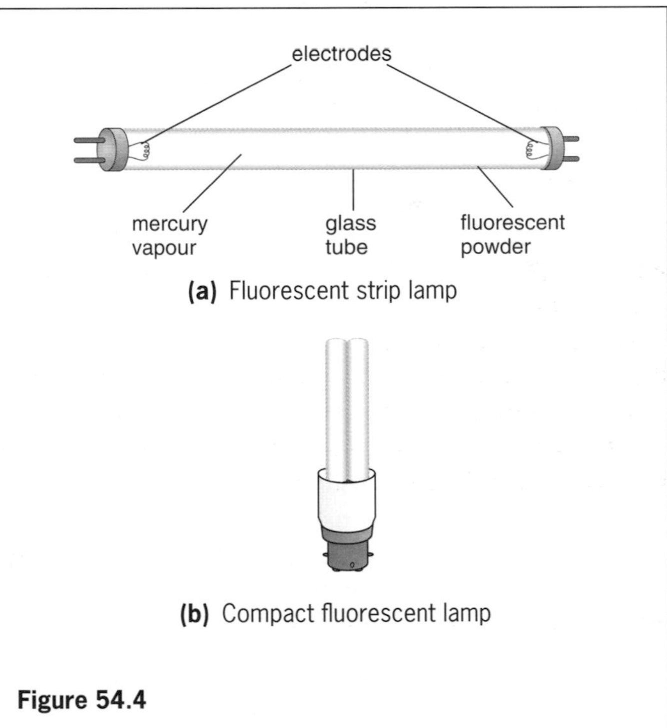

(a) Fluorescent strip lamp

(b) Compact fluorescent lamp

Figure 54.4

Electric heating

(a) Heating elements

In domestic appliances such as electric fires, cookers, kettles and irons the 'elements', Figure 54.5 (overleaf), are made from Nichrome wire. This is an alloy of nickel and chromium which does not oxidize (and become brittle) when the current makes it red hot.

The elements in **radiant** electric fires are at red heat (about 900 °C) and the radiation they emit is directed into the room by polished reflectors. In

convector types the element is below red heat (about 450 °C) and is designed to warm air which is drawn through the heater by natural or forced convection. In **storage** heaters the elements heat fireclay bricks during the night using 'off-peak' electricity. On the following day these cool down, giving off the stored heat to warm the room.

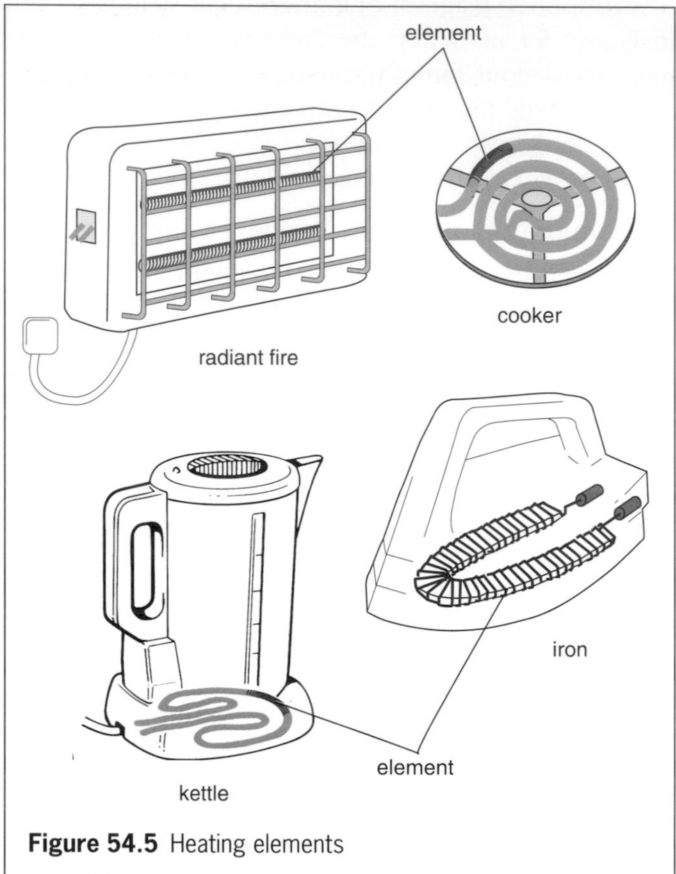

Figure 54.5 Heating elements

(b) Three-heat switch

This is sometimes used to control heating appliances. It has three settings and uses two identical elements. On 'high', the elements are in parallel across the supply voltage, Figure 54.6a; on 'medium', current only passes through one, Figure 54.6b; on 'low', they are in series, Figure 54.6c.

(c) Fuses

A fuse is a short length of wire of material with a low melting point (often tinned copper), which melts and breaks the circuit when the current through it exceeds a certain value. Two reasons for excessive currents are 'short circuits' due to worn insulation on connecting wires, and overloaded circuits. Without a fuse the wiring would become hot in these cases and could cause a fire. **A fuse should ensure that the current-carrying capacity of the wiring is not exceeded.** In general the thicker a cable is, the more current it can carry, but each size has a limit.

Two types of fuse are shown in Figure 54.7. **Always switch off before replacing a fuse,** and always replace with one of the same value as recommended by the manufacturer of the appliance.

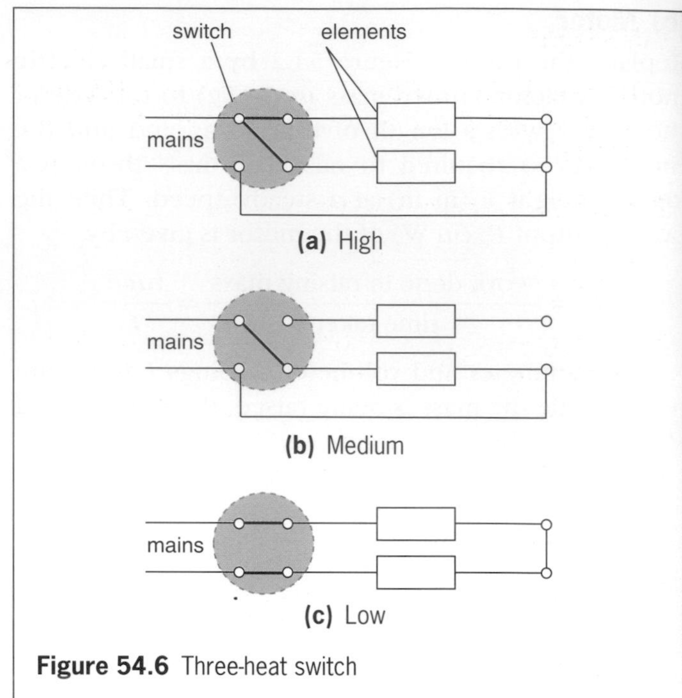

Figure 54.6 Three-heat switch

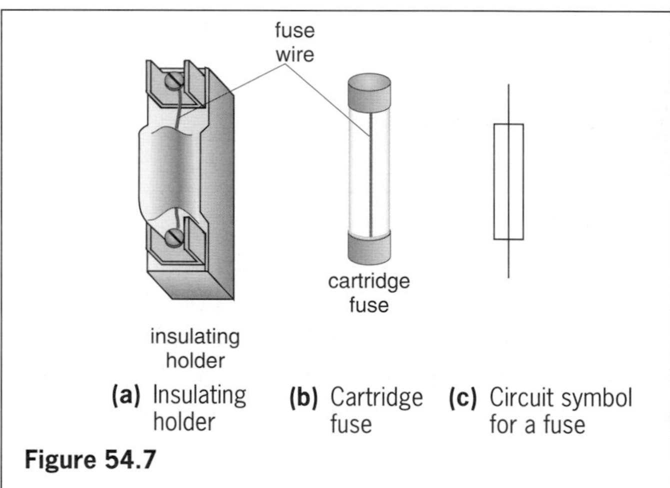

Figure 54.7

Joulemeter

Instead of using an ammeter and a voltmeter to measure the electrical energy transferred by an appliance, a **joulemeter** can be used to obtain it directly in joules. The circuit connections are shown in Figure 54.8. A household electricity meter, Figure 55.5, is a joulemeter.

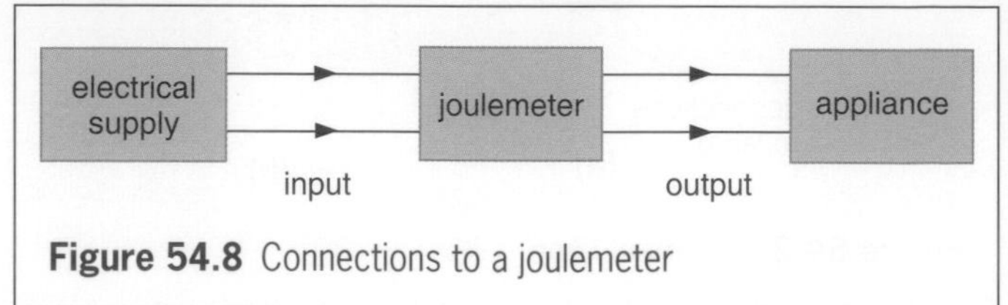

Figure 54.8 Connections to a joulemeter

QUESTIONS

1 How much electrical energy in **joules** does a 100 watt lamp transfer in **a** 1 second, **b** 5 seconds, **c** 1 minute?

2 **a** What is the power of a lamp rated at 12 V 2 A?

b How many joules of electrical energy are transferred per second by a 6 V 0.5 A lamp?

3 The largest number of 100 W bulbs which can safely be run from a 240 V supply with a 5 A fuse is

A 2 **B** 5 **C** 10 **D** 12 **E** 20

4 What is the maximum power in kilowatts of the appliance(s) that can be connected safely to a 13 A 240 V mains socket?

5 A current of 2 A passes through a resistance of 4 Ω. Calculate

a the p.d. between the ends of the resistance,

b the power used by the resistance. *(E.M.)*

6 A bulb is labelled 12 V 36 W. When used on a 12 V supply,

a what current will it take,

b what is its resistance? *(A.L.)*

7 A 3 kW electric fire is designed to be run on a 250 V supply. Assuming it to be operating at the correct voltage,

a what current will it draw from the supply,

b what is the resistance of the element of the fire? *(N.W.)*

Checklist

After studying this chapter you should be able to

- recall the relations $W = ItV$ and $P = IV$ and use them to solve simple problems on energy transfers,
- describe experiments to measure electric power,
- describe electric lamps, heating elements and fuses,
- recall that a joulemeter measures electrical energy.

55 ELECTRICITY IN THE HOME

House circuits	***Electric shock***
Paying for electricity	***Practical work*** House circuits.
Safety with electricity	

House circuits

Electricity usually comes to our homes by an underground cable containing two wires, the **live** (L) and the **neutral** (N). The neutral is earthed at the local sub-station and so there is no p.d. between it and earth. The supply is a.c. (Chapter 49) and the live wire is alternately positive and negative. Study the modern house circuit shown in Figure 55.1.

(a) Circuits in parallel

Every circuit is connected in parallel with the supply, i.e. across the live and neutral, and receives the full mains p.d. of 240 V.

(b) Switches and fuses

These are always in the live wire. If they were in the neutral, light switches and power sockets would be 'live' when switches were 'off' or fuses 'blown'. A shock (fatal) could then be obtained by, for example, touching the element of an electric fire when it was switched off.

(c) Staircase circuit

The light is controlled from two places by the two two-way switches.

(d) Ring main circuit

The live and neutral wires each run in two complete rings round the house, and the power sockets, each rated at 13 A, are tapped off from them. Thinner wires can be used since the current to each socket flows by two paths, i.e. in the whole ring. The ring has a 30 A fuse and if it has, say, ten sockets all can be used so long as the total current does not exceed 30 A, otherwise the wires overheat.

(e) Fused plug

Only one type of plug is used in a ring main circuit. It is wired as in Figure 55.2 - note the colours of the wire coverings. It has its own cartridge fuse, 3 A (red)

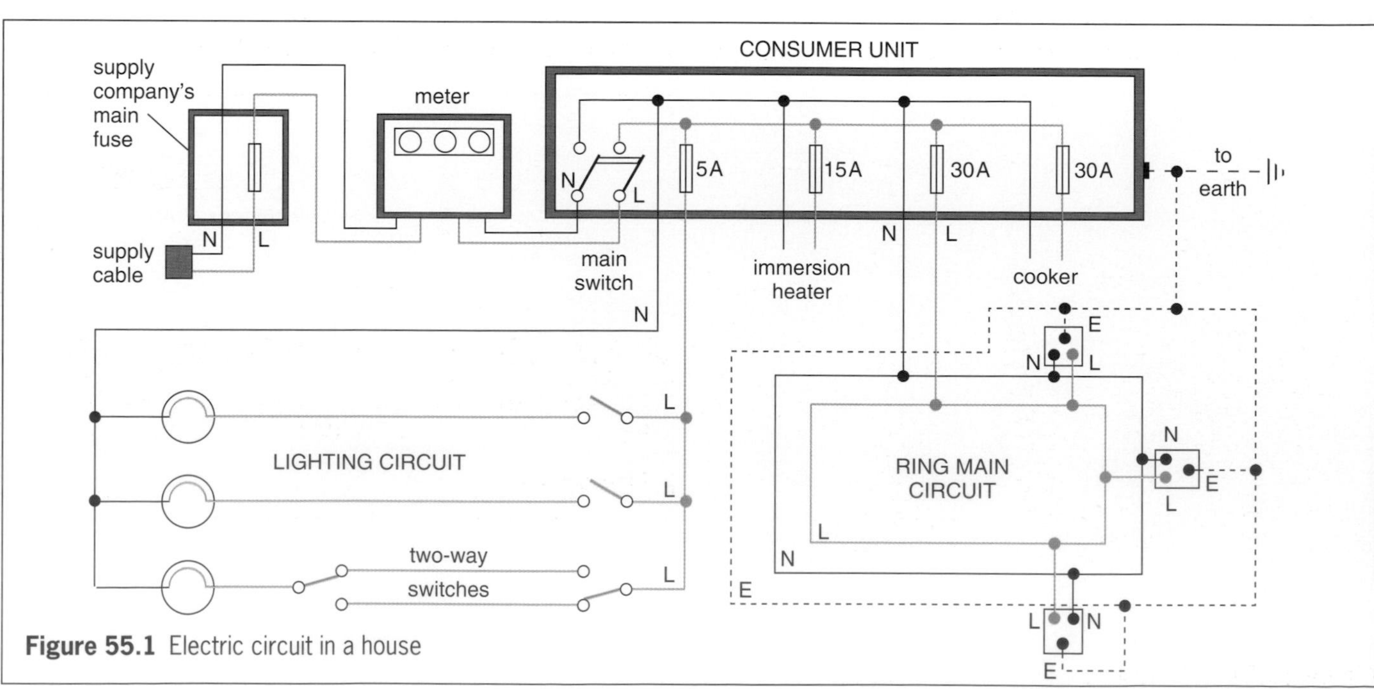

Figure 55.1 Electric circuit in a house

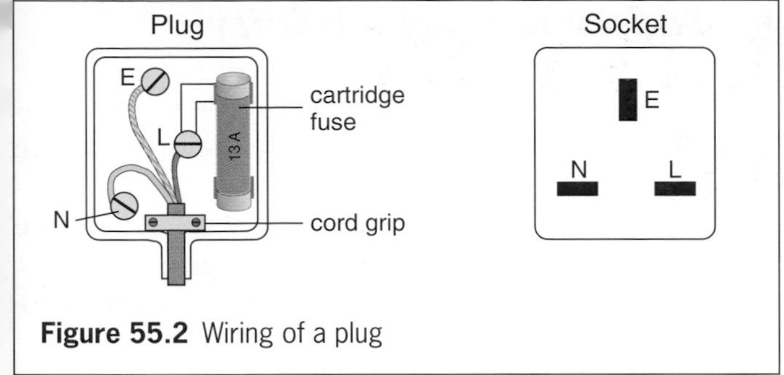

Figure 55.2 Wiring of a plug

for appliances with powers up to 720 W, or 13 A (brown) for those between 720 W and 3 kW.

(f) Earthing and safety

A ring main has a third wire which goes to the top sockets on all power points, Figure 55.1, and is earthed by being connected either to a **metal** water pipe in the house or to an earth connection on the supply cable. This third wire is a safety precaution to prevent electric shock should an appliance develop a fault.

The earth pin on a three-pin plug is connected to the metal case of the appliance which is thus joined to earth by a path of almost zero resistance. If then, for example, the element of an electric fire breaks or sags and touches the case, a large current flows to earth and 'blows' the fuse. Otherwise the case would become 'live' and anyone touching it would receive a shock which might be fatal, especially if they were 'earthed' by, say, standing in a damp environment, e.g. on a wet concrete floor.

(g) Circuit breakers, Figure 55.3

These are now used in consumer units instead of fuses. They contain an electromagnet (Chapter 57) which, when the current exceeds the rated value of the circuit breaker, becomes strong enough to separate a pair of contacts and breaks the circuit. They operate much faster than fuses and are reset by pressing a button.

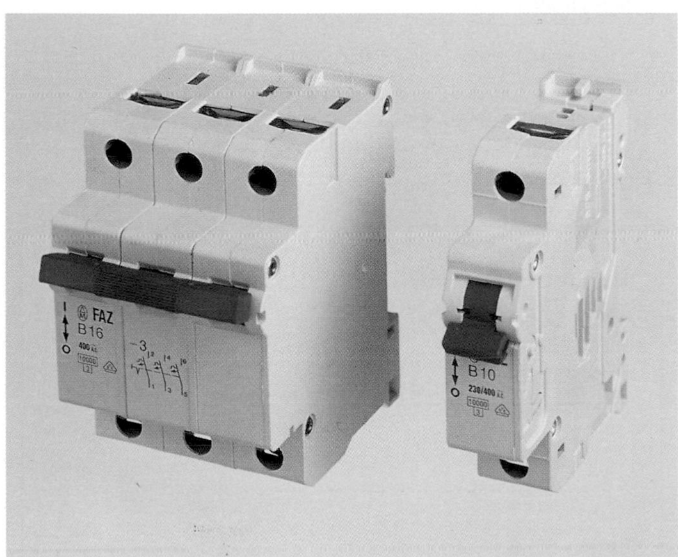

Figure 55.3 Circuit breakers

The **residual current circuit breaker** (RCCB), also called a **residual current device** (RCD), is an adapted circuit breaker which is used when the resistance of the earth path between the consumer and the sub-station is not small enough for a fault-current to blow the fuse (or circuit breaker). It works by detecting any difference between the currents in the live and neutral wires; when these become unequal due to an earth fault (i.e. some of the current returns to the sub-station via the case of the appliance and earth) it breaks the circuit before there is any danger.

An RCD should be plugged into a socket supplying power to a portable appliance such as an electric lawnmower or hedge trimmer. In these cases the risk of electrocution is greater because the user is generally making a good earth connection through the feet.

(h) Double insulation

Appliances such as vacuum cleaners, hairdryers and food mixers are usually double-insulated. Connection to the supply is by a two-core insulated cable, with no earth wire, and the appliance is enclosed in an insulating plastic case. Any metal attachments which the user might touch are fitted into this case so that they do not make a direct connection with the internal electrical parts, e.g. a motor. There is then no risk of a shock should a fault develop.

Practical work

House circuits

On a circuit board connect up and investigate

a the staircase circuit of Figure 55.4a,
b the ring main circuit of Figure 55.4b.

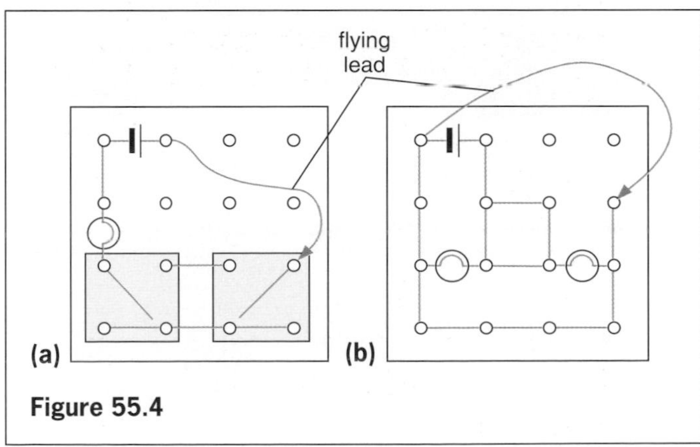

Figure 55.4

Paying for electricity

Electricity supply companies charge for the **electrical energy** they supply. A joule is a very small amount of energy and a larger unit, the **kilowatt-hour** (kWh) is used.

A kilowatt-hour is the electrical energy used by a 1 kW appliance in 1 hour.

A 3 kW electric fire working for 2 hours uses 6 kWh of electrical energy – usually called 6 'units'. Electricity meters, which are joulemeters (Chapter 54), are marked in kWh: the latest have digital read-outs like the one in Figure 55.5. At present a 'unit' costs about 8p.

Typical powers of some appliances are:

Lamps	60, 100 W	Fire	1, 2, 3 kW
Fridge	150 W	Kettle	2–3 kW
Television	200 W	Immersion heater	3 kW
Iron	750 W	Cooker	8 kW

Note that

$$\begin{aligned} 1\text{ kWh} &= 1000\text{ J/s} \times 3600\text{ s} \\ &= 3\,600\,000\text{ J} = 3.6\text{ MJ} \end{aligned}$$

Figure 55.5 Electricity meter with digital display

Safety with electricity

1 Switch off the electrical supply before starting repairs.
2 Use plugs that have a rubber or plastic case, the correct fuse, an earth pin and a cord grip.
3 Do not overload circuits by using too many adaptors.
4 Do not have long cables trailing across a room, or under a carpet that is walked over regularly.
5 Do not let appliances or cables come into contact with water, e.g. holding a hairdryer with wet hands in a bathroom can be dangerous.
6 Do not connect appliances that use large amounts of power (e.g. an electric fire) to a lighting circuit.

Electric shock

Electric current passing through the heart can be fatal. The following action should be taken in cases of electric shock.

1 **Switch off the supply** if the shocked person is still touching live equipment.

2 **Send for qualified medical assistance.**

3 **If breathing has stopped apply the 'kiss of life'.**

4 **If the heart has stopped try to restart it** by striking the chest smartly three times over the heart.

QUESTIONS

1 The circuits of Figure 55.6a, b show 'short circuits' between the live (L) and neutral (N) wires. In both, the fuse has blown but whereas (a) is now safe, (b) is still dangerous even though the lamp is out which suggests the circuit is safe. Explain.

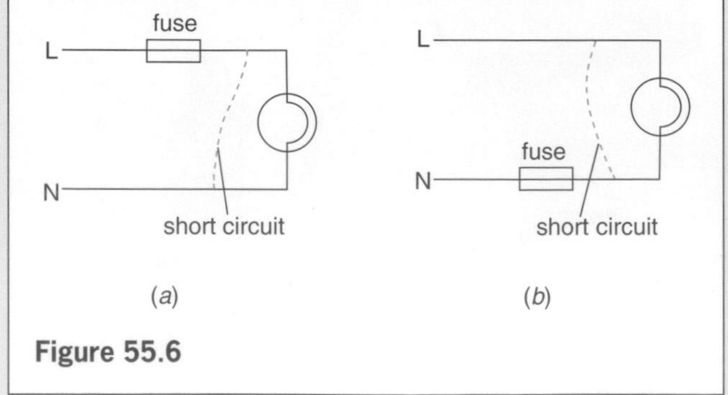

Figure 55.6

2 What steps should be taken before replacing a blown fuse in a plug?

3 What size fuse (3 A or 13 A) should be used in a plug connected to **a** a 150 W television, **b** a 750 W iron, **c** a 2 kW kettle, if the supply is 240 V?

4 An electric cooker has an oven rated at 3 kW, a grill rated at 2 kW and two rings each rated at 500 W. The cooker operates from 240 V mains.

a Are the different heating elements connected in series or in parallel? Give a reason for your answer.

b Would a 30 A fuse be suitable for the cooker, assuming that all parts are switched on? Show clearly how you obtain your answer.

c What is the cost of operating all the parts for 30 minutes if electricity costs 10p per unit? *(E.A.)*

5 If a 2 kW electric heater is used for 10 hours and electricity costs 10p per unit,

a how many units (kWh) are used,

b what is the cost? *(A.L.)*

6 What is the cost of heating a tank of water with a 3000 W immersion heater for 80 minutes if electricity costs 10p per kWh?

7 **a** A lamp is marked 240 V 60 W. What exactly does this tell you about the bulb?

b A boy has a large number of 240 V 60 W coloured bulbs he wishes to use for decorations so that the bulbs operate normally.
(i) How many can he connect to a 240 V supply through a 5 A fuse?
(ii) What would be the total power of the circuit?
(iii) Draw a diagram of the circuit he would employ.

c If electrical energy costs 10p per unit, what will be the cost of running the above circuit for 5 hours a night for 14 nights? *(S.E.)*

Checklist

After studying this chapter you should be able to

- describe with the aid of diagrams a house wiring system and explain the functions and positions of switches, fuses, circuit breakers and earth,
- wire a mains plug and recall the international insulation colour code,
- perform calculations of the cost of electrical energy in joules and kilowatt-hours,
- recall safety precautions for domestic wiring and appliances,
- recall the procedure for dealing with someone who has received an electric shock.

ADDITIONAL QUESTIONS

The first questions under each topic heading are 'basic' questions intended for all students; the following questions marked with a stripe are 'higher' questions for those seeking grades E to A*.

Static electricity

1 In the process of induction in electrostatics

A a conductor is rubbed with an insulator
B a charge is produced by friction
C negative and positive charges are separated
D a positive charge induces a positive charge
E electrons are 'sprayed' into an object.

(J.M.B.)

Electric current; potential difference; resistance

2 If two resistors of 5 Ω and 15 Ω are joined together in series and then placed in parallel with a 20 Ω resistor, the effective resistance in ohms of the combination is

A 0.1 **B** 10 **C** 20 **D** 40 **E** 400 *(N.I.)*

3 V_1, V_2, V_3 are the p.d.s across the 2 Ω, 3 Ω and 4 Ω resistors respectively in the diagram and the current is 5 A. Which one of the columns **A** to **E** shows the correct values of V_1, V_2 and V_3 measured in volts?

	A	B	C	D	E
V_1	2.0	10	0.4	2.5	3.0
V_2	3.0	15	0.6	1.6	2.0
V_3	4.0	20	0.8	1.25	1.0

(J.M.B.)

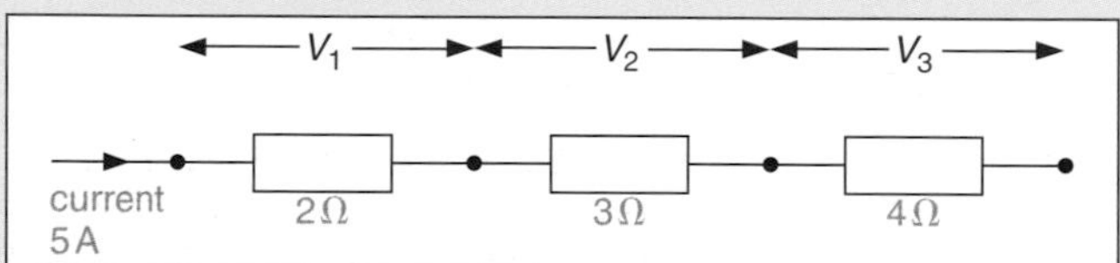

4 Two 4 Ω resistors are connected to a 2 V cell and an ammeter as shown in (a), (b), (c) below. Find the reading on the ammeter in each case. The resistance of the connecting wires, the meter and the cell may be neglected. *(E.A.)*

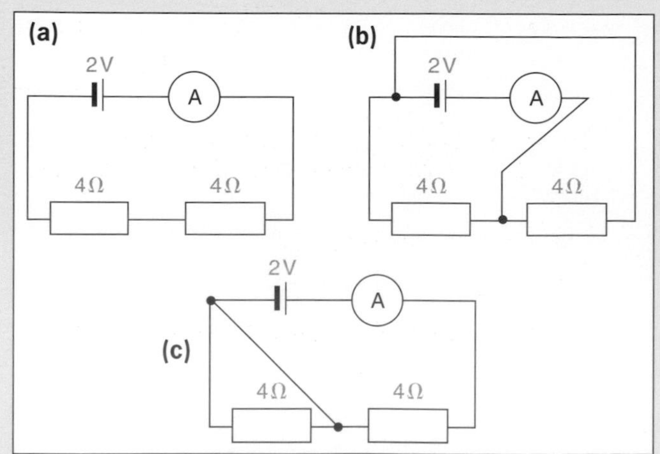

5 **a** The graph illustrates how the p.d. across the ends of a conductor is related to the current flowing through it.
(i) What law may be deduced from the graph?
(ii) What is the resistance of the conductor?

b Draw diagrams to show how six 2 V lamps could be lit to normal brightness when using a (i) 2 V supply, (ii) 6 V supply and (iii) 12 V supply.

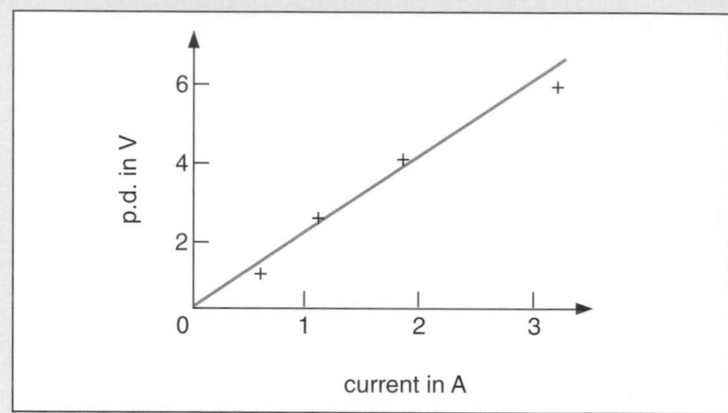

6 When a 6 Ω resistor is connected across the terminals of a 12 V battery, the number of coulombs passing through the resistor per second is

A 0.5 **B** 2 **C** 6 **D** 12 **E** 72 *(N.I.)*

7 In the circuit below, what is the voltmeter reading if the ammeter reads 2 A and the resistance of the voltmeter is 900 Ω?

A 50 **B** 180 **C** 200 **D** 450 **E** 1800 *(N.I.)*

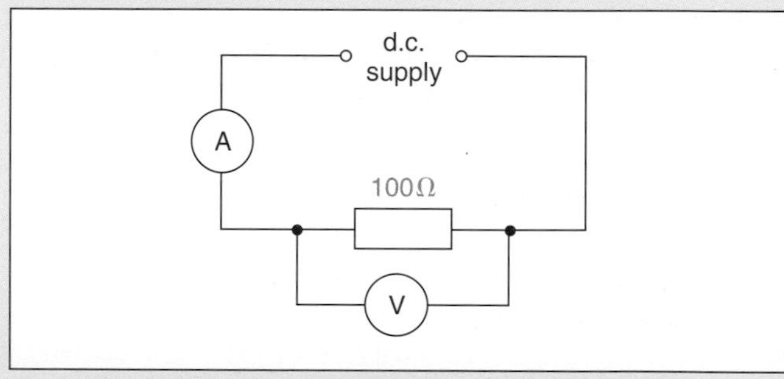

8 The diagram below shows a 2 V cell connected to an arrangement of resistors, part in series and part in parallel.

a What is the total resistance of the two 6 Ω resistors in parallel?

b What is the current flowing in the 1 Ω resistor?

c What is the current in one of the 6 Ω resistors?

(N.W.)

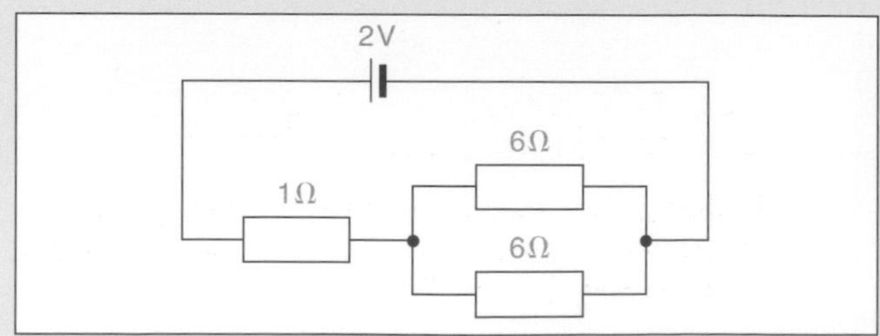

9 A 4 Ω coil and a 2 Ω coil are connected in parallel. What is their combined resistance? A total current of 3 A passes through the coils. What current flows through the 2 Ω coil?

10 In the circuits below the circles are either ammeters or voltmeters. State what each is and the reading it would show. *(E.A.)*

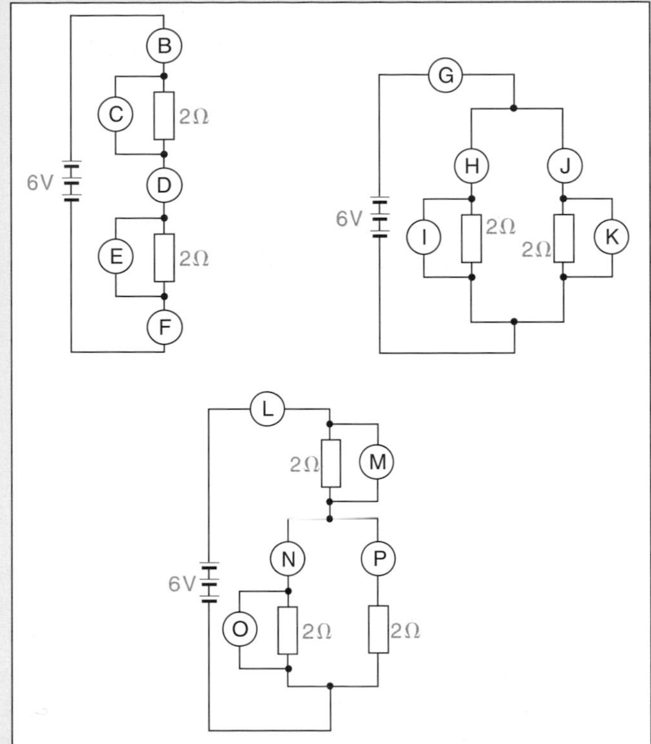

11 Resistors are connected as below and a p.d. of 12 V is applied across them. S1 and S2 are switches. Calculate

a the total resistance when both switches S1 and S2 are closed,

b the total current supplied when both switches S1 and S2 are closed,

c the current in the 3 Ω resistor when switch S1 is closed and switch S2 open,

d the current supplied when both switches are open. *(W.)*

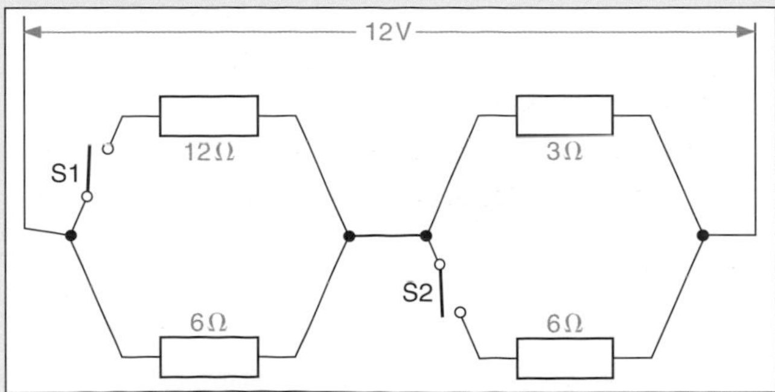

12 What is the resistance of a wire of length 300 m and cross-section area 1.0 mm^2 made of material of resistivity 1.0×10^{-7} Ω m?

13 Calculate the resistance of an aluminium cable of length 10 km and diameter 2.0 mm if the resistivity of aluminium is 2.7×10^{-8} Ω m.

Electric power; electricity in the home

14 a Below is a list of wattages of various appliances. State which is most likely to be the correct one for each of the appliances named.

60 W 250 W 750 W 1.5 kW 3 kW

(i) immersion heater (for a bath)
(ii) table lamp
(iii) iron

b What current will be taken by a 1 kW appliance if the supply voltage is 250?

(J.M.B./A.L./N.W.16+)

15 A heating coil has a resistance of 4 Ω and is connected to a 12 V supply. It is placed in a container holding 100 g of water at 20 °C. Calculate

a the current,

b the power of the circuit,

c the energy produced in 5 minutes,

d the rise in the temperature of the water, assuming no heat is lost. (Specific heat capacity of water is 4200 J/(kg °C) or 4.2 J/(g °C).) *(E.A.)*

16 A 2 kW electric fire is used for 10 hours each week and a 100 W lamp is used for 10 hours each day. Find the total energy consumed each week and the total cost for each week if 1 kWh of electricity costs 10p. *(J.M.B.)*

17 a Diagram (a) represents a simple lighting circuit.
(i) What is wrong with this circuit?
(ii) Explain why this is dangerous.

b Diagram (b) represents an incomplete two-way lighting circuit. Copy and complete the circuit in such a way that the lamp would be lit with the circuit working correctly. *(J.M.B./A.L./N.W.16+)*

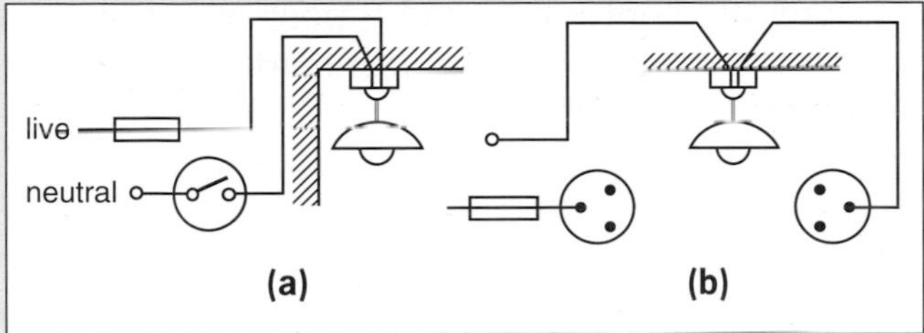

18 Two precautions usually taken when using electrical equipment connected to the mains are to

a earth the case (if metal),

b put a fuse in the circuit.

Explain clearly why each of these precautions is desirable. *(L.)*

56 MAGNETIC FIELDS

Properties of magnets
Magnetization of iron and steel
Magnetic fields
Earth's magnetic field
Practical work
Plotting lines of force.

Properties of magnets

(a) Magnetic materials

Magnets only attract strongly certain materials such as iron, steel, nickel, cobalt, which are called 'ferromagnetics'.

(b) Magnetic poles

These are the places in a magnet to which magnetic materials are attracted, e.g. iron filings. They are near the ends of a bar magnet and occur in pairs of equal strength.

(c) North and south poles

If a magnet is supported so that it can swing in a horizontal plane it comes to rest with one pole, the north-seeking or N pole, always pointing roughly towards the Earth's north pole. A magnet can therefore be used as a compass.

(d) Law of magnetic poles

If the N pole of a magnet is brought near the N pole of a suspended magnet repulsion occurs. Two S poles also repel. By contrast, N and S poles always attract. The law of magnetic poles summarizes these facts and states:

Like poles repel, unlike poles attract.

The force between magnetic poles decreases as their separation increases.

Magnetization of iron and steel

Chains of small iron nails and steel paper clips can be hung from a magnet, Figure 56.1. Each nail or clip magnetizes the one below it and the unlike poles so formed attract.

If the iron chain is removed by pulling the top nail away from the magnet, the chain collapses, showing

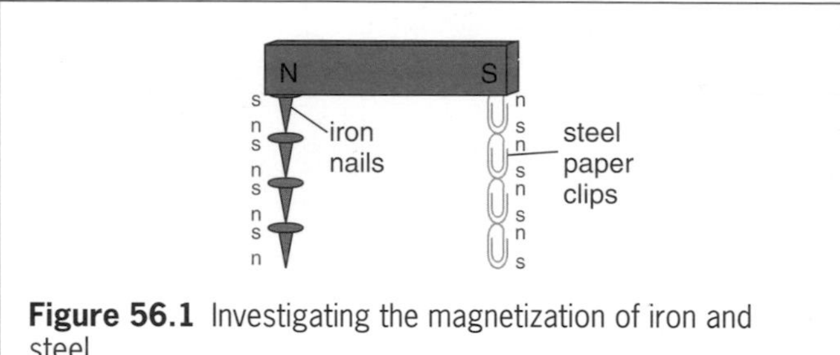

Figure 56.1 Investigating the magnetization of iron and steel

that **magnetism induced in iron is temporary**. When the same is done with the steel chain, it does not collapse; **magnetism induced in steel is permanent**.

Magnetic materials like iron which magnetize easily but do not keep their magnetism are said to be **soft**. Those like steel which are harder to magnetize than iron but stay magnetized are **hard**. Both types have their uses; very hard ones are used to make permanent magnets.

Magnetic fields

The space surrounding a magnet where it produces a magnetic force is called a **magnetic field**. The force around a bar magnet can be detected and shown to vary in direction using the apparatus in Figure 56.2. If the floating magnet is released near the N pole of the bar magnet, it is repelled to the S pole and moves along a curved path known as a **line of force** or a **field line**. It moves in the opposite direction if its south pole is uppermost.

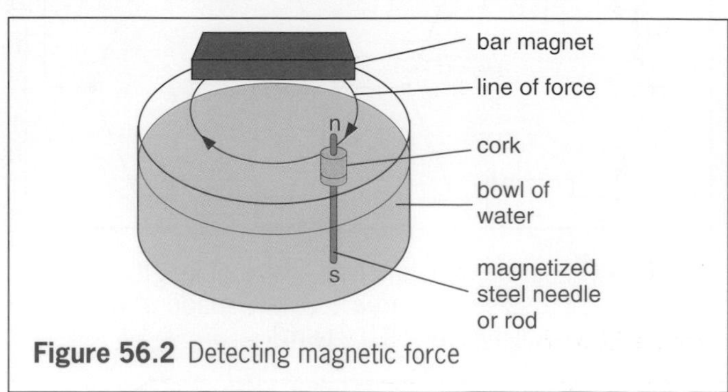

Figure 56.2 Detecting magnetic force

It is useful to consider that a magnetic field has a direction and to represent the field by lines of force. It has been decided that **the direction of the field at any point should be the direction of the force on a N pole**. To show the direction, arrows are put on the lines of force and point away from a N pole towards a S pole.

PRACTICAL WORK

Plotting lines of force

(a) Plotting compass method

A plotting compass is a small pivoted magnet in a glass case with non-magnetic metal walls, Figure 56.3a.

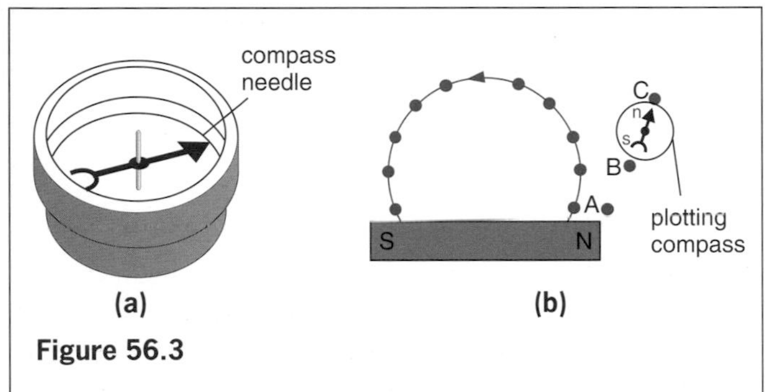

Figure 56.3

Lay a bar magnet on a sheet of paper. Place the plotting compass at a point such as A, Figure 56.3b, near one pole of the magnet. Mark the position of the poles n, s of the compass by pencil dots B, A. Move the compass so that pole s is exactly over B, mark the new position of n by dot C.

Continue this process until the S pole of the bar magnet is reached. Join the dots to give one line of force and show its direction by putting an arrow on it. Plot other lines by starting at different points round the magnet.

A typical field pattern is shown in Figure 56.4.

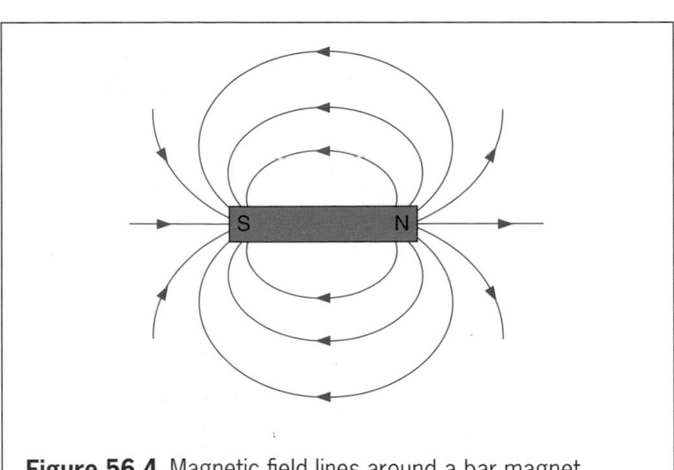

Figure 56.4 Magnetic field lines around a bar magnet

The combined field due to two neighbouring magnets can also be plotted to give patterns like those in Figure 56.5a, b. In (a), where two like poles are facing each other, the point X is called a **neutral point**. At X the field due to one magnet cancels out that due to the other and there are no lines of force.

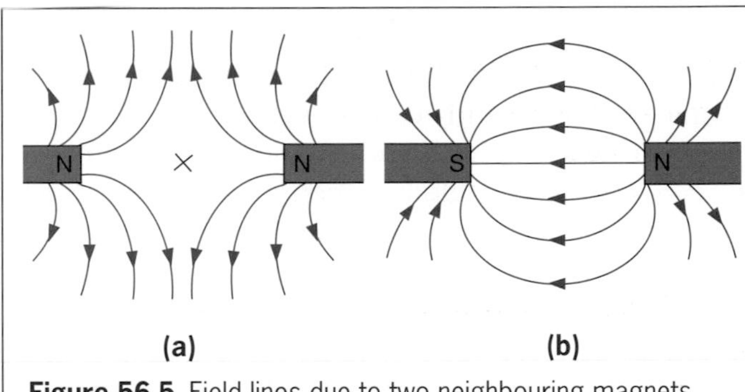

Figure 56.5 Field lines due to two neighbouring magnets

(b) Iron filings method

Place a sheet of paper *on top of* a bar magnet and sprinkle iron filings *thinly and evenly* on to the paper from a 'pepper pot'.

Tap the paper gently with a pencil and the filings should form patterns of the lines of force. Each filing turns in the direction of the field when the paper is tapped.

This method is quick but no use for weak fields. Figures 56.6a, b show typical patterns with two magnets. Why are they different?

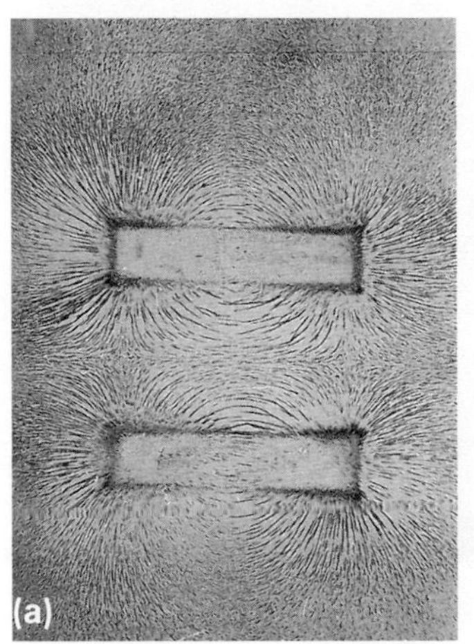

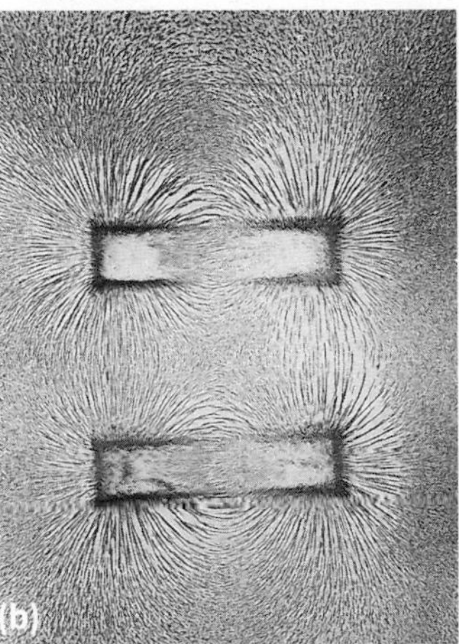

Figure 56.6 Field lines round two bar magnets shown by iron filings

Earth's magnetic field

If lines of force are plotted on a sheet of paper with no magnets near, a set of parallel straight lines is obtained. They run roughly from S to N geographically, Figure 56.7 (overleaf), and represent a small part of the Earth's magnetic field in a horizontal plane.

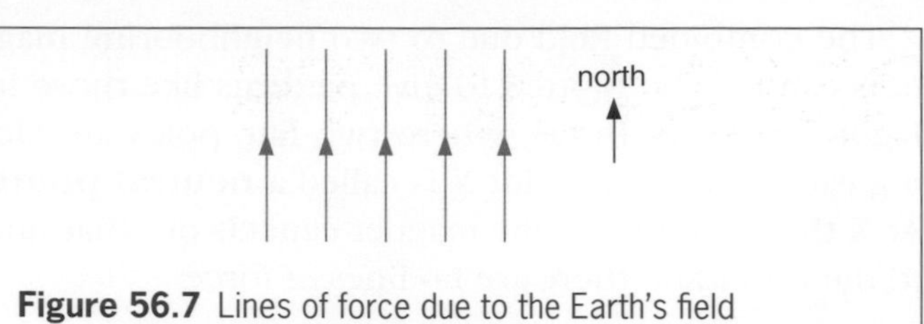

Figure 56.7 Lines of force due to the Earth's field

At most places on the Earth's surface a magnetic compass points slightly east or west of true north, i.e. the Earth's geographical and magnetic north poles do not coincide. The angle between magnetic north and true north is called the **declination**, Figure 56.8. In London at present (1994) it is 7° W of N and is decreasing. By about the year 2140 it should be 0°.

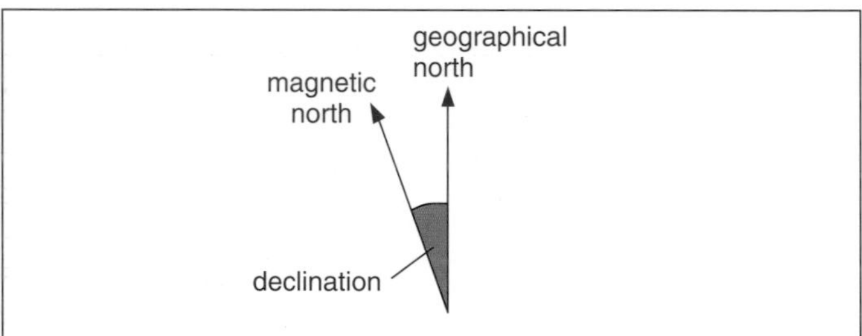

Figure 56.8 The Earth's geographical and magnetic poles do not coincide

QUESTIONS

1 A magnet attracts

A plastics **B** any metal **C** iron and steel
D aluminium **E** carbon.

2 Copy Figure 56.9 which shows a plotting compass and a magnet. Label the N pole of the magnet and draw the field line on which the compass lies.

Figure 56.9

3 The three diagrams in Figure 56.10 show the lines of force (field lines) between the poles of two magnets. Identify the poles A, B, C, D, E, F. *(E.M.)*

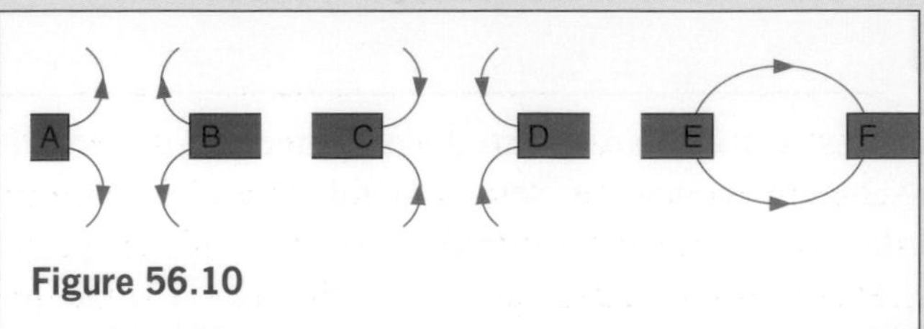

Figure 56.10

Checklist

After studying this chapter you should be able to

- state the properties of magnets,
- explain what is meant by **soft** and **hard** magnetic materials,
- recall that a magnetic field is the region round a magnet where a magnetic force is exerted and is represented by lines of force whose direction at any point is the direction of the force on a N pole,
- map magnetic fields (by the plotting compass and iron filings methods) round (a) one magnet, (b) two magnets,
- recall that at a neutral point the field due to one magnet cancels that due to another,
- define **declination**.

57 ELECTROMAGNETS

Oersted's discovery

In 1819 Oersted accidentally discovered the magnetic effect of an electric current. His experiment can be repeated by holding a wire over and parallel to a compass needle which is pointing N and S, Figure 57.1. The needle moves when the current is switched on. Reversing the current causes the needle to point in the opposite direction.

Evidently around a wire carrying a current there is a magnetic field. As with the field due to a permanent magnet, we represent the field due to a current by field lines or lines of force. Arrows on the lines show the direction of the field, i.e. the direction in which a N pole points.

Different field patterns are given by differently shaped conductors.

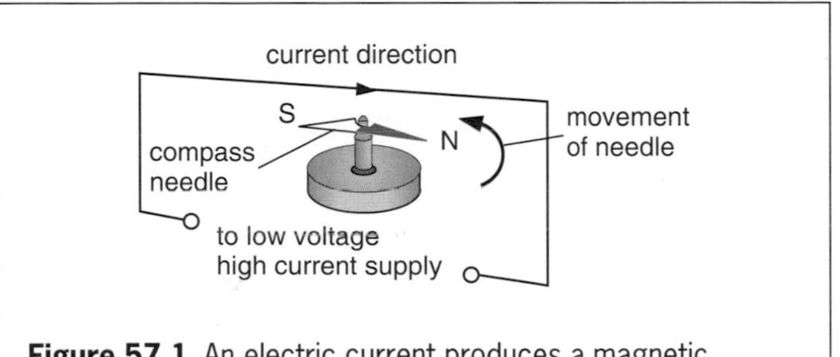

Figure 57.1 An electric current produces a magnetic effect

Field due to a straight wire

If a straight vertical wire passes through the centre of a piece of card held horizontally and a current is passed through the wire, iron filings sprinkled on the card set in concentric circles when the card is tapped, Figure 57.2. The field can be found by the **right-hand screw rule:**

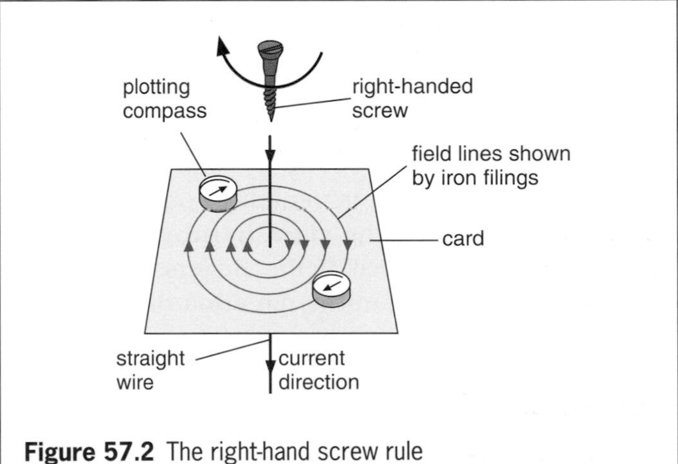

Figure 57.2 The right-hand screw rule

Plotting compasses placed on the card set along the field lines and show the direction of the field at different points. When the current direction is reversed, the compasses point in the opposite direction showing that the direction of the field reverses when the current reverses.

If the current direction is known, the direction of the field can be predicted by the **right-hand screw rule**:

> If a right-handed screw moves forwards in the direction of the current (conventional), the direction of rotation of the screw gives the direction of the field.

Field due to a circular coil

The field pattern is shown in Figure 57.3 (overleaf). At the centre of the coil the field lines are straight and at right angles to the plane of the coil. The right-hand screw rule again gives the direction of the field at any point.

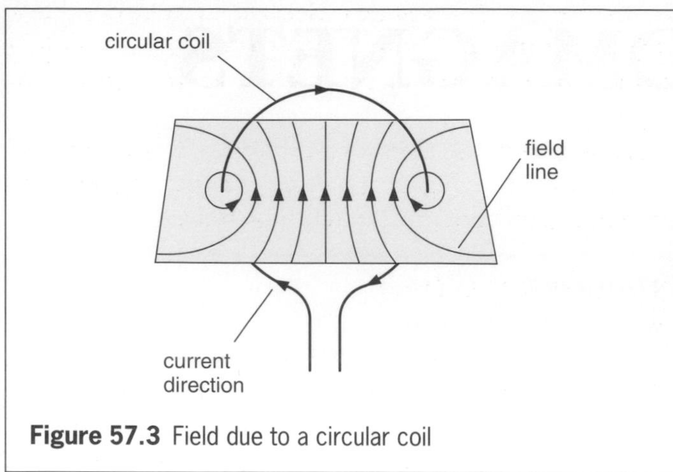

Figure 57.3 Field due to a circular coil

Field due to a solenoid

A solenoid is a long cylindrical coil. It produces a field similar to that of a bar magnet; in Figure 57.4a, end A behaves like a N pole and end B like a S pole. The polarity is found as before by applying the right-hand screw rule to a short length of one turn of the solenoid. Alternatively the **right-hand grip rule** can be used. This states that if the fingers of the right hand grip the solenoid in the direction of the current (conventional), the thumb points to the N pole, Figure 57.4b.

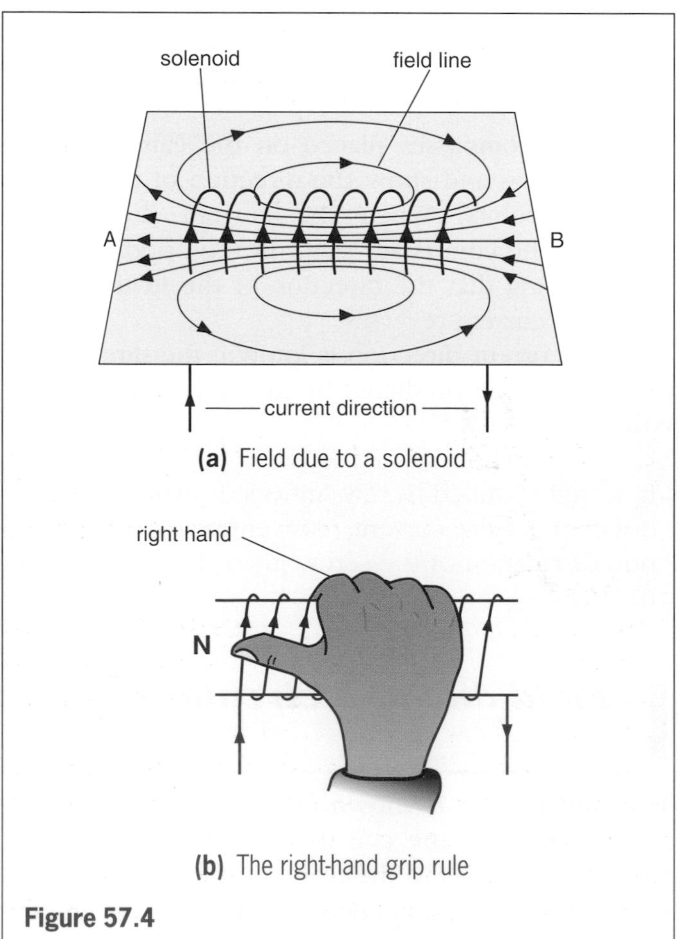

(a) Field due to a solenoid

(b) The right-hand grip rule

Figure 57.4

The field inside a solenoid can be made very strong if it has a large number of turns or a large current. Permanent magnets can be made by allowing molten ferromagnetic metal to solidify in such fields.

PRACTICAL WORK

Simple electromagnet

An electromagnet is a coil of wire wound on a soft iron core. A 5 cm iron nail and 3 m of rayon-covered copper wire (SWG 26) are needed.

(a) Leave about 25 cm at one end of the wire (for connecting to the circuit) and then wind about 50 cm as a single layer on the nail. **Keep the turns close together and always wind in the same direction.** Connect the circuit of Figure 57.5, setting the rheostat at its maximum resistance.

Find the number of paper clips the electromagnet can support when known currents between 0.2 A and 2.0 A pass through it. Record the results in a table. How does the 'strength' of the electromagnet depend on the current?

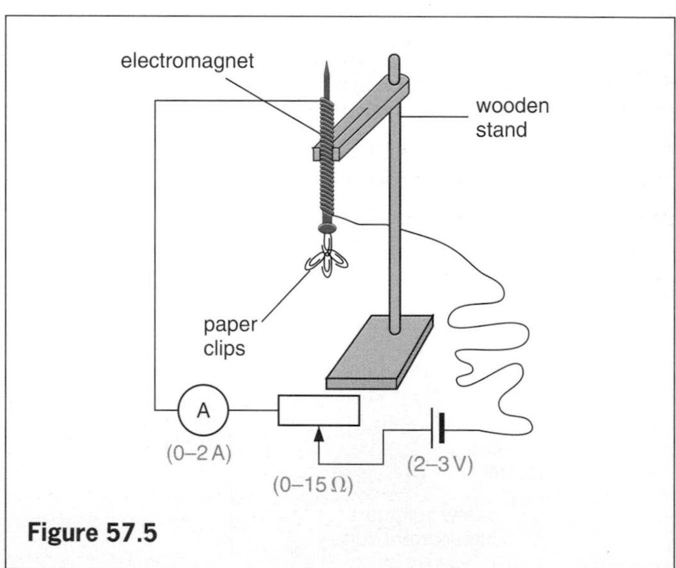

Figure 57.5

(b) Wind another two layers of wire on the nail in the *same direction* as the first layer. Repeat the experiment. What can you say about the 'strength' of an electromagnet and the number of turns of wire?

(c) Place the electromagnet on the bench and under a sheet of paper. Sprinkle iron filings on the paper, tap it gently and observe the field pattern. How does it compare with that given by a bar magnet?

(d) Use the right-hand screw (or grip) rule to predict which end of the electromagnet is a N pole. Check with a plotting compass.

Electromagnets

The magnetism of an electromagnet is *temporary* and can be switched on and off, unlike that of a permanent magnet. It has a core of soft iron which is magnetized only when current flows in the surrounding coil.

The strength of an electromagnet increases if

(i) the current in the coil increases,
(ii) the number of turns on the coil increases,
(iii) the poles are closer together.

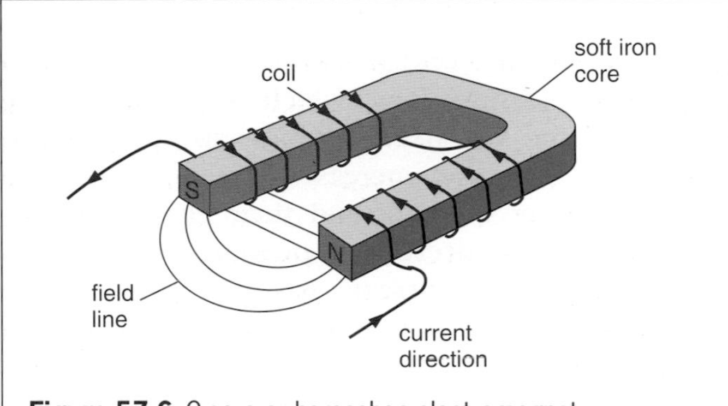

Figure 57.6 C-core or horseshoe electromagnet

Figure 57.7 Electromagnet being used to lift scrap metal

In C-core (or horseshoe) electromagnets condition (iii) is achieved, Figure 57.6. Note that the coil is wound in *opposite* directions on each limb of the core.

As well as being used as a crane to lift iron objects, scrap iron, etc., Figure 57.7, an electromagnet is an essential part of many electrical devices.

Electric bell

When the circuit in Figure 57.8 is completed, current flows in the coils of the electromagnet which becomes magnetized and attracts the soft iron bar (the armature). The hammer hits the gong but the circuit is now broken at the point C of the contact screw.

The electromagnet loses its magnetism and no longer attracts the armature. The springy metal strip is then able to pull the armature back, remaking contact at C and so completing the circuit again. This cycle is repeated so long as the bell push is depressed and continuous ringing occurs.

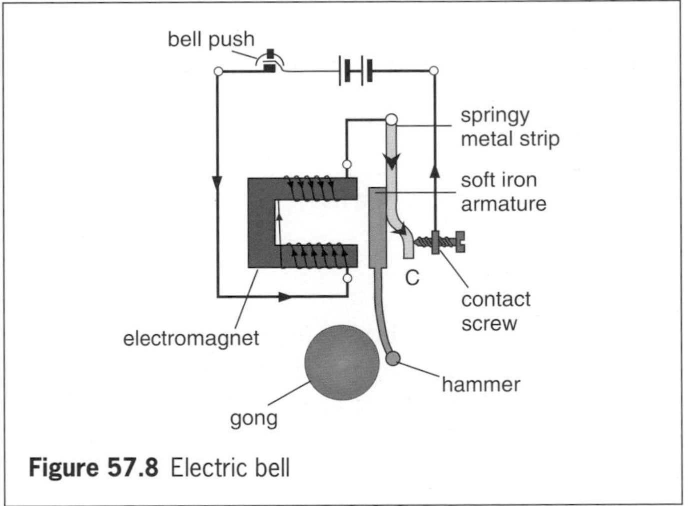

Figure 57.8 Electric bell

Relay and reed switch

(a) Relay

This is a switch based on the principle of an electromagnet. It is useful if we want one circuit to control another, especially if the current and power are larger in the second circuit (see question 3, p. 245). Figure 57.9 (overleaf) shows a typical relay. When a current is in the coil from the circuit connected to AB, the soft iron core is magnetized and attracts the L-shaped iron armature. This rocks on its pivot and closes the contacts at C in the circuit connected to DE. The relay is then 'energized' or 'on'.

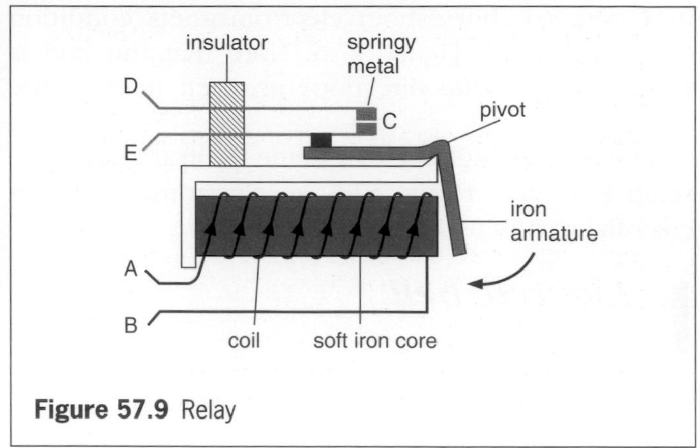

Figure 57.9 Relay

The current needed to operate a relay is called the **pull-on** current, and the **drop-off** current is the smaller current in the coil when the relay just stops working. If the coil resistance R of a relay is 185 Ω and its operating p.d. V is 12 V, the pull-on current $I = V/R = 12/185 = 0.065$ A = 65 mA. The symbols for relays with normally open and normally closed contacts are given in Figure 57.10a, b.

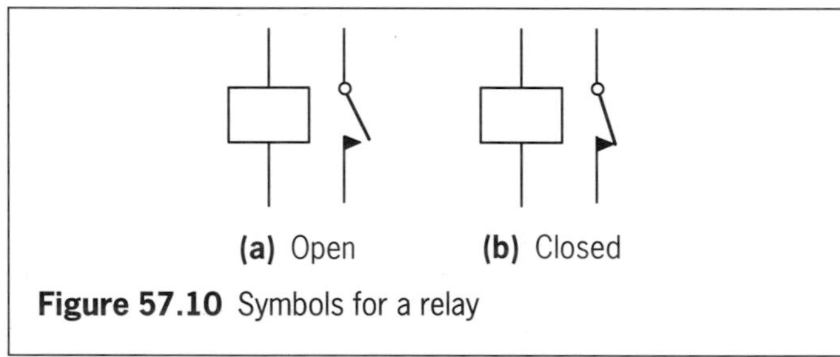

(a) Open **(b)** Closed

Figure 57.10 Symbols for a relay

(b) Reed switch

One such switch is shown in Figure 57.11a. When current flows in the coil, the magnetic field produced magnetizes the strips (called 'reeds') of magnetic material. The ends become opposite poles and one reed is attracted to the other so completing the circuit connected to AB. The reeds separate when the current in the coil is switched off.

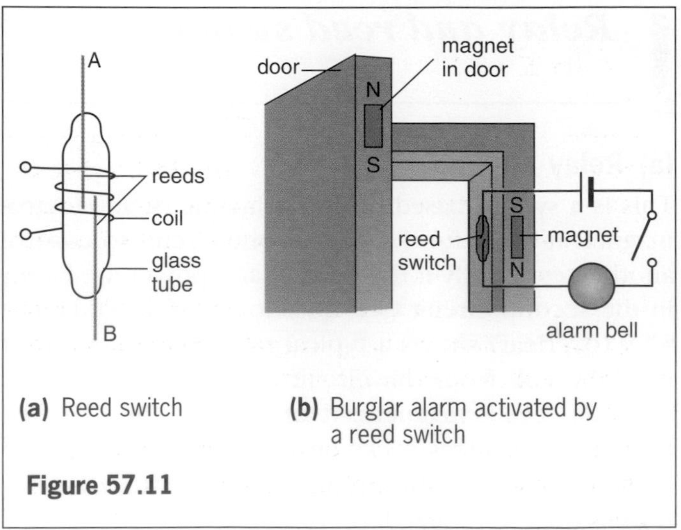

(a) Reed switch **(b)** Burglar alarm activated by a reed switch

Figure 57.11

Reed switches are also operated by permanent magnets. Figure 57.11b shows the use of a normally open reed switch as a burglar alarm. How does it work?

Telephone

A telephone contains a microphone at the speaking end and a receiver at the listening end.

(a) Carbon microphone, Figure 57.12

When someone speaks, sound waves cause the diaphragm to move backwards and forwards. This varies the pressure on the carbon granules between the movable carbon dome which is attached to the diaphragm and the fixed carbon cup at the back. When the pressure increases, the granules are squeezed closer together and their electrical resistance decreases. A decrease of pressure has the opposite effect. The current passing through the microphone varies in a similar way to the sound wave variations.

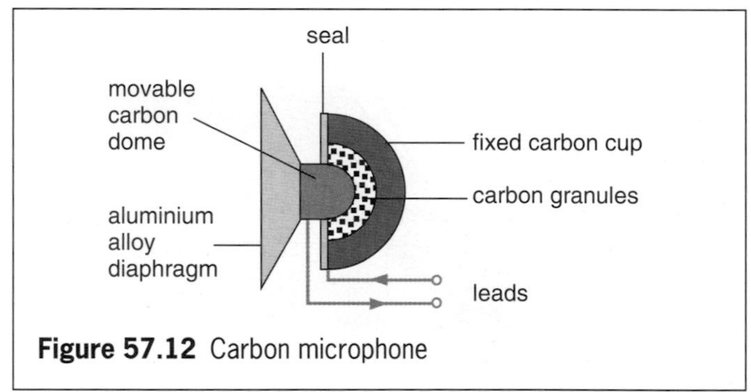

Figure 57.12 Carbon microphone

(b) Receiver

The coils are wound in opposite directions on the *two* S poles of a magnet as in Figure 57.13. If the current goes round one in a clockwise direction, it goes round the other anticlockwise, so making one S pole stronger and the other weaker. This causes the iron armature to rock on its pivot towards the stronger S pole. When the current reverses, the armature rocks the other way due to the S pole which was the

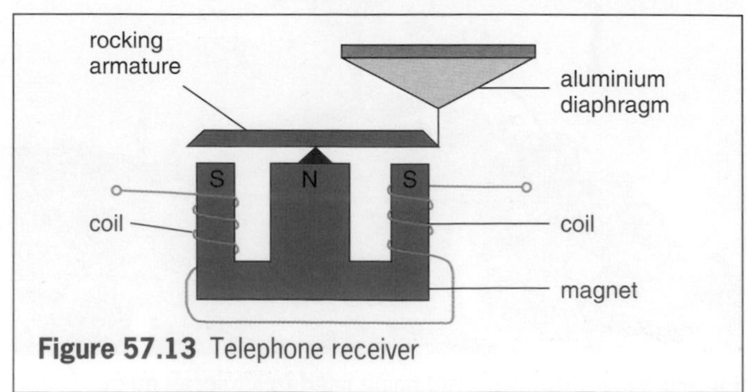

Figure 57.13 Telephone receiver

stronger before becoming the weaker. These armature movements are passed on to the diaphragm, making it vibrate and produce sound of the same frequency as the alternating current in the coil (received from the microphone).

QUESTIONS

1 The vertical wire in Figure 57.14 is at right angles to the card. In what direction will a plotting compass at A point when

a there is no current in the wire,

b current flows upwards?

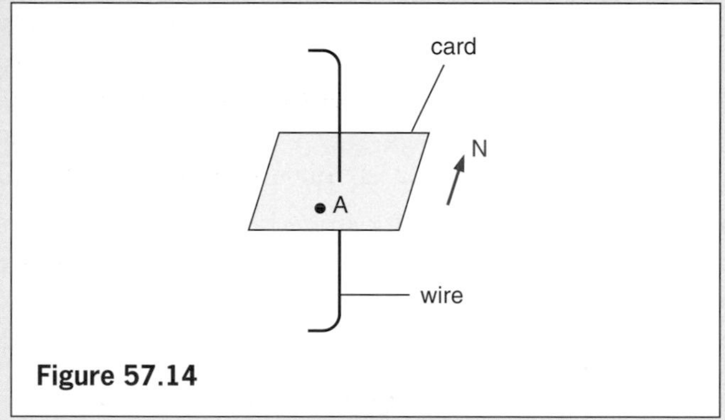

Figure 57.14

2 Figure 57.15 shows a solenoid wound round a core of soft iron. Will the end A be a N pole or S pole when the current (conventional) flows in the direction shown? *(J.M.B./A.L./N.W.16+)*

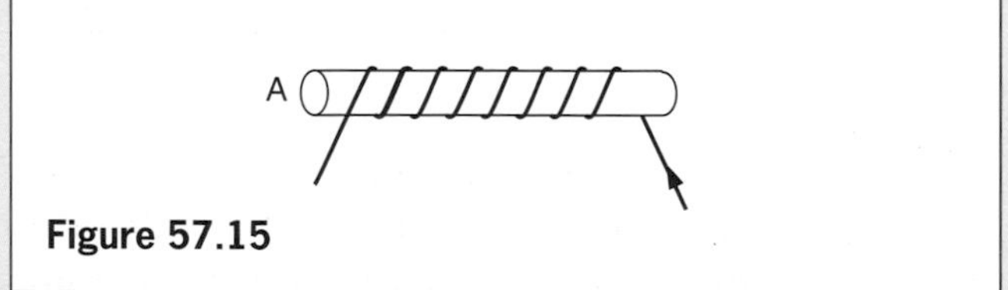

Figure 57.15

3 A small electromagnet, used for lifting and then releasing a small steel ball, is made in the laboratory, Figure 57.16.

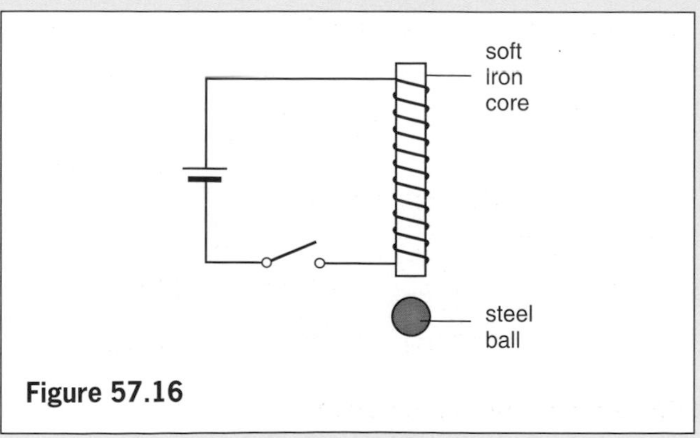

Figure 57.16

a Explain why soft iron is a better material than steel to use for the core.

b In order to lift a slightly larger ball it is necessary to make a stronger electromagnet. State *two* ways in which an electromagnet could be made more powerful.

4 You are asked to carry out an experiment to find how the lifting power of a horseshoe electromagnet depends upon the current flowing through the coils. Draw a clear diagram of the circuit you would use showing clearly the direction of the current, the direction of the winding of the coils of the electromagnet and the polarity of one end of the magnet. *(E.M. part qn.)*

5 Figure 57.17 shows an arrangement for lighting three lamps, A, B and C, only one of which is controlled directly by the switch.

a Which of the lamps is directly controlled by the switch?

b What is the name given to this use of an electromagnet?

c Which lamps can be on at once?

d Explain how lamp C comes on. *(S.R.)*

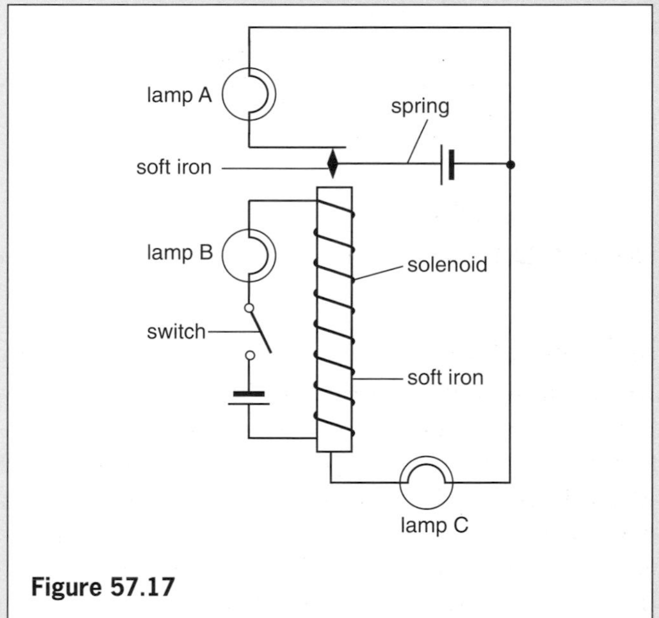

Figure 57.17

Checklist

After studying this chapter you should be able to

- describe the magnetic fields round current-carrying, straight and circular conductors and solenoids and draw sketches of them,
- recall the right-hand screw and grip rules for relating current direction and magnetic field direction,
- make a simple electromagnet,
- describe uses of electromagnets,
- explain the action of an electric bell, a relay, a reed switch, and the microphone and receiver of a telephone.

58 ELECTRIC MOTORS

The motor effect	***Moving-coil loudspeaker***
Fleming's left-hand rule	***Practical work*** A model motor.
Simple d.c. motor	
Practical motors	

Electric motors form the heart of a whole host of electrical devices ranging from domestic appliances such as vacuum cleaners and washing machines to electric trains and lifts. In a car the windscreen wipers are usually driven by one and the engine is started by another.

The motor effect

A wire carrying a current in a magnetic field experiences a force. If the wire can move it does so.

(a) Demonstration

In Figure 58.1 the flexible wire is loosely supported in the strong magnetic field of a C-shaped magnet (permanent or electro). When the switch is pressed, current flows in the wire which jumps upwards as shown. If either the direction of the current or the direction of the field is reversed, the wire moves downwards. **The force increases if the strength of the field increases and if the current increases.**

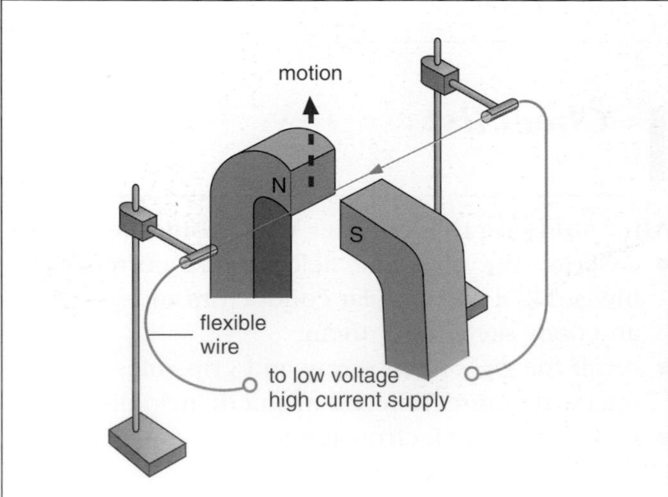

Figure 58.1 A wire carrying a current in a magnetic field experiences a force

(b) Explanation

Figure 58.2a is a side view of the magnetic field lines due to the wire and the magnet. Those due to the wire are circles and we will suppose their direction is as shown. The dotted lines represent the field lines of the magnet and their direction is to the right.

The resultant field obtained by combining both fields is shown in Figure 58.2b. There are more lines below than above the wire since both fields act in the same direction below but in opposition above. If we *suppose* the lines are like stretched elastic, those below will try to straighten out and in so doing will exert an upwards force on the wire.

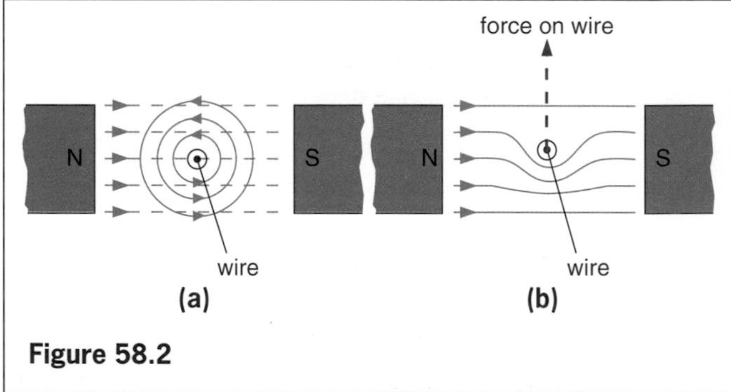

Figure 58.2

Fleming's left-hand rule

The direction of the force or thrust on the wire can be found by this rule which is also called the 'motor rule', Figure 58.3.

> Hold the thumb and first two fingers of the left hand at right angles to each other with the **F**irst finger pointing in the direction of the **F**ield and the se**C**ond finger in the direction of the **C**urrent, then the **T**humb points in the direction of the **T**hrust.

If the wire is not at right angles to the field, the force is smaller and is zero if it is parallel to the field.

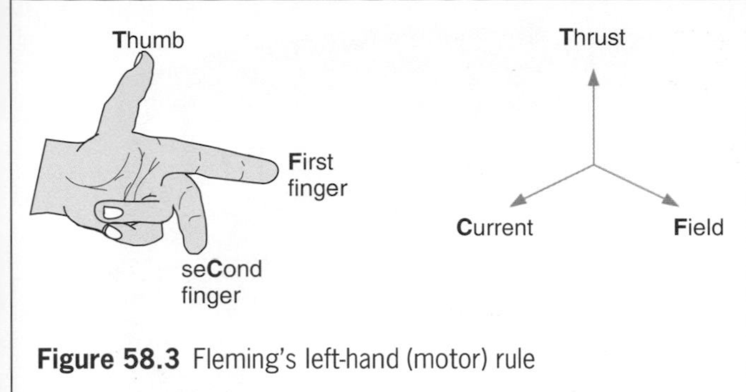

Figure 58.3 Fleming's left-hand (motor) rule

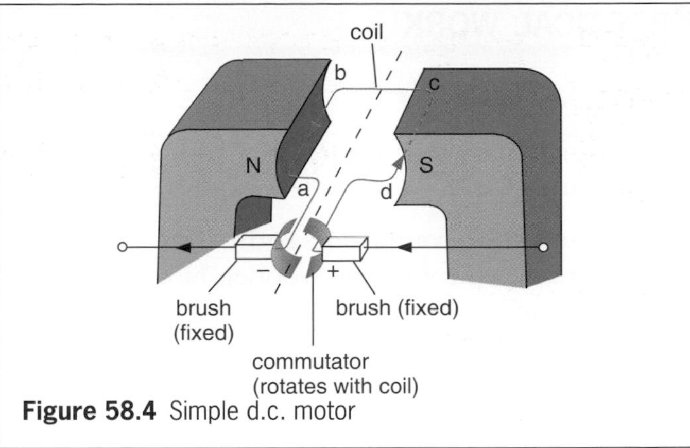

Figure 58.4 Simple d.c. motor

Simple d.c. electric motor

A simple motor to work from direct current (d.c.) consists of a rectangular coil of wire mounted on an axle which can rotate between the poles of a C-shaped magnet, Figure 58.4. Each end of the coil is connected to half of a split ring of copper, called the **commutator**, which rotates with the coil. Two carbon blocks, the **brushes**, are pressed lightly against the commutator by the springs. The brushes are connected to an electrical supply.

If Fleming's left-hand rule is applied to the coil in the position shown, we find that side **ab** experiences an upward force and side **cd** a downward force. (No forces act on **ad** and **bc** since they are parallel to the field.) These two forces form a **couple** which rotates the coil in a clockwise direction until it is vertical.

The brushes are then in line with the gaps in the commutator and the current stops. However, because of its inertia, the coil overshoots the vertical and the commutator halves change contact from one brush to the other. This reverses the current through the coil and so also the directions of the forces on its sides. Side **ab** is on the right now, acted on by a downward force, whilst **cd** is on the left with an upward force. The coil thus carries on rotating clockwise.

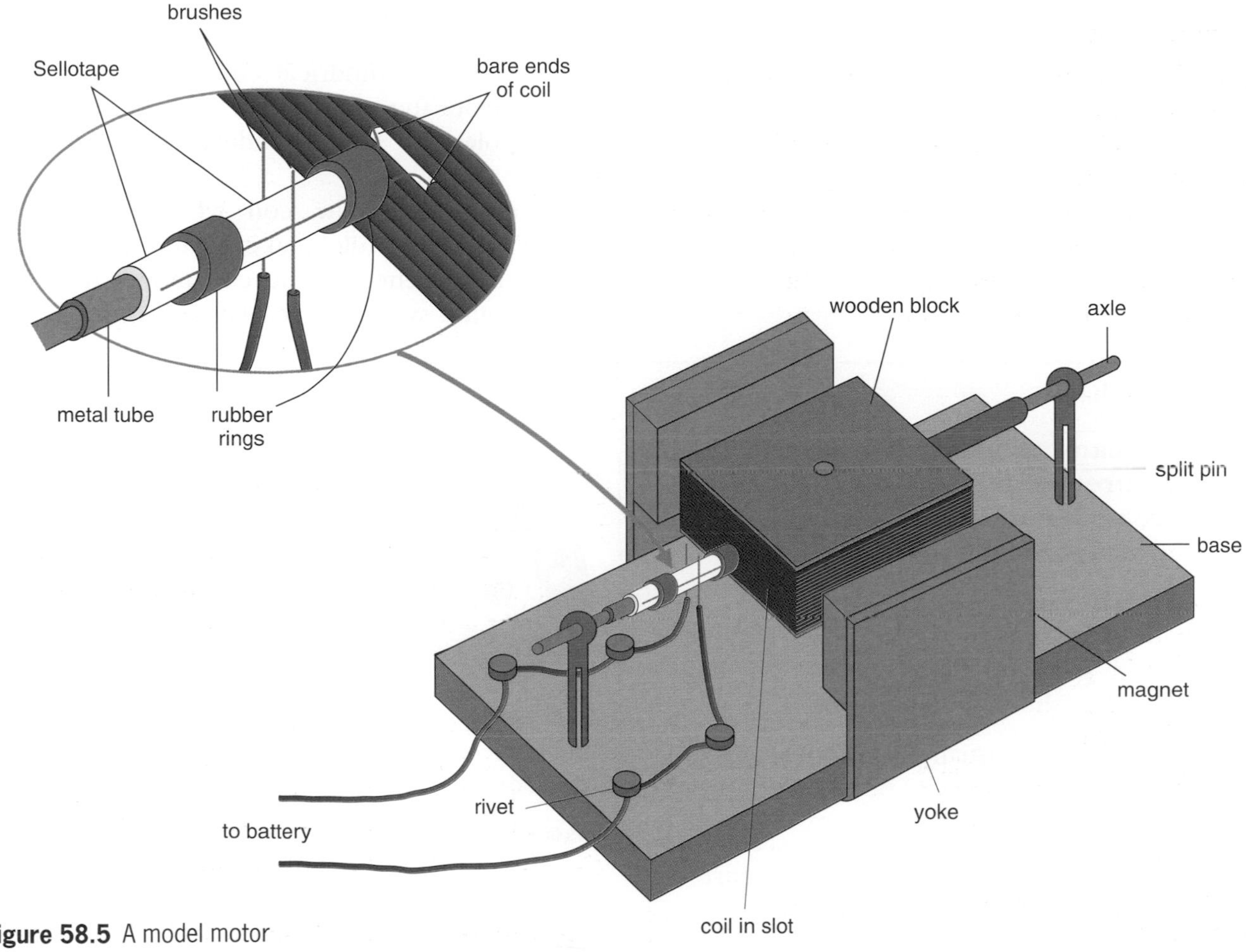

Figure 58.5 A model motor

Practical work

A model motor

The motor is shown in Figure 58.5 and is made from a kit of parts.

1 Wrap Sellotape round one end of the metal tube which passes through the wooden block.
2 Cut two rings off a piece of narrow rubber tubing; slip them onto the Sellotaped end of the metal tube.
3 Remove the insulation from one end of a $1\frac{1}{2}$ metre length of SWG 26 PVC-covered copper wire and fix it under both rubber rings so that it is held tight against the Sellotape. This forms one end of the coil.
4 Wind 10 turns of the wire in the slot in the wooden block and finish off the second end of the coil by removing the PVC and fixing this too under the rings but on the *opposite* side of the tube from the first end. The bare ends act as the **commutator**.
5 Push the axle through the metal tube of the wooden base so that the block spins freely.
6 Arrange two $\frac{1}{2}$ metre lengths of wire to act as **brushes** and leads to the supply, as shown. Adjust the brushes so that they are vertical and each touches one bare end of the coil when the plane of the coil is horizontal. **The motor will not work if this is not so.**
7 Slide the base into the magnet with *opposite poles facing*. Connect to a 3 V battery (or other low voltage d.c. supply) and a slight push of the coil should set it spinning at high speed.

Practical motors

Practical motors have:

(a) a coil of many turns wound on a soft iron cylinder or core which rotates with the coil. This makes it more powerful. The coil and core together are called the **armature**.

(b) several coils each in a slot in the core and each having a pair of commutator segments. This gives increased power and smoother running. The motor of an electric drill is shown in Figure 58.6.

(c) an electromagnet (usually) to produce the field in which the armature rotates.

Most electric motors used in industry are **induction motors**. They work off a.c. (alternating current) on a different principle to the d.c. motor.

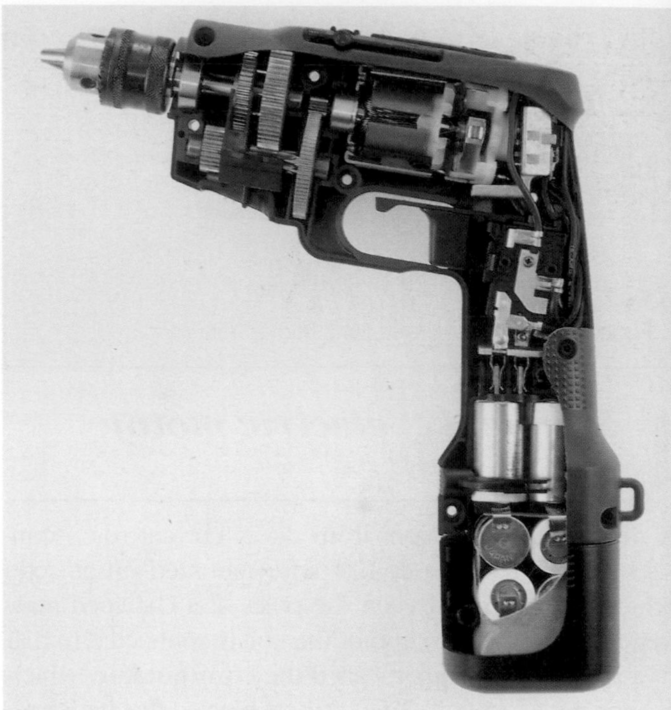

Figure 58.6 Motor inside an electric drill

Moving-coil loudspeaker

Varying currents from a radio, disc player, etc. pass through a short cylindrical coil whose turns are at right angles to the magnetic field of a magnet with a central pole and a surrounding ring pole, Figure 58.7a.

A force acts on the coil which, according to Fleming's left-hand rule, makes it move in and out. A paper cone attached to the coil moves with it and sets up sound waves in the surrounding air, Figure 58.7b.

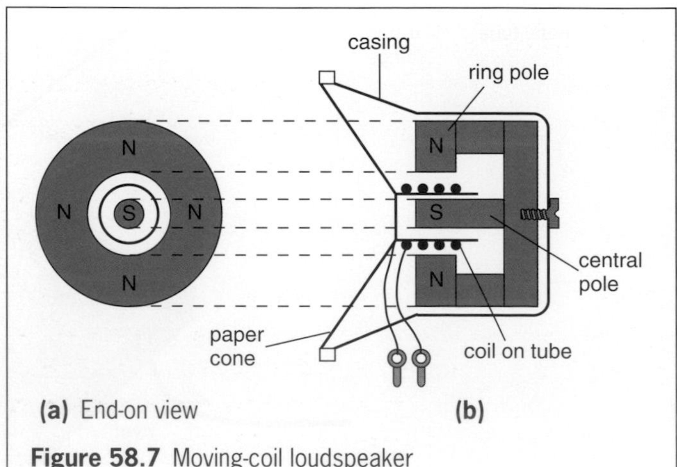

Figure 58.7 Moving-coil loudspeaker

QUESTIONS

1 A wire carries a current horizontally between the magnetic poles N and S which face each other on a table, Figure 58.8. The direction of the force on the wire due to the magnets is

A from N to S
B from S to N
C opposite to the current direction
D in the direction of the current
E vertically upwards. *(O.L.E./S.R.16+)*

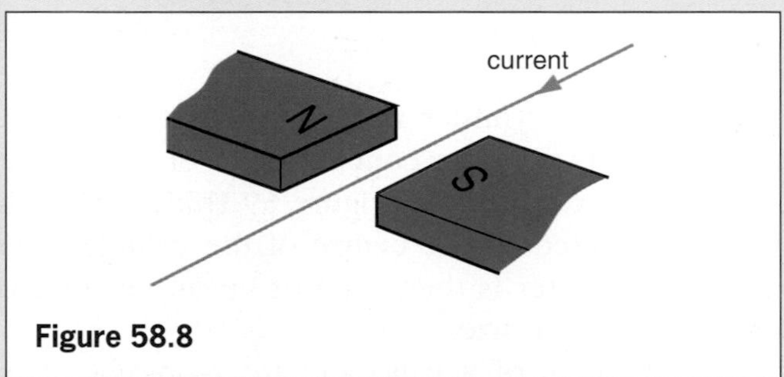

Figure 58.8

2 In Figure 58.9 AB is a copper wire hanging from a pivot at A and dipping in mercury in a copper dish at B so that it hangs between the poles of a powerful magnet.

a Copy the diagram and mark in the lines of force representing the field between the poles.

b Mark in the direction of the conventional current flow in AB when the switch is closed.

c What else will happen when the switch is closed? *(S.E.)*

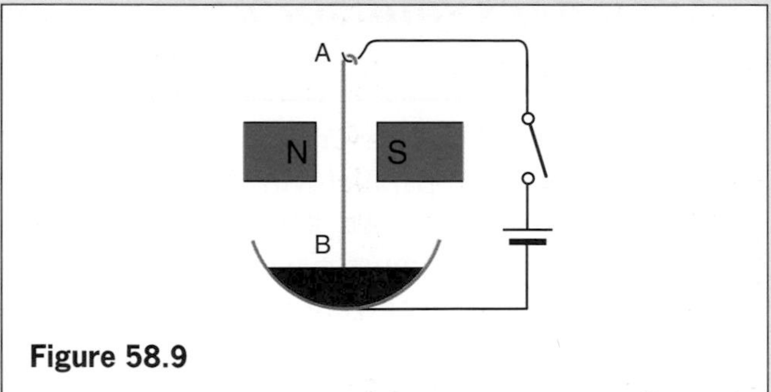

Figure 58.9

3 In the simple electric motor of Figure 58.10, will the coil rotate clockwise or anticlockwise as seen by the eye from the position X? The arrows on the diagram indicate the direction of the conventional current flow. *(J.M.B./A.L./N.W.16+)*

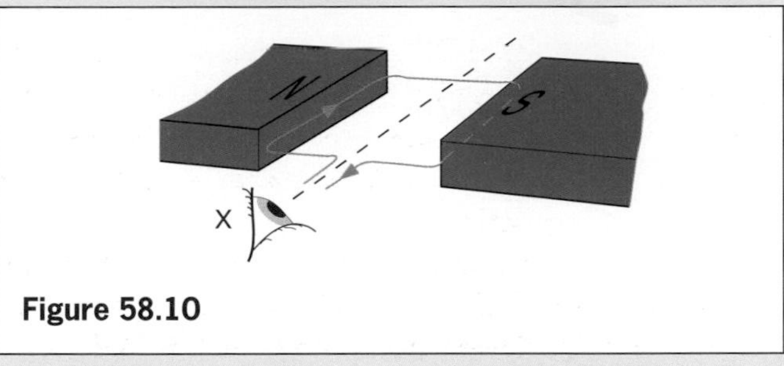

Figure 58.10

4 An electric motor is a device which transfers

A mechanical energy to electrical energy
B heat energy to electrical energy
C electrical energy to heat only
D heat to mechanical energy
E electrical energy to mechanical energy and heat.

Checklist

After studying this chapter you should be able to

- describe a demonstration to show that a force acts on a current-carrying conductor in a magnetic field and recall that it increases with the strength of the field and the size of the current,
- draw the resultant field pattern for a current-carrying conductor which is at right angles to a uniform magnetic field,
- explain why a rectangular current-carrying coil experiences a couple in a uniform magnetic field,
- draw a diagram of a simple d.c. electric motor and explain how it works,
- describe a practical d.c. motor,
- draw a diagram of a moving-coil loudspeaker and explain how it works.

59 ELECTRIC METERS

Moving-coil galvanometer
Ammeters and shunts
Voltmeters and multipliers

Moving-coil galvanometer

A galvanometer detects small currents or small p.d.s, often of the order of milliamperes (mA) or millivolts (mV).

In the moving-coil **pointer-type** meter, a coil is pivoted on jewelled bearings between the poles of a permanent magnet, Figure 59.1a. Current enters and leaves the coil by hair springs above and below it. When current flows, a couple acts on the coil (as in an electric motor), causing it to rotate until stopped by the springs. The greater the current the greater the deflection which is shown by a pointer attached to the coil.

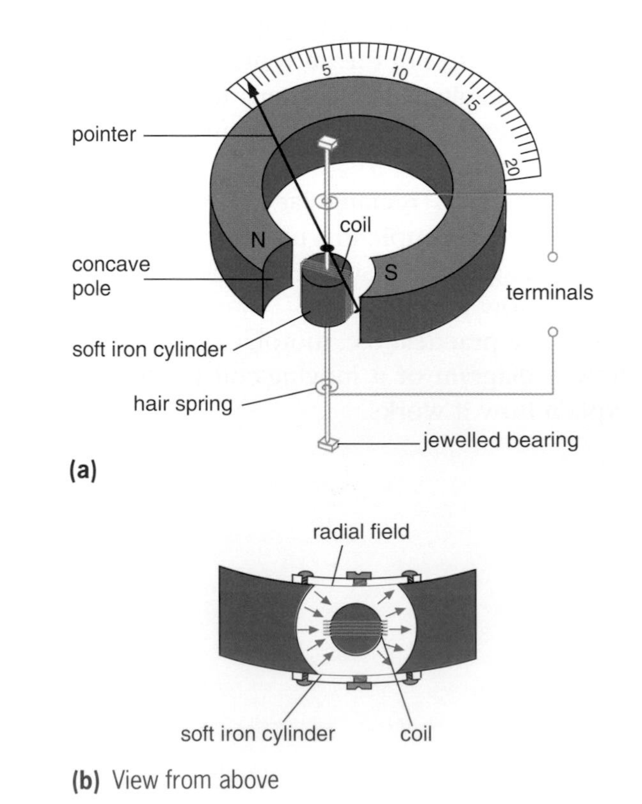

Figure 59.1 Moving-coil pointer-type galvanometer

The soft iron cylinder at the centre of the coil is fixed and along with the concave poles of the magnet it produces a **radial** field, Figure 59.1b, i.e. the field lines are directed to the centre of the cylinder. The scale on the meter is then even or linear, i.e. all divisions are the same size.

The sensitivity of a galvanometer is increased by having

(i) more turns on the coil,
(ii) a stronger magnet,
(iii) weaker hair springs or a wire suspension,
(iv) as a pointer, a long beam of light reflected from a mirror on the coil.

The last two are used in **light-beam** meters which have a full-scale deflection of a few microamperes (μA). (1 μA = 10^{-6} A)

Ammeters and shunts

An ammeter is a galvanometer having a known low resistance (a **shunt**) in parallel with it to take most of the current, Figure 59.2. An ammeter is placed in *series* in a circuit and must have a *low resistance* otherwise it changes the current to be measured.

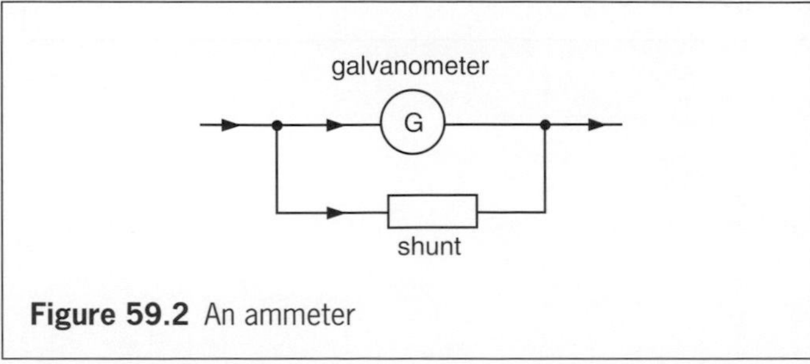

Figure 59.2 An ammeter

Voltmeters and multipliers

A voltmeter is a galvanometer having a known high resistance (a **multiplier**) in series with it, Figure 59.3. A voltmeter is placed in *parallel* with the part

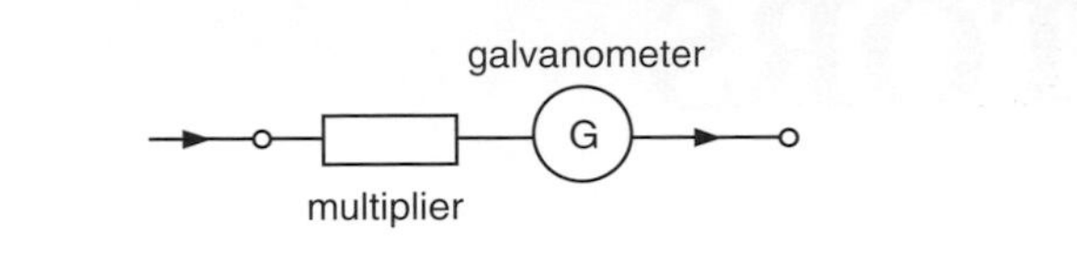

Figure 59.3 A voltmeter

of the circuit across which the p.d. is to be measured and must have a *high resistance* – otherwise the total resistance of the whole circuit is reduced so changing the current and the p.d. required.

QUESTIONS

1 What does a galvanometer do?

2 Why should the resistance of

a an ammeter be very small,

b a voltmeter be very large?

Checklist

After studying this chapter you should be able to

- draw a diagram of a simple moving-coil galvanometer and explain how it works,
- explain how a moving-coil galvanometer can be modified for use as (a) an ammeter and (b) a voltmeter,
- explain why (a) an ammeter should have a very low resistance and (b) a voltmeter should have a very high resistance.

60 GENERATORS

Electromagnetic induction	*Simple a.c. generator (alternator)*
Faraday's law	*Simple d.c. generator (dynamo)*
Lenz's law	*Practical generators*

An electric current creates a magnetic field. The reverse effect of producing electricity from magnetism was discovered in 1831 by Faraday and is called **electromagnetic induction**. It led to the construction of generators for producing electrical energy in power stations.

Electromagnetic induction

Two ways of investigating the effect follow.

(a) Straight wire and U-shaped magnet, Figure 60.1

First the wire is held at rest between the poles of the magnet and the galvanometer observed. It is then moved in each of the six directions shown. Only *when it is moving upwards* (direction 1) or *downwards* (direction 2) is there a deflection on the galvanometer, indicating an induced current in the wire. The deflection is in opposite directions in each case and only lasts while the wire is in motion.

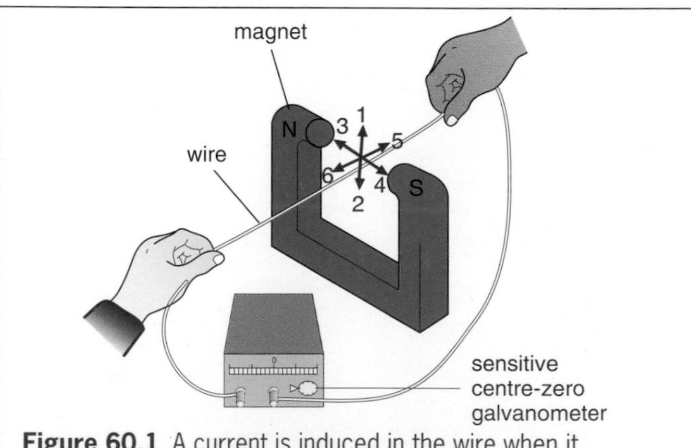

Figure 60.1 A current is induced in the wire when it moves up or down between the magnet poles

(b) Bar magnet and coil, Figure 60.2

The magnet is pushed into the coil, one pole first, then held still inside it. It is next withdrawn. The galvanometer shows that current is induced in the coil in one direction as the magnet *moves in* and in the opposite direction as it is *removed*. There is no deflection when the magnet is at rest. The results are the same if the coil is moved instead of the magnet, i.e. only **relative motion** is needed.

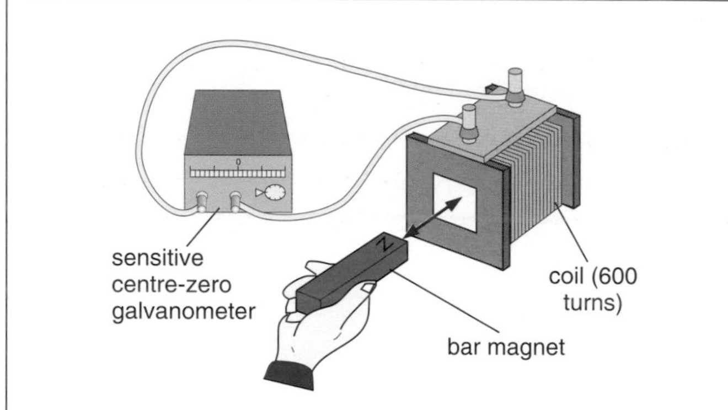

Figure 60.2 A current is induced in the coil when the magnet is moved in or out

Faraday's law

To 'explain' electromagnetic induction Faraday suggested that a voltage is induced in a conductor whenever it 'cuts' magnetic field lines, i.e. moves *across* them, but not when it moves along them or is at rest. If the conductor forms part of a complete circuit, an induced current is also produced.

Faraday found, and it can be shown with apparatus like that in Figure 60.2, that the induced p.d. or voltage increases with increases of

(i) the speed of motion of the magnet or coil,
(ii) the number of turns on the coil,
(iii) the strength of the magnet.

These facts led him to state a law:

> The size of the induced p.d. is directly proportional to the rate at which the conductor cuts magnetic field lines.

Lenz's law

The direction of the induced current can be found by a law due to the Russian scientist, Lenz.

The direction of the induced current is such as to oppose the change causing it.

In Figure 60.3a the magnet approaches the coil, north pole first. According to Lenz's law the induced current should flow in a direction which makes the coil behave like a magnet with its top a north pole. The downward motion of the magnet will then be opposed.

When the magnet is withdrawn, the top of the coil should become a south pole, Figure 60.3b, and attract the north pole of the magnet, so hindering its removal. The induced current is thus in the opposite direction to that when the magnet approaches.

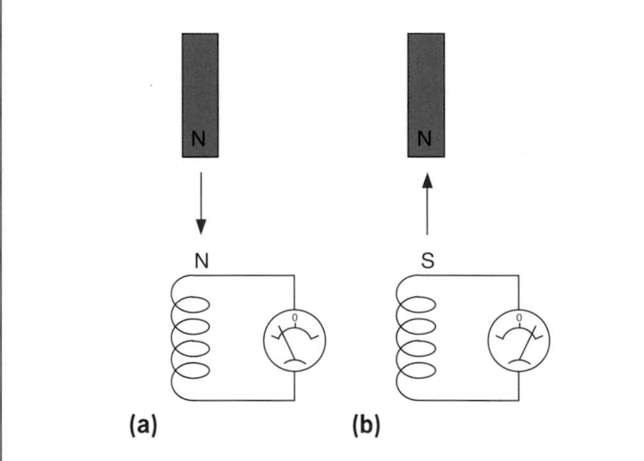

Figure 60.3 The induced current opposes the motion of the magnet

Lenz's law is an example of the principle of conservation of energy. If the currents caused opposite poles to those that they do, electrical energy would be created from nothing. As it is, mechanical energy is provided, by whoever moves the magnet, to overcome the forces that arise.

For a straight wire moving at right angles to a magnetic field a more useful form of Lenz's law is **Fleming's right-hand rule** (the 'dynamo rule'), Figure 60.4.

Hold the thumb and first two fingers of the right hand at right angles to each other with the **F**irst finger pointing in the direction of the **F**ield and the thu**M**b in the direction of **M**otion of the wire, then the se**C**ond finger points in the direction of the induced **C**urrent.

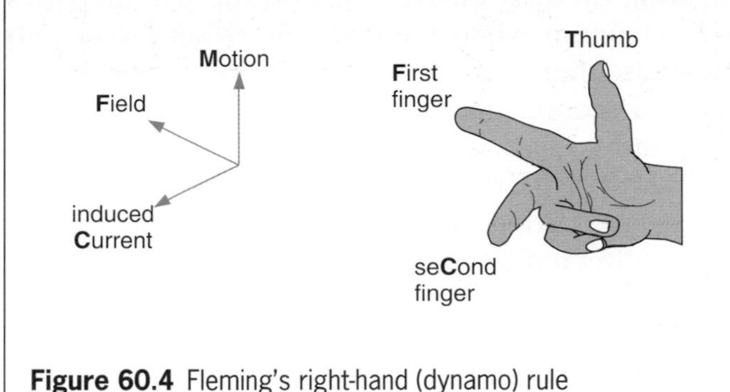

Figure 60.4 Fleming's right-hand (dynamo) rule

Simple a.c. generator (alternator)

The simplest alternating current (a.c.) generator consists of a rectangular coil between the poles of a C-shaped magnet, Figure 60.5a. The ends of the coil are joined to two **slip rings** on the axle and against which carbon **brushes** press.

When the coil is rotated it cuts the field lines and a voltage is induced in it. Figure 60.5b shows how the voltage varies over one complete rotation.

As the coil moves through the vertical position with **ab** uppermost, **ab** and **cd** are moving along the lines (**bc** and **da** do so always) and no cutting occurs. The induced voltage is zero.

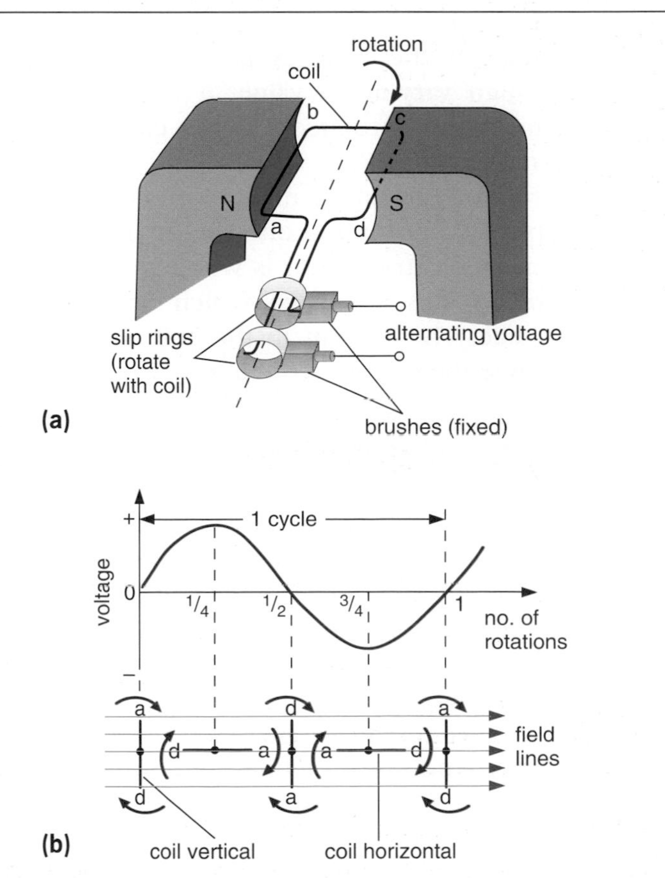

Figure 60.5 A simple a.c. generator and its output

During the first quarter rotation the p.d. increases to a maximum when the coil is horizontal. Sides **ab** and **dc** are then cutting the lines at the greatest rate.

In the second quarter rotation the p.d. decreases again and is zero when the coil is vertical with **dc** uppermost. After this, the direction of the p.d. reverses because, during the next half rotation, the motion of **ab** is directed upwards and **dc** downwards.

An alternating voltage is generated which acts first in one direction and then the other; it would cause a.c. to flow in a circuit connected to the brushes. The **frequency** of an a.c. is the number of complete cycles it makes each second (c/s) and is measured in **hertz** (Hz), i.e. 1 c/s = 1 Hz. If the coil rotates twice per second, the a.c. has frequency 2 Hz. The mains supply is a.c. of frequency 50 Hz.

Simple d.c. generator (dynamo)

An a.c. generator becomes a direct current (d.c.) one if the slip rings are replaced by a **commutator** (like that in a d.c. motor), Figure 60.6a.

The brushes are arranged so that as the coil goes through the vertical, changeover of contact occurs from one half of the split ring of the commutator to the other. In this position the voltage induced in the coil reverses and so one brush is always positive and the other negative.

The voltage at the brushes is shown in Figure 60.6b; although varying in value, it never changes direction and would produce a direct current (d.c.) in an external circuit.

In construction the simple d.c. dynamo is the same as the simple d.c. motor and one can be used as the other. When an electric motor is working it acts as a dynamo and creates a voltage which opposes the applied voltage. The current in the coil is therefore much less once the motor is running.

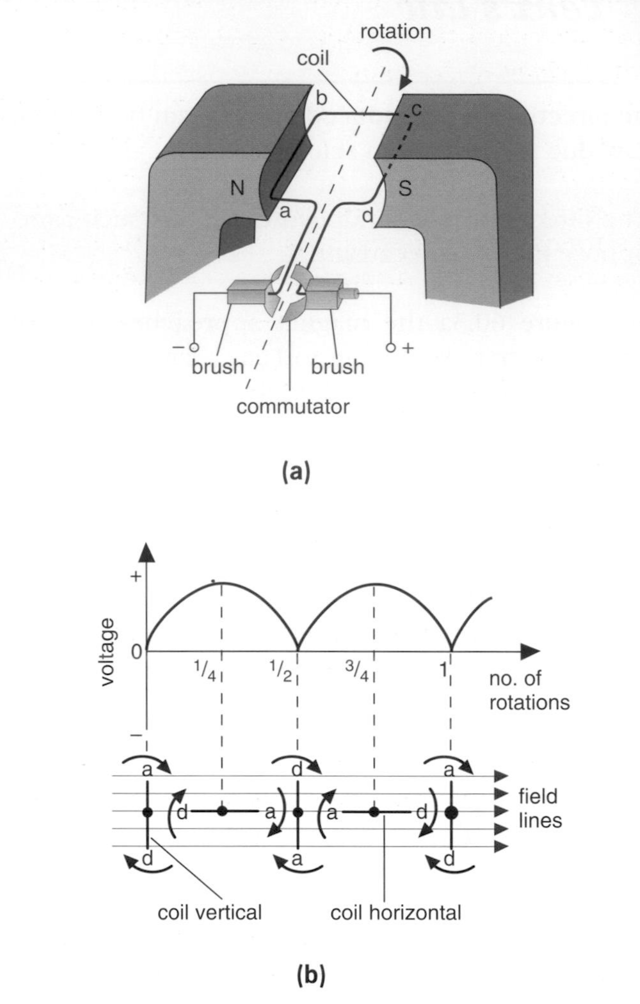

Figure 60.6 A simple d.c. generator and its output

Practical generators

In actual generators several coils are wound in evenly spaced slots in a soft iron cylinder and electromagnets usually replace permanent magnets.

(a) Power stations

In power station alternators, Figure 60.7, the electromagnets rotate (the **rotor**) while the coils and their iron core are at rest (the **stator**). The large p.d.s and currents (e.g. 25 kV at several thousand amperes) induced in the stator are led away through stationary cables, otherwise they would quickly destroy the slip rings by sparking. Instead the relatively small d.c.

Figure 60.7 General view of the alternator at a power station

required by the rotor is fed via the slip rings from a small dynamo (the **exciter**) which is driven by the same turbine as the rotor.

In a thermal power station, the turbine is rotated by high-pressure steam obtained by heating water in a coal- or oil-fired boiler or in a nuclear reactor (or by hot gas in a gas-fired power station). A block diagram of a thermal power station is shown in Figure 60.8. The energy transfer diagram was given in Figure 46.4.

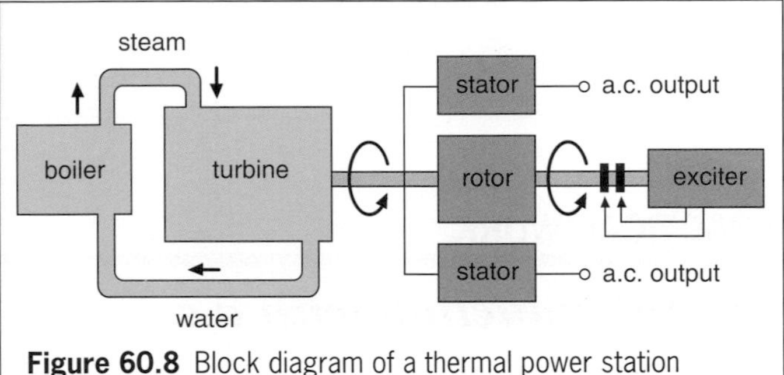

Figure 60.8 Block diagram of a thermal power station

(b) Cars

Most are now fitted with alternators because they give a greater output than dynamos at low engine speeds.

(c) Bicycles

The rotor is a permanent magnet and the voltage is induced in the coil which is at rest, Figure 60.9.

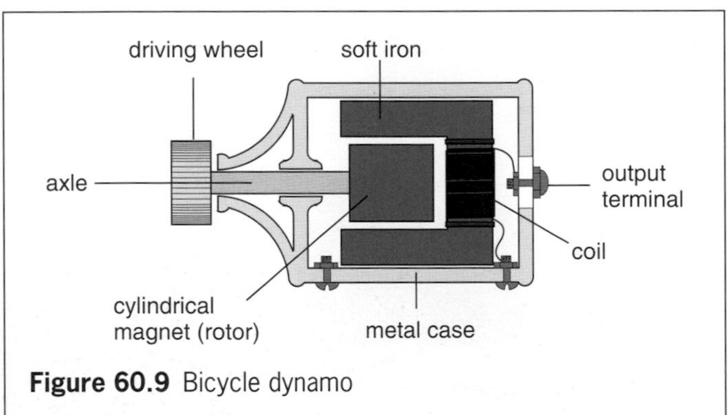

Figure 60.9 Bicycle dynamo

QUESTIONS

1 a A coil of copper wire is connected to a centre-zero galvanometer. Explain what would be observed if a bar magnet was pushed into the coil so that the north pole of the magnet entered first.

b What would be observed if the magnet was
(i) held at rest inside the coil?
(ii) pulled out again?
(iii) pushed in faster than before?
(iv) pushed in so that the south pole entered first?
(E.A.)

2 A simple generator is shown in Figure 60.10.

a What are A and B called and what is their purpose?

b What changes can be made to increase the p.d. generated?

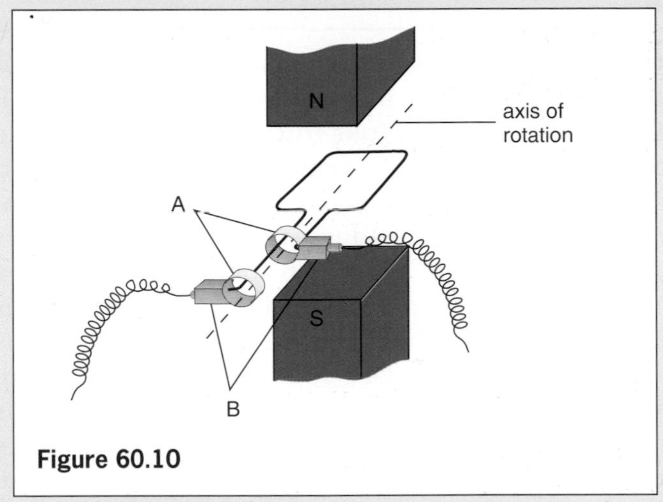

Figure 60.10

Checklist

After studying this chapter you should be able to

- describe experiments to show electromagnetic induction,
- recall Faraday's explanation of electromagnetic induction,
- predict the direction of the induced current using Lenz's law or Fleming's right-hand rule,
- draw a diagram of a simple a.c. generator and sketch a graph of its output,
- draw a diagram of a simple d.c. generator and sketch a graph of its output,
- describe practical generators.

61 TRANSFORMERS

Mutual induction	***Car ignition system***
Transformer equation	***Applications of eddy currents***
Energy losses in a transformer	***Practical work*** Mutual induction with a.c.
Transmission of electrical power	

Mutual induction

When the current in a coil is switched on or off or changed, a voltage is induced in a neighbouring coil. The effect, called **mutual induction**, is an example of electromagnetic induction and can be shown with the arrangement of Figure 61.1. Coil A is the **primary** and coil B the **secondary**.

Switching on the current in the primary sets up a magnetic field and as its field lines 'grow' outwards from the primary they 'cut' the secondary. A p.d. is induced in the secondary until the current in the primary reaches its steady value. When the current is switched off in the primary, the magnetic field dies away and we can imagine the field lines cutting the secondary as they collapse, again inducing a p.d. in it. Changing the primary current by *quickly* altering the rheostat has the same effect.

The induced p.d. is increased by having a soft iron rod in the coils or, better still, by using coils wound on a complete iron ring. More field lines then cut the secondary due to the magnetization of the iron.

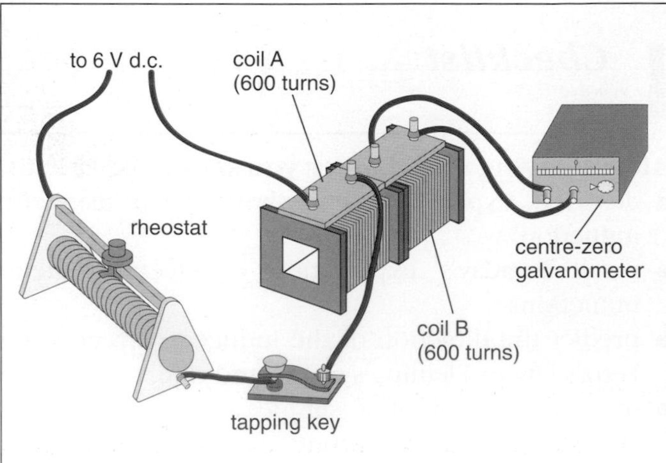

Figure 61.1 A changing current in a primary coil (A) induces a current in a secondary coil (B)

PRACTICAL WORK

Mutual induction with a.c.

An alternating current is changing all the time and if it flows in a primary coil, an alternating voltage and current are induced in a secondary coil.

Connect the circuit of Figure 61.2. The 1 V high current power unit supplies a.c. to the primary and the lamp detects the secondary current.

Find the effect on the brightness of the lamp of

(i) pulling the C-cores apart slightly,
(ii) increasing the secondary turns to 15,
(iii) decreasing the secondary turns to 5.

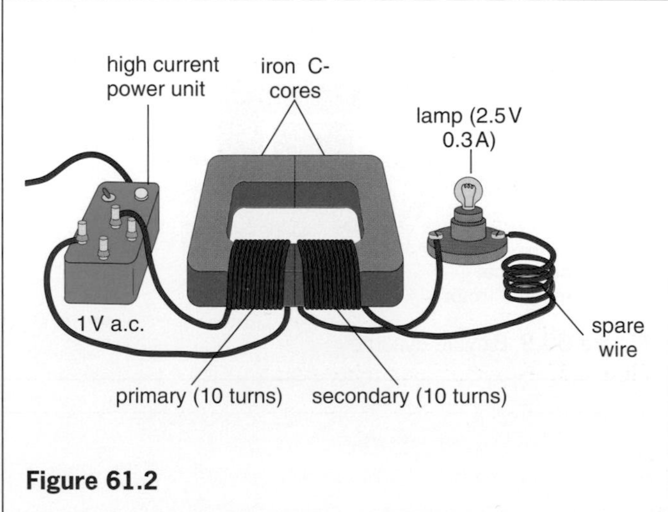

Figure 61.2

Transformer equation

A **transformer** transforms (changes) an *alternating* voltage from one value to another of greater or smaller value. It has primary and secondary coils wound on a complete soft iron core, either one on top of the other, Figure 61.3a, or on separate limbs of the core, Figure 61.3b.

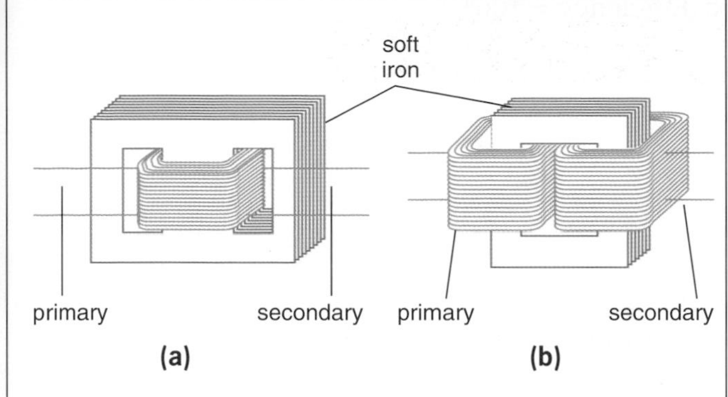

Figure 61.3 Primary and secondary coils of a transformer

An alternating voltage applied to the primary induces an alternating voltage in the secondary whose value can be shown, for an ideal transformer, to be given by

$$\frac{\text{secondary voltage}}{\text{primary voltage}} = \frac{\text{secondary turns}}{\text{primary turns}}$$

In symbols

$$\frac{V_s}{V_p} = \frac{N_s}{N_p}$$

A 'step-up' transformer has more turns on the secondary than the primary and V_s is greater than V_p, Figure 61.4a. For example, if the secondary has twice as many turns as the primary, V_s is about twice V_p. In a 'step-down' transformer there are fewer turns on the secondary than the primary and V_s is less than V_p, Figure 61.4b.

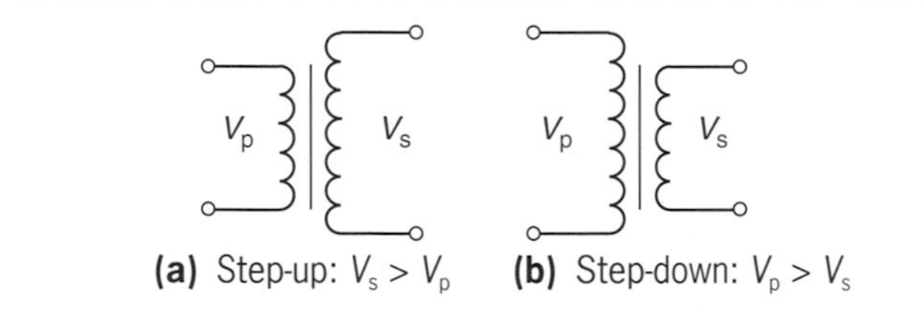

Figure 61.4 Symbols for a transformer

Energy losses in a transformer

If the p.d. is stepped up in a transformer the current is stepped down in proportion. This must be so if we assume that all the electrical energy given to the primary appears in the secondary, i.e. that energy is conserved and the transformer is 100% efficient (many approach this). Then

$$\text{power in primary} = \text{power in secondary}$$
$$V_p \times I_p = V_s \times I_s$$

where I_p and I_s are the primary and secondary currents respectively.

$$\therefore \qquad \frac{I_s}{I_p} = \frac{V_p}{V_s}$$

So, for the ideal transformer, if the p.d. is doubled the current is halved. In practice, it is more than halved due to small energy losses in the transformer arising from three causes.

(a) Resistance of windings

The windings of copper wire have some resistance and heat is produced by the current in them.

(b) Eddy currents

The iron core is in the changing magnetic field of the primary, and currents, called eddy currents, are induced in it which cause heating. These are reduced by using a **laminated** core made of sheets, insulated from each other to have a high resistance.

(c) Leakage of field lines

All the lines produced by the primary may not 'cut' the secondary, especially if the core has an air gap or is badly designed.

A large transformer like that in Figure 61.5 has to be oil-cooled to prevent overheating.

Figure 61.5 Large transformer at a power station

The stepping-up of current can be demonstrated with the 100:1 step-down transformer of Figure 61.6 (overleaf) in which the nail melts spectacularly due to the very large secondary current.

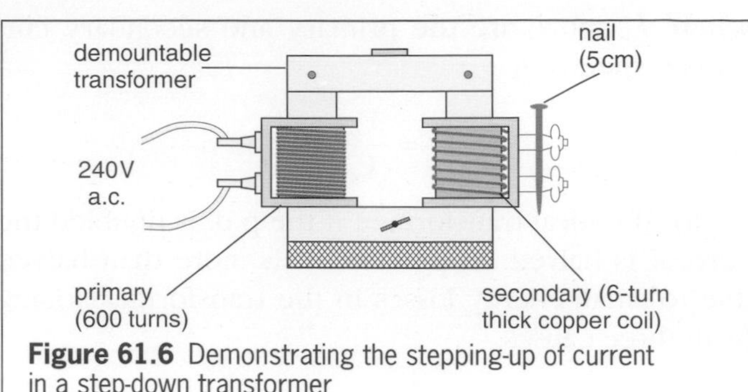

Figure 61.6 Demonstrating the stepping-up of current in a step-down transformer

Worked example

A transformer steps down the mains supply from 240 V to 12 V to operate a 12 V lamp.

a What is the turns ratio of the transformer windings?

b How many turns are on the primary if the secondary has 100 turns?

c What is the current in the primary if the transformer is 100% efficient and the current in the lamp is 2 A?

a Primary voltage = V_p = 240 V

Secondary voltage = V_s = 12 V

Turns ratio = $N_s/N_p = V_s/V_p$ = 12 V/240 V
= 1/20

b Secondary turns = N_s = 100

From **a**, $$\frac{N_s}{N_p} = \frac{1}{20}$$

$\therefore$ $$N_p = 20\,N_s = 20 \times 100 = 2000 \text{ turns}$$

c Efficiency = 100%

$\therefore$ power in primary = power in secondary
$$V_p \times I_p = V_s \times I_s$$

$\therefore$ $$I_p = \frac{V_s \times I_s}{V_p} = \frac{12\text{ V} \times 2\text{ A}}{240\text{ V}}$$
$$= 1/10\text{ A} = 0.1\text{ A}$$

Note. In this ideal transformer the current is stepped up in the same ratio as the voltage is stepped down.

Transmission of electrical power

(a) Grid system

This is a network of cables, mostly supported on pylons, which connect over 100 power stations throughout Britain to consumers. In the largest modern stations, electricity is generated at 25 000 V (25 kilovolts = 25 kV) and stepped up at once in a transformer to 275 or 400 kV to be sent over long distances on the Supergrid. Later, the p.d. is reduced by substation transformers for distribution to local users, Figure 61.7.

At the National Control Centre engineers direct the flow and re-route it when breakdown occurs. This makes the supply more reliable, and cuts costs by enabling smaller, less efficient stations to be shut down at off-peak periods.

(b) Use of high alternating p.d.s

If 400 000 W of electrical power has to be sent through cables it can be done as 400 000 V at 1 A or 400 V at 1000 A (since watts = amperes × volts). But the amount of electrical energy transferred to

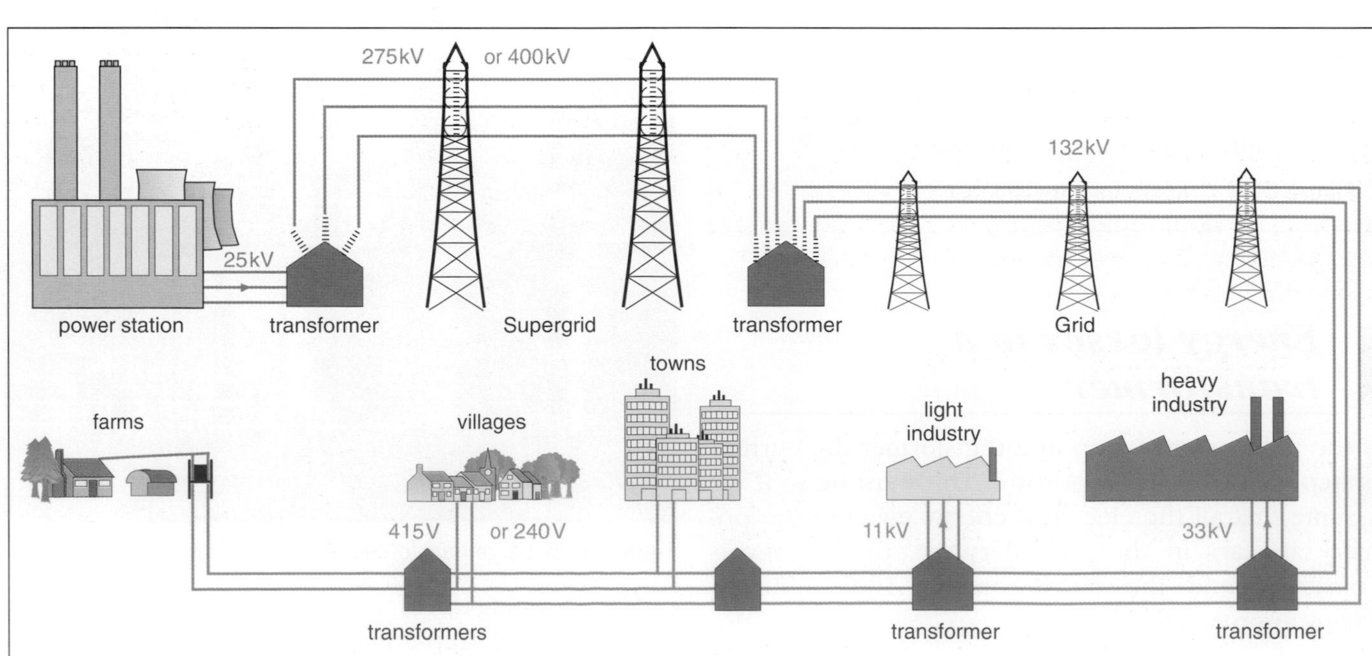

Figure 61.7 The National Grid transmission system

unwanted heat (due to resistance of the cables) is proportional to the **square of the current** and so the power loss (I^2R) is less if transmission occurs at high voltage (high tension) and low current. On the other hand, high p.d.s need good insulation. The efficiency with which transformers step alternating p.d.s up and down accounts for the use of a.c. rather than d.c.

The advantages of 'high' alternating voltage power transmission may be shown using the apparatus in Figure 61.8, **so long as the power line is well insulated** (using insulated eureka wire). The eureka resistance wires represent long transmission cables. Without the transformer at each end the lamp at the 'village' end glows dimly due to the power loss in the wires.

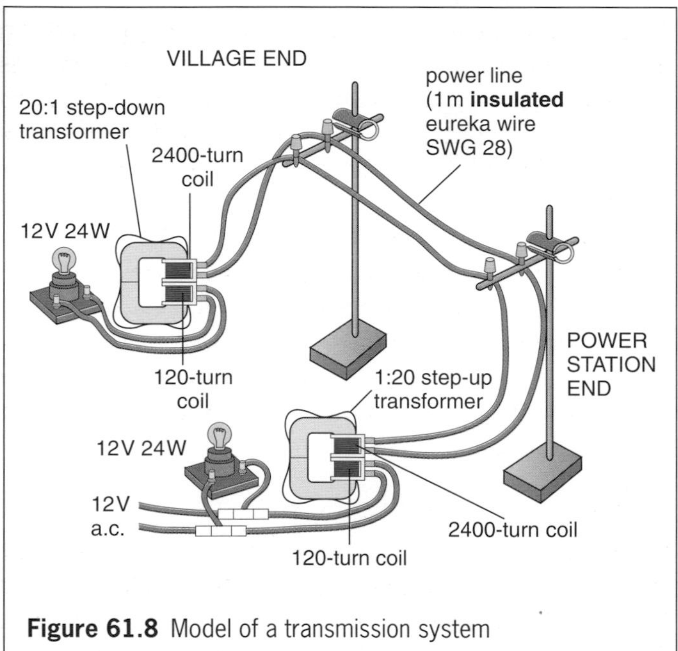

Figure 61.8 Model of a transmission system

When the transformers are connected as shown, the 12 V a.c. supply voltage is stepped up to about 240 V (since $N_p/N_s = 1/20$) at the 'power station' end, then stepped down at the 'village' end to 12 V (since $N_s/N_p = 20/1$). The current in the wires is then much less than without the transformers and so the lamp at the 'village' end is fully lit due to the smaller power loss during transmission.

Car ignition system

A transformer does not work off d.c. but the **induction coil**, which uses the same principle, does. It is used in a car to produce the high voltage that causes the sparking plug to spark and ignite the mixture of petrol vapour and air.

The primary circuit is broken when the rotating cam, Figure 61.9, separates the 'points'. A high p.d. is induced in the secondary which is applied at the right time to each plug in turn by the rotor arm of the distributor. When the 'points' open, the capacitor acts as a reservoir and stops sparking at them which would cause 'pitting'. It also ensures that the primary current falls rapidly, so giving a larger induced p.d. in the secondary.

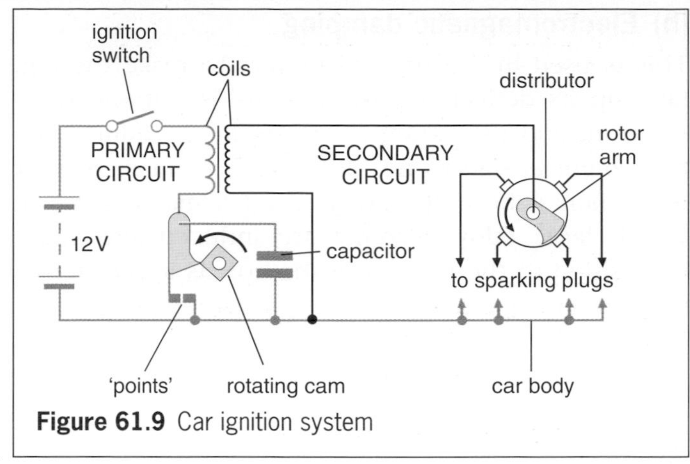

Figure 61.9 Car ignition system

Applications of eddy currents

Eddy currents are the currents induced in a piece of metal when it cuts magnetic field lines. They can be quite large due to the low resistance of the metal. They have their uses as well as their disadvantages.

(a) Car speedometer

The action depends on the eddy currents induced in a thick aluminium disc when a permanent magnet, near it but *not touching it*, is rotated by a cable driven from the gearbox of the car, Figure 61.10. The eddy currents in the disc make it rotate in an attempt to reduce the relative motion between it and the magnet (see Chapter 60). The extent to which the disc can turn however is controlled by a spring. The faster the magnet rotates the more the disc turns before it is stopped by the spring. A pointer fixed to the disc moves over a scale marked in m.p.h. (or km/h) and gives the speed of the car.

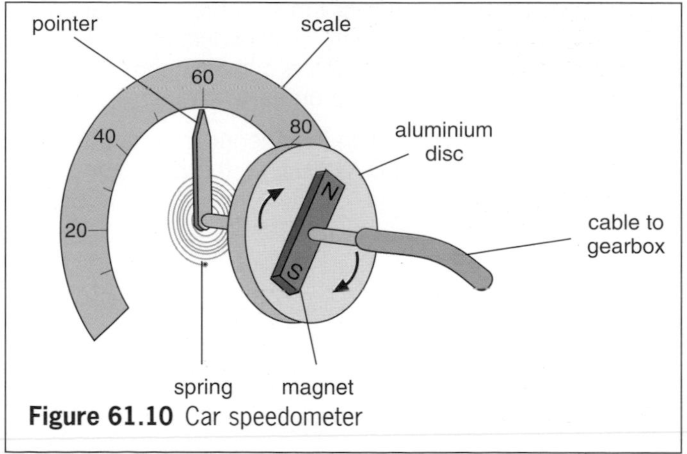

Figure 61.10 Car speedometer

(b) Electromagnetic damping

This is used in moving-coil meters to make the coil take up its deflected position quickly without overshooting and oscillating about its final reading. The movement is said to be 'dead-beat'. In most pointer instruments the coil is wound on a metal frame in which large eddy currents are induced and cause opposition to the motion of the coil as it cuts across the radial magnetic field of the permanent magnet.

QUESTIONS

1 Two coils of wire, A and B, are placed near one another, Figure 61.11. Coil A is connected to a switch and battery. Coil B is connected to a centre-reading moving-coil galvanometer.

a If the switch connected to coil A were closed for a few seconds and then opened, the galvanometer connected to coil B would be affected. Explain and describe, step by step, what would actually happen.

b What changes would you expect if a bundle of soft iron wires were placed through the centre of the coils? Give a reason for your answer.

c What would happen if more turns of wire were wound on the coil B?

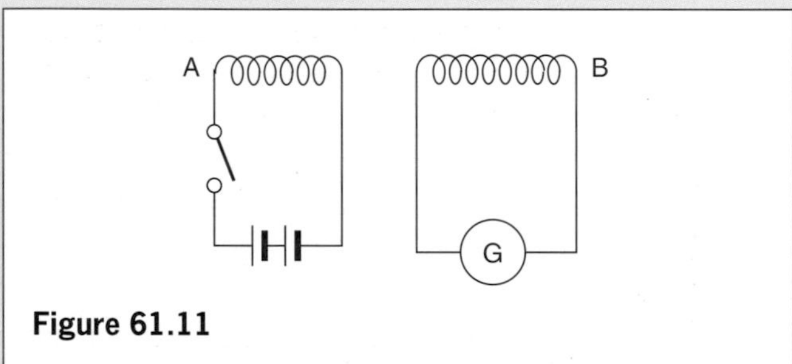

Figure 61.11

2 The main function of a step-up transformer is to

A increase current
B increase voltage
C change a.c. to d.c.
D change d.c. to a.c.
E increase the resistance of a circuit. *(S.E.)*

3 **a** Calculate the number of turns on the secondary of the step-down transformer, which would enable a 12 V bulb to be used with a 240 V a.c. mains power, if there are 480 turns on the primary.

b What current will flow in the secondary when the primary current is 0.50 A? Assume there are no energy losses. *(N.W.)*

4 The block diagram in Figure 61.12 represents the generation and distribution of electricity.

a T is a 'step-up' transformer which has a core made up of 'laminations'.
(i) Explain what is meant by a 'step-up' transformer.
(ii) What is meant by 'laminations'?
(iii) What material is used for the laminations?
(iv) Why is the core built in this way?

b If the transformer T has 1000 turns on its primary coil, how many turns must it have on its secondary coil? (Assume that there are no losses in the transformer.)

c Why is it an advantage to transmit electric power at high voltage?

d What will be the main item of equipment in the substations A and B?

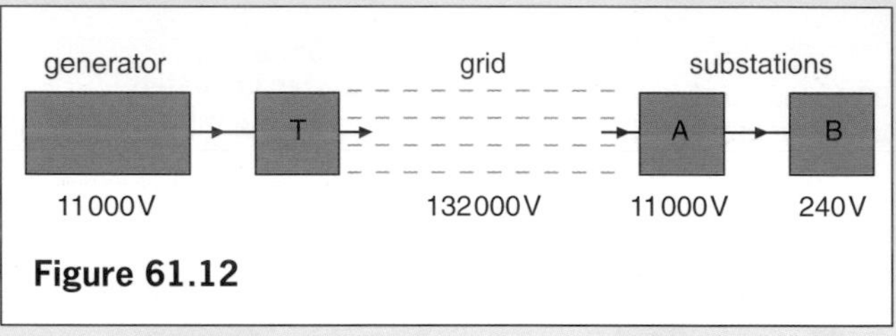

Figure 61.12

Checklist

After studying this chapter you should be able to

- explain the principle of the transformer,
- recall the transformer equation $V_s/V_p = N_s/N_p$ and use it to solve problems,
- recall that for an ideal transformer $V_p \times I_p = V_s \times I_s$, and use the relation to solve problems,
- recall the causes of energy losses in practical transformers,
- explain why high voltage a.c. is used for transmitting electrical power,
- describe the car ignition system,
- explain how eddy currents arise and how they are used in a car speedometer and in electromagnetic damping.

ADDITIONAL QUESTIONS

The first questions under each topic heading are 'basic' questions intended for all students; the following questions marked with a stripe are 'higher' questions for those seeking grades E to A*.

Magnetic fields

1 A bar magnet with north and south poles marked N and S respectively and a bar of iron PQ are fixed a short distance apart as shown. A compass needle is brought in turn to ends P, Q, N and S. Its N pole will point towards

A both ends of the iron but only to the N pole of the magnet
B both ends of the iron but only to the S pole of the magnet
C end P only of the iron and only to the S pole of the magnet
D end Q only of the iron and only to the N pole of the magnet
E end Q only of the iron and only to the S pole of the magnet. *(O.L.E./S.R.16+)*

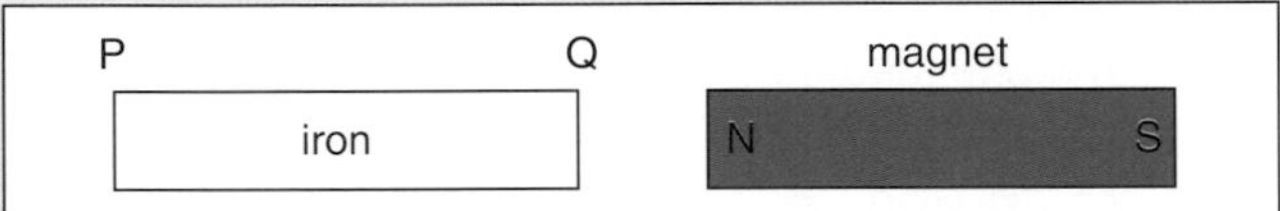

Electromagnets

2 Diagrams (a) to (d) show identical vertical insulated coils with many turns of wire and the same large direct current passing through each. The pieces of iron and steel are attached as shown. When the current is switched off what will happen to each piece of iron and steel? *(W.M.)*

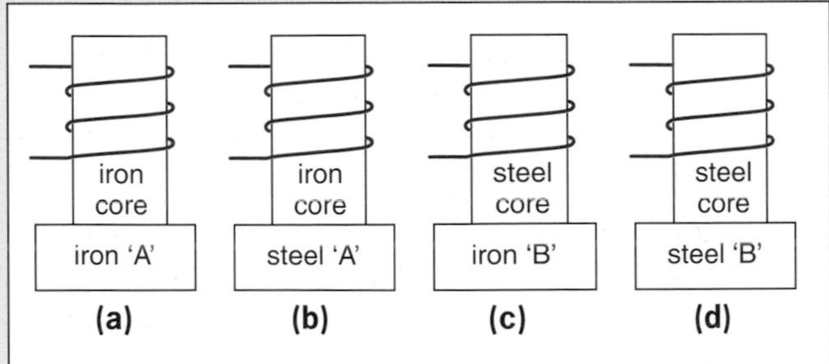

3 Part of the electrical system of a car is shown in the diagram.

a Why are connections made to the car body?

b There are *two* circuits in parallel with the battery. What are they?

c Why is wire A thicker than wire B?

d Why is a relay used?

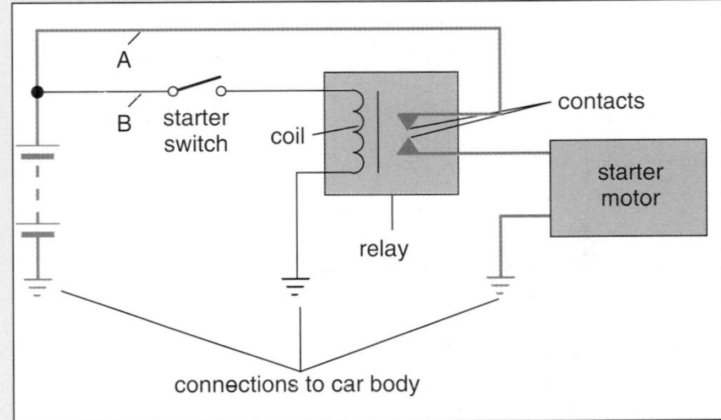

4 The pick-up in an electric guitar, diagram (a), consists of a row of small **magnets** with several thousand turns of fine **copper wire** wound round them, (b). The guitar strings are made of **steel**. Explain *briefly* how the vibrations of the strings produce a voltage in the pick-up for feeding to the amplifier.

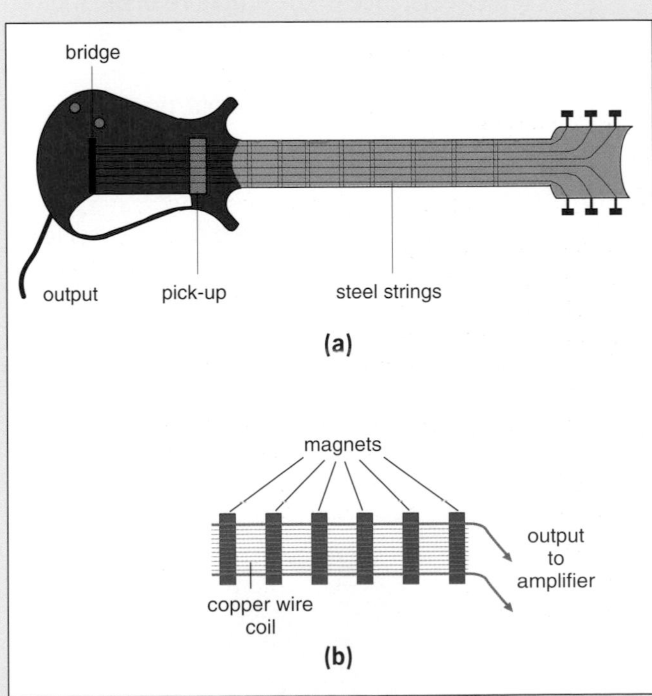

Electric motors

5 a Draw a labelled diagram of the essential components of a simple direct current electric motor. Explain clearly how continuous rotation is produced and show how the direction of rotation is related to the direction of the current.

b State what would happen to the direction of rotation of the motor you have described if (i) the current were reversed, (ii) the magnetic field were reversed, (iii) both current and field were reversed simultaneously.

c Why is a lead-acid accumulator preferred to a Leclanché (dry) battery in motor cars? *(J.M.B.)*

6 a In the diagram, AB is a copper rod resting on two horizontal copper bars between the poles of a horseshoe magnet. Describe and account for what will happen to AB when the key K is closed.

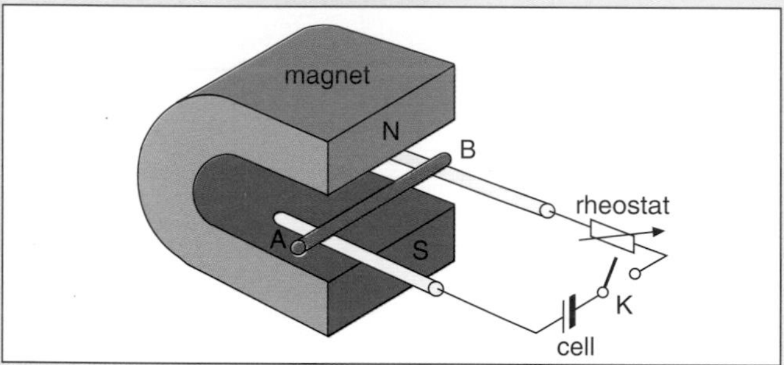

b A working electric motor takes a current of 1.5 A when the p.d. across its terminals is 250 V. Determine the efficiency of the motor if the power output is 200 W.

If the resistance of the armature of the motor is about 1 Ω why is the current as little as 1.5 A when the motor is running? *(L.)*

(**Note.** Efficiency = power output/power input. For last part, see p. 238.)

Generators

7 Describe the deflections observed on the sensitive, centre-zero galvanometer G when the copper rod XY is connected to its terminals and is made to vibrate up and down (as shown by the arrows), between the poles of a U-shaped magnet so that it is at right angles to the magnetic field.

Explain what is happening.

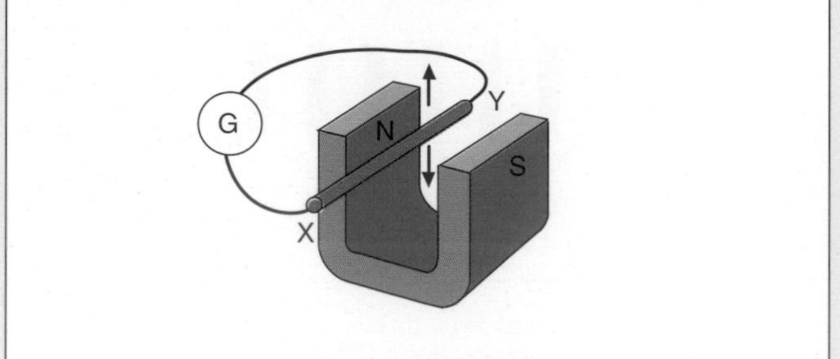

8 The power output of a generator which supplies power to a consumer along cables of total resistance 16 Ω is 30 kW at 6 kV. What is

a the current in the cables,

b the power loss in the cables,

c the drop of voltage between the ends of the cables?

9 Describe, with the aid of a labelled diagram, a simple form of generator. Explain how it may be used to generate **a** an alternating current, **b** a direct current.

In the case of **a**, sketch a graph of the variation with time of the p.d. between the ends of the coil. Use the same axes to show clearly the effects of:
(i) doubling the number of turns of the coil and keeping the rotation speed and the magnetic field constant,
(ii) doubling the speed of rotation and keeping the number of turns and the magnetic field constant. *(W.)*

Transformers

10 A transformer has 2000 turns on the primary coil. The voltage applied to the primary coil is 240 V a.c. How many turns are on the secondary coil if the output voltage is 48 V a.c.?

A 40 **B** 400 **C** 4000 **D** 5000 **E** 10 000 *(O.L.E./S.R.16+)*

11 a (i) Draw a labelled diagram showing the essential features of a voltage step-down transformer. Explain how the transformer works.
(ii) State one source of energy loss in the transformer and say how it may be reduced.

b A transformer designed to operate a 12 V lamp from a 240 V supply has 1200 turns on the primary coil. Assuming that the transformer is 100% efficient, calculate (i) the number of turns on the secondary coil, (ii) the current passing through the primary coil when the 12 V lamp has a current of 2 A flowing through it. *(J.M.B.)*

12 A transformer is used on the 240 V a.c. supply to deliver 9.0 A at 80 V to a heating coil. If 10% of the energy taken from the supply is dissipated in the transformer itself, what is the current in the primary winding? *(O. and C.)*

ELECTRONS AND ATOMS

62 ELECTRONS

Thermionic emission	***X-rays***
Cathode rays	Production.
Deflection of an electron beam	Properties and nature.
By a magnetic field.	***Photoelectric effect***
By an electric field.	***Photons***
Cathode ray oscilloscope (CRO)	***Waves or particles?***
Uses of the CRO	

The discovery of the electron was a landmark in physics and led to great technological advances.

Thermionic emission

The evacuated bulb in Figure 62.1 contains a small coil of wire, the **filament**, and a metal plate called the **anode** because it is connected to the positive of the 400 V d.c. power supply. The negative of the supply is joined to the filament which is also called the **cathode**. The filament is heated by current from a 6 V supply (a.c. or d.c.).

With the circuit as shown, the meter deflects, indicating current flow in the circuit containing the gap between anode and cathode. The current stops if *either* the 400 V supply is reversed to make the anode negative, *or* the filament is not heated.

This demonstration supports the view that negative charges, in the form of electrons, escape from the filament when it is hot because they have enough energy to get free from the metal surface. The process is known as **thermionic emission** and the bulb as a thermionic diode (since it has two electrodes). There is a certain minimum **threshold energy** (depending on the metal) which the electrons must have to escape. Also, the higher the temperature of the metal, the greater the number of electrons emitted. The electrons are attracted to the anode if it is positive and are able to reach it because there is a vacuum in the bulb.

Cathode rays

Beams of electrons moving at high speed are called **cathode rays**. Their properties can be studied using the 'Maltese cross tube', Figure 62.2.

Electrons emitted by the hot cathode are accelerated towards the anode but most pass through the hole in it and travel on along the tube. Those that miss the cross cause the screen to fluoresce with green or blue light and cast a shadow of the cross on it. The cathode rays evidently travel in straight lines.

If the N pole of a magnet is brought up to the neck of the tube, the rays (and the fluorescent shadow) move upwards. The rays are clearly deflected by a magnetic field and, using Fleming's left-hand rule (Chapter 58),

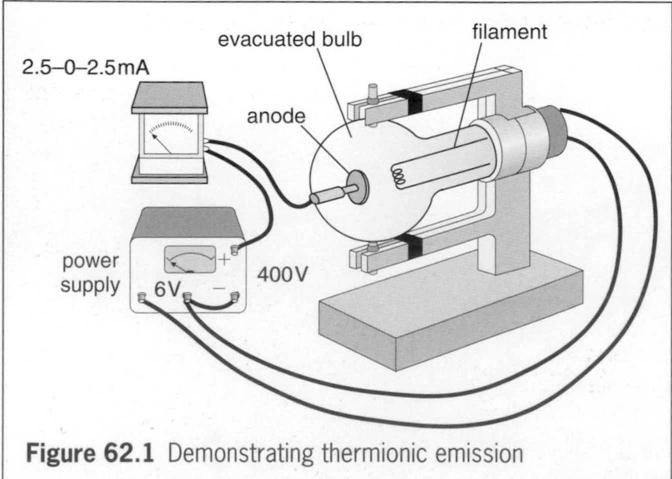

Figure 62.1 Demonstrating thermionic emission

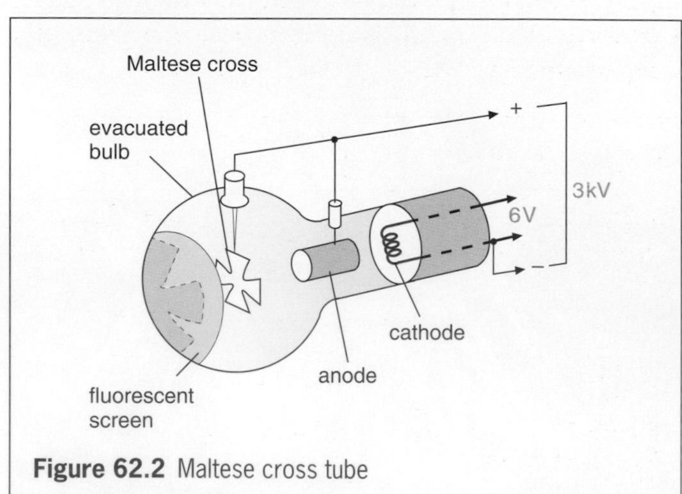

Figure 62.2 Maltese cross tube

we see that they behave like conventional current (positive charge flow) travelling from anode to cathode.

There is also an optical shadow of the cross, due to light emitted by the cathode. This is unaffected by the magnet.

Deflection of an electron beam

(a) By a magnetic field

In Figure 62.3 the evenly spaced dots represent a uniform magnetic field (i.e. one of the same strength throughout the area shown) acting into and perpendicular to the paper. An electron beam entering the field at right angles to the field experiences a force due to the motor effect (Chapter 58), whose direction is given by Fleming's left-hand rule. This indicates that the force acts inwards at right angles to the direction of the beam and makes it follow a **circular** path as shown (the beam being treated as conventional current in the opposite direction).

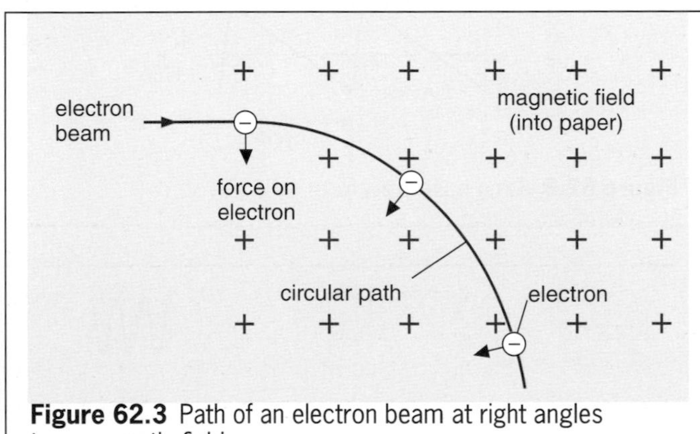

Figure 62.3 Path of an electron beam at right angles to a magnetic field

(b) By an electric field

An electric field is a region where an electric charge experiences a force due to other charges. In Figure 62.4 the two metal plates behave like a capacitor which has been charged by connection to a voltage supply. If the charge is evenly spread over the plates, a uniform electric field is created between them and is represented by parallel, equally spaced lines.

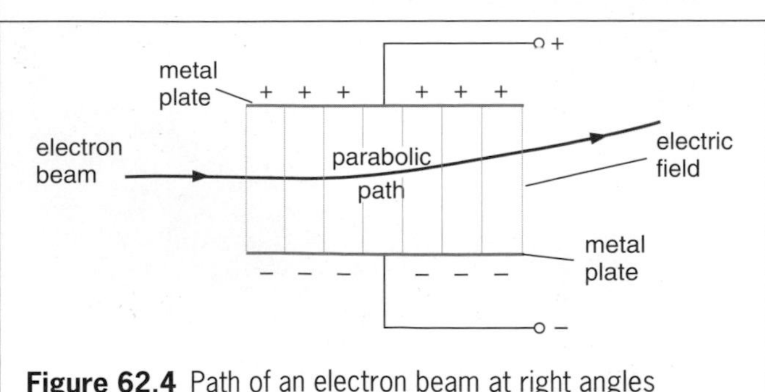

Figure 62.4 Path of an electron beam at right angles to an electric field

If an electron beam enters the field at right angles to it, the beam is attracted towards the positively charged plate and follows a **parabolic** path, as shown. In fact its behaviour is not unlike that of a projectile (Chapter 33) in which the horizontal and vertical motions can be treated separately.

(c) Demonstration

The deflection tube in Figure 62.5 can be used to show the deflection of an electron beam in electric and magnetic fields. Electrons from a hot cathode strike a fluorescent screen S set at an angle. A p.d. applied across two horizontal metal plates Y_1Y_2 creates a *vertical* electric field which deflects the rays upwards if Y_1 is positive (as shown) and downwards if it is negative.

When current flows in the two coils X_1X_2 (in series) outside the tube, a *horizontal* magnetic field is produced across the tube. It can be used instead of a magnet to deflect the rays, or to cancel the deflection due to an electric field.

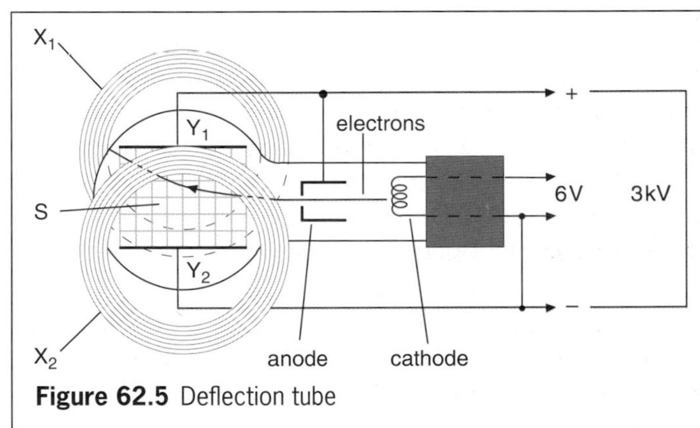

Figure 62.5 Deflection tube

Cathode ray oscilloscope (CRO)

The CRO is one of the most important scientific instruments ever to be developed. It contains, like a television and a computer monitor, a cathode ray tube which has three main parts, Figure 62.6.

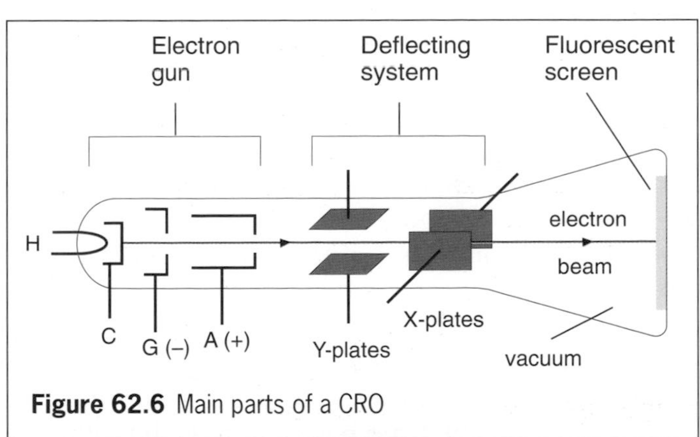

Figure 62.6 Main parts of a CRO

(a) Electron gun

This consists of a **heater** H, a **cathode** C, another electrode called the **grid** G and two or three **anodes** A. G is at a negative voltage with respect to C and controls the number of electrons passing through its central hole from C to A; it is the **brilliance** or **brightness** control. The anodes are at high positive voltages relative to C; they accelerate the electrons along the highly evacuated tube and also **focus** them into a narrow beam.

(b) Fluorescent screen

A bright spot of light is produced on the screen where the beam hits it.

(c) Deflecting system

Beyond A are two pairs of deflecting plates to which p.d.s can be applied. The **Y-plates** are horizontal but create a vertical electric field which deflects the beam vertically. The **X-plates** are vertical and deflect the beam horizontally.

The p.d. to create the electric field between the Y-plates is applied to the **Y-input** terminals (often marked 'high' and 'low') on the front of the CRO. The input is usually amplified by an amount which depends on the setting of the **Y-amp gain** control, before it is applied to the Y-plates. It can then be made large enough to give a suitable vertical deflection of the beam.

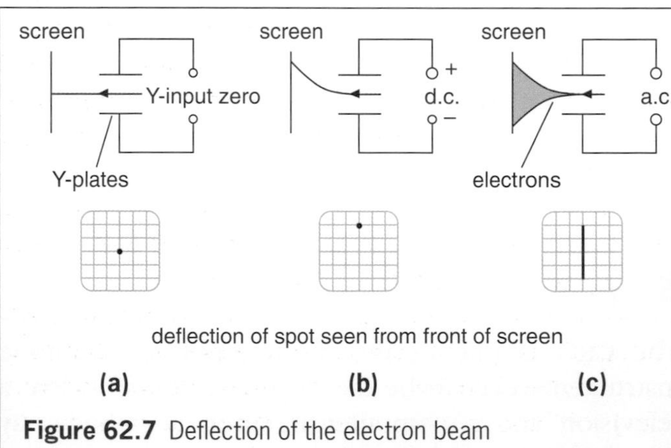

Figure 62.7 Deflection of the electron beam

In Figure 62.7a the p.d. between the Y-plates is zero as is the deflection. In (b) the d.c. input p.d. makes the upper plate positive and attracts the beam of negatively charged electrons upwards. In (c) the 50 Hz a.c. input makes the beam move up and down so rapidly that it produces a continuous vertical line (whose length increases if the Y-amp gain is turned up).

The p.d. applied to the X-plates is also via an amplifier, the X-amplifier, and can either be from an external source connected to the **X-input** terminal or, more commonly, from the **time base** circuit in the CRO.

The time base deflects the beam horizontally in the X-direction and makes the spot sweep across the screen from left to right at a steady speed determined by the setting of the time base controls (usually 'coarse' and 'fine'). It must then make the spot 'fly' back very rapidly to its starting point, ready for the next sweep. The p.d. from the time base should therefore have a sawtooth waveform like that in Figure 62.8. Since AB is a straight line, the distance moved by the spot is directly proportional to time and the horizontal deflection becomes a measure of time, i.e. a time axis or base.

In Figures 62.9a, b, c, the time base is on. In (a) the Y-input p.d. is zero, in (b) the Y-input is d.c. which makes the upper Y-plate positive. In both cases the spot traces out a horizontal line which appears to be continuous if the flyback is fast enough. In (c) the Y-input is a.c., i.e. the Y-plates are alternately positive and negative and the spot moves accordingly.

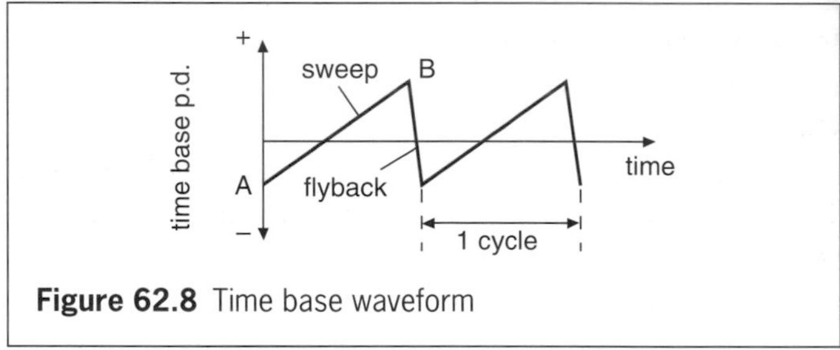

Figure 62.8 Time base waveform

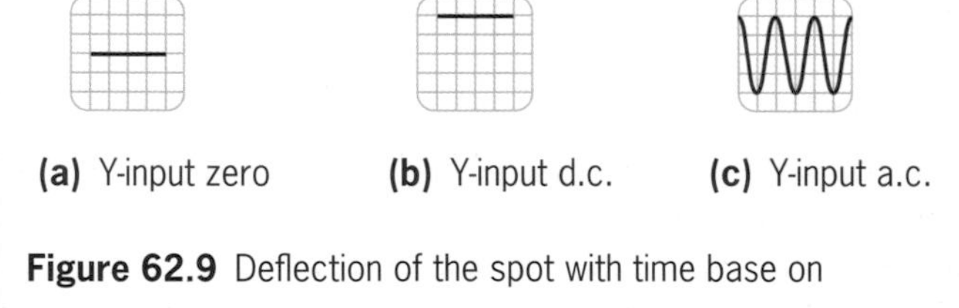

Figure 62.9 Deflection of the spot with time base on

Uses of the CRO

A small CRO is shown in Figure 62.10.

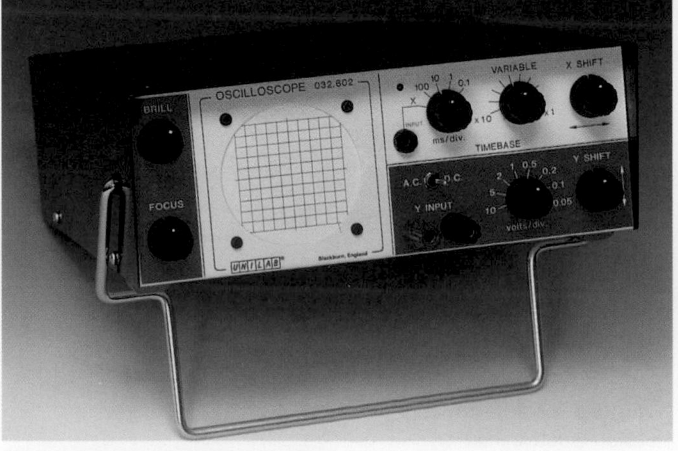

Figure 62.10 Single-beam CRO

(a) Practical points

The **brilliance** control, which is usually the **on/off** switch as well, should be as low as possible when there is just a spot on the screen. Otherwise screen 'burn' occurs which damages the fluorescent material. If possible it is best to defocus the spot or draw it into a line by running the time base.

When preparing the CRO for use, set the **brilliance**, **focus**, **X-** and **Y-shift** controls (which allow the spot to be moved 'manually' over the screen in the X and Y directions respectively) to their mid-positions. The **time base** and **Y-amp gain** controls can then be adjusted to suit the input.

When the **a.c./d.c. selector** switch is in the 'd.c.' (or 'direct') position, both d.c. and a.c. can pass to the Y-input. In the 'a.c.' (or 'via C') position, a capacitor blocks d.c. in the input but allows a.c. to pass.

(b) Measuring p.d.s

A CRO can be used as a d.c./a.c. voltmeter if the p.d. to be measured is connected across the Y-input terminals; **the deflection of the spot is proportional to the p.d.**

For example, if the **Y-amp gain** control is on, say, 1 V/div, a deflection of 1 vertical division on the screen graticule (like graph paper with squares for measuring deflections) would be given by a 1 V d.c. input. A line 1 division long (time base off) would be produced by an a.c. input of 1 V peak-to-peak, i.e. peak p.d. = 0.5 V.

(c) Displaying waveforms

In this widely used role, the time base is on and the CRO acts as a 'graph-plotter' to show the waveform, i.e. the variation with time, of the p.d. applied to its Y-input. The displays in Figures 62.11a, b are of alternating p.d.s with sine waveforms. In (a), the time base frequency *equals* that of the input and one complete wave is obtained. In (b) it is *half* that of the input and two waves are formed.

Sound waveforms can be displayed if a microphone is connected to the Y-input terminals (see Chapter 16).

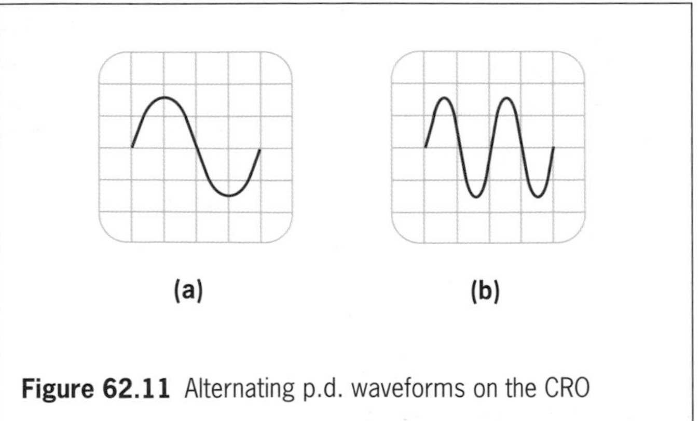

Figure 62.11 Alternating p.d. waveforms on the CRO

(d) Measuring time intervals and frequency

These can be measured if the CRO has a calibrated time base. For example, when the time base is set on 10 ms/div, the spot takes 10 milliseconds to move 1 division horizontally across the screen graticule. If this is the time base setting for the waveform in Figure 62.11b then since 1 complete wave occupies 2 horizontal divisions, we can say

$$\begin{aligned}\text{time for 1 complete wave} &= 2\ \text{divs} \times 10\ \text{ms/div}\\ &= 20\ \text{ms}\\ &= 20/1000 = 1/50\ \text{s}\end{aligned}$$

$\therefore$ number of complete waves per second = 50

$\therefore$ frequency of a.c. applied to Y-input = 50 Hz

X-rays

X-rays are produced when high-speed electrons are stopped by matter.

(a) Production

In an X-ray tube, Figure 62.12, electrons from a hot filament are accelerated across a vacuum to the anode by a large p.d. (up to 100 kV). The anode is a copper block with a 'target' of a high melting-point metal such as tungsten on which the electrons are focused by the concave cathode. The tube has a lead shield with a small exit for the X-rays. Less than $\frac{1}{2}$% of the k.e. of the electrons becomes X-rays, the rest heats the anode which has to be cooled.

High p.d.s give short wavelength, very penetrating (**hard**) X-rays. Less penetrating (**soft**) rays, of longer wavelength, are obtained with lower p.d.s. The absorption of X-rays by matter is greatest by materials of high density having a large number of outer electrons in their atoms, i.e. of high atomic number (Chapter 64). A more intense beam of rays is produced if the rate of emission of electrons is raised by increasing the filament current.

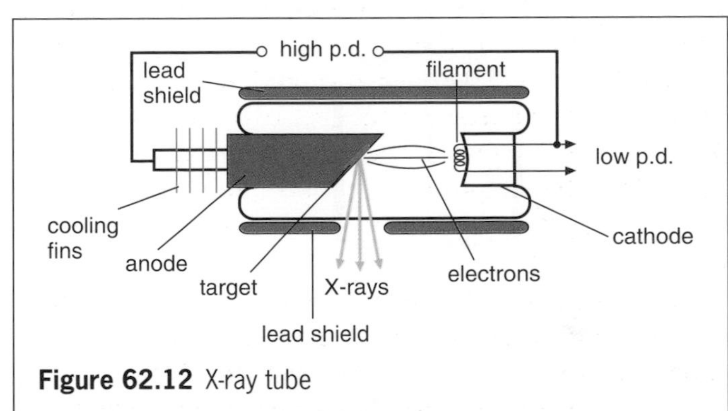

Figure 62.12 X-ray tube

(b) Properties and nature

X-rays

(i) readily penetrate matter – up to 1 mm of lead,
(ii) are not deflected by electric or magnetic fields,
(iii) ionize a gas, making it a conductor, e.g. a charged electroscope discharges when X-rays pass through the surrounding air,
(iv) affect a photographic film,
(v) cause fluorescence,
(vi) give interference and diffraction effects.

These facts (and others) suggest that X-rays are electromagnetic waves of very short wavelength.

(c) Uses

These were considered earlier (Chapter 14).

Photoelectric effect

Electrons are emitted by certain metals when electromagnetic radiation of small enough wavelength falls on them. The effect is called **photoelectric emission** and is given by zinc exposed to ultraviolet, as may be shown with the apparatus of Figure 62.13a. The zinc plate must be *cleaned thoroughly* with emery cloth. If the electroscope is charge positively, it is unaffected when the zinc is exposed to UV. If charged negatively it discharges but a sheet of glass between the lamp and the plate halts the discharge.

When the plate and electroscope are negatively charged the plate repels the emitted electrons and the electroscope loses its excess negative charge, Figure 62.13b. With a positive charge on the plate, the emitted electrons are attracted back to the plate which therefore keeps its original charge, Figure 62.13c. The glass stops most of the UV but transmits violet light (also emitted by the lamp), so the demonstration shows that zinc does not give the effect with this longer wavelength (lower frequency) radiation.

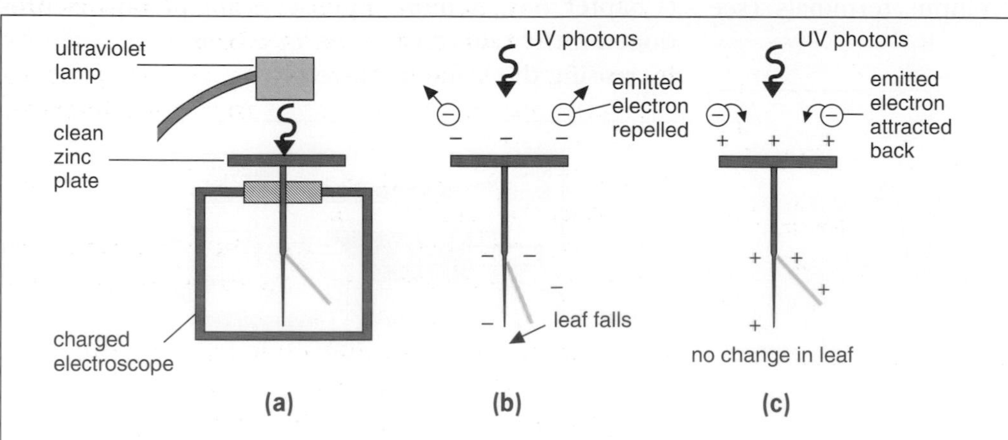

Figure 62.13 Demonstrating the photoelectric effect

Photons

The photoelectric effect only occurs for a given metal if the frequency of the incident electromagnetic radiation exceeds a certain **threshold frequency**.

We can explain this by assuming that

(i) all electromagnetic radiation is emitted and absorbed as packets of energy, called **photons**, and
(ii) the energy of a photon is directly proportional to its frequency.

Ultraviolet photons would therefore have more energy than light photons since UV has a higher frequency than light. The behaviour of zinc (and most other substances) in not giving photoelectric emission with light but with UV would therefore be explained: a photon of light has less than the minimum energy required to emit an electron.

The absorption of a photon by an atom results in the electron gaining energy and the photon disappearing. If the photon has more than the minimum amount of energy required to enable an electron to escape, the excess appears as k.e. of the emitted electron.

$$\text{energy of photon} = \frac{\text{energy needed for}}{\text{electron to escape}} + \text{k.e. of electron}$$

The photoelectric effect is the process by which X-ray photons are absorbed by matter; in effect it causes **ionization** (Chapter 63) since electrons are ejected and positive ions remain. Photons not absorbed pass through with unchanged energy.

Waves or particles?

The wave theory of electromagnetic radiation can account for properties such as interference, diffraction and polarization which the photon theory cannot. On the other hand it does not explain the photoelectric effect which the photon theory does.

It would seem that electromagnetic radiation has a dual nature and has to be regarded as waves on some occasions and as 'particles' (photons) on others.

QUESTIONS

1 Cathode rays consist of

A beams of fast moving electrons
B fluorescent particles
C light rays from a screen
D light rays from a hot filament
E infrared rays from a hot filament.

(O.L.E./S.R. 16+)

2 a In Figure 62.14a, to which terminals on the power supply must plates A and B be connected to deflect the cathode rays downwards?

b In Figure 62.14b, in which direction will the cathode rays be deflected?

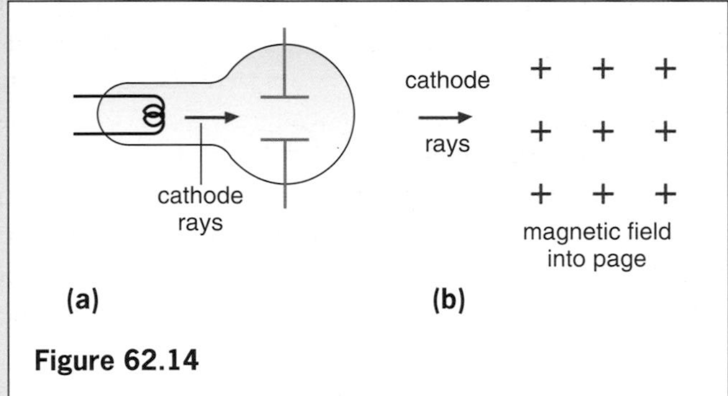

Figure 62.14

3 a Figure 62.15a shows a cathode ray tube fitted with one pair of plates.
(i) What name is given to part P?
(ii) What does part P produce?

b With switch S off a spot is seen in the centre of the screen as shown in Figure 62.15b.
(i) What happens to this spot when switch S is closed?
(ii) What would happen to the spot if the p.d. of the battery were increased?
(iii) What would you *see* on the screen if a low frequency alternating p.d. of about 1 Hz were connected to the plates Q and R?
(iv) What would you *see* on the screen if a higher frequency alternating p.d. of about 50 Hz were connected to the plates Q and R?

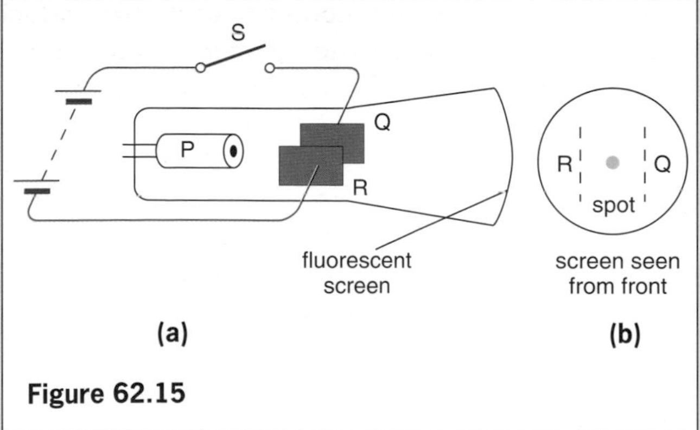

Figure 62.15

c Describe another way in which the spot can be made to behave exactly as it does when the switch S is closed as in **b**(i). Illustrate your answer with a diagram. *(S.E.)*

4 In an X-ray tube what is the effect on the X-ray beam of (i) increasing the accelerating voltage, (ii) decreasing the filament current?

5 a What is the photoelectric effect?

b Why do ultraviolet but not violet light photons give the photoelectric effect with zinc?

Checklist

After studying this chapter you should be able to

- explain the terms **thermionic emission** and **cathode rays**,
- describe experiments to show that cathode rays are deflected by magnetic and electric fields,
- describe with the aid of diagrams the jobs done in a CRO by the electron gun, the X- and Y-plates, the fluorescent screen and the time base,
- describe how the CRO is used to measure p.d.s, to display waveforms and to measure time intervals and frequency,
- describe with the aid of a diagram how X-rays are produced in an X-ray tube,
- list six properties of X-rays which indicate that they are electromagnetic waves of short wavelength,
- explain the term **photoelectric emission** and describe a demonstration of the effect,
- explain the term **threshold frequency** and why this leads to the idea of **photons** of electromagnetic energy.

63 RADIOACTIVITY

Ionizing effect of radiation
Geiger–Müller (GM) tube
Alpha, beta, gamma rays
Cloud chambers
Radioactive decay: half-life
Uses of radioactivity
Dangers and safety

The discovery of radioactivity in 1896 by the French scientist Becquerel was accidental. He found that uranium compounds emitted radiation which (i) affected a photographic plate even when wrapped in black paper and (ii) ionized a gas. Soon afterwards Marie Curie discovered the radioactive element radium. Today radioactivity is used widely in industry, medicine and research.

We are all exposed to **background radiation** caused partly by radioactive materials in rocks, the air and our bodies, and partly by cosmic rays from outer space.

Ionizing effect of radiation

A charged electroscope discharges when a lighted match or a radium source (**held in forceps**) is brought near the cap, Figures 63.1a, b.

In the first case the flame knocks electrons out of surrounding air molecules leaving them as positively charged air **ions**, i.e. air molecules which have lost one or more electrons, Figure 63.2; in the second case radiation causes the same effect, called **ionization**. The positive air ions are attracted to the cap if it is negatively charged; if it is positively charged the electrons are attracted. As a result in either case the charge on the electroscope is neutralized, i.e. it loses its charge.

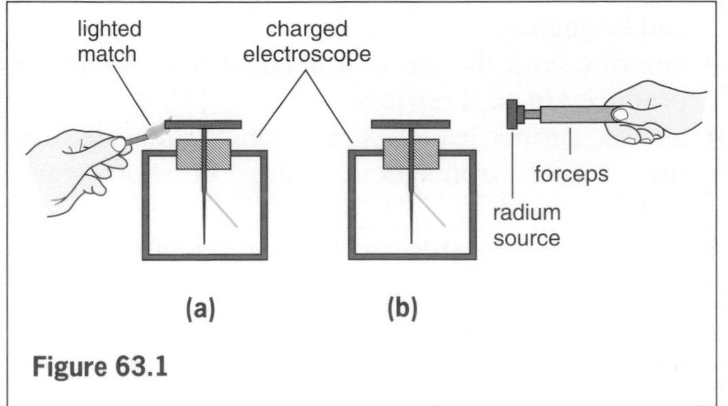

Figure 63.1

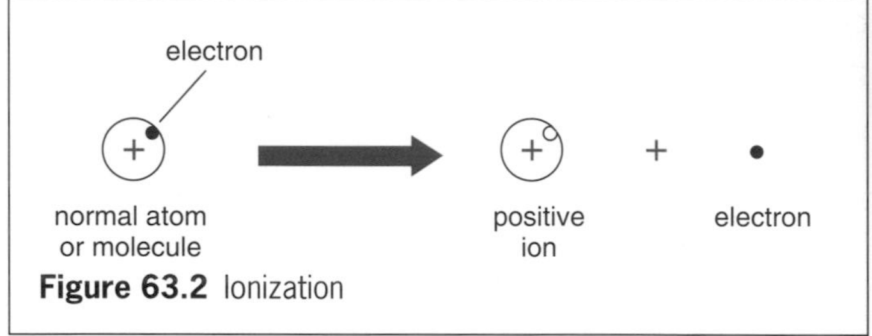

Figure 63.2 Ionization

Geiger–Müller (GM) tube

The ionizing effect is used to detect radiation.

When radiation enters a GM tube, Figure 63.3, either through a thin end-window made of mica, or, if it is very penetrating, through the wall, it creates argon ions and electrons. These are accelerated towards the electrodes and cause more ionization by colliding with other argon atoms.

On reaching the electrodes, the ions produce a current pulse which is amplified and fed either to a **scaler** or a **ratemeter**. A scaler counts the pulses and shows the total received in a certain time. A ratemeter has a meter marked in 'counts per second (or minute)' from which the average pulse-rate can be read. It usually has a loudspeaker which gives a 'click' for each pulse.

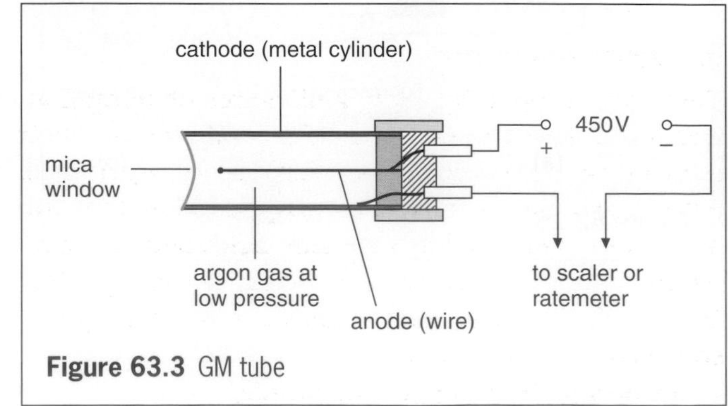

Figure 63.3 GM tube

Alpha, beta, gamma rays

Experiments to study the penetrating power, ionizing ability and behaviour in magnetic and electric fields, show that a radioactive substance emits one or more of three types of radiation – called alpha (α), beta (β) and gamma (γ) rays.

Penetrating power can be investigated as in Figure 63.4 by observing the effect on the count-rate of placing in turn between the GM tube and the lead, a sheet of (i) thick paper (the radium source, lead and tube must be close together for this part), (ii) aluminium 2 mm thick and (iii) lead 2 cm thick. Other sources can be tried, e.g. americium, strontium and cobalt.

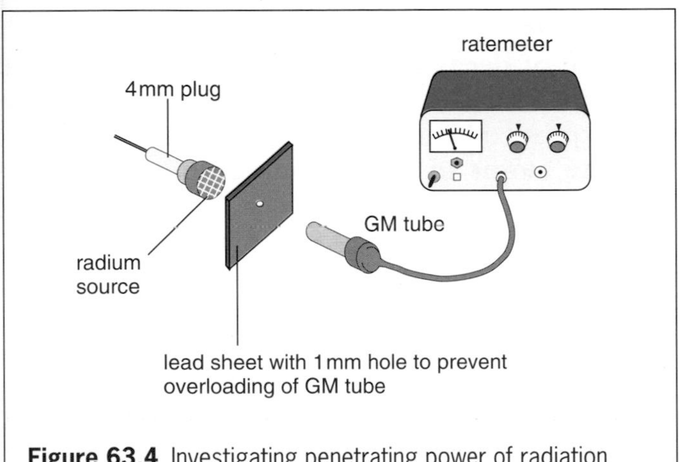

Figure 63.4 Investigating penetrating power of radiation

(a) Alpha rays

These are stopped by a thick sheet of paper and have a range in air of only a few centimetres since they cause intense ionization in a gas due to frequent collisions with gas molecules. They are deflected by electric and *strong* magnetic fields in a direction and by an amount which suggests they are helium atoms minus two electrons, i.e. **helium ions with a double positive charge**. From a particular substance, they are all emitted with the same speed (about 1/20th of that of light).

Americium (Am 241) may be used as a pure α source.

(b) Beta rays

These are stopped by a few millimetres of aluminium and some have a range in air of several metres. Their ionizing power is much less than that of α-particles. As well as being deflected by electric fields, they are more easily deflected by magnetic fields and measurements show they are streams of **high-energy electrons**, like cathode rays, emitted with a range of speeds up to that of light.

Strontium (Sr 90) emits β-rays only.

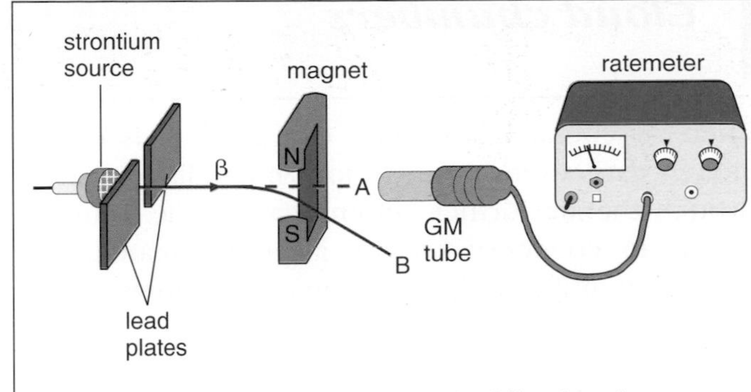

Figure 63.5 Demonstrating magnetic deflection of β-particles

The magnetic deflection of β-particles can be shown as in Figure 63.5. With the GM tube at A and without the magnet, the count-rate is noted. Inserting the magnet reduces the count-rate but it increases again when the GM tube is at B.

(c) Gamma rays

These are the most penetrating and are stopped only by many centimetres of lead. They ionize a gas even less than β-particles and are not deflected by electric and magnetic fields. They give interference and diffraction effects and are **electromagnetic radiation** travelling at the speed of light. Their wavelengths are those of very short X-rays, from which they differ only because they arise in atomic nuclei whereas X-rays come from energy changes in the electrons outside the nucleus.

Cobalt (Co 60) is a pure γ source. Radium (Ra 226) emits α, β and γ-rays.

A GM tube detects β-particles and γ-photons and energetic α-particles; a charged electroscope detects α only. All three types of rays cause fluorescence.

The behaviour of the three kinds of radiation in a magnetic field is summarized in Figure 63.6. The deflections (not to scale) are found from Fleming's left-hand rule, taking negative charge moving to the right as equivalent to positive (conventional) current to the left.

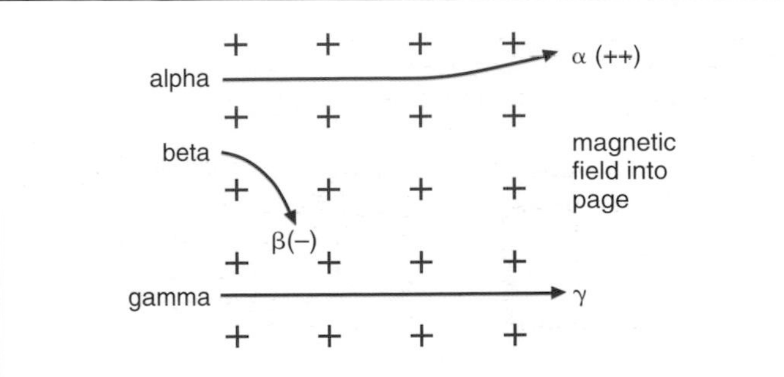

Figure 63.6 Deflection of α, β and γ-rays in a magnetic field

Cloud chambers

When air containing vapour, e.g. alcohol, is cooled enough, saturation occurs. If ionizing radiation passes through the air, further cooling causes the saturated vapour to condense on the air ions created. The resulting white line of tiny liquid drops shows up as a track when illuminated.

In a **diffusion cloud chamber**, Figure 63.7, vapour from alcohol in the felt ring diffuses downwards, is cooled by the 'dry ice' (solid carbon dioxide at $-78\,°C$) in the lower section and condenses near the floor on air ions formed by radiation from the source in the chamber. Tracks are produced continuously which are sharp if an electric field is created by frequently rubbing the plastic lid of the chamber with a cloth.

α-particles give straight, thick tracks, Figure 63.8a. Very fast β-particles produce thin, straight tracks; slower ones give short, twisted, thicker tracks, Figure 63.8b. γ-rays eject electrons from air molecules; the electrons behave like β-particles and produce their own tracks spreading out from the γ-rays.

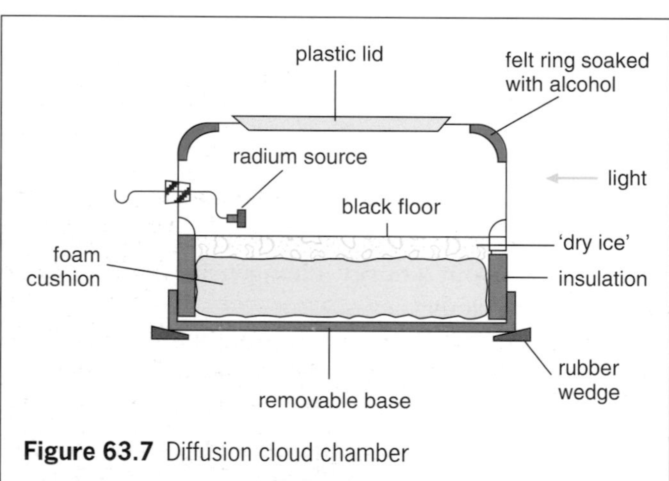

Figure 63.7 Diffusion cloud chamber

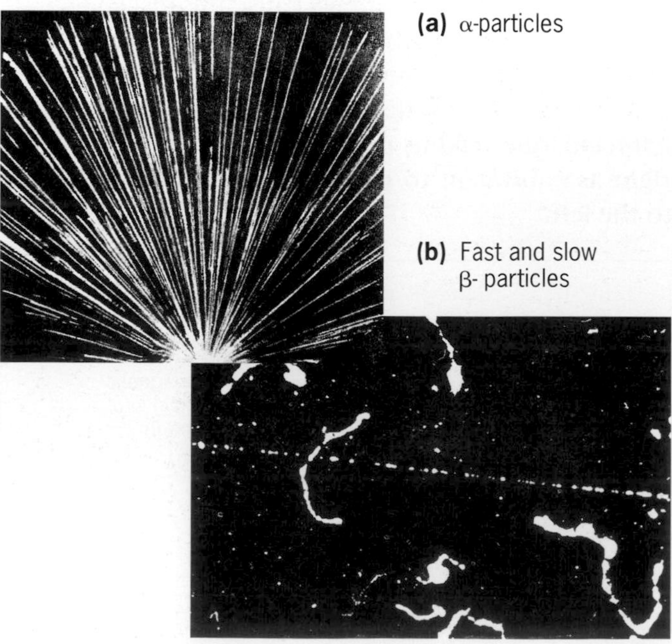

Figure 63.8 Tracks in a cloud chamber

The **bubble chamber**, in which the radiation leaves a trail of bubbles in liquid hydrogen, has now replaced the cloud chamber in research work.

Radioactive decay: half-life

Radioactive atoms have unstable nuclei and 'decay' into atoms of different elements with more stable nuclei when they emit α or β-particles. These changes are spontaneous and cannot be controlled; also, it does not matter whether the material is pure or combined chemically with something else.

(a) Half-life

The **rate of decay** is unaffected by temperature but every radioactive element has its own definite decay rate, expressed by its **half-life**. This is **the time for half the atoms in a given sample to decay**. It is difficult to know when a substance has lost all its radioactivity, but the time for its activity to fall to half its value can be found more easily.

(b) Decay curve

The average number of disintegrations (i.e. decaying atoms) per second of a sample is its **activity**. If it is measured at different times (e.g. by finding the count-rate using a GM tube and ratemeter), a decay curve of activity against time can be plotted. The one in Figure 63.9 shows that the activity decreases by the *same* fraction in successive equal time intervals. It falls from 80 to 40 disintegrations per second in 10 minutes, from 40 to 20 in the next 10 minutes, from 20 to 10 in the third 10 minutes and so on. The half-life is 10 minutes.

Half-lives vary from millionths of a second to millions of years. For radium it is 1600 years.

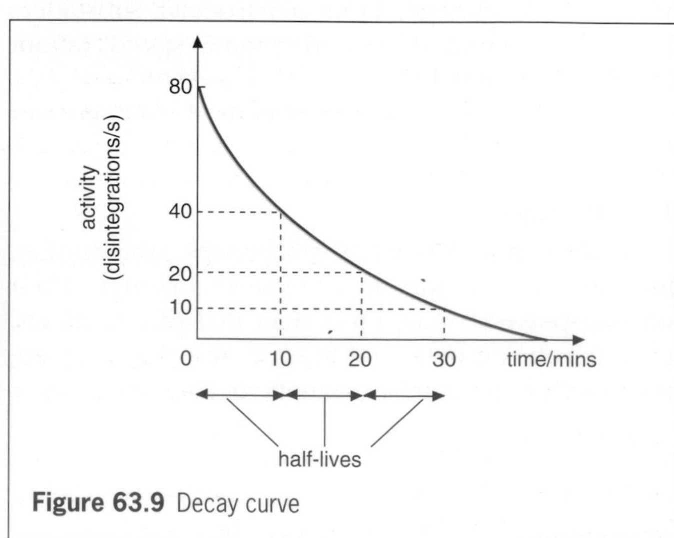

Figure 63.9 Decay curve

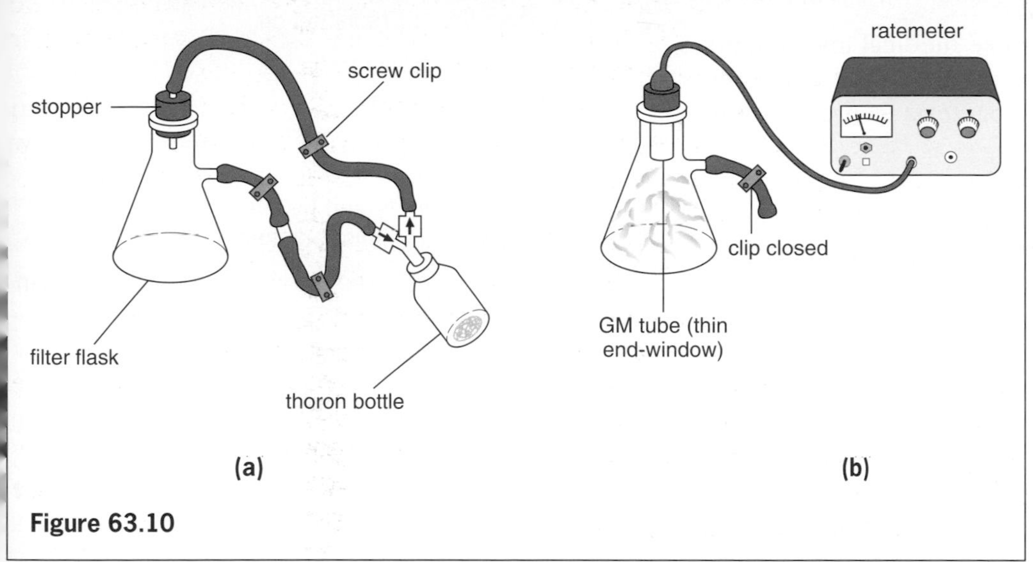

Figure 63.10

(c) Experiment

The half-life of the α-emitting gas **thoron** can be found as in Figure 63.10. The thoron bottle is squeezed 3 or 4 times to transfer some thoron to the flask, (a). The clips are then closed, the bottle removed and the stopper replaced by a GM tube so that it seals the top, (b).

When the ratemeter reading has reached its maximum and started to fall, the count-rate is noted every 15 s for 2 minutes and then every 60 s for the next few minutes. (The GM tube is left in the flask for at least 1 hour until the radioactivity has decayed.)

A graph of count-rate (from which the background count-rate, found separately, has been subtracted) against time is plotted and the half-life (52 s) estimated from it.

(d) Random nature

During the previous experiment it becomes evident that the count-rate varies irregularly: the loudspeaker of the ratemeter 'clicks' erratically, not at a steady rate. This is because radioactive decay is a **random** process, in that it is a matter of pure chance whether or not a particular nucleus will decay during a certain period of time. All we can say is that about half the nuclei in a sample will decay during the half-life. We cannot say which nuclei these will be, nor can we influence the process in any way.

Uses of radioactivity

Radioactive substances, called **radioisotopes**, are now made in nuclear reactors and have many uses.

(a) Thickness gauge

If a radioisotope is placed on one side of a moving sheet of material and a GM tube on the other, the count-rate decreases if the thickness increases. This technique is used to control automatically the thickness of paper, plastic and metal sheets during manufacture, Figure 63.11.

Figure 63.11 Quality control in the manufacture of paper using a radioactive gauge

(b) Tracers

The progress of a small amount of a weak radioisotope injected into a system can be 'traced' by a GM tube or other detector. The method is used in medicine to detect brain tumours, in agriculture to study the uptake of fertilizers by plants, and in industry to measure fluid flow in pipes.

(c) Radiotherapy

Gamma rays from strong cobalt radioisotopes are replacing X-rays in the treatment of cancer.

(d) Sterilization

Gamma rays are used to sterilize medical instruments etc. by killing bacteria. They are also used to 'irradiate' some foods, again killing bacteria to preserve the food for longer.

(e) Archaeology

A radioisotope of carbon, present in the air, is taken in by plants and trees and its rate of decay is used to date archaeological remains, e.g. of wood and linen.

The ages of radioactive rocks have been estimated in a similar way (see Chapter 71).

Dangers and safety

The danger from α-particles (due, as with all radiation, to their ionizing effect) is small unless the source enters the body. β and γ-rays can cause radiation burns (i.e. redness and sores on the skin) and delayed effects such as cancer and eye cataracts. Fall-out from atomic explosions contains highly active elements (e.g. strontium) with long half-lives, which are absorbed by the bones.

The weak sources used at school **should always be lifted with forceps, never held near the eyes and should be kept in their boxes when not in use**. In industry sources are handled by long tongs and transported in thick lead containers. Workers are protected by lead and concrete walls, and wear radiation dose badges which keep a check on the amount of radiation they have been exposed to.

Worked example

A radioactive source has a half-life of 20 minutes. What fraction is left after 1 hour?

After 20 minutes, fraction left = $\frac{1}{2}$
After 40 minutes, fraction left = $\frac{1}{2} \times \frac{1}{2} = \frac{1}{4}$
After 60 minutes, fraction left = $\frac{1}{2} \times \frac{1}{4} = \frac{1}{8}$

QUESTIONS

1 Which type of radiation from radioactive materials

a has a positive charge?

b is the most penetrating?

c is easily deflected by a magnetic field?

d consists of waves?

e causes the most intense ionization?

f has the shortest range in air?

g has a negative charge?

h is not deflected by an electric field?

2 What is the effect of radiation on

a a charged electroscope?

b a photographic film?

c a fluorescent screen?

3 **a** What do you understand by 'background radiation'? State two sources of this radiation.

b State three safety precautions to be observed when using radioactive sources.

c Describe briefly two uses of radioactive sources.

d How would you test to distinguish between two radioactive sources, one of which emits only alpha particles and the other which emits only beta particles? *(A.L.)*

4 In an experiment to find the half-life of radioactive iodine, the count-rate falls from 200 counts per second to 25 counts per second in 75 minutes. What is its half-life?

5 If the half-life of a radioactive gas is 2 minutes, then after 8 minutes the activity will have fallen to a fraction of its initial value. This fraction is

A $\frac{1}{4}$ **B** $\frac{1}{6}$ **C** $\frac{1}{8}$ **D** $\frac{1}{16}$ **E** $\frac{1}{32}$

Checklist

After studying this chapter you should be able to

- recall that the radiation emitted by a radioactive substance can be detected by its ionizing effect,
- explain the principle of operation of a Geiger–Müller tube and a diffusion cloud chamber,
- recall the nature of α, β and γ-rays,
- describe experiments to compare the range and penetrating power of α, β and γ-rays in different materials,
- recall the ionizing abilities of α, β and γ-rays and relate them to their ranges,
- predict how α, β and γ-rays will be deflected in magnetic and electric fields,
- define the term **half-life**,
- describe an experiment from which a radioactive decay curve can be obtained,
- show from a graph that radioactive decay processes have a constant half-life,
- solve simple problems on half-life,
- recall that radioactivity is (a) a random process, (b) due to nuclear instability, (c) independent of external conditions,
- recall some uses of radioactivity,
- discuss the dangers of radioactivity and safety precautions necessary.

64 ATOMIC STRUCTURE

Nuclear model	***Models of the atom***
Protons and neutrons	Rutherford–Bohr model. Modern model: energy levels.
Isotopes and nuclides	***Nuclear energy***
Radioactive decay	

The discovery of the electron and of radioactivity seemed to indicate that atoms contained negatively and positively charged particles and were not indivisible as was previously thought. The questions then were 'How are the particles arranged inside an atom?', and 'How many are there in the atom of each element?'

An early theory, called the 'plum-pudding' model, regarded the atom as a positively charged sphere in which the negative electrons were distributed all over it (like currants in a pudding) and in sufficient numbers to make the atom electrically neutral. Doubts arose about this model.

Nuclear model

Whilst investigating radioactivity, the physicist Rutherford noticed that not only could α-particles pass straight through very thin metal foil as if it wasn't there but also that some were deflected from their initial direction. With the help of Geiger (of tube fame) and Marsden, Rutherford investigated this in detail at Manchester University using the arrangement in Figure 64.1. The fate of the α-particles after striking the gold foil was detected by the scintillations (flashes of light) they produced on a glass screen coated with zinc sulphide and fixed to a rotatable microscope.

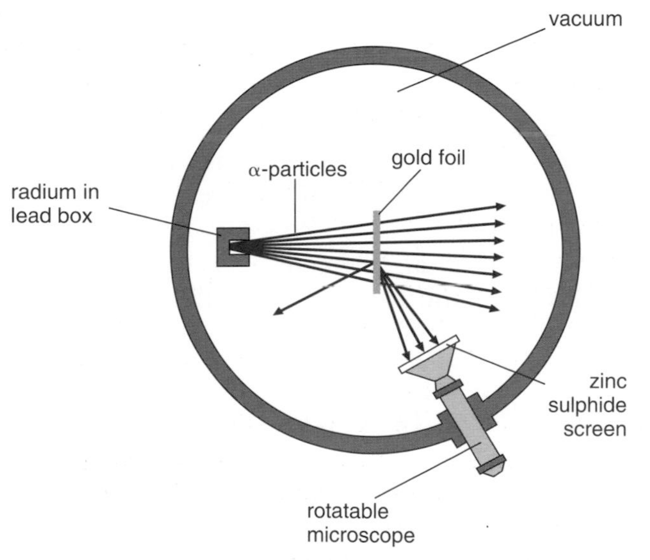

Figure 64.1 Geiger and Marsden's scattering experiment

They found that most of the α-particles were undeflected, some were scattered by appreciable angles and a few (about 1 in 8000) surprisingly 'bounced' back. To explain these results Rutherford proposed in 1911 a 'nuclear' model of the atom in which **all the positive charge and most of the mass of an atom** formed a dense core or **nucleus**, of very small size compared with the whole atom. The electrons surrounded the nucleus some distance away.

He derived a formula for the number of α-particles deflected at various angles, assuming that the electrostatic force of repulsion between the positive charge on an α-particle and the positive charge on the nucleus of a gold atom obeyed an inverse square law (i.e. the force increases four times if the separation is halved). Geiger and Marsden's experimental results completely confirmed Rutherford's formula and supported the view that an atom is mostly empty space. In fact the nucleus and electrons occupy about one million millionth of the volume of an atom. Putting it another way, the nucleus is like a sugar lump in a very large hall and the electrons a swarm of flies.

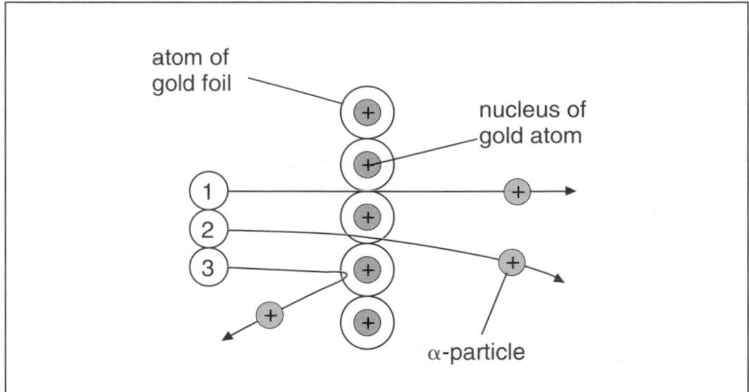

Figure 64.2 Electrostatic scattering of α-particles

The paths of three α-particles are shown in Figure 64.2 (on the previous page).

① is clear of all nuclei and passes straight through the gold atoms.
② suffers some deflection.
③ approaches a gold nucleus so closely as to be violently repelled by it and 'rebounds', appearing to have had a head-on 'collision'.

Protons and neutrons

We now believe as a result of other experiments, in some of which α and other high-speed particles were used as 'atomic probes', that atoms contain three basic particles – protons, neutrons and electrons.

A **proton** is a hydrogen atom minus an electron, i.e. a positive hydrogen ion. Its charge is equal in size but opposite in sign to that of an electron but its mass is about 2000 times greater.

A **neutron** is uncharged with almost the same mass as a proton.

Protons and neutrons are in the nucleus and are called **nucleons**. Together they account for the mass of the nucleus (and most of that of the atom); the protons account for its positive charge. These facts are summarized in Table 64.1.

Table 64.1

Particle	Relative mass	Charge	Location
Proton	1836	$+e$	In nucleus
Neutron	1839	0	In nucleus
Electron	1	$-e$	Outside nucleus

In a neutral atom the number of protons equals the number of electrons surrounding the nucleus. Table 64.2 shows the particles in some atoms. Hydrogen is simplest with 1 proton and 1 electron. Next is the inert gas helium with 2 protons, 2 neutrons and 2 electrons. The soft white metal lithium has 3 protons and 4 neutrons.

Table 64.2

	Hydrogen	Helium	Lithium	Oxygen	Copper
Protons	1	2	3	8	29
Neutrons	0	2	4	8	34
Electrons	1	2	3	8	29

The atomic or proton number Z of an atom is the number of protons in the nucleus.

It is also the number of electrons in the atom. The electrons determine the chemical properties of an atom and when the elements are arranged in order of atomic number in the Periodic Table, they fall into chemical families.

The mass or nucleon number A of an atom is the number of nucleons in the nucleus.

In general

$$A = Z + N$$

where N is the **neutron number** of the element.

Atomic **nuclei** are represented by symbols. Hydrogen is written ^1_1H, helium ^4_2He, lithium ^7_3Li and in general atom X is written ^A_ZX where A is the nucleon number and Z the proton number.

Isotopes and nuclides

Isotopes of an element are atoms which have the same number of protons but different numbers of neutrons. That is, their proton numbers are the same but not their nucleon numbers.

Isotopes have identical chemical properties since they have the same number of electrons and occupy the same place in the Periodic Table. (In Greek, *isos* means same and *topos* means place.)

Few elements consist of identical atoms; most are mixtures of isotopes. Chlorine has two isotopes; one has 17 protons and 18 neutrons (i.e. $Z = 17$, $A = 35$) and is written $^{35}_{17}\text{Cl}$, the other has 17 protons and 20 neutrons (i.e. $Z = 17$, $A = 37$) and is written $^{37}_{17}\text{Cl}$. They are present in ordinary chlorine in the ratio of three atoms of $^{35}_{17}\text{Cl}$ to one atom of $^{37}_{17}\text{Cl}$, giving chlorine an average atomic mass of 35.5.

Hydrogen has three isotopes: ^1_1H with 1 proton, **deuterium** ^2_1D with 1 proton and 1 neutron and **tritium** ^3_1T with 1 proton and 2 neutrons. Ordinary hydrogen contains 99.99 per cent of ^1_1H atoms. Water made from deuterium is called 'heavy water' (D_2O); it has a density of 1.108 g/cm^3, it freezes at 3.8 °C and boils at 101.4 °C.

Each form of an element is called a **nuclide**. Nuclides with the same Z but different A are isotopes.

Radioactive decay

The emission of an α or a β-particle from a nucleus produces an atom of a different element, which may itself be unstable. After a series of changes a stable end-element is formed.

(a) Alpha decay

An α-particle is a helium nucleus having 2 protons and 2 neutrons and when an atom decays by α emission, its nucleon number decreases by 4 and its proton number by 2. For example, when radium of nucleon number 226 and proton number 88 emits an α-particle, it decays to radon of nucleon number 222 and proton number 86. We can write:

$$^{226}_{88}\text{Ra} \rightarrow {}^{222}_{86}\text{Rn} + {}^{4}_{2}\text{He}$$

The values of A and Z must balance on both sides of the equation since nucleons and charge are conserved.

(b) Beta decay

Here a neutron changes to a proton and an electron. The proton remains in the nucleus and the electron is emitted as a β-particle. The new nucleus has the same nucleon number, but its proton number increases by one since it has one more proton. Radioactive carbon, called carbon 14, decays by β emission to nitrogen:

$$^{14}_{6}\text{C} \rightarrow {}^{14}_{7}\text{N} + {}^{0}_{-1}\text{e}$$

(c) Gamma emission

After emitting an α or a β-particle some nuclei are left in an 'excited' state. Rearrangement of the protons and neutrons occurs and a burst of γ-rays is released.

Models of the atom

(a) Rutherford–Bohr model

Shortly after Rutherford proposed his nuclear model of the atom, Bohr, a Danish physicist, developed it to explain how an atom emits light. He suggested that the electrons circled the nucleus at high speed being kept in **certain orbits** by the electrostatic attraction of the nucleus for them. He pictured atoms as miniature solar systems. Figure 64.3 shows the models for three elements.

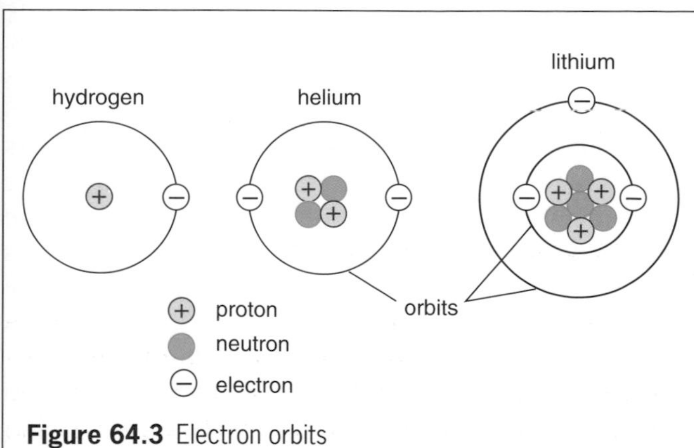

Figure 64.3 Electron orbits

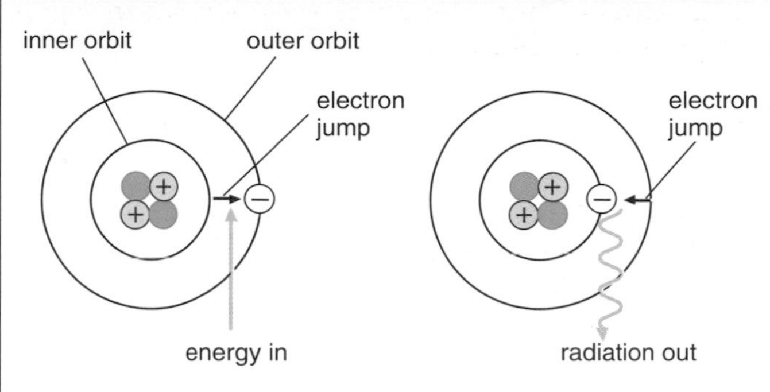

Figure 64.4 Bohr's explanation of energy changes in an atom

Normally the electrons remain in their orbits but if the atom is given energy, e.g. by being heated, electrons may jump to an outer orbit. The atom is then said to be **excited**. Very soon afterwards the electrons return to an inner orbit and as they do, they emit energy in the form of bursts of electromagnetic radiation (called **photons**), e.g. as infrared light, ultraviolet or X-rays, Figure 64.4. The wavelength of the radiation emitted depends on the two orbits between which the electrons jump.

If an atom gains enough energy for an electron to escape altogether, the atom becomes an ion and the energy needed to achieve this is called the **ionization energy** of the atom.

(b) Modern model

Although still useful for some purposes, the Rutherford–Bohr model has now been replaced by a **mathematical model** which is not easy to picture. The best we can do, without using advanced mathematics, is to say that the atom consists of a nucleus surrounded by a hazy cloud of electrons. Regions of the atom where the mathematics predicts electrons are likely to be found are represented by dense shading, Figure 64.5.

The modern theory does away with the idea of electrons moving in definite orbits and replaces them by **energy levels** that are different for each element. When an electron 'jumps' from one level, say E_3 in

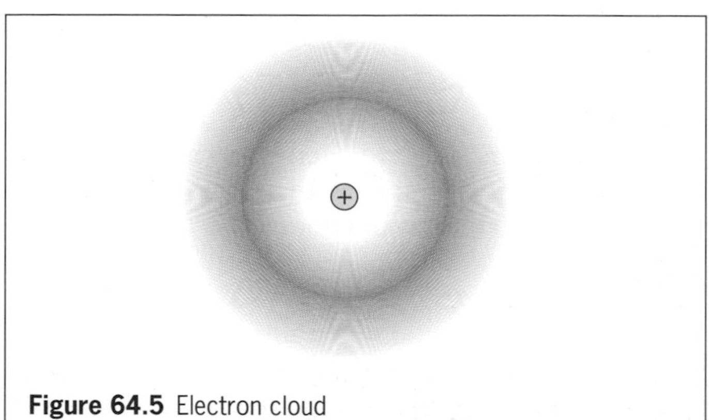

Figure 64.5 Electron cloud

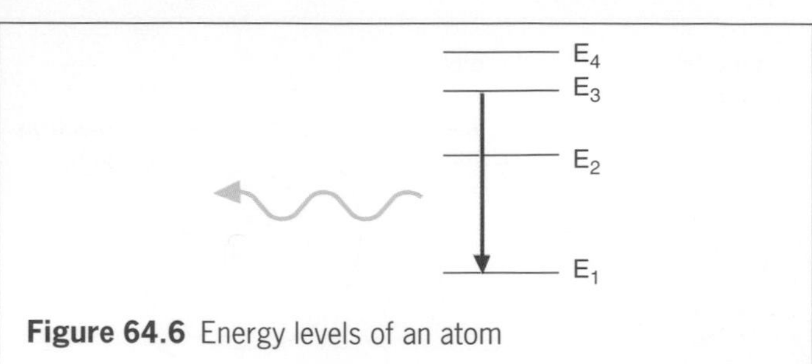

Figure 64.6 Energy levels of an atom

Figure 64.6, to a lower one E_1, a photon of electromagnetic radiation is emitted with energy equal to the difference in energy of the two levels. The frequency (and wavelength) of the radiation emitted by an atom is thus dependent on the arrangement of energy levels. For an atom emitting visible light, the resulting spectrum (produced for example by a prism) is a series of coloured **lines** which is unique to each element. Sodium vapour in a gas discharge tube (or street light) gives two adjacent yellow-orange lines, Figure 64.7. Light from the Sun is due to energy changes in many different atoms and the resulting spectrum is a **continuous** one with all colours, see Figure 9.2c.

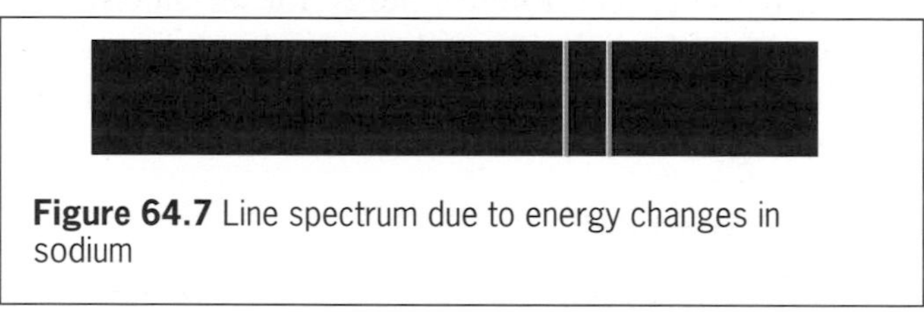
Figure 64.7 Line spectrum due to energy changes in sodium

Nuclear energy

(a) $E = mc^2$

Einstein predicted that if the energy of a body changes by an amount E, its mass changes by an amount m given by the equation

$$E = mc^2$$

where c is the speed of light (3×10^8 m/s). The implication is that any reaction in which there is a decrease of mass, called a **mass defect**, is a source of energy. The energy and mass changes in physical and chemical changes are very small; those in some nuclear reactions, e.g. radioactive decay, are millions of times greater. It appears that mass (matter) is a very concentrated form of energy.

(b) Fission

The heavy metal uranium is a mixture of isotopes of which $^{235}_{92}$U, called uranium 235, is the most important. Some atoms of this isotope decay quite naturally, emitting high-speed neutrons. If one of these hits the nucleus of a neighbouring uranium 235 atom (being uncharged the neutron is not repelled by the nucleus), this may break (**fission**) into two nearly equal radioactive nuclei, often of barium and krypton, with the production of two or three more neutrons:

$$^{235}_{92}\text{U} + \underset{\text{neutron}}{^{1}_{0}\text{n}} \rightarrow \underbrace{^{144}_{56}\text{Ba} + ^{90}_{36}\text{Kr}}_{\text{fission fragments}} + \underset{\text{neutrons}}{2^{1}_{0}\text{n}}$$

The mass defect is large and appears mostly as k.e. of the fission fragments. These fly apart at great speed colliding with surrounding atoms and raising their average k.e., i.e. their temperature, so producing heat.

If the fission neutrons split other uranium 235 nuclei, a **chain reaction** is set up, Figure 64.8. In practice some fission neutrons are lost by escaping from the surface of the uranium before this happens. The ratio of those causing fission to those escaping increases as the mass of uranium 235 increases. This must exceed a certain **critical** value to sustain the chain reaction.

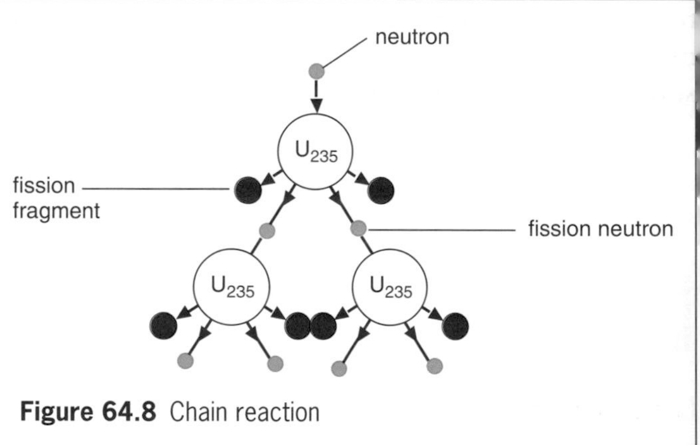

Figure 64.8 Chain reaction

(c) Nuclear reactor

In a nuclear power station a nuclear reactor produces the steam for the turbines instead of a coal- or oil-burning furnace. Figure 64.9 is a simplified diagram of a reactor.

The chain reaction occurs at a steady rate which is controlled by inserting or withdrawing neutron-absorbing rods of boron among the uranium rods. The graphite core is called the **moderator** and slows down the fission neutrons; fission of uranium 235 occurs more readily with slow than with fast neutrons. Carbon dioxide gas is pumped through the core and carries off heat to the heat exchanger where steam is produced. The concrete shield gives workers protection from γ-rays and escaping neutrons. The radioactive fission fragments must be removed periodically if the nuclear fuel is to be used efficiently.

In an **atomic bomb** an increasing uncontrolled chain reaction occurs when two pieces of uranium 235 come together and exceed the critical mass.

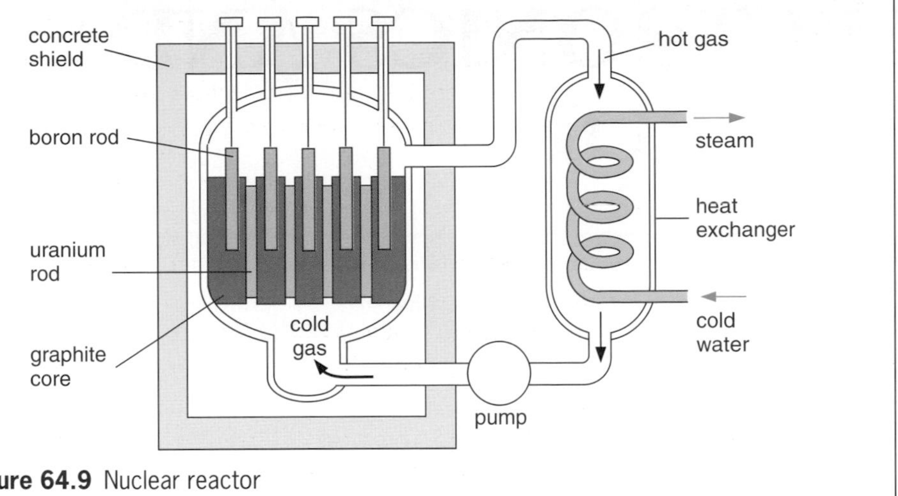

Figure 64.9 Nuclear reactor

(d) Fusion

The union of light nuclei into heavier ones can also lead to a loss of mass and, as a result, the release of energy. At present, research is being done on the controlled fusion of isotopes of hydrogen (deuterium and tritium) to give helium. Temperatures of over 100 million °C are required. Fusion is believed to be the source of the Sun's energy.

QUESTIONS

1 What are the particles you would expect to find in an atom? Give some idea of their relative masses and state what electrical charge, if any, each kind has. *(S.E.)*

2 **a** An atom of cobalt has an atomic number of 27 and a mass number of 59. Describe simply the structure of the cobalt atom.

b What are isotopes?

c Why are isotopes difficult to separate by chemical methods? *(W.M.)*

3 Uranium 238 and uranium 235 are 'isotopes' of uranium and have the same atomic number, 92.

a What do the numbers 238 and 235 represent?

b (i) What does the number 92 tell you about the nucleus of either of these two atoms?
(ii) What else does the number 92 tell you about the atom as a whole?

c In what way does the nucleus of uranium 238 differ from the nucleus of uranium 235? *(J.M.B./A.L./N.W.16+)*

4 What changes, if any, occur in the atomic number of a radioactive atom if the nucleus emits **a** an α-particle, **b** a β-particle, **c** a γ-ray? *(E.A.)*

5 Atomic bombs and atomic reactors both provide large quantities of energy through chain reactions.

a Explain what is meant by a chain reaction.

b Explain how the reaction, which is violent in the case of a bomb, is slowed down in the reactor. *(E.A.)*

Checklist

After studying this chapter you should be able to

- describe how Rutherford and Bohr contributed to views about the structure of the atom,
- describe the Geiger–Marsden experiment which established the nuclear model of the atom,
- recall the charge, relative mass and location in the atom of protons, neutrons and electrons,
- define the terms **proton number** (Z), **neutron number** (N) and **nucleon number** (A) and use the equation $A = Z + N$,
- explain the terms **isotope** and **nuclide** and use symbols to represent them, e.g. $^{35}_{17}Cl$,
- write equations for radioactive decay and interpret them,
- outline the modern view of the atom and explain in terms of **energy levels** how line spectra are produced,
- connect the release of energy in a nuclear reaction with a change of mass according to the equation $E = mc^2$,
- describe the process of **fission**,
- describe a nuclear reactor,
- outline the process of **fusion**.

65 ELECTRONICS AND CONTROL

Electronics is being used more and more in our homes, factories, offices, banks, shops and hospitals. The development of semiconductor devices such as transistors and integrated circuits ('chips') has given us, among other things, pocket calculators that are also clocks and diaries, Figure 65.1a, heart pacemakers, Figure 65.1b, microcomputers, robots, computer games, machines for teaching spelling and arithmetic and even for recognizing signatures.

Semiconductors

Semiconductors, of which silicon and germanium are the two best known, are insulators if they are very pure, especially at low temperatures. Their conductivity increases at higher temperatures. This is the opposite to the case with metals. In metals the number of electrons available for conduction, i.e. to carry current, is fixed. As the temperature increases the resistance *increases* due to increased vibration of the atoms making electron flow more difficult. In semiconductors (and carbon) this effect is more than offset by the 'freeing' of more charges for conduction as the temperature increases. As a result the resistance *decreases* at higher temperatures.

The conductivity of semiconductors can be greatly increased by adding tiny but controlled amounts of certain other substances (called 'impurities') by a process known as 'doping'. They can then be used to make diodes, transistors and integrated circuits.

Semiconductor diode

A diode is a two-terminal, one-way device which lets current pass through it in one direction only. One is shown in Figure 65.2 with its symbol. The wire nearest the band is the **cathode** and the one at the other end is the **anode**.

The diode conducts when the anode goes to the + terminal of the voltage supply and the cathode to the – terminal, Figure 65.3a. It is then **forward biased**, its resistance is small and conventional current passes in the direction of the arrow on its symbol. If the connections are the other way round, it does not conduct, its resistance is large and it is **reverse biased**, Figure 65.3b.

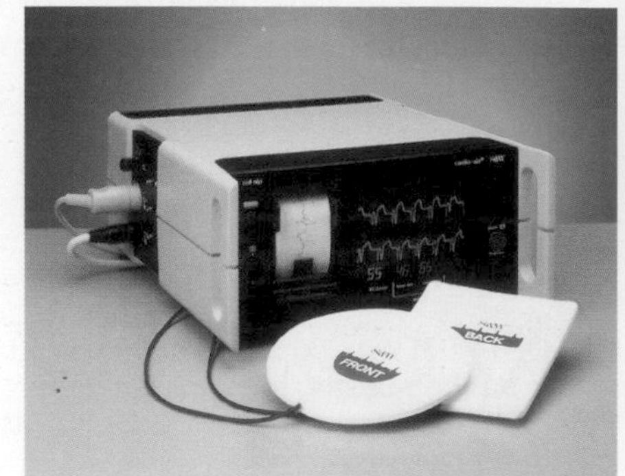

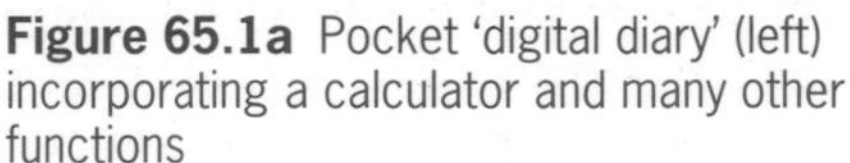

Figure 65.1a Pocket 'digital diary' (left) incorporating a calculator and many other functions

Figure 65.1b Heart pacemaker (right)

The lamp in the circuit shows when the diode conducts by lighting up. It also acts as a resistor to limit the current when the diode is forward biased. Otherwise the diode might overheat and be damaged.

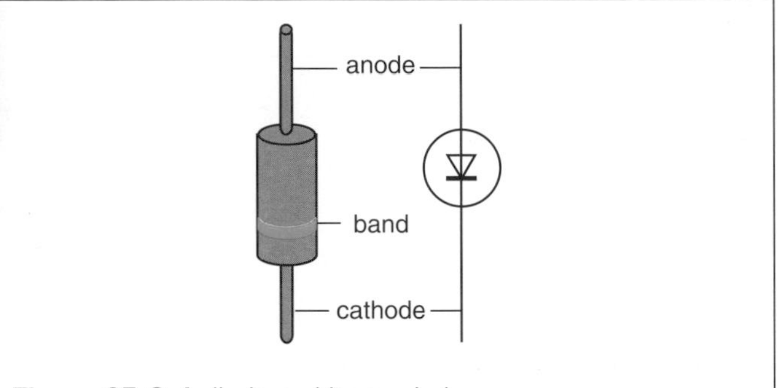

Figure 65.2 A diode and its symbol

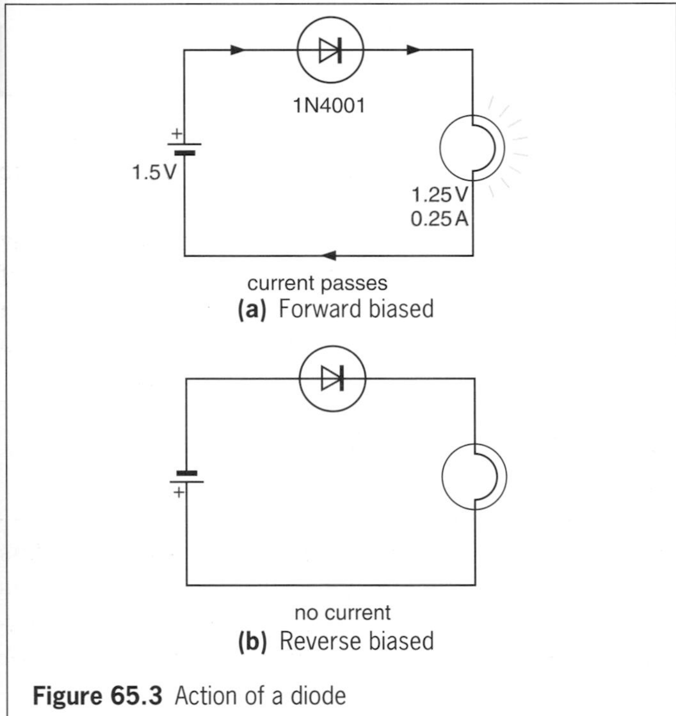

Figure 65.3 Action of a diode

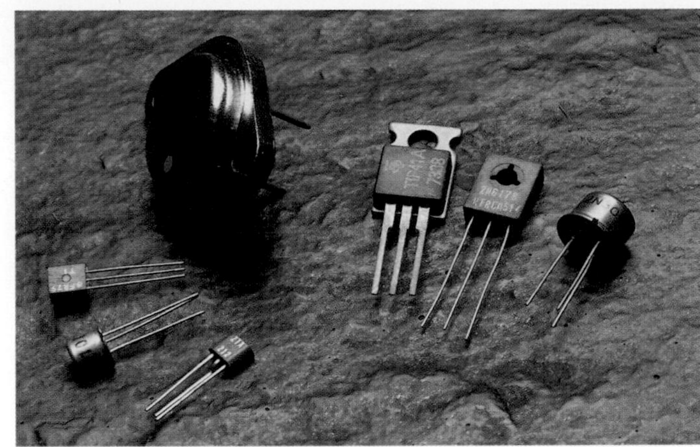

Figure 65.4a Transistor components

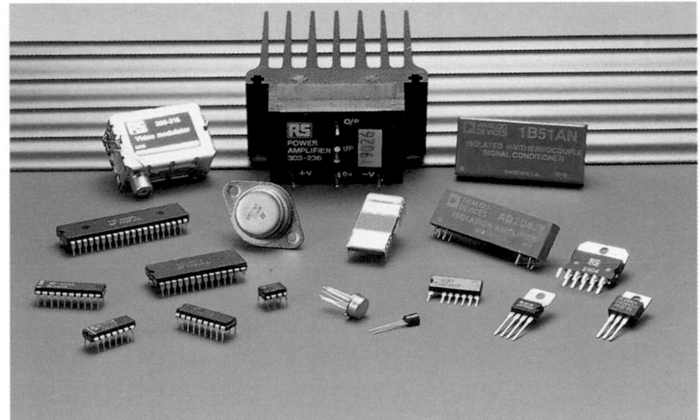

Figure 65.4b Integrated circuits which may each contain millions of transistors

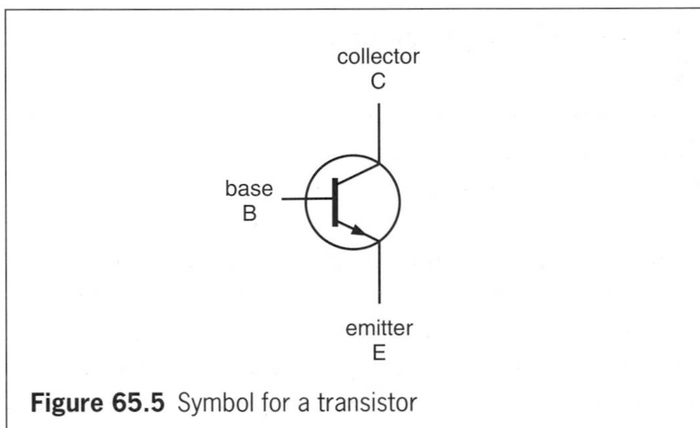

Figure 65.5 Symbol for a transistor

Transistor

Transistors are the small semiconductor devices which have revolutionized electronics. They are made as separate components, like those in Figure 65.4a in their cases, and also as parts of **integrated circuits** (ICs) in which millions may be 'etched' on a 'chip' of silicon, Figure 65.4b.

Transistors have three connections called the **base** (B), the **collector** (C) and the **emitter** (E). In the transistor symbol shown in Figure 65.5, the arrow indicates the direction in which conventional current flows through it when C and B are connected to battery +, and E to battery −.

There are two current paths through a transistor. One is the **base-emitter path** and the other is the **collector-emitter** (via base) **path**. The transistor's usefulness arises from the fact that it can link circuits connected to each path so that the current in one controls that in the other, just like a relay (Chapter 57).

Its action can be shown using the circuit of Figure 65.6 (overleaf). When **S is open**, the base current I_B is zero and neither L_1 nor L_2 light up, showing that the collector current I_C is also zero even though the battery is correctly connected across the C-E path.

When **S is closed**, B is connected through R to battery +, and L_2 lights up but not L_1. This shows there is now collector current (which passes through L_2) and that it is much greater than the base current (which passes through L_1 but is too small to light it).

Therefore, **in a transistor the base current I_B switches on and controls the much greater collector current I_C.**

Resistor R must be in the circuit to limit the base current which would otherwise create so large a collector current as to destroy the transistor by overheating.

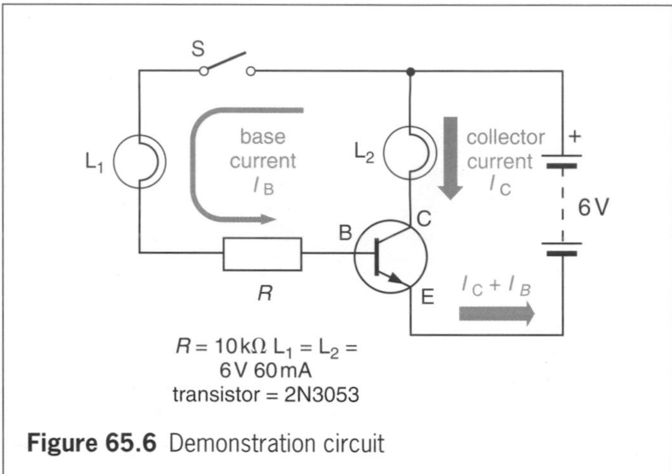

Figure 65.6 Demonstration circuit

Transistor as a switch

(a) Advantages

Transistors have many advantages over other electrically operated switches such as relays and reed switches. They are small, cheap, reliable, have no moving parts, their life is almost indefinite (in well-designed circuits) and they can switch on and off millions of times a second.

(b) 'On' and 'off' states

A transistor is considered to be 'off' when the collector current is zero or very small. It is 'on' when the collector current is much larger. The resistance of the collector-emitter path is large when the transistor is 'off' (as it is for an ordinary mechanical switch) and small (ideally it should be zero) when it is 'on'.

To switch a transistor 'on' requires the base voltage (and therefore the base current) to exceed a certain minimum value (about +0.6 V).

(c) Basic switching circuits

Two are shown in Figures 65.7a, b. The 'on' state is shown by the lamp in the collector circuit becoming fully lit.

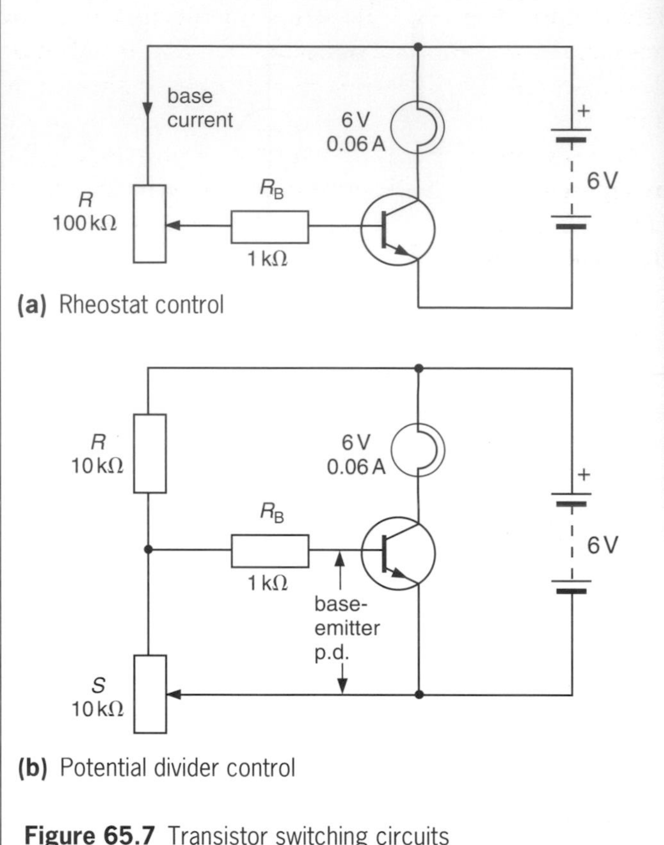

Figure 65.7 Transistor switching circuits

Rheostat control is used in (a) where 'switch-on' occurs by reducing R until the base current is large enough to make the collector current light the lamp. (The base resistor R_B is *essential* in case R is made zero and results in +6 V from the battery being applied directly to the base. This would produce very large base and collector currents and destroy the transistor by overheating.)

Potential divider control is used in (b). Here 'switch-on' is obtained by adjusting the variable resistor S until the p.d. across S (which is the base-emitter p.d. and depends on the value of S compared with that of R) exceeds +0.6 V or so.

Note. In a potential divider the p.d.s across the resistors are in the ratio of their resistances. For example, in Figure 65.7b if $R = 10\ \text{k}\Omega$ and $S = 5\ \text{k}\Omega$ then p.d. across $R = V_R = 4$ V and p.d. across $S = V_S = 2$ V, i.e. $V_R/V_S = 4\text{ V}/2\text{ V} = 2/1$ and $V_R + V_S = 6\text{ V}$ = p.d. across R and S in series where $R/S = 10\ \text{k}\Omega/5\ \Omega = 2/1$.

In general

$$\frac{V_R}{V_S} = \frac{R}{S} \quad \text{and} \quad V_S = (V_R + V_S)S/(R + S)$$

Also see question 2 on p. 275.

Other semiconductor devices

(a) Light dependent resistor (LDR)

The action of an LDR depends on the fact that the resistance of the semiconductor cadmium sulphide decreases as the intensity of the light falling on it increases.

An LDR and a circuit showing its action are shown in Figures 65.8a, b. Note the circuit symbol for an LDR. When light from a lamp falls on the 'window' of the LDR, its resistance decreases and the increased current lights the lamp.

LDRs are used in photographic exposure meters.

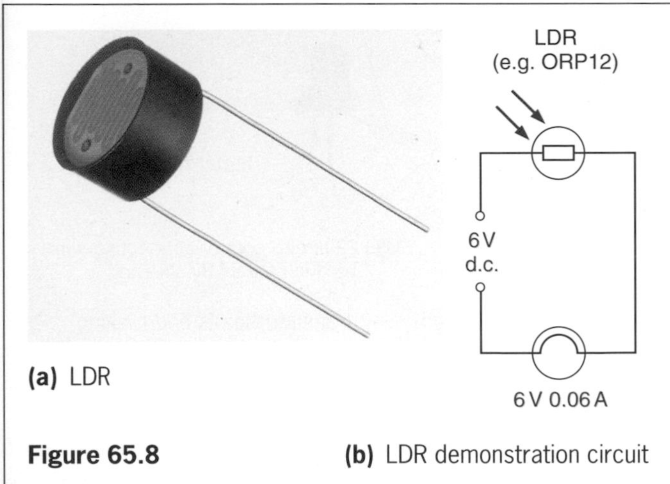

(a) LDR

Figure 65.8 **(b)** LDR demonstration circuit

(b) Thermistor

This contains semiconducting metallic oxides whose resistance decreases markedly when the temperature rises either due to heating the thermistor directly or to passing a current through it.

Two are shown in Figure 65.9a, and the symbol for a thermistor is shown in a circuit to demonstrate its action in Figure 65.9b. Heating the thermistor with a match lights the bulb.

A thermistor in series with a meter marked in °C can measure temperatures.

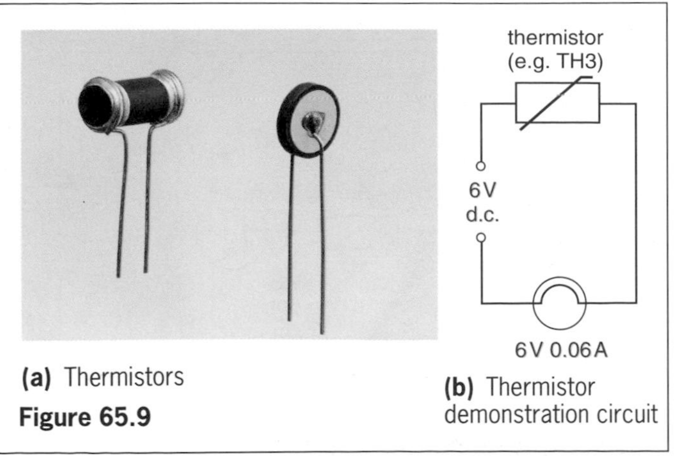

(a) Thermistors

Figure 65.9 **(b)** Thermistor demonstration circuit

(c) Light emitting diode (LED)

An LED, shown in Figure 65.10a, is a diode made from the semiconductor gallium arsenide phosphide. When forward biased, the current through it makes it emit red, yellow or green light. No light is emitted on reverse bias which, if it exceeds 5 V, may cause damage.

In use an LED must have a suitable resistor R in series with it (e.g. 300 Ω on a 5 V supply) to limit the current (typically 10 mA). Figure 65.10b shows the symbol for an LED in a demonstration circuit.

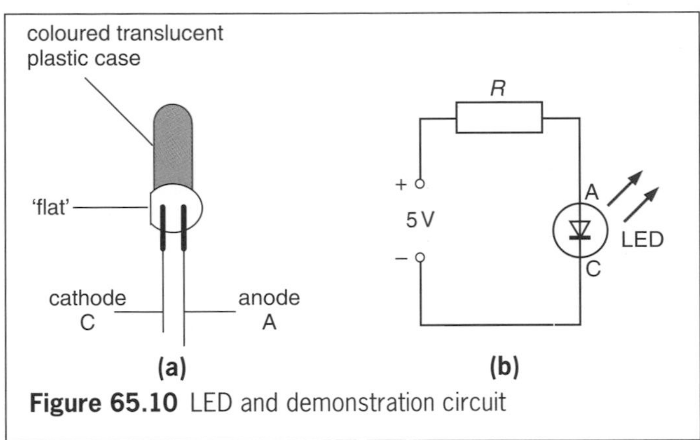

Figure 65.10 LED and demonstration circuit

LEDs are used as indicator lamps on computers, music centres, radio receivers and other electronic equipment. Many clocks, calculators, cash registers and measuring instruments have seven-segment red or green numerical displays, Figure 65.11a. Each segment is an LED and, depending on which have a voltage across them, the display lights up the numbers 0 to 9, as in Figure 65.11b.

LEDs are small, reliable, have a long life, their operating speed is high and their current requirements very modest.

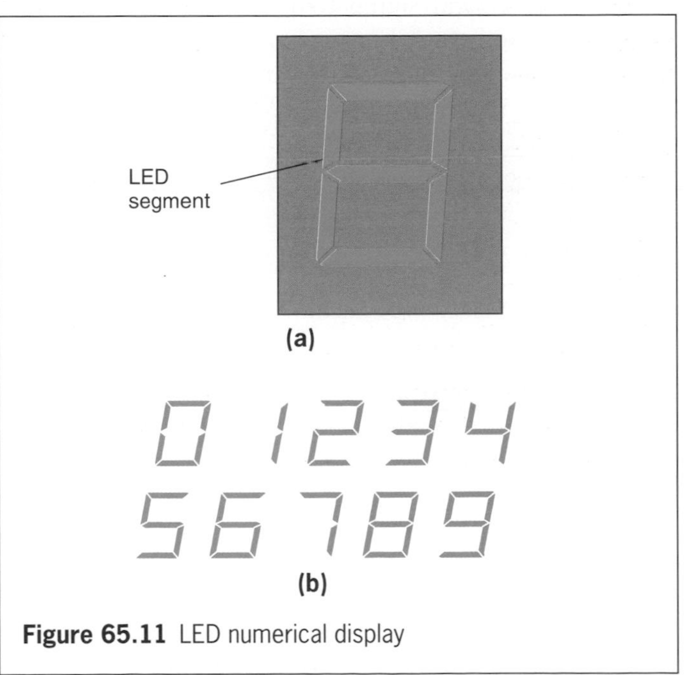

Figure 65.11 LED numerical display

(d) Photodiode

This is a normal semiconductor diode with a 'window' through which light can enter. It works in reverse bias and the tiny reverse current increases with the light intensity. Photodiodes are used as fast counters which produce a current pulse every time a beam of light is interrupted. Its symbol is the same as for an LED but with the arrows reversed.

Electronic systems

Any electronic system can be considered to consist of the three parts shown in the block diagram of Figure 65.12, i.e.

(i) an **input sensor** or **transducer**,
(ii) a **processor**, and
(iii) an **output transducer**.

A 'transducer' is a device for converting a non-electrical input into an electrical signal or vice versa.

The **input sensor** detects changes in the environment and converts them from their present form of energy into electrical energy. Input sensors or transducers include LDRs, thermistors, microphones and switches which respond, for instance, to pressure changes.

The **processor** decides on what action to take on the electrical signal it receives from the input sensor. It may involve an operation such as counting, amplifying, timing or storing.

The **output transducer** converts the electrical energy supplied by the processor into another form. Output transducers include lamps, LEDs, loudspeakers, motors, heaters, relays and cathode ray tubes.

In a radio, the input sensor is the aerial which sends an electrical signal to processors in the radio which, among other things, amplify the signal so that it can enable the output transducer, in the form of a loudspeaker, to produce sound.

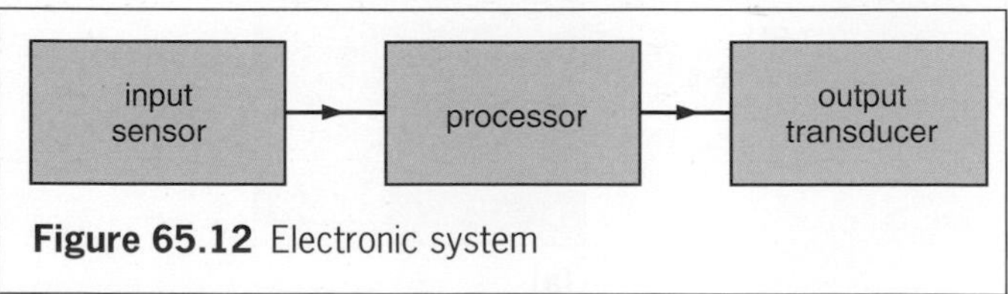

Figure 65.12 Electronic system

PRACTICAL WORK

Transistor switching circuits

The components can be mounted on a circuit board, e.g. an S-DeC, Figure 65.13a. The diagrams in Figure 65.13b show how to lengthen transistor leads and also how to make connections (without soldering) to parts that have 'tags', e.g. variable resistors.

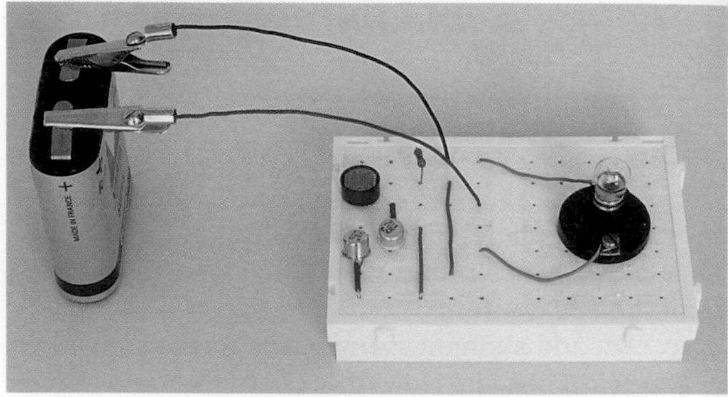

Figure 65.13a Partly built transistor switching circuit on an S-DeC

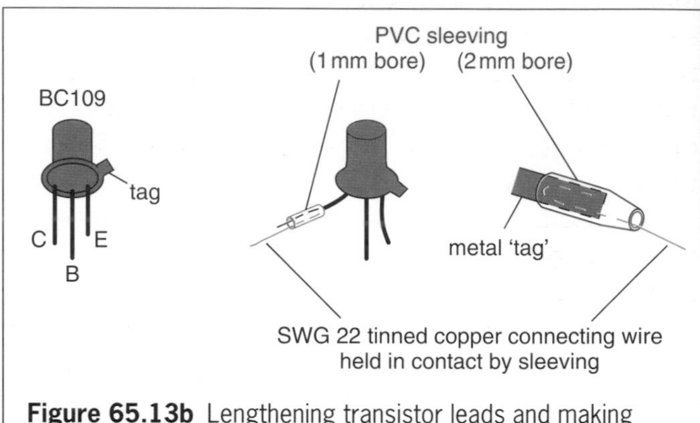

Figure 65.13b Lengthening transistor leads and making connections to tags

In many control circuits, devices such as LDRs and thermistors are used in potential divider arrangements to detect small changes of light intensity and temperature respectively. These changes then enable a transistor to act as a simple processor by controlling the current to an output transducer, e.g. a lamp or bell.

(a) Light-operated

In the circuit of Figure 65.14 the LDR is part of a potential divider. The lamp comes on when the LDR is shielded, due to more of the battery p.d. being dropped across the increased resistance of the LDR (i.e. more than 0.6 V) and less across *R*. In the dark, the base-emitter p.d. increases as does the base current and so also the collector current.

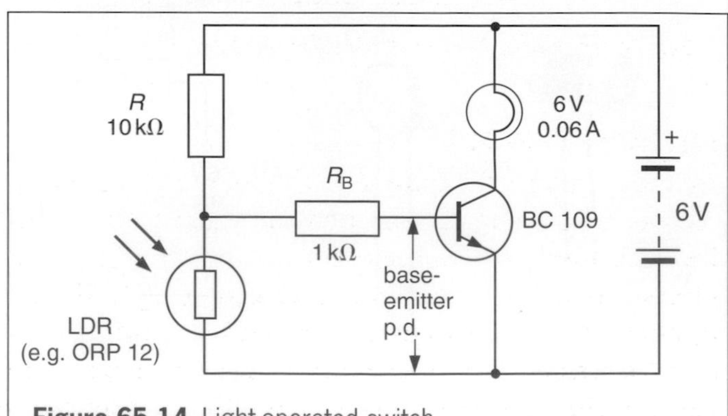

Figure 65.14 Light-operated switch

If the LDR and R are interchanged the lamp goes off in the dark and the circuit could act as a light-operated intruder alarm.

If a variable resistor is used for R, the light level at which switching occurs can be changed.

(b) Temperature-operated

In the high-temperature alarm circuit of Figure 65.15 a thermistor and a resistor R form a potential divider across the 6 V supply.

When the temperature of the thermistor rises, its resistance decreases and a larger fraction of the 6 V supply is dropped across R, i.e. the base-emitter p.d. increases. When it exceeds 0.6 V or so the transistor switches on and collector current (too small to ring the bell directly) goes through the relay coil. The relay contacts close, enabling the bell to obtain directly from the 6 V supply the larger current it needs.

The diode D protects the transistor from damage by the large p.d. induced in the relay coil (due to its inductance) when the collector current falls to zero at switch off. The diode is forward biased by the induced p.d. (which tries to maintain the current through the relay coil) and, because of its low forward resistance (e.g. 1 Ω), offers an easy path for the current produced. To the 6 V supply the diode is reverse biased and its high resistance does not short-circuit the relay coil when the transistor is on.

If the thermistor and R are interchanged, the circuit could act as a frost-warning device. If R is variable the temperature at which switching occurs can be changed.

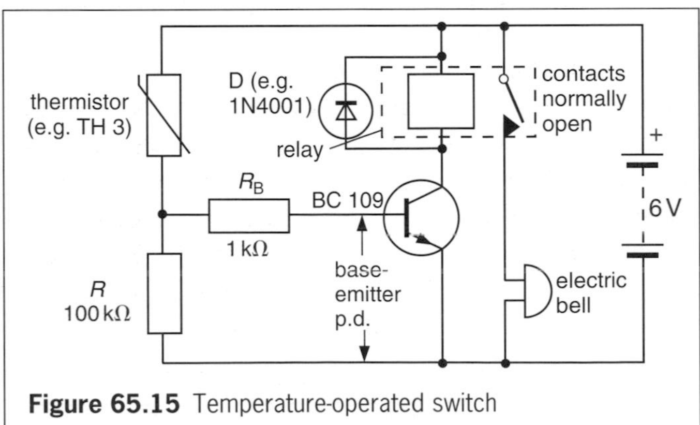

Figure 65.15 Temperature-operated switch

(c) Time-operated

In the circuit of Figure 65.16 when S_1 and S_2 are closed, the lamp is on and the transistor is off because the base-emitter p.d. is zero (due to S_2 short-circuiting C and stopping it charging up). If S_2 is opened, C starts to charge through R, and, *after a certain time*, the base-emitter p.d. exceeds 0.6 V causing the transistor to switch on. This operates the relay whose contacts open and switch off the lamp. The time delay between opening S_2 and the lamp going off increases if either C or R are increased.

The circuit is reset by opening S_1 and closing S_2 to let C discharge. It could be used as a timer to control a lamp in a photographic dark room.

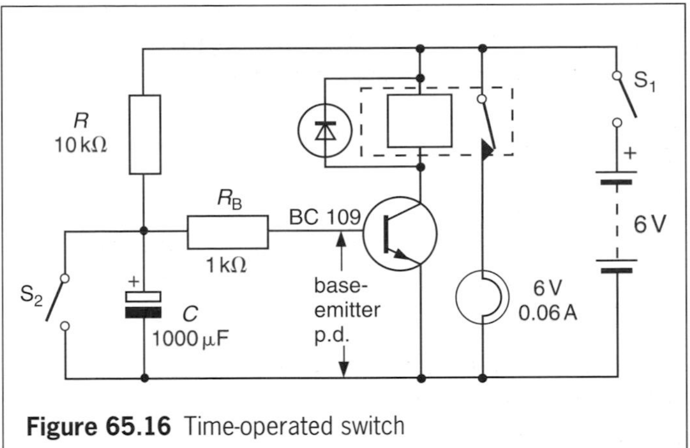

Figure 65.16 Time-operated switch

Logic gates

Logic gates are switching circuits used in computers and other electronic systems. They 'open' and give a 'high' output voltage, i.e. a signal (e.g. 5 V), depending on the combination of voltages at their inputs, of which there is usually more than one.

There are five basic types, all made from transistors in integrated circuit form. The behaviour of each is described by a **truth table** showing what the output is for all possible inputs. 'High' (e.g. 5 V) and 'low' (e.g. near 0 V) outputs and inputs are represented by 1 and 0 respectively and are referred to as **logic levels** 1 and 0.

(a) NOT gate or inverter

This is the simplest gate, with one input and one output. It produces a 'high' output if the input is 'low', i.e. NOT high, and vice versa. Whatever the input, the gate inverts it. The symbol and truth table are given in Figure 65.17.

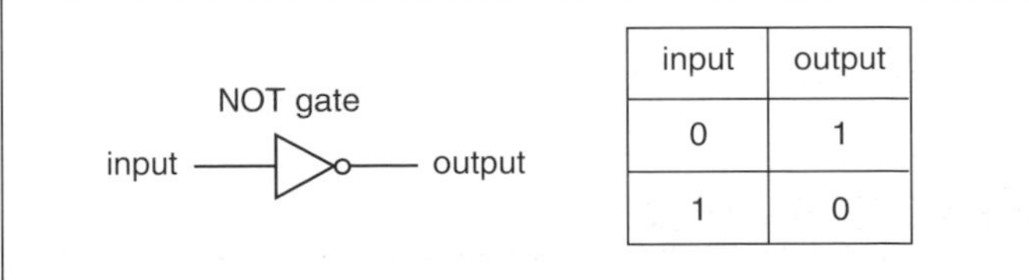

input	output
0	1
1	0

Figure 65.17 NOT gate symbol and truth table

(b) OR, NOR, AND, NAND gates

All these have two or more inputs and one output. The truth tables and symbols for 2-input gates are shown in Figure 65.18. Try to remember the following.

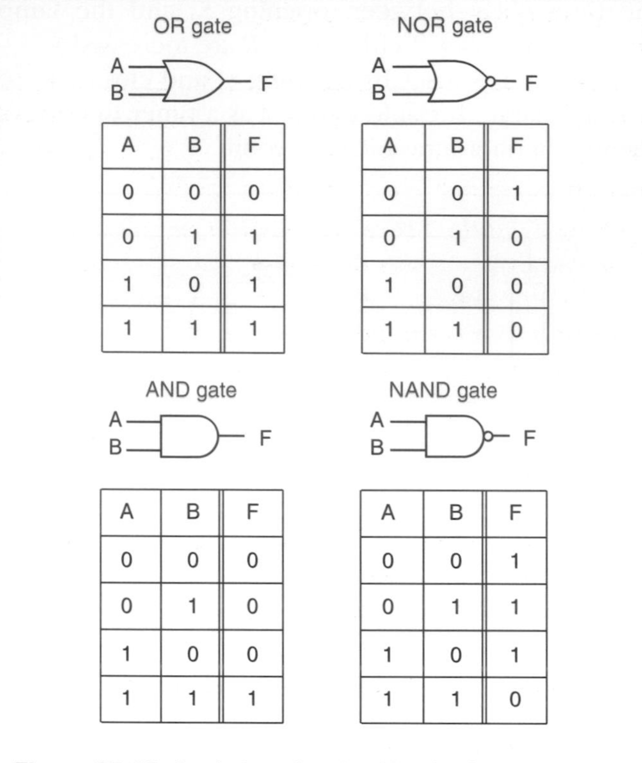

OR gate

A	B	F
0	0	0
0	1	1
1	0	1
1	1	1

NOR gate

A	B	F
0	0	1
0	1	0
1	0	0
1	1	0

AND gate

A	B	F
0	0	0
0	1	0
1	0	0
1	1	1

NAND gate

A	B	F
0	0	1
0	1	1
1	0	1
1	1	0

Figure 65.18 Symbols and truth tables for 2-input gates

OR: output is 1 if input A **OR** input B **OR** both are 1
NOR: output is 1 if neither input A **NOR** input B is 1
AND: output is 1 if input A **AND** input B are 1
NAND: output is 1 if input A **AND** input B are **NOT** both 1

Note from the truth tables that the outputs of the NOR and NAND gates are those of the OR and AND gates respectively inverted. They have a small circle at the output end of their symbols to show this inversion.

(c) Testing logic gates

The truth tables for the various gates can be conveniently checked by having the logic gate IC mounted on a small board with sockets for the power supply, inputs A and B and output F, Figure 65.19. A 'high' input (i.e. logic level 1) is obtained by connecting the input socket to the positive of the power supply, e.g. +5 V, and a 'low' one (i.e. logic level 0) to 0 V.

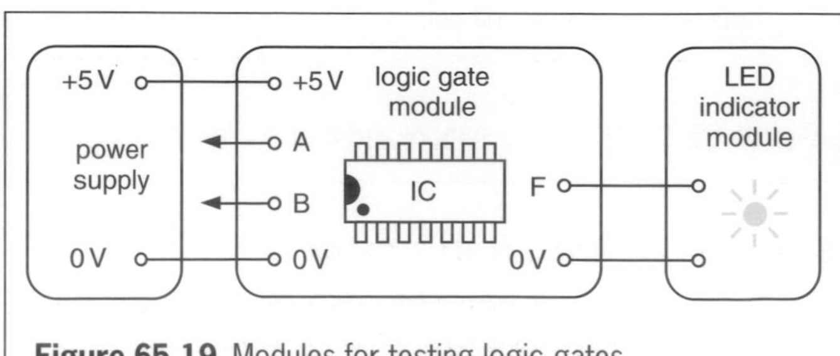

Figure 65.19 Modules for testing logic gates

The output can be detected using an indicator module containing an LED which lights up for a 1 and stays off for a 0.

Logic gate control systems

Logic gates can be used as processors in electronic control systems. Many of these can be demonstrated by connecting together commercial modules like those in Figure 65.21b.

(a) Security system

The block diagram for a simple system that might be used by a jeweller to protect an expensive clock is shown in Figure 65.20. The clock sits on a push switch which sends a 1 to the NOT gate unless the clock is lifted when a 0 is sent. In that case the output from the NOT gate is a 1 which rings the bell.

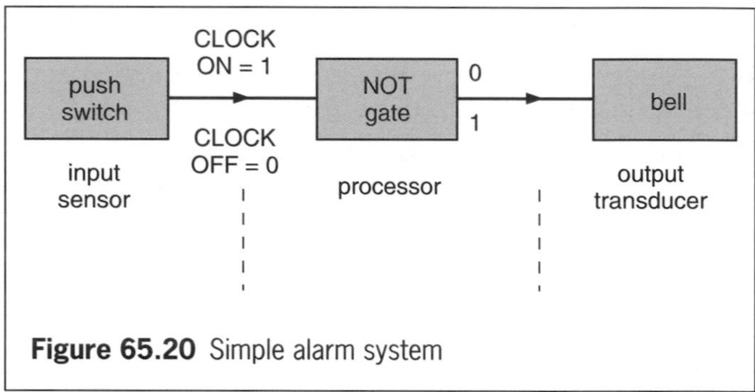

Figure 65.20 Simple alarm system

(b) Safety system for a machine operator

The system could prevent a machine (e.g. an electric motor) being switched on before another switch had been operated by a protective safety guard being in the correct position. In Figure 65.21a, when switches A *and* B are pressed they supply a 1 to each input of the AND gate which can then start the motor. Figure 65.21b shows the system built from modules with a transducer driver included to supply the large current required by the motor.

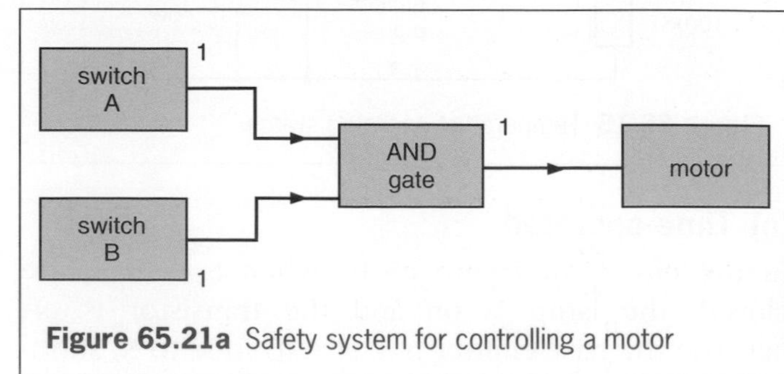

Figure 65.21a Safety system for controlling a motor

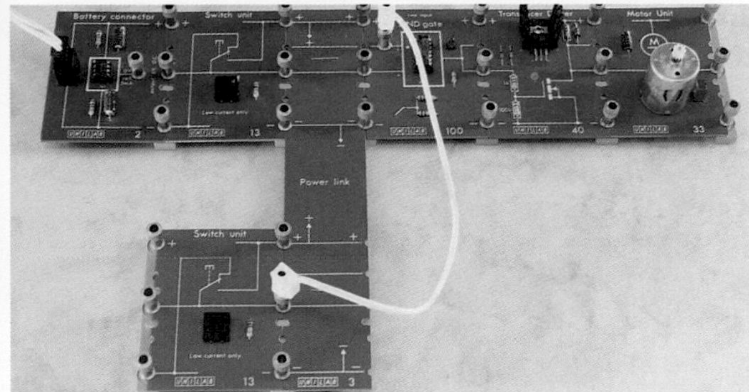

Figure 65.21b Modules for demonstrating the safety system

(c) Heater control system

The heater control has to switch on the heating system when it is

(i) **cold**, i.e. the temperature is below a certain value and the output from the temperature sensor is 0, and
(ii) **daylight**, i.e. the light sensor output is 1.

With these outputs from the sensors applied to the processor in Figure 65.22, the AND gate has two 1 inputs. The output from the AND gate is then 1 needed to turn on the heater control. Any other combination of sensor outputs produces a 0 output from the AND gate, as you can check.

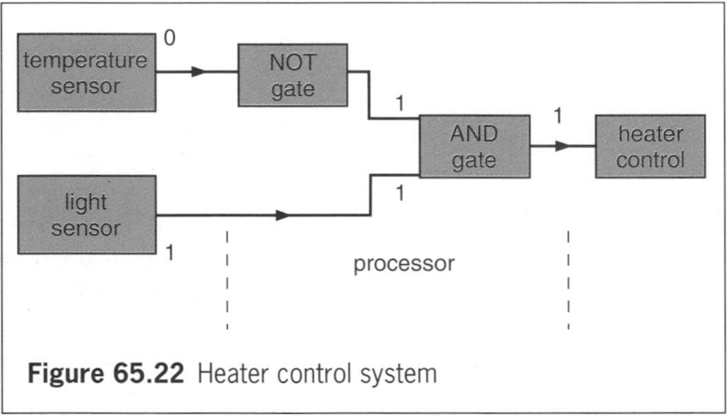

Figure 65.22 Heater control system

(d) Street lights

A system is required which allows the street lights either to be turned on manually by a switch at any time, or automatically by a light sensor when it is dark. The arrangement in Figure 65.23 achieves this since the OR gate gives a 1 output when either or both of its inputs are 1.

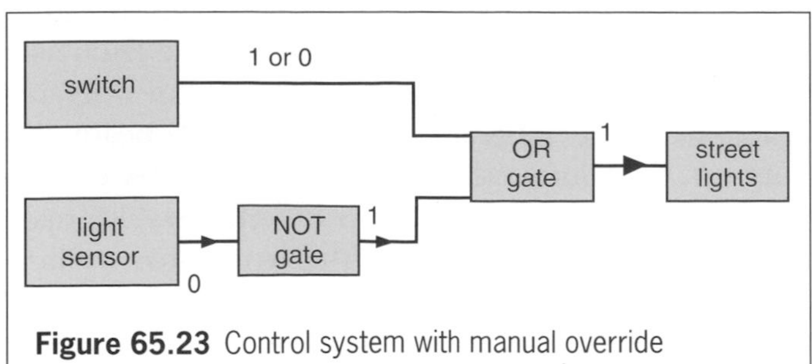

Figure 65.23 Control system with manual override

Problems to solve

Design and draw block diagrams for logic control systems to indicate how the following jobs could be done. If possible build them using modules.

1 Allow a doorbell to work only during the **day**.
2 Give warning when the temperature of a domestic hot water system is too **high** or when a switch is pressed to **test** the alarm.
3 Switch on a bathroom heater when it is **cold** and **light**.
4 Sound an alarm when it is **cold** or a switch is **pressed**.
5 Give warning if the temperature of a room **falls** during the **day** and also allow a **test** switch to check the alarm works.
6 Give warning of **frosty** conditions at **night** to a gardener who is sometimes very tired after a hard day and may want to switch the alarm off.

Feedback and control

In many kinds of system, electronic and other, all or part of the output (or information about it) is **fed back** to the input and affects the output.

Feedback is **positive** if it acts in the same direction as the input and increases the output. It is used to produce a.c. in electrical oscillators and in bistables (considered in the next section) which are the basic memory elements in computers. Too much positive feedback can cause instability in a system. This may happen in a public address system if the sound (output) from a loudspeaker is picked up by the microphone and fed back to the amplifier and then on to the speaker where it produces 'howls', Figure 64.24.

Feedback is **negative** if it acts in the opposite direction to the input and reduces the output. In an amplifier, though it decreases the gain, it makes for stability, better control and less distortion.

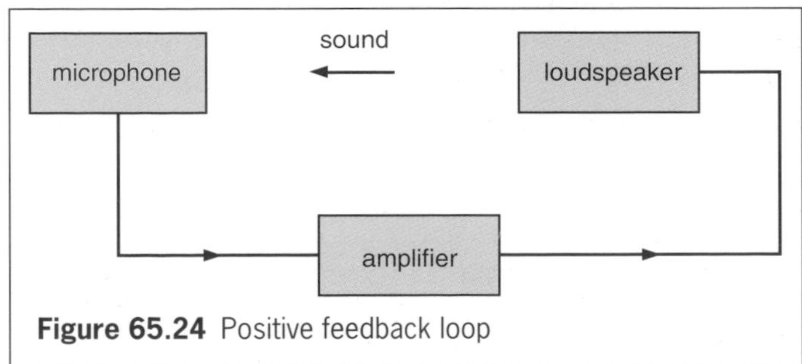

Figure 65.24 Positive feedback loop

(a) Control of body temperature

Feedback plays an important part in this process. If the body temperature rises above normal, the blood and temperature sensors in the skin feed back information to a part of the brain which responds in two ways. First it increases the rate of flow of blood to skin capillaries, so increasing the rate at which the blood is cooled. Second, sweating starts, producing cooling of the body by evaporation. If the temperature falls below normal, feedback reduces the blood flow to the skin and no sweating occurs, and shivering is activated.

(b) Control of a heater

A block diagram of a system to regulate the temperature of a room is shown in Figure 65.25. The required temperature is set by hand on the **thermostat**. If the setting is above room temperature, the thermostat switches on the **electrical supply** to the **heater**. The room warms up and when its temperature reaches the one selected, the thermostat responds to this information, which is fed back to it from the **air in the room**, by switching off the electrical supply.

If the room cools and its temperature falls below the setting, the thermostat again responds and switches on the heater. In this way, temperature control by feedback is achieved.

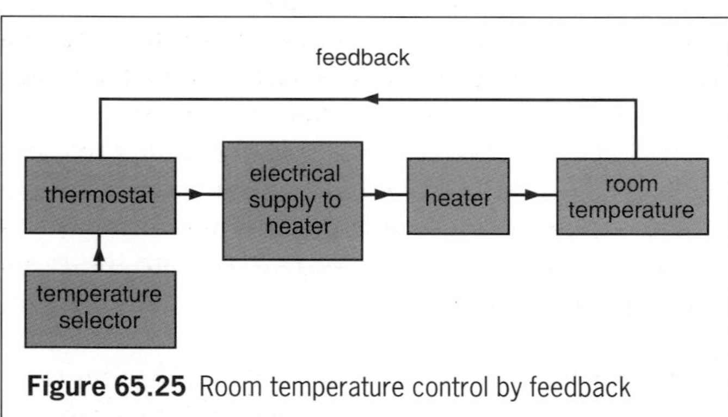

Figure 65.25 Room temperature control by feedback

(c) Control of a car exhaust

Many cars are now fitted with a feedback system to minimize and control exhaust gases. In Figure 65.26 the **oxygen sensor** in the exhaust pipe feeds back information about the emission gases to the **electronic control unit** (a microprocessor). This adjusts the air/fuel mixture to the **cylinder** to optimize the efficiency of the **catalytic converter** in changing harmful gases into carbon dioxide, nitrogen and water.

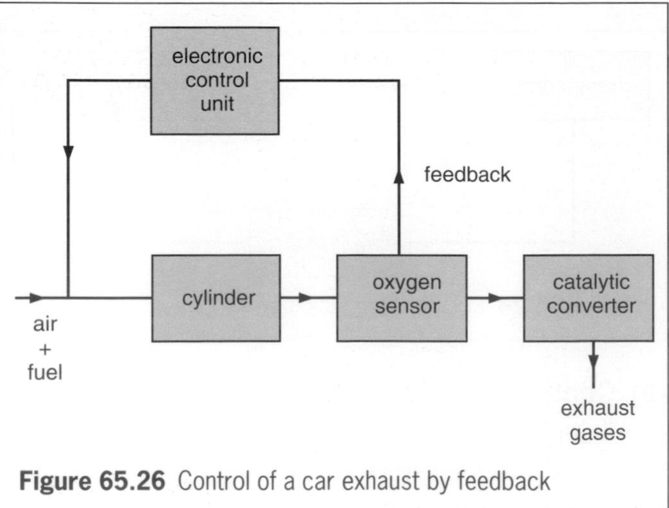

Figure 65.26 Control of a car exhaust by feedback

SR bistable

(a) Action

A **bistable** or **flip-flop** is a switching circuit with two stable states and a kind of built-in memory that forms the basis of memory systems in computers. It has two inputs, SET (S) and RESET (R) and one output (Q). In the SET stable state Q = 1, i.e. the output is 'on' or 'high', Figure 65.27a. In the RESET stable state Q = 0, i.e. the output is 'off' or 'low', Figure 65.27b. Normally the S and R inputs are kept 'high'.

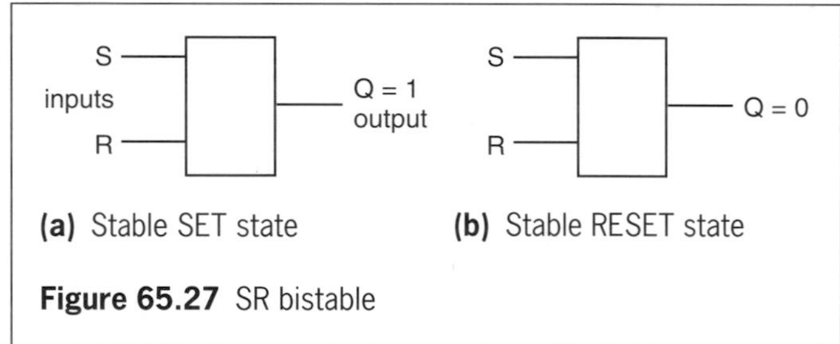

(a) Stable SET state **(b)** Stable RESET state

Figure 65.27 SR bistable

When the circuit is first switched on, it can settle in either state. Suppose it is in the RESET state with Q = 0. It switches to the SET state if the S input is changed (from 1 to 0). Q then stays at 1, even if S is changed back again. The circuit 'remembers' the first change to the S input and is now **latched** in the SET state *with further changes at the* S *input having no effect on* Q.

To switch the circuit to the RESET state, the R input has to be changed (from 1 to 0). Q then stays at 0 *no matter how often the R input is changed*. The circuit is now latched in the RESET state.

(b) Burglar alarm

The block diagram for the system is shown in Figure 65.28: it demonstrates the use of a bistable latch as a **processor** in a control system. The alarm operates when an intruder breaks a light (infrared) beam and cuts off the illumination to the **light sensor**. This changes the input to the **latch** (an SR bistable) making it go 'high' and bring on the **transistor switch** which operates the **alarm**. The latter is kept on by the latch, *even when the light beam is no longer interrupted*, until the latch is reset.

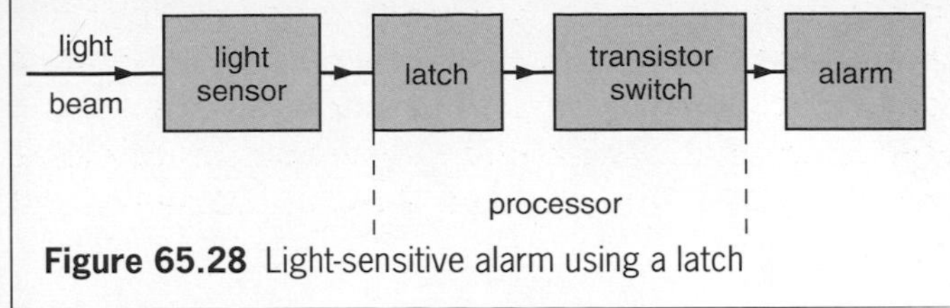

Figure 65.28 Light-sensitive alarm using a latch

(c) NAND gate bistable

Two 2-input NAND gates connected as in Figure 65.29a behave as an SR bistable. The output from each gate is **fed back** to one of the inputs (not S or R) of the other and this is the reason for the circuit having a 'memory'.

The sequence of events described in **(a)** can be checked, either using a ready-made bistable or by wiring up an IC, e.g. CMOS 4011B which contains four 2-input NAND gates, on a prototype board as in Figure 65.29b. The LED lights up when the output is 1 and the 'high' and 'low' inputs for S and R are obtained from the +9 V and 0 V battery supply lines respectively.

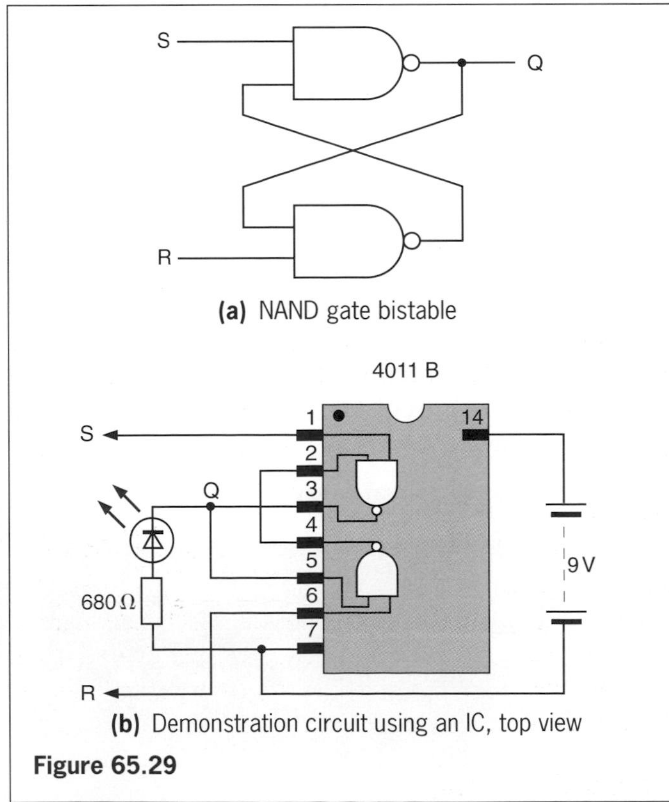

(a) NAND gate bistable

(b) Demonstration circuit using an IC, top view

Figure 65.29

Analogue and digital electronics

There are two main types of electronic circuits, devices or systems – analogue and digital.

In **analogue circuits**, voltages (and currents) can have any value within a certain range over which they can be varied smoothly and continuously, Figure 65.30a. They include amplifier-type circuits.

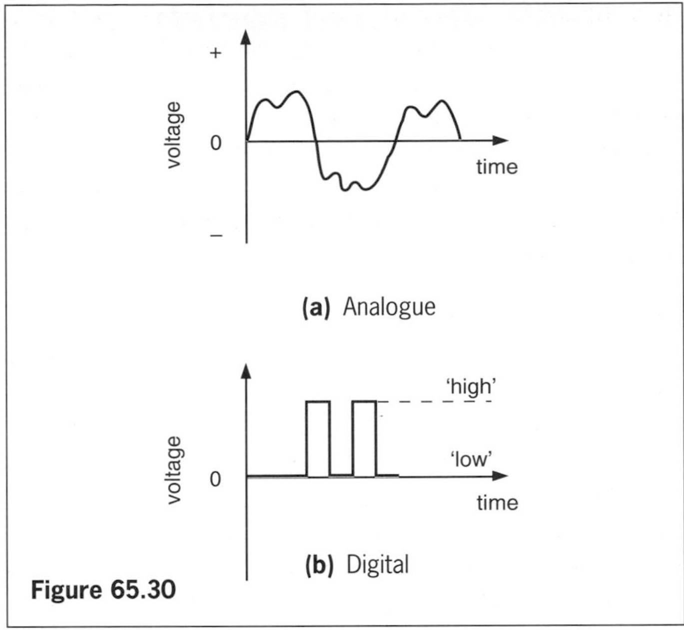

(a) Analogue

(b) Digital

Figure 65.30

In **digital circuits**, voltages have only one of two values, either 'high' (e.g. 5 V) or 'low' (e.g. near 0 V), Figure 65.30b. They include switching-type circuits and are what we have been considering.

A **variable resistor** is an analogue device which, in a circuit with a lamp, allows the lamp to have a wide range of light levels. A **switch** is a digital device which allows a lamp to be either 'on' or 'off'.

Analogue meters display their readings by the deflection of a pointer over a continuous scale, Figure 65.31a. **Digital meters** display their readings as digits, i.e. numbers, which change by one digit at a time, Figure 65.31b.

Figure 65.31a Analogue meter

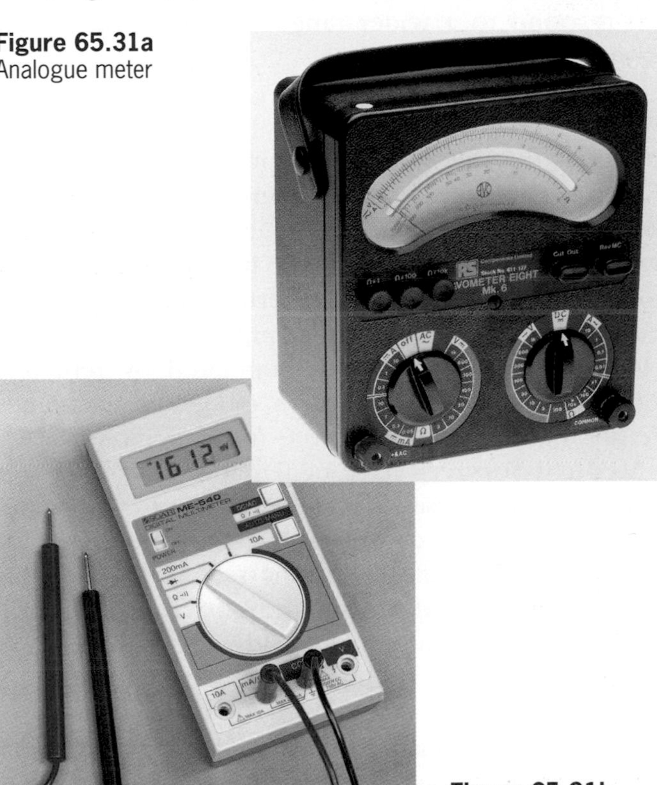

Figure 65.31b Digital meter

Electronics and society

Electronics is having an ever-increasing impact on all our lives. Work and leisure are changing as a result of the social, economic and environmental influences of new technology.

(a) Reasons for the impact

Why is electronics having such a great impact? Some of the reasons are listed below.

(i) **Mass production** of large quantities of semiconductor devices (e.g. ICs) allows them to be made very cheaply.
(ii) **Miniaturization** of components means that even complex systems can be quite compact.
(iii) **Reliability** of electronic components is a feature of well-designed circuits. There are no moving parts to wear out and systems can be robust.
(iv) **Energy consumption** and use of natural resources is often much less than for their non-electronic counterparts, for example the transistor uses less power than a relay.
(v) **Speed of operation** can be millions of times greater than for other alternatives (e.g. mechanical devices).
(vi) **Transducers** of many different types are available for transferring information in and out of an electronic system.

To sum up, electronic systems tend to be cheaper, smaller, more reliable, less wasteful, much faster and can respond to a wider range of signals than other systems.

(b) Some areas of impact

At home devices such as washing machines, burglar alarms, telephones, cookers and sewing machines now contain electronic components. Central heating systems and garage doors may have automatic electronic control. For home entertainment, video cassette recorders (VCRs), compact disc players, television sets with Teletext operated by remotely controlled keypads, computers and electronic games are finding their way into more and more homes.

Medical services have benefited greatly in recent years from the use of electronic instruments and appliances. Electrocardiograph (ECG) recorders for monitoring the heart, ultrasonic scanners for checks during pregnancy, deaf aids, heart pacemakers, artificial kidneys, limbs and hands with electronic control, Figure 65.32a, and talking newspapers for the blind are some examples.

In industry microprocessor-controlled equipment is taking over. Robots are widely used for car assembly work, and to do dull, routine, dirty jobs such as welding and paint spraying. In many cases production lines and even whole factories, e.g. sugar refineries and oil refineries, are almost entirely automated. Computer-aided design (CAD) of products is increasing, Figure 65.32b, even in the clothing industry.

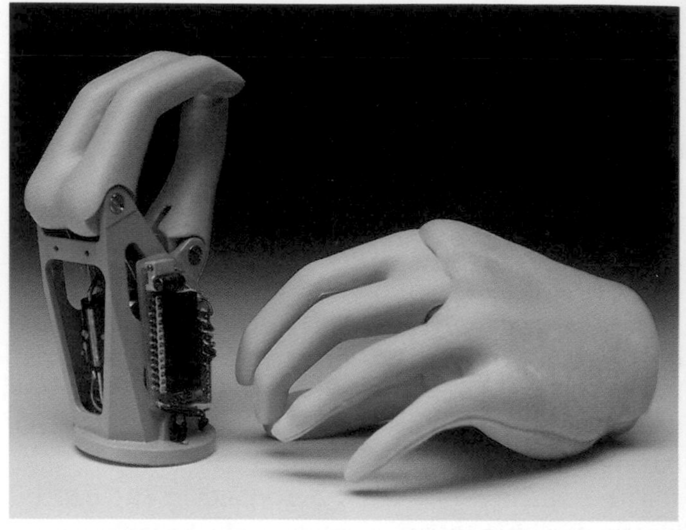

Figure 65.32a Electronically controlled artificial hands

In offices, banks and shops word processors are efficient and more versatile replacements for typewriters. Mail in the form of text, numbers and pictures can be transmitted by electronic means. Cash dispensers at banks and building society offices are a great convenience for their customers. Bar codes (like the one on the back cover of this book) on tins and packets are used by supermarkets for stock control in conjunction with a bar code reader (which uses a laser) and a data recorder, possibly connected to a computer. A similar system is operated by some libraries to record the issue and return of books. Shop cash registers are almost universally electronic today.

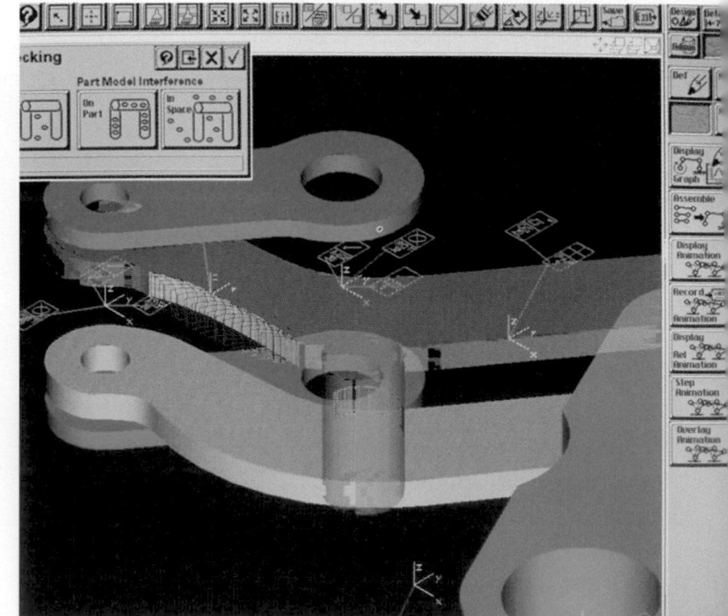

Figure 65.32b Computer-aided design of car components

Communications have been transformed. Satellites enable events on one side of the world to be seen and heard on the other side, as they happen. Electronic telephone exchanges like System X are the order of the day. Car telephone systems, called cellular radio, permit people in cars, trains and ferries to make calls on the existing telephone network, computers being used to switch them when they leave one area (cell) and enter another.

Leisure activities have been affected by electronic developments. For some people, leisure means participating in or attending sporting activities and here the electronic scoreboard may be much in evidence. For the golf enthusiast, electronic machines claim to analyse 'swings' and reduce handicaps. For others, leisure means listening to music, whose production, recording and listening facilities have been transformed. Electronically synthesized music has become the norm for popular recordings. The lighting and sound effects in modern musical shows are programmed by computer. The availability of home computers in recent years has enabled a huge market in computer games to develop.

(c) Consequences of the impact

Most of the social and economic consequences of electronics are beneficial but a few cause problems.

An improved quality of life has resulted from the greater convenience and reliability of electronic systems, increased life expectancy and leisure time, and fewer dull, repetitive jobs.

Better communication has made the world a smaller place. The speed with which news can be reported to our homes by radio and television, and the influence of information systems such as Teletext, enable the public to be better informed.

Databases have been developed. These are memories which can store huge amounts of information for rapid transmission from one place to another. For example, the police can obtain in seconds, by radio, details of a car they are following. Databases raise questions, however, about invasion of privacy and security.

Employment is affected by the demand for new equipment - new industry and jobs are created to make and maintain it - but when electronic systems replace mechanical ones, redundancy and/or retraining needs arise. Conditions of employment and long-term job prospects can also be affected for many people, especially certain manual and clerical workers, e.g. typists. One industrial robot replaces four factory workers.

The public attitude to the electronics revolution is not always positive. Modern electronics is a 'hidden' technology with parts that are enclosed in a tiny package (or 'black box') and do not move. It is also a 'throwaway' technology in which the whole lot is discarded and replaced - by an expert - if a part fails, and rapid advances in design technology cause equipment to quickly become obsolete. For these reasons it may be regarded as mysterious and unfriendly - people feel they do not understand what makes it tick.

(d) The future

The only certain prediction about the future is that new technologies will be developed and these, like present ones, will continue to have a considerable influence on our lives.

Today the development of 'intelligent' Fifth Generation computers is being pursued with great vigour, and optical systems which are more efficient than electronic ones for transferring (by optical fibres), storing (by holographic memories) and processing information are starting to appear.

QUESTIONS

1 Figure 65.33a shows a lamp, a semiconductor diode and a cell connected in series. The lamp lights when the diode is connected in this direction. Say what happens to each of the lamps in (b), (c) and (d). Give reasons for your answer. *(S.E.)*

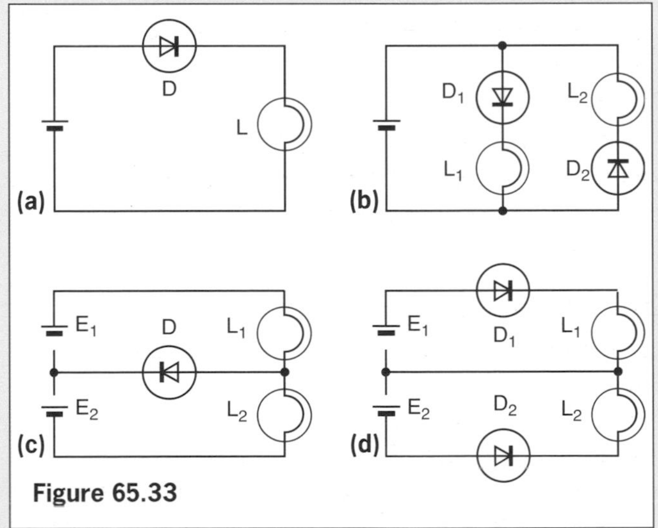

Figure 65.33

2 What are the readings V_1 and V_2 on the high-resistance voltmeters in the potential divider circuit of Figure 65.34 if **a** $R_1 = R_2 = 10\ \text{k}\Omega$, **b** $R_1 = 10\ \text{k}\Omega$, $R_2 = 50\ \text{k}\Omega$, **c** $R_1 = 20\ \text{k}\Omega$, $R_2 = 10\ \text{k}\Omega$?

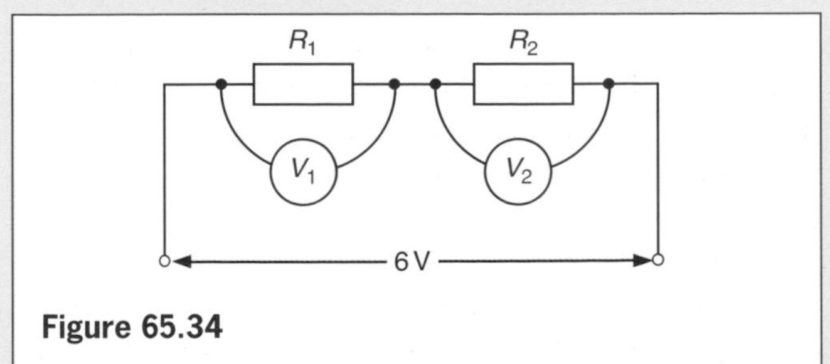

Figure 65.34

3 Figure 65.35 shows an incomplete electronic circuit.

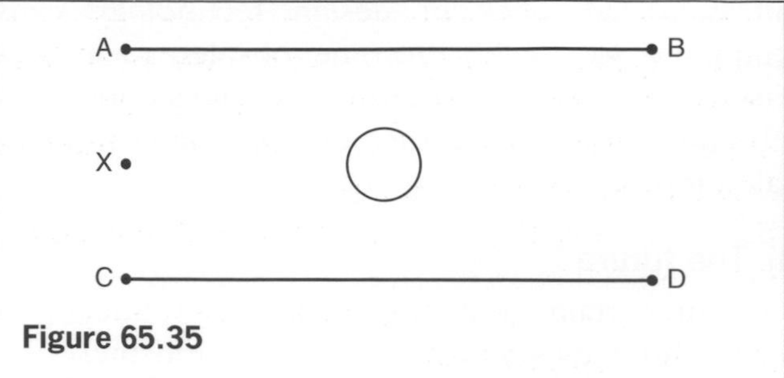

Figure 65.35

a Copy the diagram and make the following additions.
(i) In the circle draw the symbol for a transistor.
(ii) Add the symbol for a battery, correctly connected between B and D.
(iii) Draw the symbol for a thermistor between points A and X.
(iv) Draw the symbol for a variable resistor between points X and C.
(v) Draw the symbol for a lamp in the collector circuit.
(vi) Connect the base and emitter of the transistor to the correct places in the circuit.

b As the circuit is arranged the lamp will not light. If there is a rise in temperature the lamp will light. Explain this.

c A simple rearrangement of this circuit would make the lamp light when the temperature falls. Explain what changes should be made, and how they will affect the working of the circuit.

d In the original circuit, instead of a variable resistor, a fixed resistor could have been used between points X and C. Explain why it might have been considered desirable to use a variable resistor for these two circuits. *(J.M.B./A.L./N.W.16+)*

4 A simple moisture-warning circuit is shown in Figure 65.36 in which the moisture detector is two closely spaced copper rods.

a Describe how the circuit works when the detector gets wet.

b Warning lamps are often placed in the collector circuit of a transistor. Why is a relay used here?

c What is the function of D?

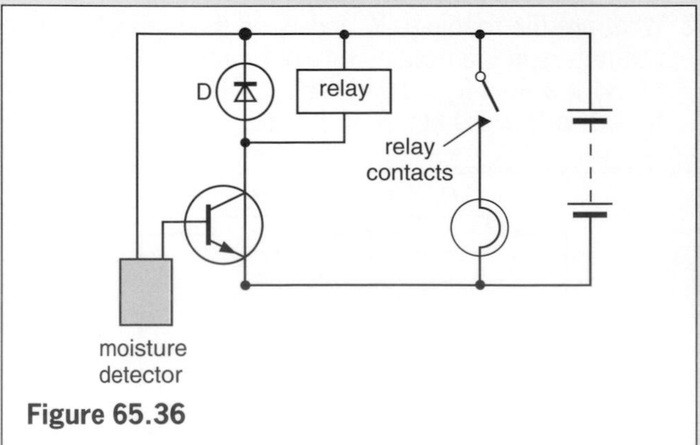

Figure 65.36

5 The combined truth tables for four logic gates A, B, C, D are given below. State what kind of gate each one is.

Inputs		Outputs			
		A	B	C	D
0	0	0	0	1	1
0	1	0	1	1	0
1	0	0	1	1	0
1	1	1	1	0	0

6 What do the symbols represent in Figure 65.37a, b, c, d, e?

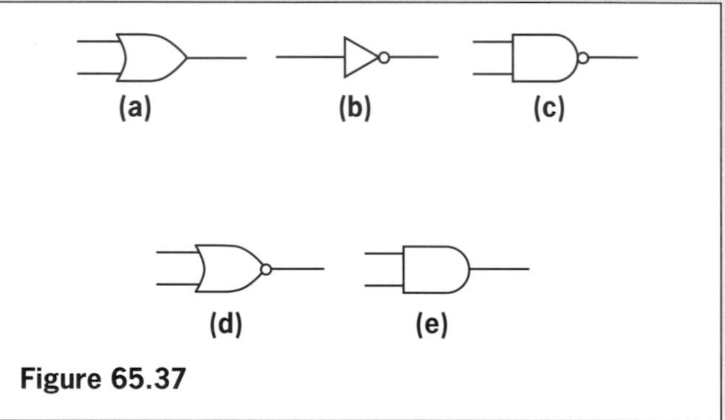

Figure 65.37

7 Combinations of switches can act as logic gates. Which type are those in Figures 65.38a, b? Treat A and B as the inputs and Q as the output.

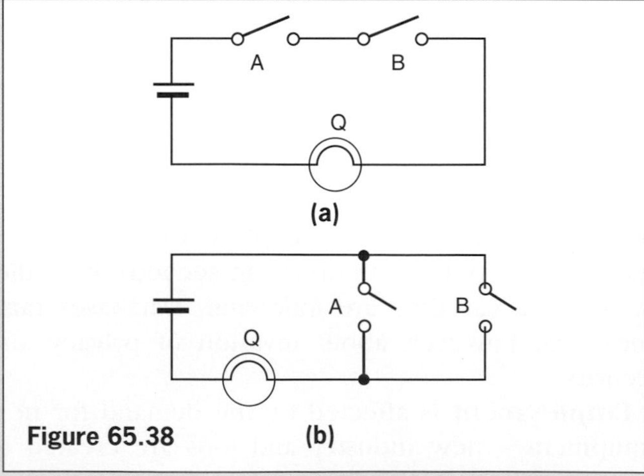

Figure 65.38

8 Design and draw the block diagrams for logic control systems to:

a wake you at the crack of dawn and which you can also switch off,

b protect the contents of a drawer which you can still open without setting off the alarm.

9 The logic levels are shown in Figure 65.39 for an SR bistable.

a Which state is it in?

b What happens to Q if the S input is changed to 0?

c What happens to Q if the S input is returned to 1?

d What must be done to return Q to its original state?

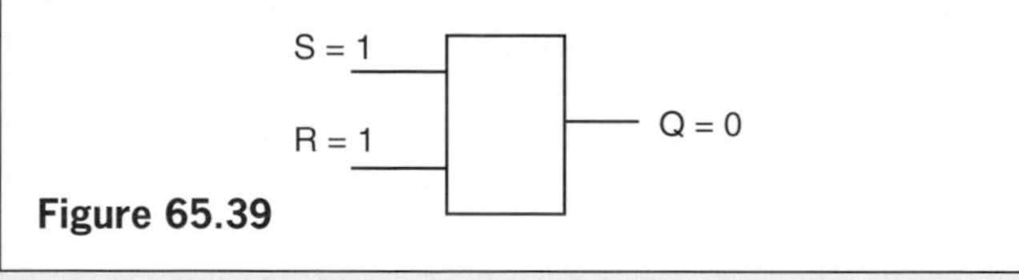

Figure 65.39

Checklist

After studying this chapter you should be able to

- explain the term **doping** of semiconductors,
- explain what is meant by a diode being **forward biased** and **reverse biased**,
- describe the action of a transistor with the aid of a circuit diagram,
- describe the action of an **LDR**, a **thermistor**, an **LED** and a **photodiode,**
- describe how a transistor can be used as a switch,
- recall the functions of the **input sensor**, **processor** and **output transducer** in an electronic system and give some examples,
- explain the operation of light-, temperature- and time-operated transistor switching circuits with the aid of diagrams,
- describe the action of **NOT**, **OR**, **NOR**, **AND** and **NAND** logic gates and recall their truth tables,
- design and draw block diagrams of logic control systems for given requirements,
- explain with the aid of a sketch the term **feedback** in electronic systems and how it can be positive or negative, and give examples of the use of feedback,
- describe the action of an **SR bistable** and explain how it can be used as a **latch**,
- distinguish between **analogue** and **digital** circuits and devices,
- show an appreciation of the impact of electronics on society.

66 TELECOMMUNICATIONS

Telecommunications is concerned with sending and receiving information over a distance. In the broadest sense, information can be in the form of written or spoken words, numbers, diagrams, pictures, music or computer data.

Early history

The earliest line-of-sight methods involved sending smoke signals from hill top to hill top. It was the forerunner of flag signalling by semaphore and ship-to-ship signalling by Aldis lamp.

The mid-19th century saw the invention of the electric telegraph by **Wheatstone**, Figure 66.1a, in which messages were sent in Morse code as currents in cables. Shortly afterwards **Bell** developed the telephone, Figure 66.1b, making it possible to transmit speech electrically.

In 1895, after **Hertz** had shown how to produce the electromagnetic waves predicted mathematically by **Maxwell**, **Marconi** succeeded in transmitting them over a mile. In 1901 he established that wireless telegraphy signals could be sent and received across the Atlantic; shortly afterwards radio telephony became possible. Picture transmission, i.e. television, came in the 1930s after much of the early development work had been done by **Baird**.

Representing information

Information can be represented electrically in two ways.

(a) Digital method

In this method electricity is switched on and off and the information is in the form of electrical pulses. For example, in the simple circuit of Figure 66.2a, data can be sent by the 'dots' and 'dashes' of Morse code by closing the switch for a short or a longer time. In Figure 66.2b the letter A (· –) is shown.

Computers use the simpler **binary code** with 1 and 0 represented by 'high' and 'low' voltages respectively. They can only handle numbers (i.e. 1 and 0)

Figure 66.1a Telegraph transmitting key c.1890 (left)
Figure 66.1b Early telephone c.1890 (centre)
Figure 66.1c Radio set 1946 (right)

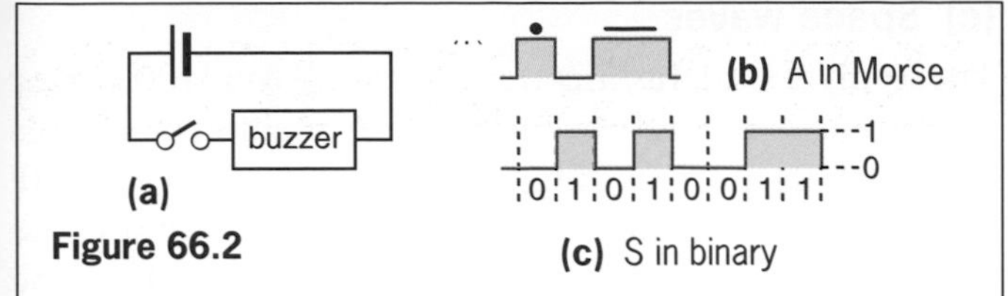

Figure 66.2

and so there is a pattern of 1s and 0s for each of the 26 letters of the alphabet and for other symbols etc. Figure 66.2c represents capital S in this code.

(b) Analogue method

In this case the information is changed to a voltage or current that varies continuously and smoothly over a range of values. The waveform of the voltage or current, i.e. its variation with time, represents the information and is an analogue of it, Figure 66.3. For example, the loudness and frequency (pitch) of a sound determines the amplitude and frequency of the waveform produced by a microphone on which the sound falls.

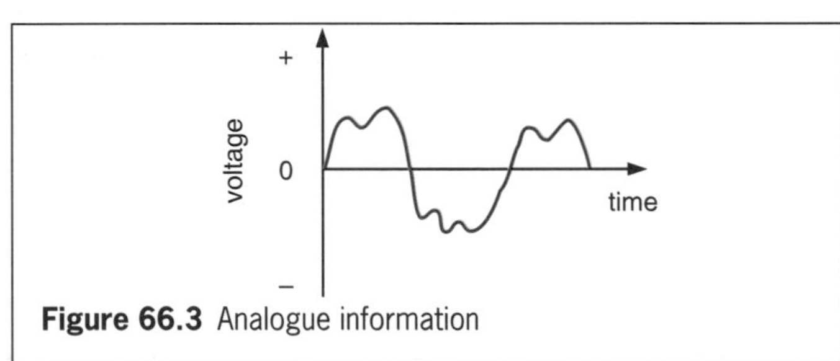

Figure 66.3 Analogue information

Basic communication system

Electrical signals representing 'information' from a microphone, a television camera, a computer, etc., can be sent from place to place using either cables or radio waves. While some signals can be sent directly through cables, in general, and certainly in radio and television, as we will see later, a 'carrier' wave is required to transport them.

The basic building blocks of any communication system are shown in Figure 66.4. Signals from the **input transducer** are added to the **carrier** in the **modulator** or **encoder** by the process of **modulation**. 'Modulation' is the process of impressing one wave system (the signal) upon another, higher frequency, system (the carrier). The modulated signal is then sent by the **transmitter** into the **propagating medium** (i.e. cable or radio wave).

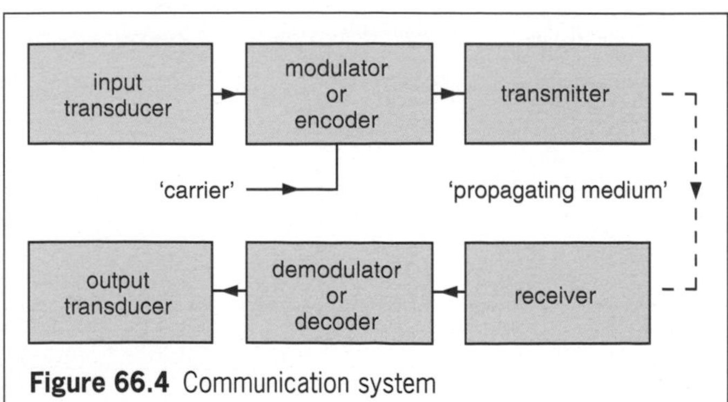

Figure 66.4 Communication system

At the receiving end, the **receiver** may have to select and perhaps **amplify** the modulated signal before the **demodulator** or **decoder** extracts from it the information signal for delivery to the **output transducer**. Some form of **storage** by a tape or disc may also be required.

Basically the transmitting and receiving ends each consist of an electronic system comprising an input transducer, a processor consisting of several subsystems, and an output transducer.

Radio waves

Radio waves are members of the family of electromagnetic radiation considered in Chapter 14. They are energy carriers which travel at the speed of light c, their frequency f and wavelength λ being related, as for any wave motion, by the equation

$$v = f\lambda$$

where $v = c = 3.0 \times 10^8$ m/s in a vacuum (or air).

Radio waves can be classified either by their frequency or their wavelength; if one is large the other is small. Frequency is more fundamental since, unlike λ (and v), f does not change when the waves travel from one medium to another. Table 66.1 shows how they are grouped into different frequency bands.

Table 66.1

Frequency band	Typical wavelength	Some uses
Low (LF) 30 kHz–300 kHz	Long 1500 m	Long-wave radio and communication over large distances
Medium (MF) 300 kHz–3 MHz	Medium 300 m	Medium-wave local and distant radio
High (HF) 3 MHz–30 MHz	Short 30 m	Short-wave radio and communication, amateur and CB radio
Very high (VHF) 30 MHz–300 MHz	Very short 3 m	FM radio, police, emergency services
Ultra high (UHF) 300 MHz–3 GHz	Ultra short 30 cm	TV (bands 4, 5)
Super high (SHF) above 3 GHz	Microwaves 3 cm	Radar, communication satellites, telephone and TV links

(1 GHz = 1000 MHz)

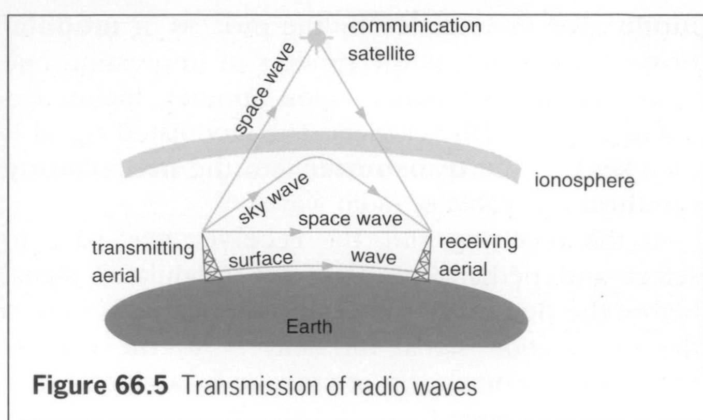

Figure 66.5 Transmission of radio waves

Radio waves can travel from a transmitting aerial in one or more of three different ways, Figure 66.5.

(a) Surface or ground waves

These follow the Earth's surface and have a limited range, being greatest (1500 km) for long waves but much less for VHF.

(b) Sky waves

These travel skywards and if they are below a certain **critical frequency** (typically 30 MHz) are returned to Earth by bouncing off the ionosphere. This consists of layers of air molecules (the D, E and F layers), stretching from about 80 km above the Earth to 500 km, which have become positively charged (i.e. ionized) by the removal of electrons due to the Sun's ultraviolet radiation. The critical frequency varies with the time of day and the seasons.

Sky waves of low, medium and high frequencies can travel thousands of kilometres.

(c) Space waves

These give straight-line transmission over 100 km or so on Earth in the absence of intervening obstacles such as hills and buildings, and are the way VHF, UHF and microwaves travel. They can also penetrate the ionosphere so satellite communication uses this type of transmission.

Radio system

Radio waves are emitted by aerials when a.c. flows in them but the length of the aerial must be comparable with the wavelength of the wave produced for the radiation to be appreciable. A 50 Hz a.c. corresponds to a wavelength of 6×10^6 m (since $v = f\lambda$ where $v = 3 \times 10^8$ m/s and $f = 50$ Hz).

Alternating currents with frequencies below about 20 kHz are called **audio frequency** (a.f.) currents; those with frequencies greater than this are **radio frequency** (r.f.) currents. Therefore, so that aerials are not too large, they are supplied with r.f. currents. However, speech and music generate a.f. currents and so some way of combining a.f. with r.f. is required if they are to be sent over a distance.

(a) Transmitter

In this an **oscillator** produces an r.f. current which would cause an aerial connected to it to send out an electromagnetic wave, called a **carrier wave**, of constant amplitude and having the same frequency as the r.f. current. If a normal receiver picked up such a signal nothing would be heard. The r.f. signal has to be modified or **modulated** so that it 'carries' the a.f. This is done in various ways.

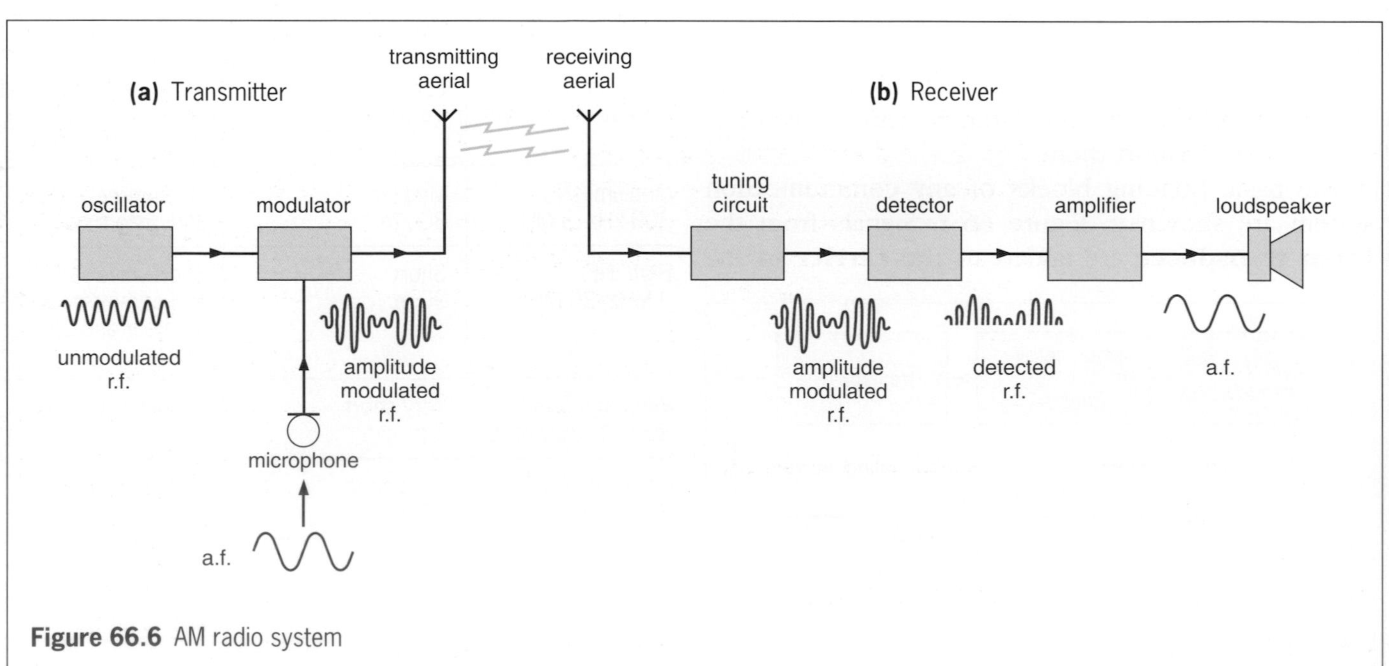

Figure 66.6 AM radio system

In **amplitude modulation** (AM) the amplitude of the r.f. is varied so that it depends on the a.f. current from the microphone, the process occurring in a **modulator**. This type of transmitter is used for medium and long-wave broadcasting in Britain: Figure 66.6a is a block diagram for one.

In **frequency modulation** (FM), which is used in VHF broadcasts, the frequency of the carrier is altered at a rate equal to the frequency of the a.f. but the amplitude remains constant, Figure 66.7. This kind of signal is fairly free from electrical interference.

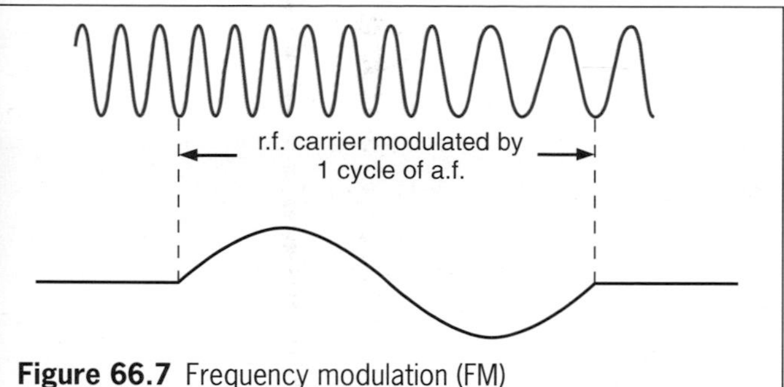

Figure 66.7 Frequency modulation (FM)

(b) Receiver

A block diagram of a simple AM receiver is shown in Figure 66.6b. The **tuning circuit** selects the wanted signal from the **aerial**. The **detector** (or demodulator) separates the a.f. (speech or music) from the r.f. carrier. The **amplifier** then boosts the a.f. which produces sound in the loudspeaker.

Electrical oscillations

The production of r.f. currents is impossible with mechanical generators but is readily achieved electrically.

(a) Oscillatory circuit

This contains a capacitor C and a coil L, Figure 66.8a. If C is charged and then discharges through L, current flows backwards and forwards round the circuit, alternately discharging and charging C, first one way then the other. The frequency of the a.c. produced depends on C and L; the smaller they are the higher is the frequency and each circuit has a **natural frequency** of oscillation. Due to the resistance of the coil, the current eventually stops and a decaying or damped oscillation is obtained, Figure 66.8b.

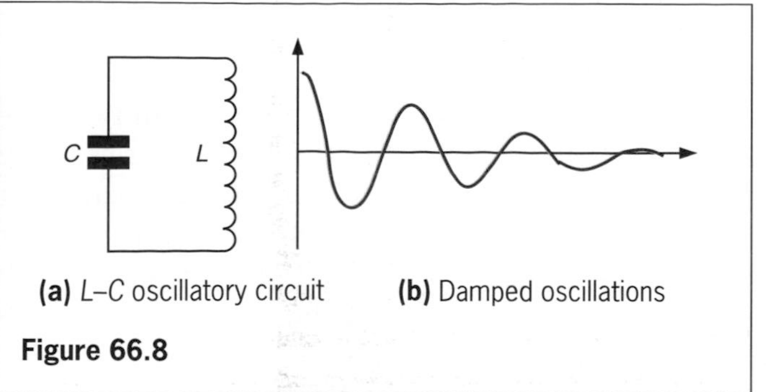

(a) *L–C* oscillatory circuit **(b)** Damped oscillations

Figure 66.8

Slow damped oscillations (about 2 Hz) can be shown using the circuit of Figure 66.9 with the CRO on its slowest time base speed.

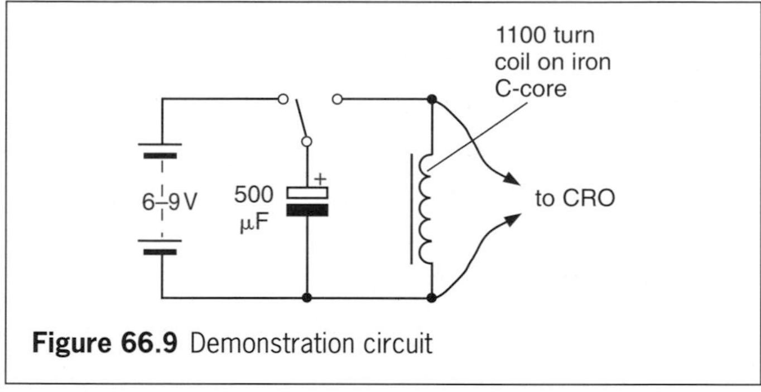

Figure 66.9 Demonstration circuit

(b) Transistor oscillator

To obtain oscillations that do not die away, energy has to be fed into the *L-C* circuit at the right time. This can be done using a transistor as in Figure 66.10. The coil L_1, by being near L, feeds the energy needed into the input (base-emitter) circuit of the transmitter.

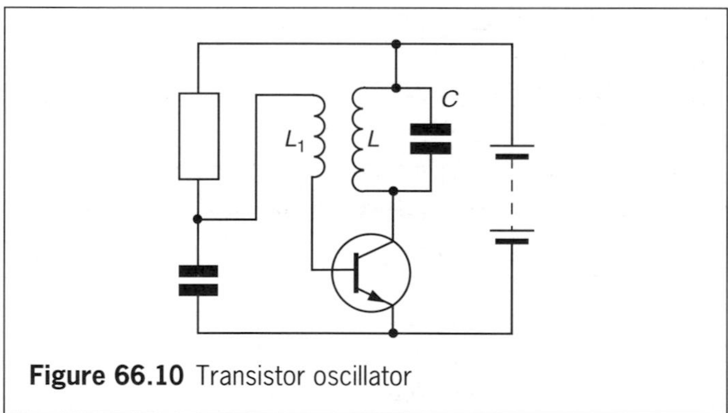

Figure 66.10 Transistor oscillator

Tuning circuit

Different transmitting stations send out radio waves of different frequencies. Each induces a signal in the aerial of a radio receiver but if the aerial is connected to an *L-C* circuit, Figure 66.11, only the signal whose frequency equals the natural frequency of the *L-C*

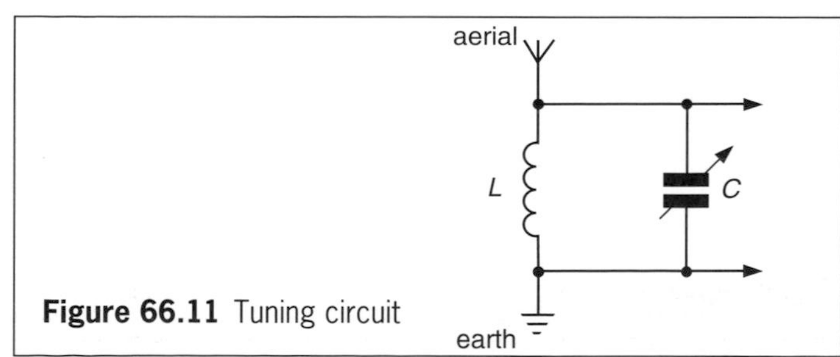

Figure 66.11 Tuning circuit

circuit is selected. If the value of C is varied different stations can be 'tuned in'.

This is an example of **electrical resonance** (Chapter 16).

Detection or demodulation

In **detection** or **demodulation** the a.f. which was 'added' to the r.f. in the transmitter is recovered in the receiver.

Suppose the amplitude modulated r.f. signal V of Figure 66.12a is applied to the detector circuit of Figure 66.12d. The diode produces rectified pulses of r.f. current I, Figure 66.12b. These charge up C during the positive half-cycles as well as flowing through the earphone. During the negative half-cycles when the diode is non-conducting, C *partly* discharges through the earphone.

The p.d. V_c across C (and the earphone) varies as in Figure 66.12c if C and R have the correct values. Apart from the slight r.f. ripple V_c has the same frequency and shape as the modulating a.f. and reproduces the original sound in the earphone.

Germanium diodes are used as detectors. If a loudspeaker is to be operated, the detector is followed by several stages of a.f. amplification. A suitable resistor then replaces the earphone.

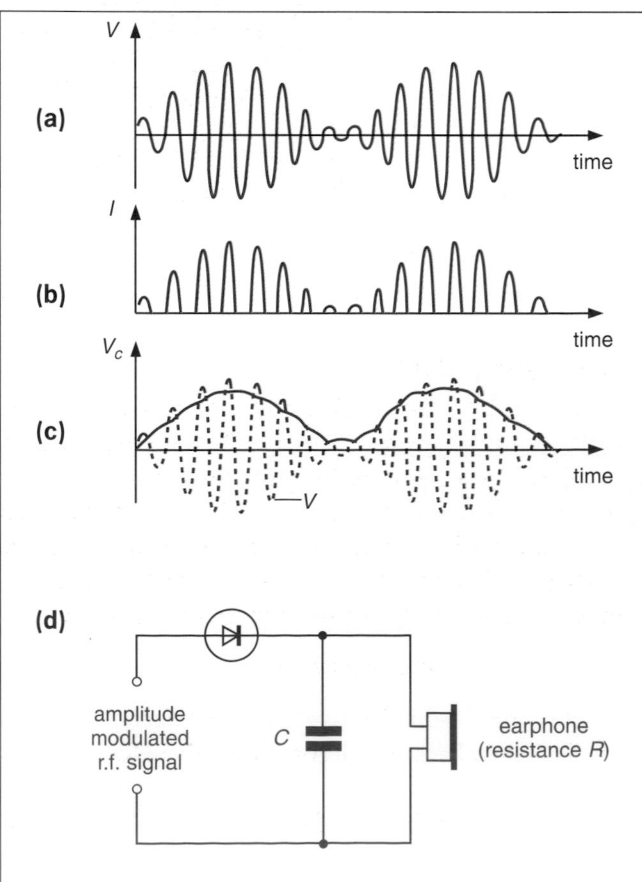

Figure 66.12 Detection of an AM signal

PRACTICAL WORK

Simple radio receiver

Connect the circuit of Figure 66.13 (opposite) on an S-DeC, Figure 66.14a. The transistor leads can be lengthened and connections made to the 'tags' on the variable capacitor as in Figure 65.13b, p. 268. The resistor colour code is given on p. 288.

For an aerial, support a length of wire (e.g. 10 m) as high as you can; making an earth connection to a water tap will improve reception, Figure 66.14b.

You should be able to tune in to one or two stations (depending on your location) by altering the variable capacitor.

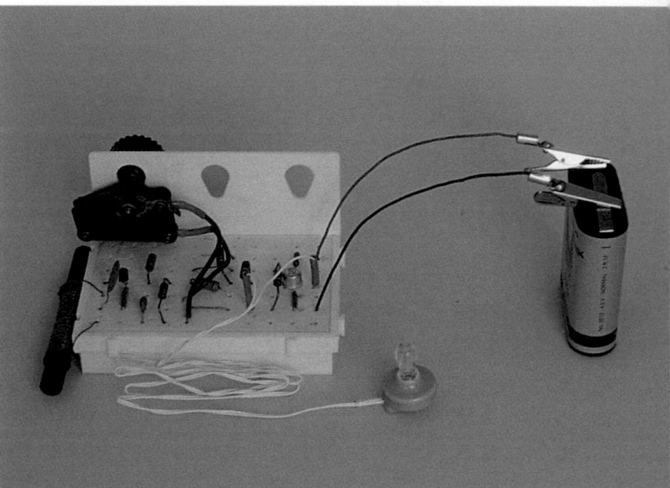

Figure 66.14a Radio receiving circuit built on an S-DeC

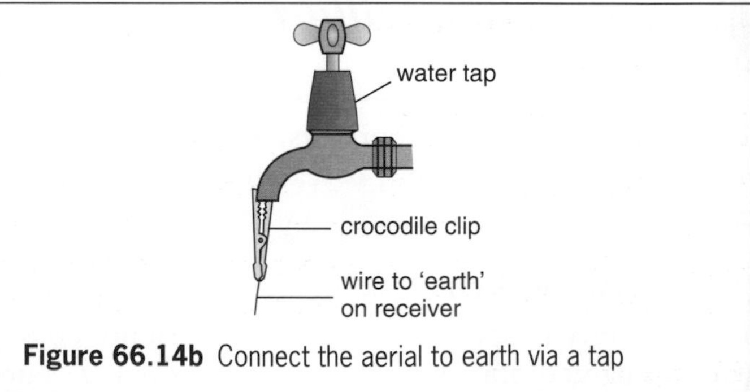

Figure 66.14b Connect the aerial to earth via a tap

Television

(a) Black and white

A television receiver is basically a CRO with two time bases. The horizontal or **line time base** acts as in the CRO. The vertical or **frame time base** operates at the same time and draws the spot at a much slower rate down to the bottom of the screen and then returns it almost at once to the top. The spot thus 'draws' a series of parallel lines of light (625) which covers the screen, Figure 66.15, and is called a **raster**.

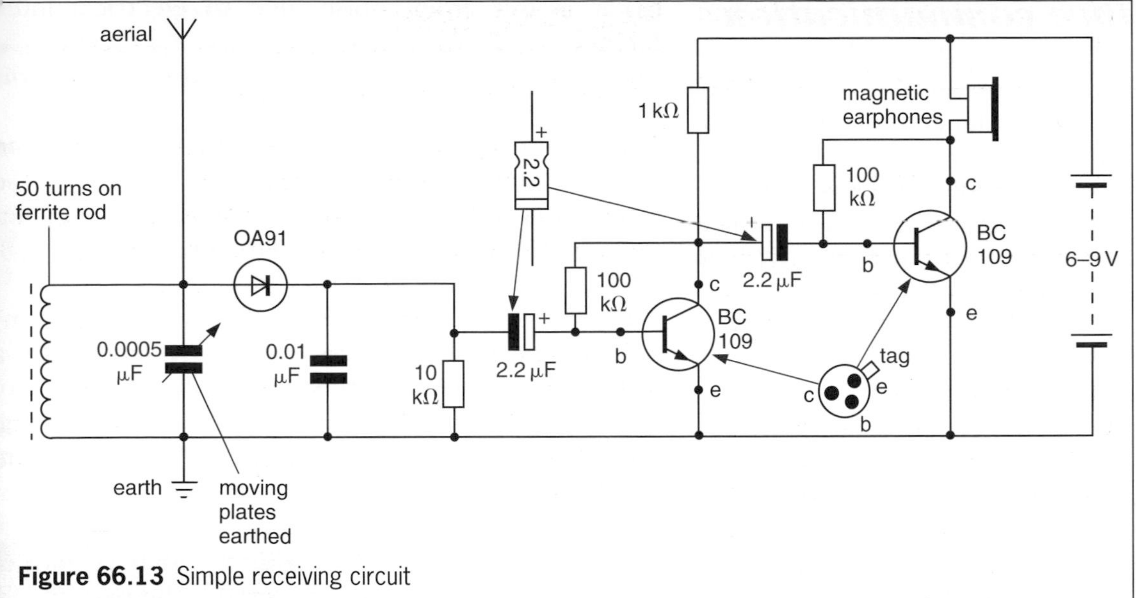

Figure 66.13 Simple receiving circuit

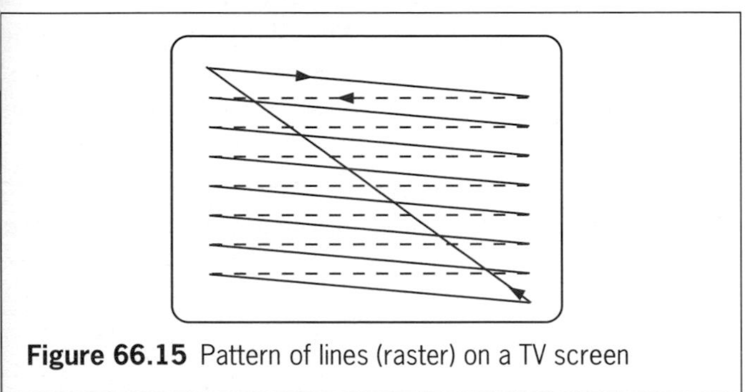
Figure 66.15 Pattern of lines (raster) on a TV screen

A picture is produced by the incoming signal altering the number of electrons which travel from the electron gun to the screen. The greater the number the brighter the spot. The brightness of the spot varies from white through grey to black as it sweeps across the screen. A complete picture appears every 1/25 s, but because of the persistence of vision we see the picture as continuous. If each picture is just slightly different from its predecessor, the resultant effect is that of a 'movie' and not a sequence of 'stills'.

(b) Colour

One type of colour television has three electron guns and the screen is coated with about a million tiny light-emitting 'dots' arranged in triangles. One 'dot' in each triangle emits red light when hit by electrons, another green light and the third blue light.

As the three electron beams scan the screen, an accurately placed 'shadow mask' consisting of a perforated metal plate with about one-third of a million holes ensures that each beam strikes only dots of one 'colour', e.g. electrons from the 'red' gun strike only 'red' dots, Figure 66.16.

When a triangle of dots is struck it may be that the red and green electron beams are intense but not the blue. The triangle will emit red and green light strongly and appear yellowish (Chapter 9). The triangles of dots are struck in turn, and since the dots are so small and the scanning so fast, we see a continuous colour picture.

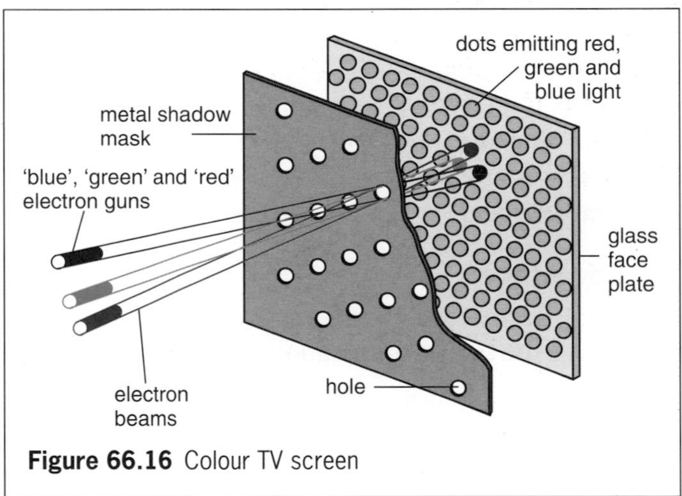

Figure 66.16 Colour TV screen

(c) Energy transfers

Electrons emitted by an electron gun in the cathode ray tube of a television receiver or a CRO are accelerated towards the screen by the high p.d. which is applied across the tube. If this is V and an electron has charge e and mass m, then an amount of electrical energy eV (from $W = QV$, Chapter 50) is transferred to kinetic energy of the electron. Therefore

$$\text{k.e.} = \tfrac{1}{2}mv^2 = eV$$

where v is the final velocity of the electron. This k.e. is transferred to heat and light when the electron hits the screen.

We can calculate the k.e. if we take $e = 1.6 \times 10^{-19}$ C and $V = 25$ kV then

$$\text{k.e.} = eV = 1.6 \times 10^{-19}\ \text{C} \times 25 \times 10^{3}\ \text{V}$$
$$= 4.0 \times 10^{-15}\ \text{J}$$

Optical fibre communication systems

For both local and long-distance communication, copper telephone cables carrying electric currents are being replaced very rapidly by optical fibres. These are very thin fibres of very pure glass in which 'light' is trapped by total internal reflection (Chapter 6) and used to carry information.

(a) Outline of system

An optical cable (showing fibres inside) is shown in Figure 66.17a and a simplified block diagram of a communication system in Figure 66.17b. Electrical signals from the **input transducer** (representing the information to be transmitted) are modulated by a process called **pulse code modulation** in a **coder**. This produces a stream of equivalent digital electrical pulses which are changed by the **optical transmitter** into 'light' pulses for transmission by the optical fibre. The transmitter is either an LED or a miniature laser. The 'light' used is infrared radiation because it suffers less absorption in the glass.

At the receiving end the **optical receiver** is a **photodiode** (a diode that conducts when light falls on it). This converts the incoming infrared signals into the corresponding electrical ones before they are processed by the **decoder** for passage to the **output transducer**.

(b) Advantages

A digital optical fibre system has important advantages over other communication systems.

(i) It has a high information-carrying capacity – at present about 30 000 telephone calls at once on a pair of fibres, whereas a copper cable can carry only about 2000.

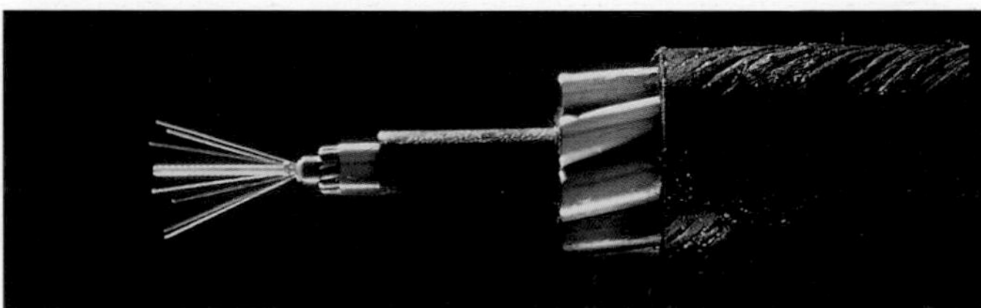

Figure 66.17a Optical cable

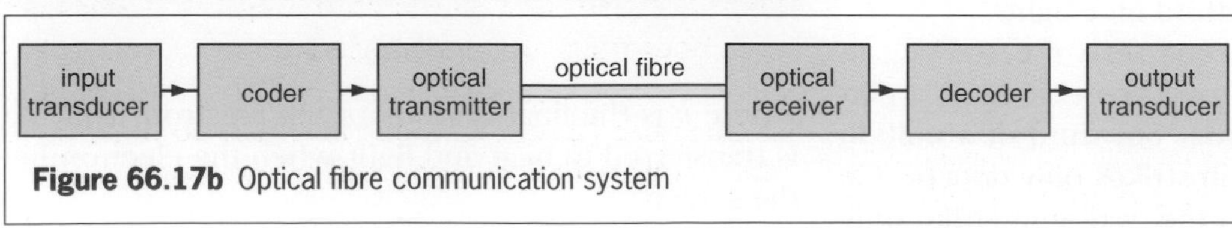

Figure 66.17b Optical fibre communication system

(ii) It is free from 'noise' due to electrical interference.

(iii) Greater distances can be covered without amplifiers than with copper cables.

(iv) A cable of optical fibres is lighter, smaller and easier to handle than a copper cable.

(v) Crosstalk between adjacent channels is negligible.

(vi) It offers greater security to the user.

(c) Pulse code modulation

In this process an analogue signal is changed into a digital one before it is transmitted. The amplitude of the analogue signal, Figure 66.18a, is 'sampled' at regular time intervals to find its value. The values are measured on a scale of equally spaced voltage levels, six in Figure 66.18. Each level is represented in binary code, Figure 66.18b, by the appropriate pattern of electrical pulses, i.e. by a certain 'bit-pattern'. A three-bit code as here can represent up to eight levels (0 to 7); four bits would allow sixteen levels to be coded. The bit-pattern is sent as a series of pulses, forming a pulse code modulated signal, Figure 66.18c.

The accuracy of the representation increases with the number of voltage levels and the sampling frequency.

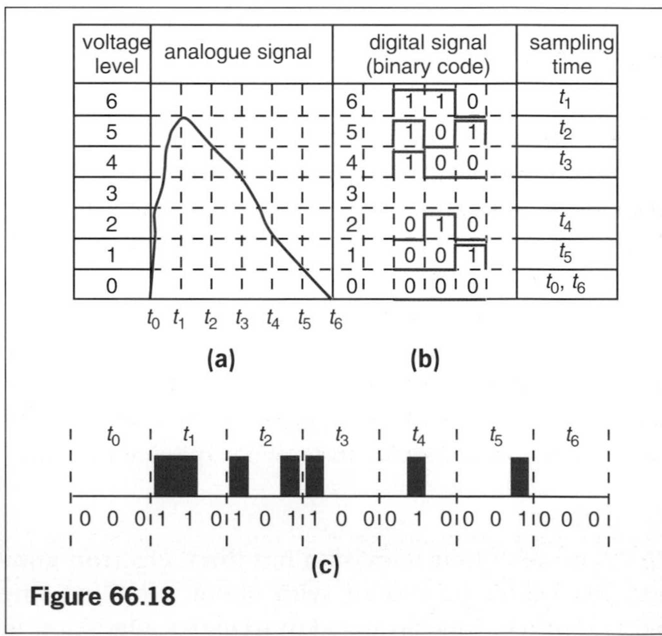

Figure 66.18

Communication satellites

More than half of all intercontinental telephone calls, as well as worldwide television broadcasts of major events, use the satellite communication network.

(a) Geostationary synchronous satellites

The idea proposed by the space science writer Arthur C. Clarke in 1945 has been realised with the advent of powerful rockets which can place satellites in orbit 36 000 km (22 500 miles) above the **equator**, where they orbit in 24 hours and appear to be stationary from the Earth. Such satellites 'see' 120° of longitude, so that the whole of the populated Earth's surface can be covered by three correctly positioned satellites.

Throughout the world there are now over 200 Earth stations which transmit signals to satellites and receive signals via satellites from other countries. The largest satellite system is managed by the 126-nation International Telecommunication Satellite Organization (INTELSAT). Satellites have been launched for many countries by American, Russian and Chinese rockets and by the European Space Agency's *Ariane* rockets. An INTELSAT VI satellite is shown in Figure 2c, p. vii. It operates at microwave frequencies of 4, 6, 11 and 14 GHz and has a capacity of 30 000 two-way telephone circuits plus three TV channels.

(b) Dish aerials

The beams of microwaves, directed to and received from communication satellites by the large saucer-shaped metal dishes at Earth stations, Figure 66.19, should be as parallel as possible to reduce energy loss. However there is some spreading to give a radiation pattern not unlike that due to diffraction of light at a single slit (Figure 13.1). It can be shown that the larger the dish and the shorter the wavelength of the microwaves, the more directional is the beam.

Figure 66.19 Dish aerials at an Earth station

(c) Orbital relationships

To put an artificial satellite of mass m into a circular orbit at a distance r from the *centre of the Earth*, it must enter the orbit with the correct velocity v. If it does not, the force of gravity, which decreases with height, will not be equal to the centripetal force F (Chapter 37) required for that orbit. It can be shown that the relationship which must exist is given by

$$F = \frac{mv^2}{r}$$

The orbital period T (i.e. time for one orbit) is then

$$T = \frac{2\pi r}{v}$$

From these two equations the various features of the satellite's motion can be calculated.

For example, for a synchronous satellite at height 36 000 km above the Earth's surface, $r = (36\,000 + 6000)\text{ km} = 42\,000\text{ km} = 42 \times 10^6\text{ m}$ (taking the Earth's radius as 6000 km) and $T = 24\text{ hours} = 24 \times 3600\text{ s}$. Hence the orbital velocity v is

$$v = \frac{2\pi r}{T} = \frac{2\pi \times 42 \times 10^6\text{ m}}{24 \times 3600\text{ s}}$$

$$= 3.1 \times 10^3\text{ m/s} = 3.1\text{ km/s}$$

If F is the gravitational force per kg to keep the satellite in orbit, then

$$F = \frac{mv^2}{r} = \frac{1.0\text{ kg}}{42 \times 10^6\text{ m}} \times (3.1 \times 10^3)^2\,\frac{\text{m}^2}{\text{s}^2}$$

$$= \frac{9.6}{42}\text{ kg m/s}^2 = 0.23\text{ N}$$

Information storage

There are several ways of storing information.

(a) Vinyl discs or records

These store sound as a wavy shape on the sides of a groove in a plastic disc, Figure 66.20. The groove is produced by a heated cutter with a fine point which is moved a little from side to side by the a.f. currents being recorded. In a stereo recording, wavy lines are cut in both walls of a right-angled groove. During playback the recording process is reversed by the pick-up. For their size (25 cm diameter) records do not store a large amount of information and are no longer made.

Figure 66.20 Groove in a record

(b) Compact discs (CDs)

These record sound in digital form as a series of tiny pits of different lengths and spacing on a plastic disc with a reflecting surface. The pick-up on the disc recorder sends a very narrow intense beam of light from a laser through a partly reflecting mirror and a lens system which focuses it on the disc, Figure 66.21. If the beam falls in a pit, it is scattered, otherwise it is reflected back to a photodiode which produces an electronic signal. The reproduction is of high quality. CDs can store pictures (and text) as well as sound.

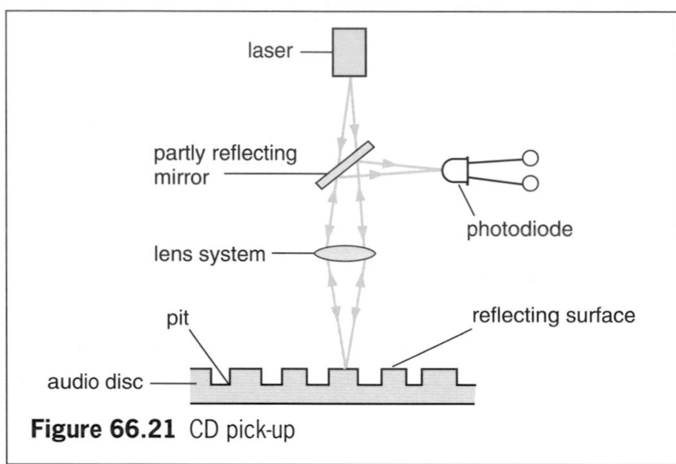

Figure 66.21 CD pick-up

(c) Audio and video tapes

Sound and pictures can be recorded on plastic tape coated with a fine powder of magnetic material (iron oxide or chromium oxide) which becomes magnetized during recording. A chain of minute permanent magnets is produced on the tape in a pattern which represents the original sound or picture. Figure 66.22 shows the simple magnet patterns for single 'high' and 'low' frequencies.

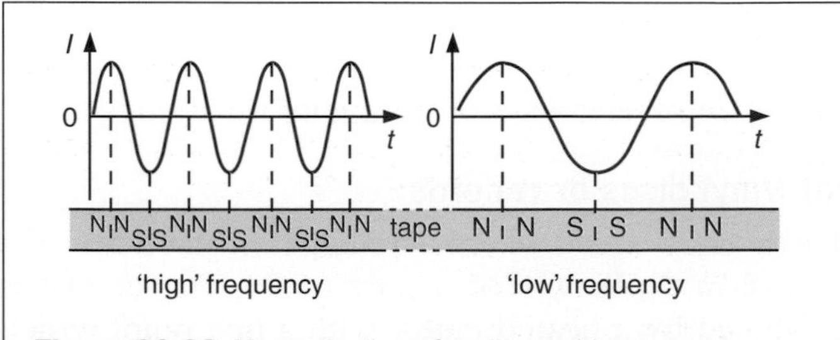

Figure 66.22 Magnetization of audio or video tape

(d) Computer disks

These are used in rigid form for large computers and as 'floppy' disks in small computers. Data is stored magnetically in binary form as 1s and 0s.

QUESTIONS

1 List the limitations of sending information by **a** a line-of-sight method, **b** telegraphy, **c** telephony.

2 **a** (i) Draw a labelled diagram of the electromagnetic spectrum showing how radio waves fit in with other types of electromagnetic radiation. What do all these radiations have in common?
(ii) A radio station broadcasts on a frequency of 100 MHz. Assuming the speed of light to be 3×10^8 m/s what will be the wavelength of the radio frequency carrier wave from this station?

b What is meant by the terms (i) audio frequencies, (ii) radio frequencies? Illustrate your answer by reference to sound waves and carrier waves. *(N.W.)*

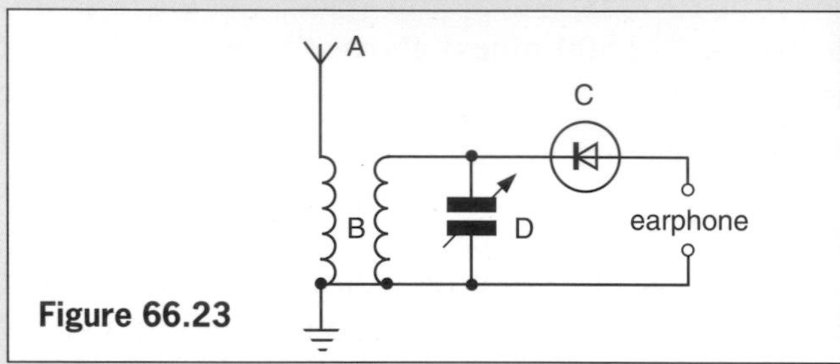

Figure 66.23

3 Figure 66.23 shows the circuit of a simple radio receiver.

a State the names of the parts labelled A, B, C.

b What is the purpose of the part labelled D?

4 State briefly how each of the following has extended the range over which communication can be made: **a** radio transmission, **b** satellites, **c** optical fibres.

Checklist

After studying this chapter you should be able to

- describe the early history of telecommunications,
- explain how information can be represented electrically in digital and analogue form,
- describe with the aid of a diagram the main parts of a communication system,
- recall that radio waves are a form of electromagnetic radiation, their range of frequency and wavelength, and the various uses of the different bands,
- recall the ways in which radio waves travel,
- outline a radio system using amplitude modulation,
- recall how electrical oscillations are produced,
- explain the action of the tuning and detector circuits in a radio receiver,
- outline how black and white and colour television receivers work,
- recall that in a cathode ray tube the k.e. of an electron of charge e, mass m, accelerated through a p.d. V to velocity v is given by $\frac{1}{2}mv^2 = eV$,
- give an outline of an optical fibre communication system and state its advantages over other systems,
- outline the process of pulse code modulation,
- describe how geostationary communication satellites operate and solve problems on their orbital speed etc.,
- recall some ways in which information can be stored.

ADDITIONAL QUESTIONS

The first questions under each topic heading are 'basic' questions intended for all students; the following questions marked with a stripe are 'higher' questions for those seeking grades E to A*.

Electrons

1 **a** How are cathode rays (i) produced in a cathode ray tube, (ii) made to travel along the tube to the screen?

b State three ways in which cathode rays differ from light rays.

c Give three uses of cathode ray tubes.

2 **a** When the Y-amp gain control on a CRO is set on 2 V/div, an a.c. input produces a vertical line 10 divisions long. What is the peak voltage of the input?

b What is the frequency of an alternating p.d. which is applied to the Y-plates of a CRO and produces five complete waves covering 10 horizontal divisions of the screen when the time base setting is 10 ms/div?

Radioactivity

3 Explain what is meant by **half-life**, **background radiation**.

At a certain instant the corrected count-rate registered on a detector placed close to an α-particle emitter is 200 per second and this falls to 50 per second in 12 minutes. Determine the half-life of the source. *(L.)*

4 Describe simple experiments (details of instruments are not expected) to show the difference between α and β radiation when

a passed through a magnetic field, and

b appropriate sheets of material are placed in their paths.

What are the essential differences in nature and properties between the above types of radiation and γ radiation?

State, with a reason, an essential precaution which is necessary when using equipment known to emit γ radiation or X-rays.

A radioactive substance has a half-life of 1 minute. By what factor would you expect the level of activity to drop after 4 minutes? *(S.)*

Atomic structure

5 Which of the following statements most correctly represents the important deductions from experiments on the scattering of alpha particles by matter?

A Alpha particles are nuclei of helium atoms.

B Atoms are composed of protons, neutrons and electrons.

C Atoms consist of positively and negatively charged matter.

D Electrons move in orbits round the nucleus of atoms.

E Atoms have a small nucleus carrying a positive charge. *(J.M.B.)*

6 The following represents part of a radioactive series in which the chemical symbols have been replaced by letters.

$$^{234}_{90}A \xrightarrow{(i)} {}^{234}_{91}B \xrightarrow{(ii)} {}^{234}_{92}C \xrightarrow{(iii)} {}^{230}_{90}D$$

Name the particles emitted in the three changes.

Write down the two letters which represent isotopes. *(W.)*

Electronics and control

7 The diagram shows a simple safety circuit.

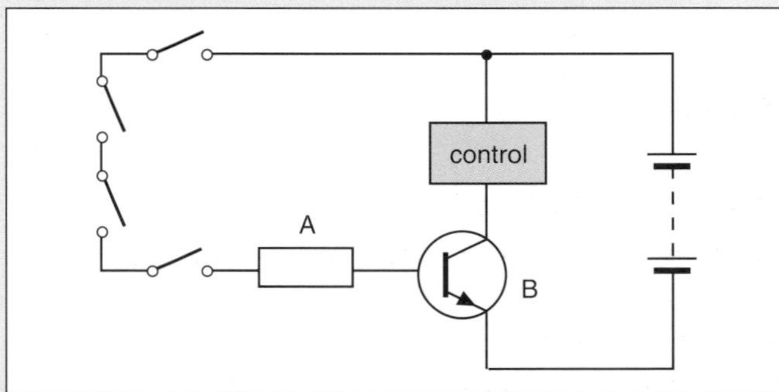

a (i) Name component A.
(ii) Name component B.

b (i) What is the name of the terminal of component B to which component A is connected?
(ii) Why is component A usually needed in this part of this type of circuit?

c (i) How many contacts must be closed to operate this circuit?
(ii) Explain exactly what happens in the circuit when the required number of contacts is closed.

8 **a** In circuit (a) what is the value of V_{out} if S is (i) open, (ii) closed?

b Repeat for circuit (b).

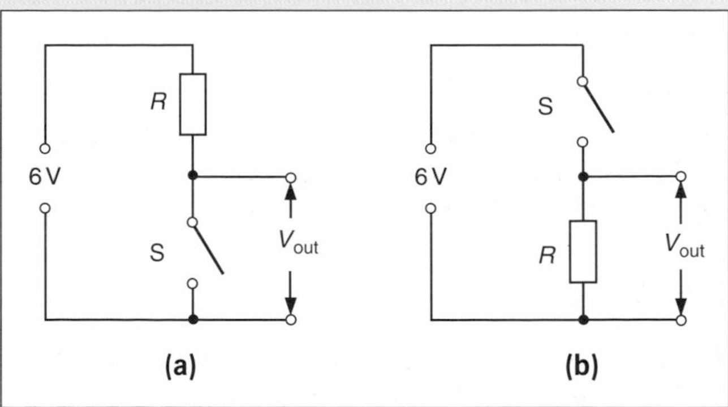

9 The diagram below shows the logic levels for an SR bistable in its RESET state.

Redraw the diagram with the new logic levels when the S input is changed to 0. Explain the changes.

Hint. You need to refer in your explanation to the truth table of a NAND gate.

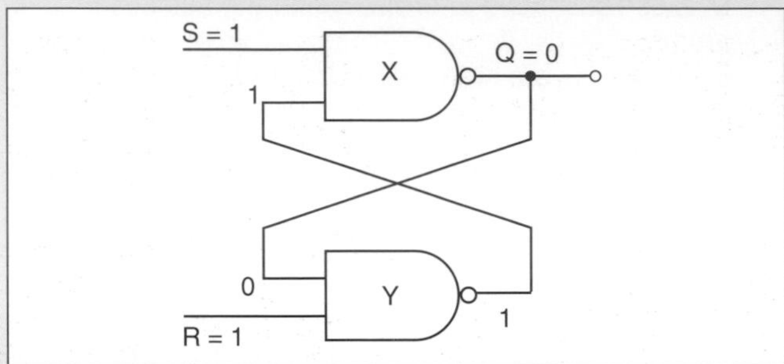

Telecommunications

10 The circuit of a very simple radio receiver is shown. It can be divided into three parts along the dotted lines.

a Name part X and state what it does.

b Name part Y and state what each of the *three* components does.

c Name part Z and state its function.

d Draw two diagrams to show the colour coding on the 10 kΩ and 100 kΩ resistors. (See the resistor colour code below the circuit diagram.)

11 An electron, charge *e* and mass *m*, is accelerated in a cathode ray tube by a p.d. of 1000 V. Calculate

a the kinetic energy gained by the electron,

b the speed it acquires.

($e = 1.6 \times 10^{-19}$ C, $m = 9.1 \times 10^{-31}$ kg)

12 a Describe a geostationary satellite orbit.

b State the conditions for a satellite to remain in a geostationary orbit.

c Explain why a geostationary orbit is necessary for effective satellite communication.

13 A satellite close to the Earth (at a height of about 200 km) has an orbit speed of 8 km/s. Taking the radius of its orbit as approximately equal to the Earth's radius of 6400 km, calculate the time it takes to make one orbit.

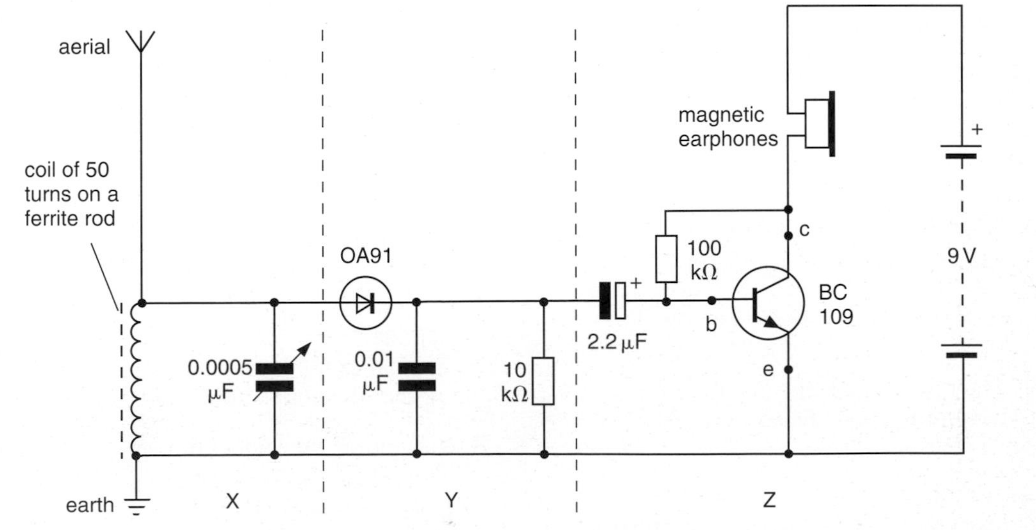

aerial
coil of 50 turns on a ferrite rod
0.0005 μF
OA91
0.01 μF
10 kΩ
2.2 μF
100 kΩ
magnetic earphones
c
b
e
BC 109
9 V
earth
X
Y
Z

Resistor colour code

black	0	green	5
brown	1	blue	6
red	2	violet	7
orange	3	grey	8
yellow	4	white	9

4th band gives tolerance:
gold ±5%
silver ±10%
no colour ±20%

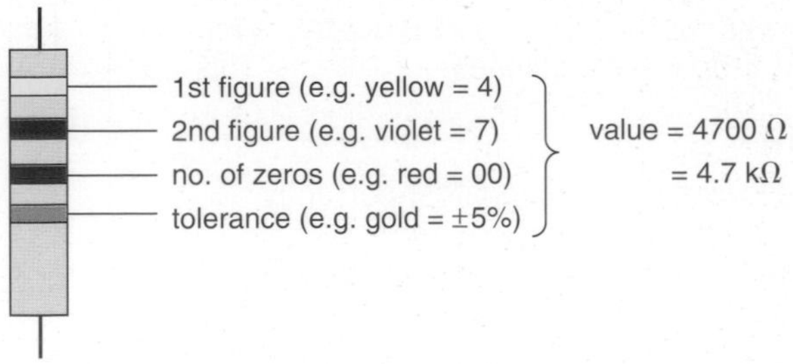

1st figure (e.g. yellow = 4)
2nd figure (e.g. violet = 7)
no. of zeros (e.g. red = 00)
tolerance (e.g. gold = ±5%)
value = 4700 Ω = 4.7 kΩ

EARTH AND SPACE PHYSICS

67 THE ATMOSPHERE AND WEATHER

To understand how our weather arises, we need to know about the nature of the Earth's atmosphere and how it receives energy from the Sun.

I *The Earth's atmosphere*

The atmosphere is a blanket of gases, about four-fifths nitrogen, one-fifth oxygen, with small but important amounts of carbon dioxide and water vapour. It stretches several thousand kilometres into space, thinning very rapidly with height. It consists of several different layers, Figure 67.1.

1 The **troposphere** occupies the first 10 km or so and is where most of our weather is formed. In it the temperature and pressure fall quickly as the altitude (height) increases; the temperature drops from an average of about 15 °C at sea-level to −50 °C and the air pressure (and density) at 10 km is about one-quarter of its sea-level value.

2 The **stratosphere** extends from 10 km to about 60 km and contains no water vapour and so no clouds. The temperature, although still low, rises with altitude. Narrow belts of fast-moving winds, called 'jet streams', always blow at certain latitudes in an easterly direction. Conditions are suitable for high-flying aircraft.

3 The **ionosphere** stretches beyond the stratosphere and consists of layers of electrically charged particles (ions), which as we saw earlier (Chapter 66) reflect radio signals back to Earth.

The composition of the atmosphere is fairly constant, but it is thought that in the early stages of the Earth's history there was more carbon dioxide and less oxygen. However, with the evolution of plants and trees, carbon dioxide was used and oxygen released in photosynthesis. This is the process in which carbon dioxide and water are changed into glucose (sugar) and oxygen by the action of solar energy.

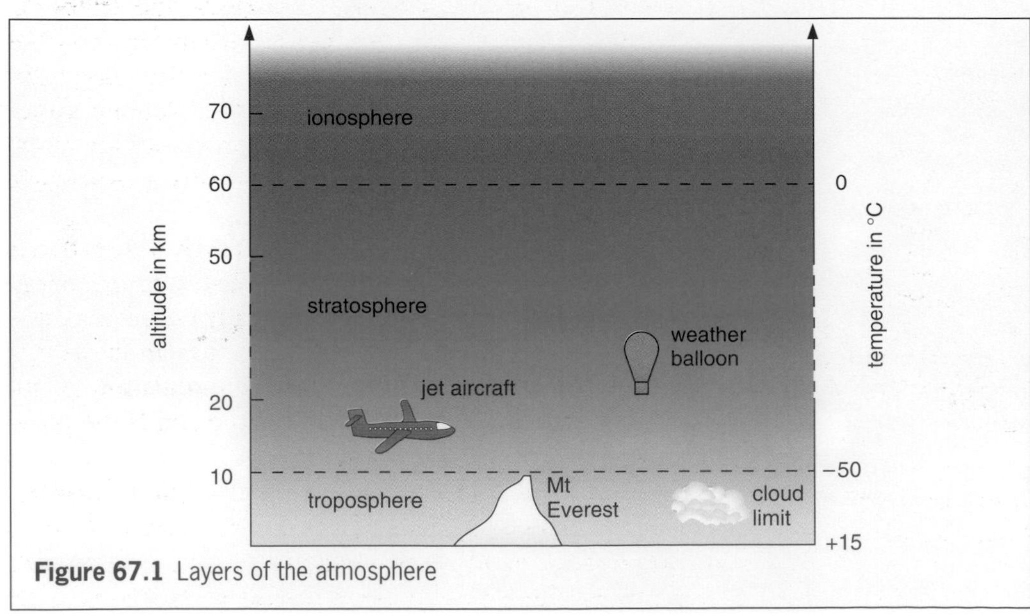

Figure 67.1 Layers of the atmosphere

How the Earth is heated

The source of the Earth's energy is the Sun, whether it is supplied directly as solar radiation or as stored energy in fossil fuels (Chapter 46).

(a) Radiation

Solar radiation is mostly light and short-wavelength infrared, about *one-third* of which is reflected back into space by the Earth's atmosphere. Most of the rest of this short-wave radiation passes through the atmosphere without warming it and is absorbed by the Earth's surface where it is changed to heat energy.

As a result the surface is warmed and, being at a comparatively low temperature, radiates *long-wave* infrared (Chapter 14). This is readily absorbed by the atmosphere, especially by water vapour and carbon dioxide. These act like the glass in a greenhouse (Chapter 45) by allowing short-wave solar radiation to pass to Earth but preventing long-wave radiation from the Earth's surface from escaping. Some re-radiation of long-wave infrared from the atmosphere back to the surface and to outer space also occurs.

The **Earth's surface** is thus warmed by both **direct** and **indirect** radiation, as shown diagrammatically in Figure 67.2. The extent to which a particular area is heated depends on factors such as altitude, latitude, nature of the surface (e.g. whether land or sea), cloud cover, time of day and year. Unequal heating results, causing changes in the temperature and pressure of the air which are responsible for many weather changes.

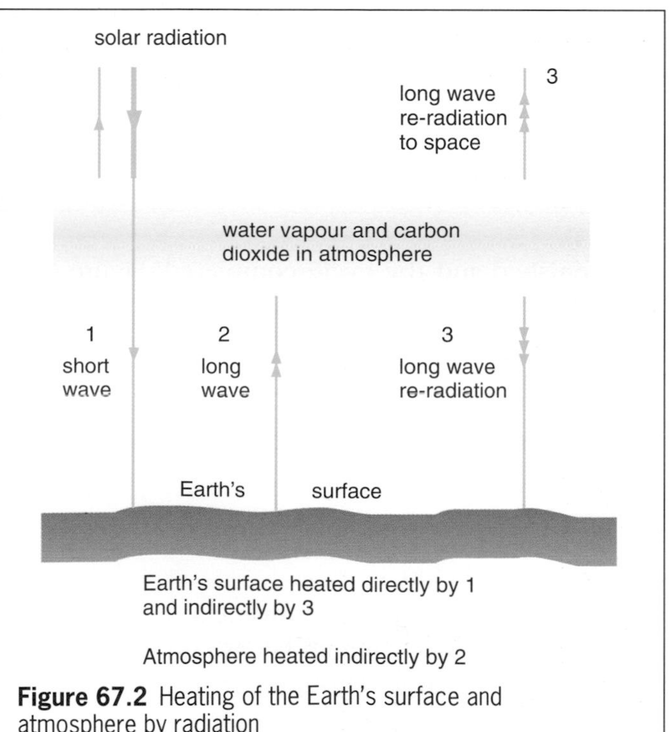

Figure 67.2 Heating of the Earth's surface and atmosphere by radiation

The **atmosphere** is warmed **indirectly** (mostly) by radiation but other processes are also involved.

(b) Conduction

Heat is conducted from the Earth's surface to the air above it but very little is passed on because air is such a poor conductor.

(c) Convection

When air is heated it expands, becomes less dense and rises. Cooler, more dense air falls to replace it and in turn is warmed and rises. Convection currents (Chapter 44) are set up, transferring heat to other parts of the atmosphere.

(d) Evaporation and condensation

Much of the radiation that falls on the Earth's surface is on the oceans where it changes water into water vapour by evaporation. The energy used in this process is stored as latent heat (Chapter 42) in the vapour. If the vapour is cooled and condenses to water (e.g. rain) or snow, its latent heat is released into the atmosphere.

(e) Compression

When air is squeezed by, for example, falling from a higher to a lower altitude where the atmospheric pressure is greater, its temperature rises as a result of the work done by compressing it. The same happens when a bicycle tyre is inflated, the rise in temperature of the air when it is squeezed making the barrel of the pump warmer.

The weather and energy transfers

The processes described in **(a)** to **(e)** of the previous section all involve **energy transfers** and for some weather phenomena they act as driving agents. For example, convection currents provide the energy for land and sea breezes (Chapter 44). Among other effects that are energized by such processes are cloud formation, fog, frost and thunderstorms – as we will see shortly.

The **unequal heating of the Earth's surface** is often the cause of convection currents and can set in motion weather changes both locally (as in land and sea breezes) and globally. Globally it has an important effect on the large-scale complex circulation of air that occurs in the Earth's atmosphere and is the background to the world's weather.

Since most solar radiation is received in the Tropics (bands either side of the Equator), warm air rises there and is carried northwards and southwards towards the Poles to be replaced at the surface by

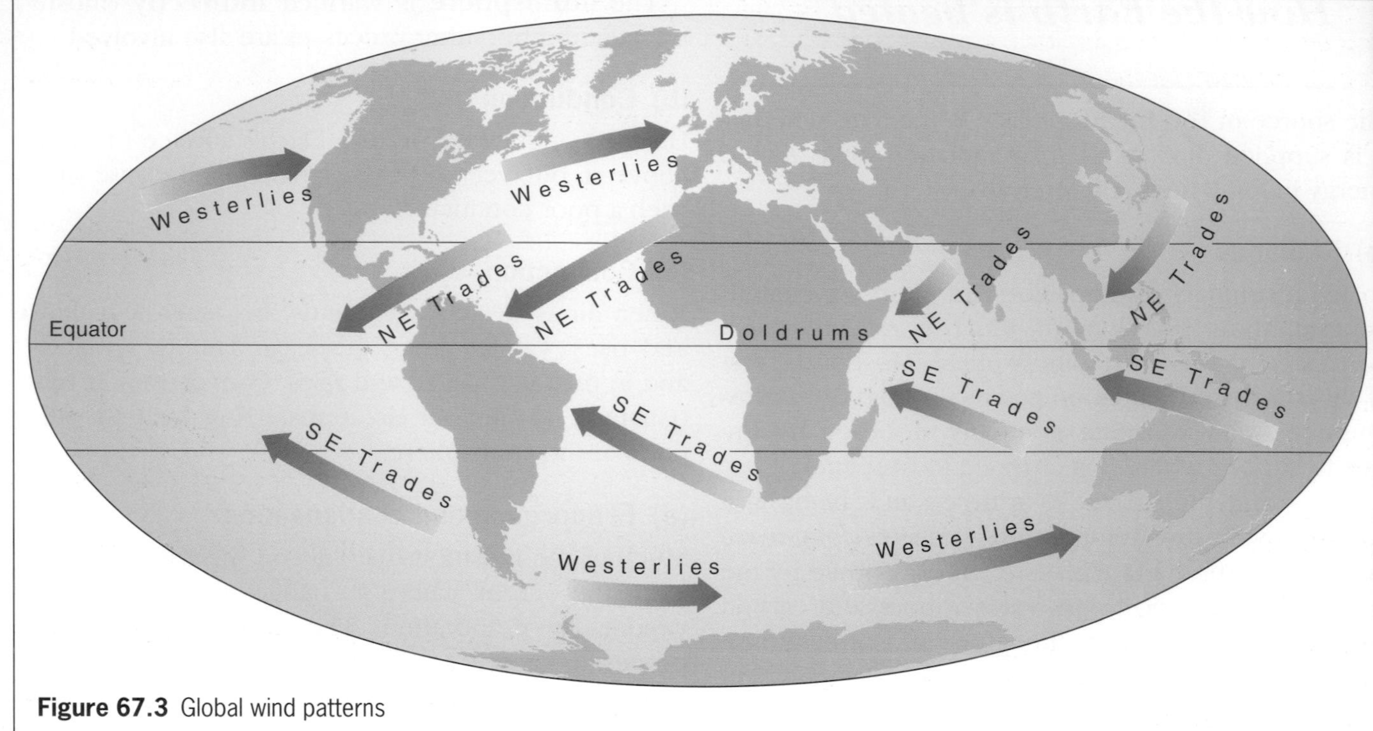

Figure 67.3 Global wind patterns

cooler air moving in. These air movements cause the pressure differences that create winds whose directions are also affected by the Earth's easterly rotation.

The resulting **wind pattern** is shown in Figure 67.3 in outline. At the Equator there are the windless **Doldrums**, where sailing ships were becalmed. In the Tropics the steady **Trade Winds** blow for most of the year from the north-east in the northern hemisphere and from the south-east in the southern hemisphere. Further north and south the prevailing **Westerlies** blow strongly in the Atlantic (bringing Britain most of its weather) and also in the Pacific on the west coast of North America.

The atmosphere can be regarded as a **huge transport system carrying energy** from warmer to colder regions using winds, local and global, to smooth out temperature and pressure differences. In this way the Tropics do not become hotter nor polar regions colder.

The water cycle

The Earth has a limited supply of water which circulates round and round between the Earth's surface and the atmosphere. It is explained by a model called the **water cycle**.

Water evaporates continuously from the seas, lakes and rivers, from plants and trees, and from the ground. If the water vapour condenses to form water droplets in clouds, it may eventually return to the Earth as rain or snow. In this way the supply of water is replenished and the cycle completed, Figure 67.4; the same processes are repeated again and again.

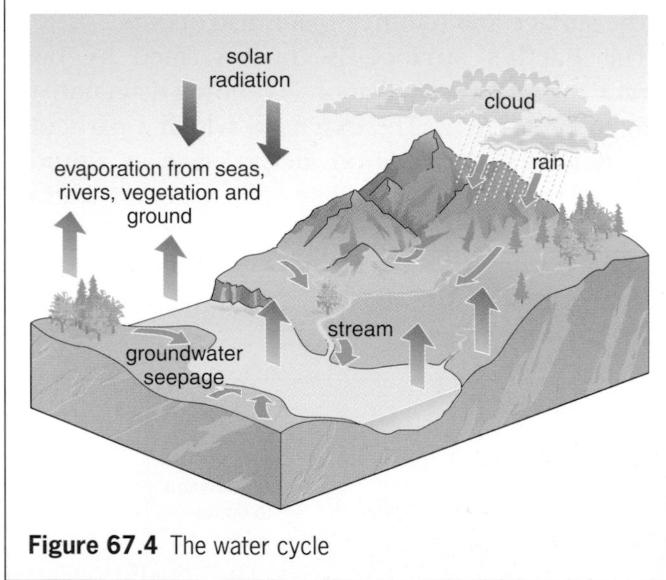

Figure 67.4 The water cycle

The water cycle is called a 'closed system'; it is powered by the Sun.

Clouds and precipitation

(a) Clouds and rain

Clouds consist of tiny water droplets or, at high altitudes, of ice particles, which fall as rain, snow or hail when they are large enough. Clouds are formed when water vapour in the atmosphere condenses or freezes as the air cools and its capacity to carry water vapour is reduced. The cooling can occur if the air is forced to rise. When it does so, it expands because the air pressure is less higher up. However it requires energy to push back the surrounding air and this is provided by its internal supply of heat energy. As a result the air temperature falls.

There are three main ways in which air is made to rise and produce clouds and possibly rain.

(i) Convection currents develop due to the unequal heating especially of those parts of the Earth's surface receiving a great deal of solar radiation. They create upward currents of warm air (as in land and sea breezes) and **convection rain** may fall.
(ii) A wind forces warm, moist air to flow up over a mountainous region where it may cause **relief rain**, mostly on the windward side of the region, Figure 67.5.
(iii) When warm air meets cold air, the warm air is driven upwards, as happens in the warm and cold fronts that accompany depressions (Chapter 68). **Depression rain** may follow.

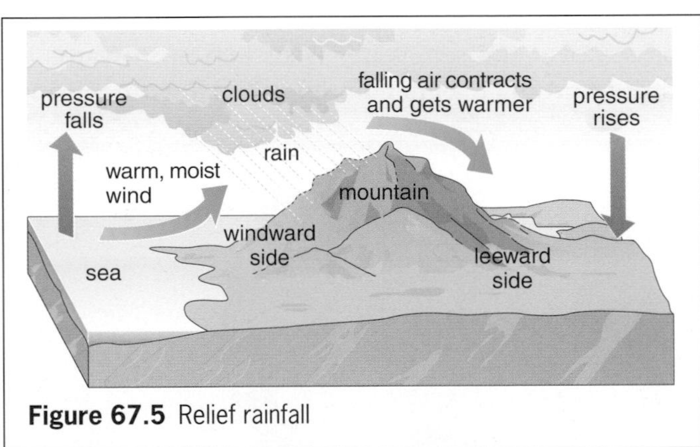

Figure 67.5 Relief rainfall

(b) Snow

When water vapour condenses directly in very cold clouds to form ice crystals which join together, snowflakes are formed.

(c) Hail

If raindrops are carried up and down rapidly in a thundercloud by convection currents, they experience temperatures alternately above and below freezing. The resulting hailstones consist of layers of ice.

Rain, snow and hail are called **precipitation**.

(d) Cloud types

The appearance, height and motion of clouds give useful clues to forecasters about weather developments in the next few hours. Some of the terms used to describe them are:

(i) **cirrus** - meaning feathery, and refers to very high clouds, Figure 67.6a,
(ii) **cumulus** - meaning heap-like with flat bases and rounded tops; such clouds occur at high, middle and low levels, Figure 67.6b,
(iii) **stratus** - used for layer-type clouds which are grey and bring drizzle; they can develop at all altitudes, Figure 67.6c,
(iv) **nimbus** - meaning rain cloud which is dark, shapeless and often brings continuous rain,
(v) **alto** - refers to mid-level clouds.

The two basic types are **cumulus** and **stratus**, but clouds which are described by combinations of the above terms occur frequently. For example, **cirrocumulus** and **nimbostratus** (p. 299), **cumulonimbus** (pp. 294 and 300).

Figure 67.6 Cloud formations

(a) Cirrus clouds

(b) Cumulus clouds

(c) Stratus clouds

Other weather phenomena

(a) Dew and frost

Dew forms on a still, clear night when the ground loses heat by radiation. If the air in contact with the ground has a high water vapour content and is cooled below the **dew point** (i.e. the temperature at which the air is saturated and can hold no more water vapour), tiny drops of water are deposited on the ground as dew.

If the dew point is below the freezing point of water (0 °C), water vapour turns to ice directly and **ground frost** results.

Both of these weather effects are driven by the energy transfer which occurs when the ground emits radiation.

(b) Fog and mist

These consist of tiny droplets of water, smaller in mists than in fogs, suspended in nearly still air. In many ways they are like very low clouds, but while clouds form when air is cooled below the dew point by rising, fogs and mists are produced when a large layer of air is cooled by contact with a cold surface (that may itself have cooled by radiation).

(c) Thunderstorms

Massive, often rain-bearing clouds (**cumulonimbus**) in which there are strong convection currents carrying raindrops and hailstones up and down continually, cause thunderstorms. As a result of this violent motion, charged particles (ions) are formed within the cloud, its top being positive and most of the rest negative, Figure 67.7. The large electrical voltage produced (millions of volts) by the separation of charges causes huge sparks, i.e. lightning, to jump between different parts of the same cloud or from one cloud to another or between the cloud and a part of the ground under it that has an opposite charge.

The accompanying clap of thunder is due to the huge current (thousands of amperes) that passes, making the air so hot that it expands and contracts violently, creating sound waves.

(d) Hurricanes, typhoons, tropical cyclones

These are all terms used in different parts of the world for an area of very deep low pressure. At the centre or 'eye' of this area it is calm but strong winds spiral upwards towards it, Figure 67.8. These can cause severe damage to property and there may also be torrential rain from the rapidly rising air currents. They form mostly over tropical seas, from which they obtain a plentiful supply of warm moisture.

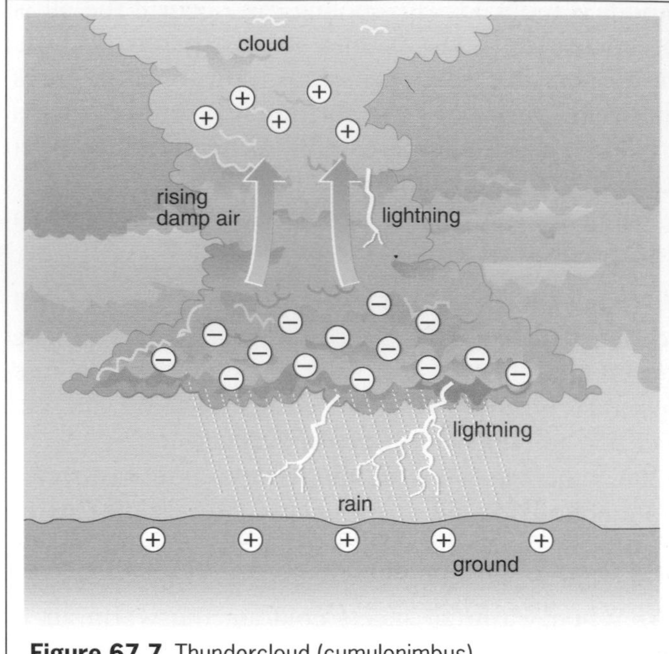

Figure 67.7 Thundercloud (cumulonimbus)

A **tornado** differs from a hurricane by forming over land. Its winds are even more destructive and if it passes over the sea, water is sucked up towards its centre to produce a tall column called a **waterspout**.

Figure 67.8 Aerial view of a hurricane centre off Scotland

(e) Local winds

These are small-scale winds, like land and sea breezes, which occur regularly for a short time in certain localities where they have particular effects. For example, in the Alps the notorious warm, dry Föhn wind is the result of air being warmed and dried by compression as it moves to lower levels; it can raise the temperature in valleys below by 10 to 15 °C in an hour.

(f) Monsoons

These are large-scale winds which affect continents and are formed in a similar way to land and sea breezes. In southern Asia they bring heavy rain.

Environmental concerns

(a) The greenhouse effect

During the last hundred years the burning of fossil fuels (coal, oil, gas) in the industrialized world has led to a steady increase of carbon dioxide in the atmosphere. Some scientists feel that the result could be a considerable rise in the Earth's surface temperature in the future due to the 'greenhouse-like' action (Chapter 45) of carbon dioxide (and other gases such as methane). The natural regulation of the Earth's temperature by the greenhouse effect would thus be upset and might lead to drastic climatic changes, e.g. rising sea levels threatening coastal regions. On the other hand, carbon dioxide is necessary for green plants to make their (and our) food by photosynthesis.

(b) The ozone layer

In the stratosphere at a height of about 20 km there is a layer of the gas ozone (O_3) which absorbs most, but not all, of the Sun's harmful ultraviolet radiation. Some years ago, it was discovered that the ozone layer had thinned, especially in the Arctic and Antarctic, leaving 'holes'. The cause is thought to be chlorofluorocarbons (CFCs) used in aerosols, refrigerators and air-conditioning systems. Light releases chlorine from CFCs which reacts with ozone to produce oxygen (O_2).

The use of CFCs is now decreasing and hopefully the 'holes' will disappear and stop ultraviolet rays causing skin cancer and genetic mutations in plants and animals.

(c) Deforestation

Plants and trees absorb carbon dioxide in photosynthesis and so help to counter the greenhouse effect. The widespread deforestation of large areas of the Earth's surface is therefore a matter of concern unless extensive tree-planting programmes are undertaken.

(d) Siting of structures

A knowledge of local climatic features is important when siting certain structures. For example, reservoirs need to be built in mountainous regions where the rainfall is high; wind generators require a site where there are strong prevailing winds. However, such locations are often in beauty spots and other environmental factors may also have to be considered, such as the need to cut down trees.

QUESTIONS

1 a About two-thirds of solar radiation is absorbed by the Earth's surface. What happens to the rest that reaches the atmosphere?

b State *three* ways in which the atmosphere is heated.

c Name *three* factors which cause unequal heating of the Earth's surface.

2 Name *three* weather phenomena that are driven by the energy transfers occurring in convection and radiation.

3 How does the atmosphere prevent the Tropics from becoming hotter and polar regions becoming colder?

4 Name the physical processes involved in the water cycle.

Checklist

After studying this chapter you should be able to

- recall the structure of the Earth's atmosphere,
- describe how the Earth and the atmosphere are heated,
- explain how some weather phenomena are driven by energy transfer processes such as unequal heating of the Earth's surface, convection and radiation, and give examples,
- describe the water cycle,
- recall how clouds and rain, snow and hail are formed,
- recognize and name the main types of cloud formation,
- explain the formation of dew and frost, fog and mist, thunderstorms, hurricanes (typhoons, tropical cyclones), tornadoes, local winds and monsoons,
- discuss environmental concerns such as the greenhouse effect, the depletion of the ozone layer, deforestation and the siting of large structures.

68 WEATHER MAPS AND FORECASTS

Weather data	***Air streams over the British Isles***
Weather symbols	***Climate in the British Isles*** Rainfall. Temperature. Jet streams.
Weather maps Isobars and winds. Anticyclones. Depressions.	***Weather forecasting today***
Fronts	

As well as being of general interest, weather forecasts are important to farmers, navigators, aviators, motorists, holiday makers, the emergency services when there is a risk of floods, hurricanes, etc., and others.

1 *Weather data*

Weather forecasting requires measurements and observations to be made at many weather stations at regular intervals, usually every six hours.

(a) Temperature

Readings are taken on a mercury thermometer (Chapter 38) in °C, *in the shade*, i.e. out of direct sunlight, preferably in a Stevenson screen, Figure 68.1. The highest and lowest temperatures over 24 hours are also recorded by a maximum (mercury) and minimum (alcohol) thermometer, but are not shown on weather maps. Humidity measurements (i.e. how wet or dry the air is) are also taken, using a wet bulb thermometer, but are not included on maps.

insulated roof
sides of wooden louvres to allow air to enter and leave
one side hinged to give access to instruments

Figure 68.1 Stevenson screen

(b) Pressure

This is an important measurement because changes in it are often a good guide to future weather. In general, a steady rise is a sign of good weather and a rapid fall a sign of bad. A mercury barometer (Chapter 28) gives accurate readings but **barographs**, Figure 68.2, based on the aneroid barometer (Chapter 28) are cheaper and more portable. They have a pen that moves over a chart on a revolving drum and a continuous visual record for a week can be obtained.

Pressure on weather maps is given in millibars (mb). Normal atmospheric pressure at sea-level is

$$1000 \text{ mb} = 1 \text{ bar} = 100\,000 \text{ Pa} = 760 \text{ mmHg}$$

(c) Wind

Wind speed is measured by an **anemometer** consisting of three metal cups fixed to a vertical spindle by arms which rotate when blown by the wind, Figure

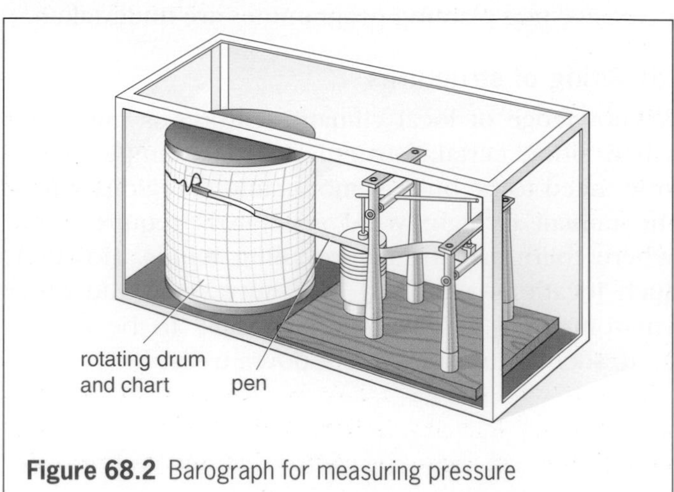

Figure 68.2 Barograph for measuring pressure

68.3a. The speed is calculated in knots from the number of rotations recorded on the meter in a certain time. (1 knot = 1.9 km/h = 1.2 mph approx.)

The **Beaufort scale** for describing wind force or speed was invented by Admiral Sir Francis Beaufort in 1805, by watching the effects that different wind speeds had on the sea. Force 1 produces slight ripples, Force 4 commonly causes 'white horses', Force 9 gives high waves with tumbling crests and thick spray and is referred to as a gale, Force 12 is a hurricane. (See the wind speed chart in Figure 68.5b.)

Wind direction is shown by a **wind vane** in the form of a freely rotating horizontal arm with a vertical flat surface, the vane at one end and a pointer at the other, Figure 68.3b. The wind blows against the vane and the pointer shows the direction from which it has come, as indicated by four fixed arms marked with the points of the compass. A west wind is one that blows *from* the west.

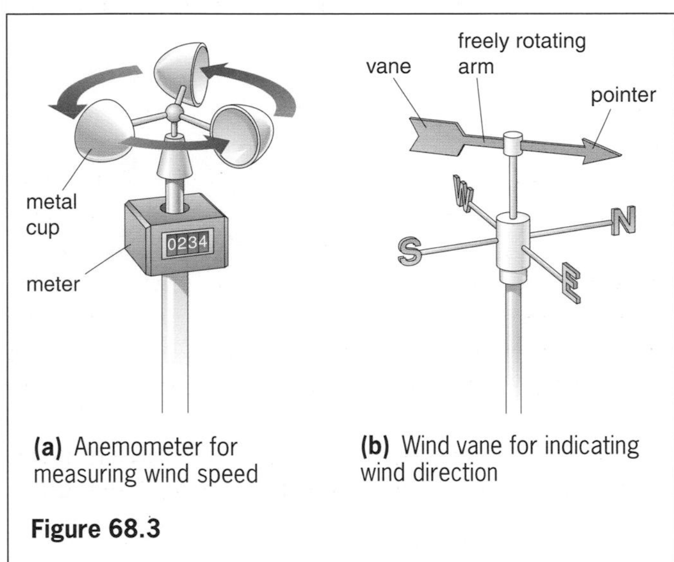

(a) Anemometer for measuring wind speed

(b) Wind vane for indicating wind direction

Figure 68.3

(d) Rainfall

A **rain gauge** is a can with a standard-sized funnel opening and a collecting bottle inside it, Figure 68.4a. Every 24 hours the bottle's contents are emptied into a measuring cylinder and the rainfall obtained in millimetres.

(e) Sunshine

A **sunshine recorder** is a glass sphere that acts as a lens and focuses the Sun's rays on to a blue sensitized card, Figure 68.4b. A brown (scorched) trace is produced from which the number of hours of sunshine can be read off.

(f) Clouds

The amount of cloud cover is estimated in fractions of eighths. Clear sky is 0/8, half-clouded is 4/8 and completely overcast is 8/8. Comments are also recorded about the type of clouds and their altitude.

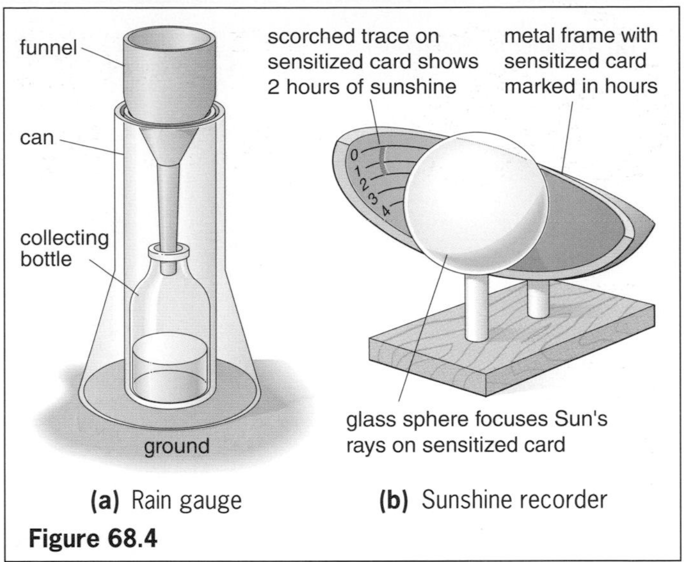

(a) Rain gauge

(b) Sunshine recorder

Figure 68.4

(g) Precipitation

The type of any precipitation is reported as well, e.g. drizzle, showers, continuous rain, hail, snow.

Weather symbols

The data gathered at every weather station is summarized in a 'station model' before it is sent on to the weather centre for plotting on a weather map. A simplified model is shown in Figure 68.5a (overleaf), along with the symbols for certain data (b). Each piece of data always occupies the same position on the model except wind direction (which may cause others to be slightly displaced).

Only the last three figures are given for the pressure reading and the decimal point (before the third figure) is omitted. This does not cause confusion because pressures are usually between 950 mb and 1050 mb. Therefore if the first two figures are between 01 and 49, 10 has to be added to the front, otherwise 9 goes in front. For example:

235 = 1023.5 mb
916 = 991.6 mb
024 = 1002.4 mb

Any **pressure change** in the previous three hours is recorded in a similar way. For example, +15 means a rise of 1.5 mb, while −35 indicates a fall of 3.5 mb.

Weather maps

(a) Isobars and winds

An essential stage in producing a weather map or 'synoptic chart', is the plotting of the pressure readings from the weather stations in the region. When this has been done, lines, called **isobars**, are drawn joining places of equal pressure (just like contours

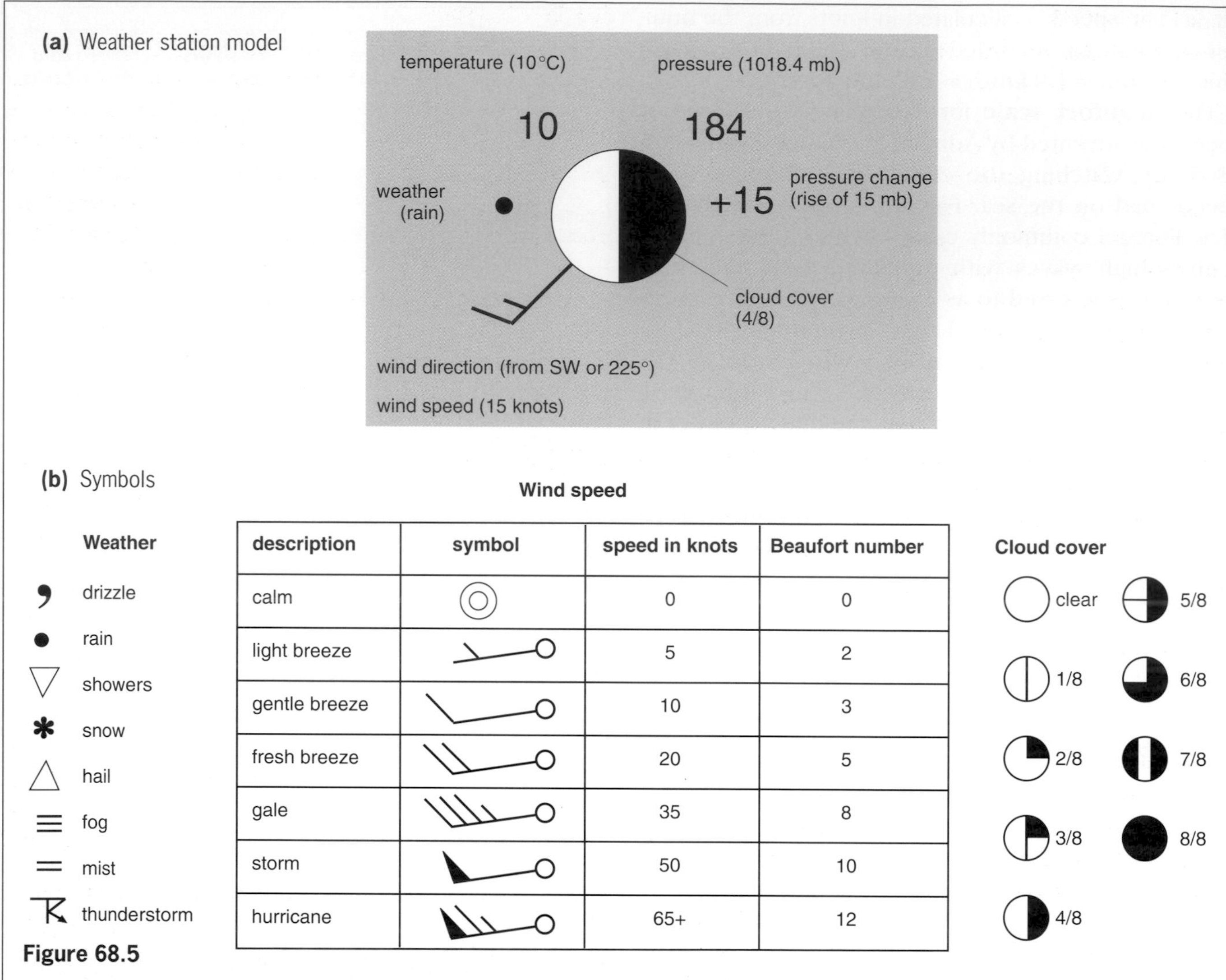

description	symbol	speed in knots	Beaufort number
calm		0	0
light breeze		5	2
gentle breeze		10	3
fresh breeze		20	5
gale		35	8
storm		50	10
hurricane		65+	12

Figure 68.5

that join places of equal altitude on a map). These are often shown at 4 mb intervals, as the lines in the simplified map of Figure 68.6. The resulting pressure pattern, along with a comparison of the readings from all the weather stations, helps to give an overall picture of the present situation and forms the basis for predicting probable developments.

Figure 68.6 Isobars

For example, closely spaced isobars indicate a big pressure difference over a short distance and suggest strong winds are likely; widely spaced isobars suggest light winds. Winds do not blow directly from places of high pressure to those of low pressure across the isobars. The rotation of the Earth makes them blow more or less *along* the isobars.

(b) Anticyclones

If the isobars are closed and roughly circular with the highest pressure at the centre, the system is termed an anticyclone or 'high'. It consists of dry, denser air which exerts a greater pressure on the ground than would a similar mass of normal air. This heralds generally settled, clear, calm weather with warm sunny days in summer. In winter there is the risk of frost and fog due to cloud-free skies allowing a rapid temperature fall near the ground. Winds are non-existent or light and circulate round the 'high' in a clockwise direction in the northern hemisphere, anticlockwise in the southern.

Once an anticyclone has formed it often does not move for several days and covers a wide area.

(c) Depressions

These are regions where the isobars are again closed and oval-shaped, but the pressure is lowest at the centre. They are also called 'lows' or cyclones and contain air that is less dense than colder air because it is warm and has expanded. The pressure acting on the ground is therefore less in this case.

'Lows' usually bring unsettled, wet, windy weather. Winds blow anticlockwise round them in the northern hemisphere, clockwise in the southern.

A change in the weather is often preceded by a shift in the direction of the wind. If it **veers**, that is moves in a clockwise direction (e.g. from east to south), high pressure and fine weather may follow. If it **backs**, by moving anticlockwise, low pressure and storms can be expected.

Associated with depressions are **fronts** which, as we will see in the next section, are a great help in forecasting once their position has been identified.

Fronts

(a) Air masses

An air mass is a large volume of air with a fairly uniform temperature and humidity. It covers a large area of the Earth's surface, such as a desert, a continent or an ocean, from which its characteristics originate. Air masses often move and when one meets another with very different properties, a 'front' develops at the boundary or leading edge of the advancing air mass, where there are marked changes of temperature, pressure, density, humidity, but little mixing.

Although fronts usually mean bad weather, what they bring depends to some extent on their recent path over land and sea. For example, rain is less likely if its path has been over a large land mass.

There are three main types of front.

(b) Warm front

A warm front is formed when a warm air mass meets a cooler one and has to *slide up over it* to continue advancing, Figure 68.7a. The boundary separating warm air from cold air slopes up gradually from the ground. As the warm air is forced to rise into regions of lower pressure, it expands, cools and clouds (**cirrocumulus** that give a 'mackerel' sky, Figure 68.8a) form by condensation of water vapour. Eventually when the clouds are thick and low enough (**nimbostratus**, Figure 68.8b), heavy rain falls continuously for several hours and is followed by drizzle and hill fog. The temperature also rises when the rain starts.

Figure 68.8 Clouds associated with a warm front

(a) Cirrocumulus

(b) Nimbostratus

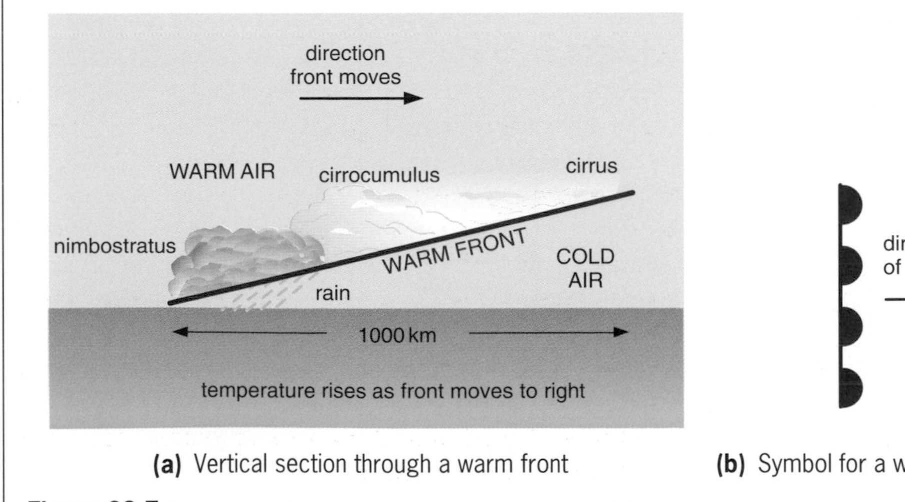

(a) Vertical section through a warm front

(b) Symbol for a warm front

Figure 68.7

The first signs of the approach of a warm front are small wisps of high-level cloud (cirrus) about 1000 km ahead of the front at ground level. The position of a warm front on a weather map is found by studying the data from weather stations. In Figure 68.7a there is a *slow temperature rise* along a line of stations (from right to left) towards the front and a *marked rise* at the front, as well as increasing cloud and rain.

The symbol for a warm front and its direction of travel are shown in Figure 68.7b.

(c) Cold front

A cold front, which often follows fairly quickly after a warm front, is formed when cold air pushes forwards *under* warm air, Figure 68.9a. The boundary separating the two air masses is much steeper than for a warm front and slopes backwards towards the direction from which the front has come. Consequently, the warm air is forced up sharply by the cold mass and thick towering clouds (**cumulonimbus**) form in the rapidly rising air as it cools. If they are tall enough their tops are cut off by the jet stream (Chapter 67) to give them the shape of a blacksmith's anvil, Figure 68.10. There is a brief spell of heavy, often thundery rain over a narrow band (since the front is narrow). It is followed by a fall in temperature with showers and sunny intervals.

A cold front is identified by a *sudden fall* in temperature along a line of weather stations where there is heavy rain from low clouds (cumulus). Stations behind the front may be having better weather. Not all cold fronts have such well-defined features as those described - some may bring only small amounts of rain.

The symbol for a cold front and its direction of travel are shown in Figure 68.9b.

Figure 68.10 Cumulonimbus clouds associated with a cold front

(d) Occluded front

Cold fronts travel about twice as fast as warm fronts (30 mph/50 kmph compared with 15 mph/25 kmph). When a cold front catches up on a warm front, it pushes it off the Earth's surface and an 'occluded front' is formed, Figure 68.11. It produces the steady rain typical of a warm front followed by showers from the cold front with no fine spell between. The depression then dies away. Occluded fronts are more difficult to detect than the other two types.

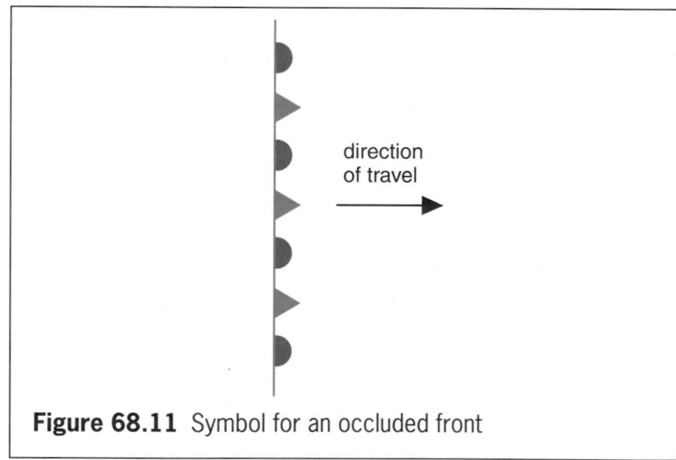

Figure 68.11 Symbol for an occluded front

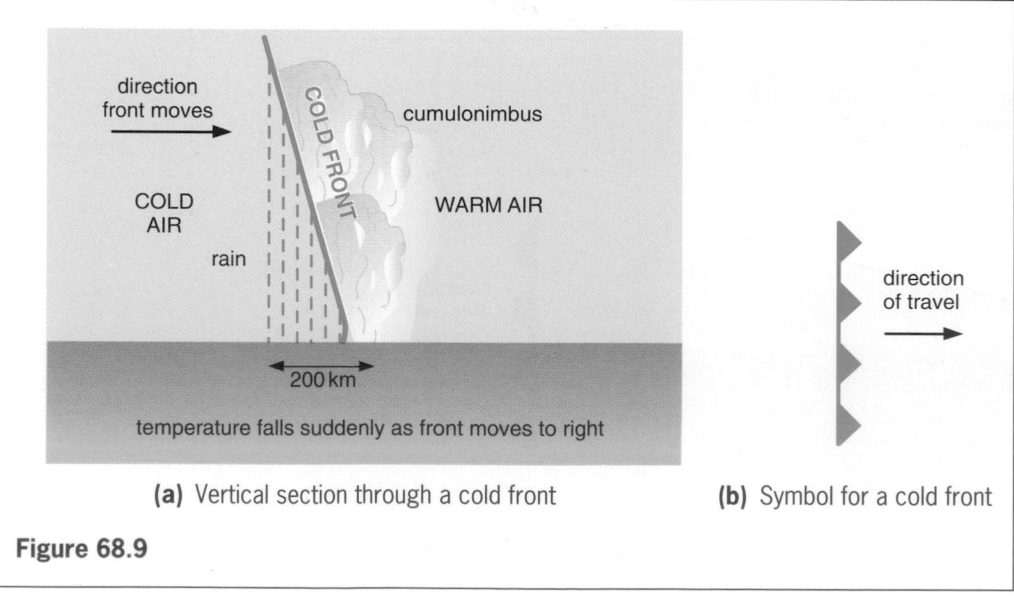

(a) Vertical section through a cold front

(b) Symbol for a cold front

Figure 68.9

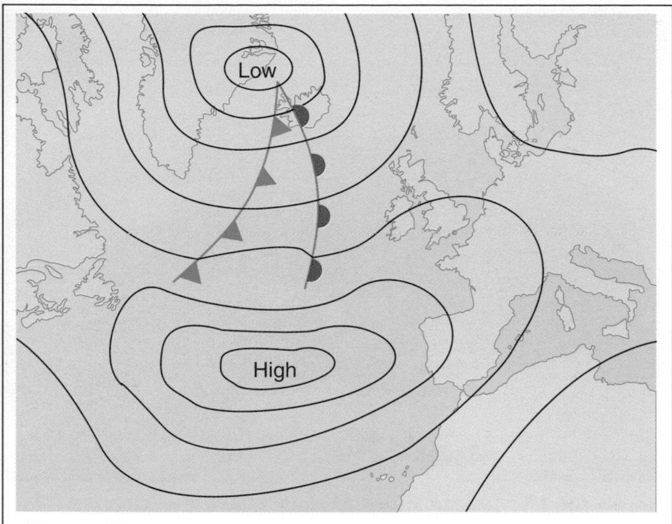

Figure 68.12 Warm and cold fronts approaching the British Isles

The weather map of Figure 68.6 is shown again in Figure 68.12 with the predicted positions of two fronts included. An easterly-moving warm front is being followed by a south-easterly cold front in a typical depression situation. The fronts meet near the centre of low pressure. Unsettled weather in the British Isles is a distinct possibility!

Air streams over the British Isles

There are *five* different air streams (moving air masses) that can affect the weather in Britain and Ireland. They are shown in Figure 68.13. The pressure pattern they give often consists not of 'highs' and 'lows' but of fairly straight and parallel isobars. The two **continental** air streams originate over land and are likely to be dry; the three **maritime** ones have come from oceans and are moist.

1 **Polar continental.** This brings very cold, *dry*, clear weather in winter and cool, *dry* weather in summer, from northern Asia.
2 **Tropical continental.** This air stream moves north from the African deserts to give *hot*, dry summers, sometimes carrying red dust with it.
3 **Tropical maritime.** This originates over the Azores and brings in from the central Atlantic and Caribbean *warm*, *moist* tropical air. This gives the south-west of Britain its typically mild, cloudy weather in winter and bright, sunny days as it dries out over Europe in summer.
4 **Polar maritime.** This arrives from the North Atlantic on a north-west wind and causes cold, showery weather with bright spells.
5 **Arctic maritime.** With its origin in the Arctic Ocean, this may bring chilly north winds accompanied by sleet or snow.

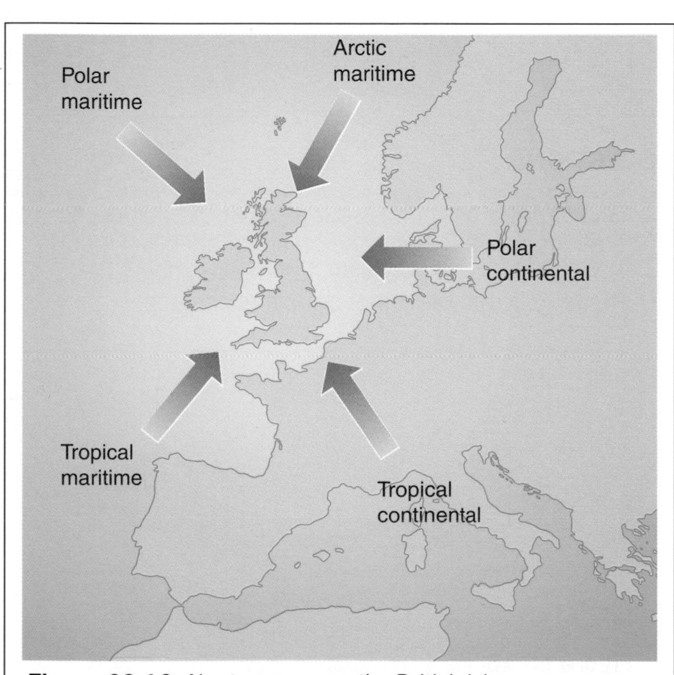

Figure 68.13 Air streams over the British Isles

Climate in the British Isles

(a) Rainfall

The south-west coast of Ireland is wetter than eastern England as the average monthly rainfall bar charts of Figures 68.14a and b show. This is because the prevailing winds from the Atlantic are moist (since the air has been over the sea) and reach western regions first, where it is also mountainous.

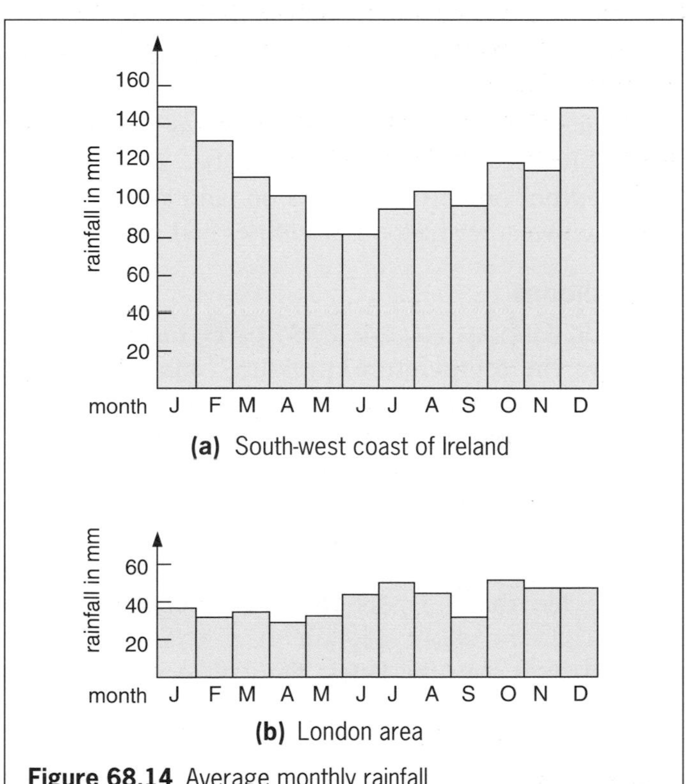

Figure 68.14 Average monthly rainfall

(b) Temperature

In winter it is warmer in the west than in the east because at this time of year Britain's temperatures are decided by the closeness of the Atlantic. The sea is then warmer than the land, making the prevailing westerly winds warm (as well as moist).

In summer it is warmer in the south than the north because then Britain gets most of its heat from the Sun and temperatures depend mainly on latitude. The Atlantic influence is much less since in summer the sea is not as warm as the land.

(c) Jet streams

These narrow belts of fast-moving air (200 km/h), which blow at mid-latitudes in an easterly direction at heights of about 10 km, were mentioned in the previous chapter (p. 290). They wobble about and the British Isles are often under one, which has an important effect on our weather.

When the stream is well to the north, warm air can move up from the Tropics and good weather results. This is what happened in the hot summers of 1976 and 1989. When it moves further south, mixtures of 'lows' and 'highs' occur to give rainy weather.

Long-range weather forecasting involves trying to predict the path of the jet stream.

Weather forecasting today

The atmosphere is a highly complex system in which conditions vary almost by the minute as the Earth, spinning on its tilted axis, travels on its yearly journey round the Sun. Forecasting the weather more than a few days ahead is therefore not easy and the more data that is available from as many sources as possible, the better are the predictions likely to be. Today, data is obtained not only from stations on land and at sea but also from weather balloons, satellites and radar.

(a) Balloons

Weather balloons (Figure 29.5) carry instruments to measure the temperature, pressure, wind speed and humidity at certain altitudes up to 35 km. The readings are sent back to Earth continuously by radio, and provide useful information about the jet streams.

(b) Polar orbiting satellites

These transmit down to Earth infrared cloud pattern pictures like that in Figure 68.15. They circle the Earth at a height of about 850 km in a 100-minute orbit which passes over the Poles. They observe a 3000 km wide band of the Earth's surface, and as the Earth rotates they achieve complete coverage in 24 hours. The pictures are relayed continuously and picked up by receiving stations in turn round the Earth.

Figure 68.15 Satellite image of cloud over Europe

(c) Geostationary satellites

These circle the Equator at a height of 36000 km. They travel at the same speed as the Earth's rotation so observe the same area at all times.

Our knowledge of world weather has received a great boost from satellites; they are particularly useful for observing regions such as the Poles, the oceans and the Tropics where there is a shortage of ground-based weather stations.

(d) Radar

This is able to detect rain and follow its motion.

(e) Computers

Using complex mathematical models of the atmosphere, computers are fed with the large amount of data now available to try to make forecasts more accurate but unfortunately, the undeniable truth is that we have no control over the weather.

QUESTIONS

1 Give reports on the present weather at each of the stations A and B using the station models in Figures 68.16a and b respectively. Reference should be made to *seven* pieces of data in each case.

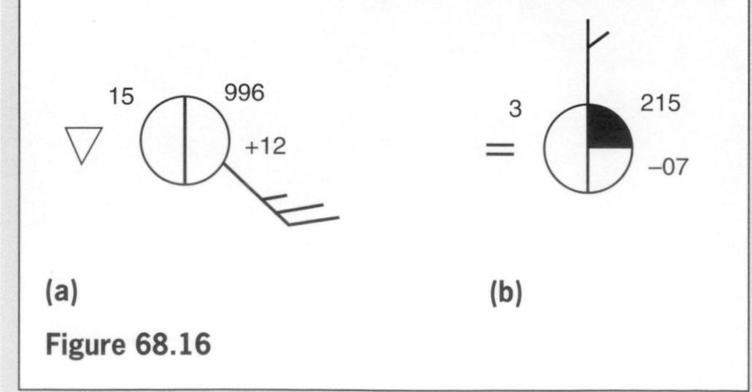

Figure 68.16

2 Draw up station models for weather stations X and Y based on the following data.

Station X
Pressure = 1039.2 mb
Pressure change = rise of 2.5 mb
Temperature = 9 °C
Cloud cover = 5/8
Wind = 10 knots from south
Drizzle

Station Y
Pressure = 997.6 mb
Pressure change = fall of 1.1 mb
Temperature = 4 °C
Cloud cover = 7/8
Wind = 30 knots from west
Fog

3 a What is an isobar?

b What kind of pressure system is represented in the small parts of the synoptic charts shown in Figures 68.17a, b by (i) A, (ii) B?

c In which system will the wind be stronger? Why?

d Why do winds blow *along* isobars rather than *across* them?

e If system A is in the northern hemisphere and B in the southern hemisphere what will be the wind direction at (i) X, (ii) Y?

f Outline the kind of weather each system brings.

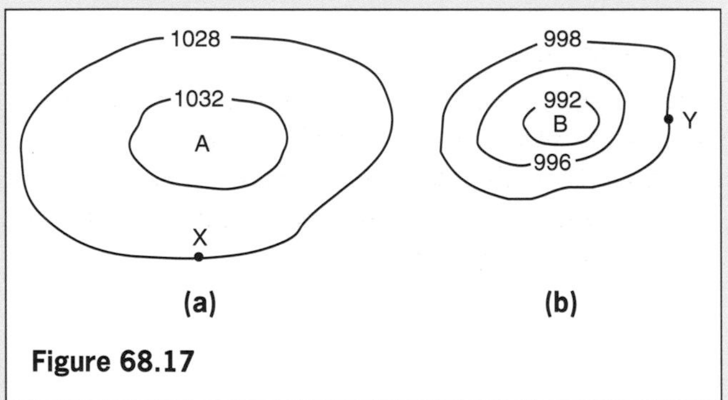

Figure 68.17

4 a What is a warm front?

b Draw the symbol used to show one on a weather map.

c State briefly a common sequence of events that accompanies a warm front.

Checklist

After studying this chapter you should be able to

- state what measurements of weather conditions are made and describe the instruments used,
- recall how weather measurements are recorded on a 'station model' and recognize the meteorological symbols used,
- use weather maps having isobars marked to identify winds (strength and direction), high and low pressure systems,
- recall that a **front** is the leading edge of a moving air mass,
- explain how warm, cold and occluded fronts are formed,
- describe the weather conditions associated with different fronts,
- predict an outline weather forecast from a weather map of present conditions,
- recall the five different air streams that can affect weather in the British Isles,
- describe the climate in different parts of the British Isles in terms of rainfall and temperature,
- outline the role played in weather forecasting by balloons, satellites, radar and computers.

69 STRUCTURE OF THE EARTH

Figure 69.1 The Earth is approximately spherical with a radius of about 6400 km or 4000 miles. Nearly 70% of its surface is covered by water

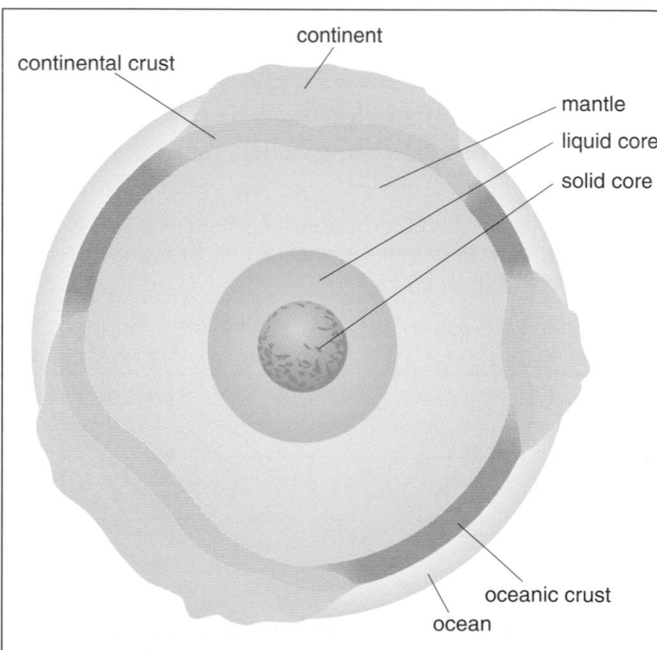

Figure 69.2 The structure of the Earth (not to scale)

I *Inside the Earth*

(a) Structure

There are three main zones or layers, Figure 69.2.

1 The **core** is a sphere of very dense material (probably iron and nickel) which stretches almost halfway to the surface of the Earth. It contains a solid inner part surrounded by a liquid outer shell. The Earth's magnetic field is thought to be caused by electric currents in the outer core.

2 The **mantle** surrounds the core and consists of semi-solid material (compounds of magnesium, silicon and oxygen) of much greater density than is found at the Earth's surface. When heat is generated within the mantle, the materials there are changed into hot, very viscous molten rock called **magma**.

3 The **crust** is the solid outer shell around the mantle. It comprises **oceanic** crust, made of dense, dark-coloured material (mostly basalt) which lies under the oceans to a depth of 5 to 10 km. Also, less dense, lighter coloured **continental** crust (mostly granite) is found under continents with a thickness ranging from 25 to 90 km. When the crust heats up, the solid materials of which it is composed become molten and can flow as magma.

The crust and the outermost part of the mantle form a firm shell round the Earth which is called the **lithosphere** (see Figure 69.6).

b) Evidence

The average density of the Earth is 5.5 g/cm^3, which is twice that of the rocks in the crust. This suggests there is a large volume of different, denser material at greater depths within the Earth. However, the main evidence for the existence of its layered structure comes from studying how shock waves from earthquakes travel through it. Such waves are called **seismic waves** and are detected by a seismograph. Basically this is a very sensitive electrical instrument with high-precision timing that responds to any vibration of the rocks in which it is embedded. The ground disturbances are recorded digitally for computer analysis or displayed as a trace (a **seismogram**) on a roll of paper on a revolving drum by a pen recorder.

c) Temperature

The Earth's temperature increases with depth and in some deep gold mines it can reach 50 °C. It is estimated that the core temperature is over 4000 °C (however the core is not gaseous because of the high pressure to which it is subjected by the weight of the surrounding rocks). The core's hotness arises from the Earth having been formed at a very high temperature (Chapter 71). The mantle and crust act as 'insulators' to heat flow so slowing down cooling. Another contributory factor is the heat produced by the decay of radioactive elements within the Earth.

Seismic waves

Seismic waves travel at different speeds in rocks of different density; **the greater the density, the greater the speed**. When a change of speed occurs, the waves change direction, i.e. are refracted (Chapter 5). By studying seismograms from widely separated seismograph stations, it has been possible to work out the depths in the Earth at which the structure changes, i.e. roughly where the layers of different density start and finish.

There are three types of seismic waves, denoted by the letters P, S and L.

The P or **primary** waves have the highest speeds (6 to 24 km/s) and are the first to be detected in an earthquake. They are longitudinal waves which can pass through **both liquids and solids** and, due to refraction, follow curved paths in the Earth. The direction of refraction shows that they travel faster in the mantle than in the crust and also that the deeper they go into the mantle and core the faster they travel.

The slower S or **secondary** waves, with transverse motion, follow closely behind the P waves and are the second type to be recorded. They also travel in curved paths through the mantle but **cannot pass through liquids** and so are stopped by the outer liquid core thus creating a 'shadow zone' on the other side of it.

The L or **long** waves travel most slowly and are largely confined to the crust. They are the main cause of earthquake damage near to a seismic event.

A typical seismogram with wave traces of the three types and their arrival times is shown in Figure 69.3. Note that the L waves have appreciably greater amplitudes than the other two types, i.e. they carry much more energy and so are much more destructive.

The paths of the seismic waves from the **epicentre** of an earthquake (i.e. the point on the Earth's surface directly above the origin of the earthquake) are shown in Figure 69.4.

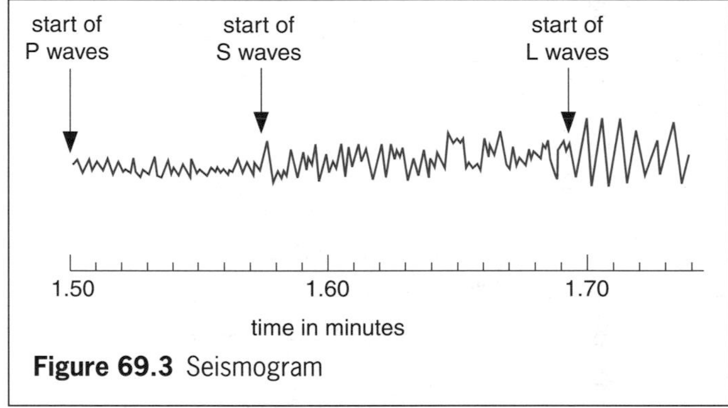

Figure 69.3 Seismogram

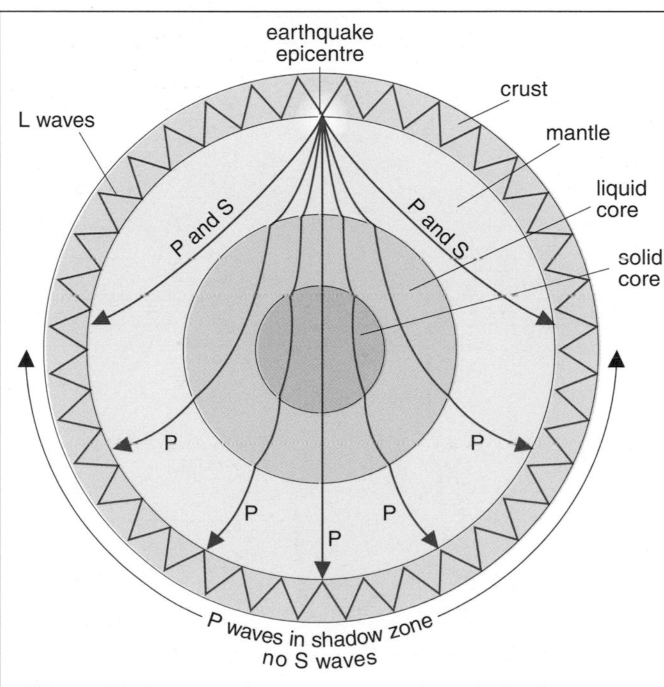

Figure 69.4 Paths of seismic waves through the Earth

Continental drift

It is now thought that all today's continents were once part of just one land mass, a super continent, which started to break up and drift apart about 200 million years ago. The fairly close fit that the east coast of South America makes with the west coast of Africa suggests that they might at one time have been joined. It is just one example of how the shapes of widely separated land masses look as if they might in the past have been part of a larger mass.

The movement is still going on. Africa and Europe are colliding as are India and central Asia. The Red Sea is widening by about 1 cm per year; northern England is rising 5 mm a year, while south-east England is sinking 1 mm or so a year. Evidence supporting the motion of continents comes from various sources.

Rocks on the floor of the oceans and those on the edges of surrounding continents have been found to have similarities. In some regions plant and animal **fossils** reveal that the climate in which they thrived was very different from the present one and continental drift seems to be a likely explanation for the change.

Paleomagnetism, which is the study of the magnetism of ancient rocks, provides the most convincing evidence for continental drift. When rocks containing magnetic materials are formed, they become magnetized in a N–S direction. Examination of rocks of different ages show there have been changes in the direction of the Earth's magnetic field over long periods. As a result the N and S magnetic poles appear to have wandered. Complete reversals of polarity have occurred at least nine times.

Theory of plate tectonics

Earthquakes, volcanoes and the forming of mountains were once thought to be due to the Earth contracting as it cooled. For the Earth to have a smaller volume, the crust would need to develop 'wrinkles' in the form of mountain ranges. Earthquakes and volcanoes were explained as being due to the crust cracking.

The Earth's crust is now considered to consist of several large slabs of rock and some smaller ones (about 15 in all), called **tectonic plates**, Figure 69.5. These are moving slowly relative to one another, by a few centimetres per year, causing among other things continental drift. The plates contain both continental and oceanic crust and move like rafts on a layer of the outer mantle, 200 to 300 km thick, called the **asthenosphere**, Figure 69.6.

It is thought that the source of energy for the movement of plates may be convection currents within the outer mantle, generated by heat from radioactive rocks and from the cooling of the Earth's core.

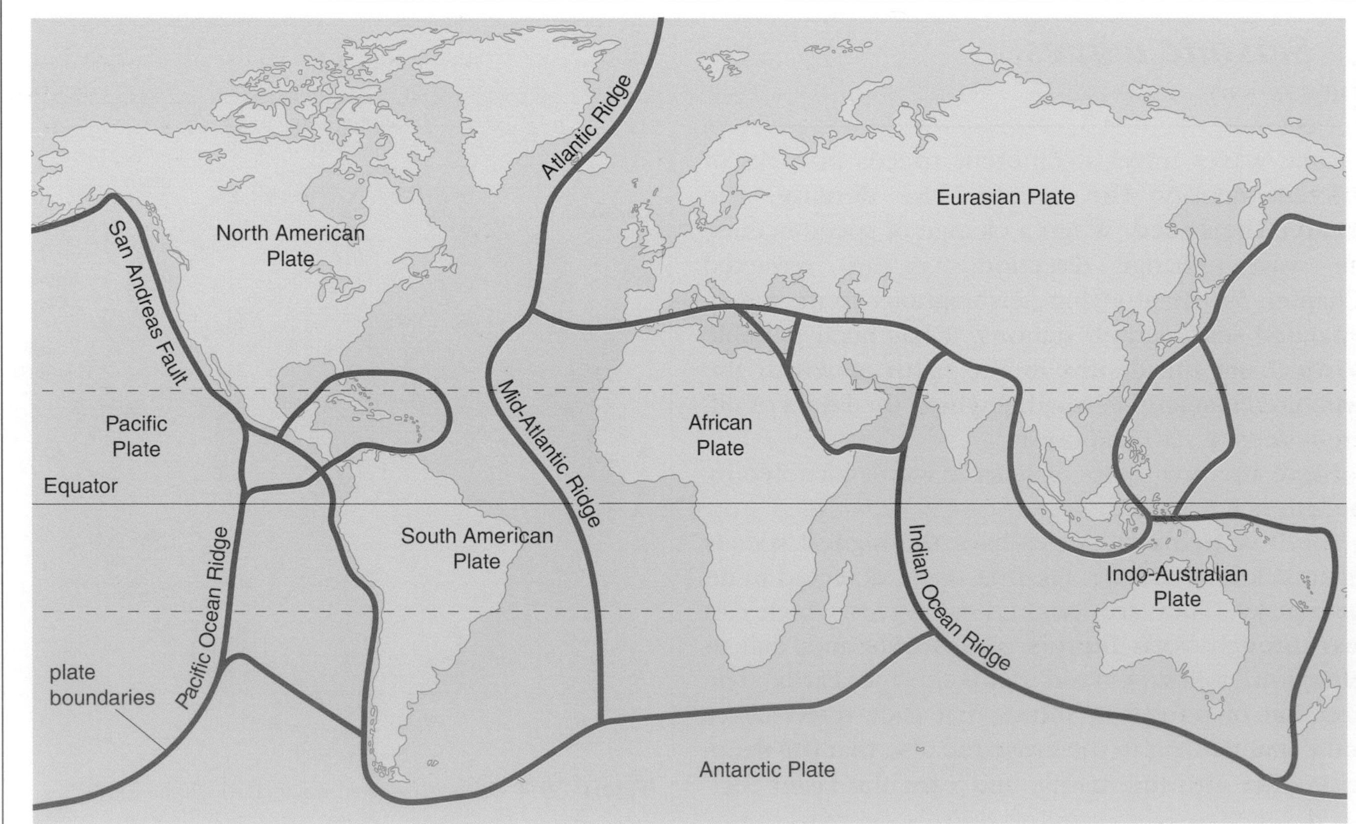

Figure 69.5 Tectonic plates

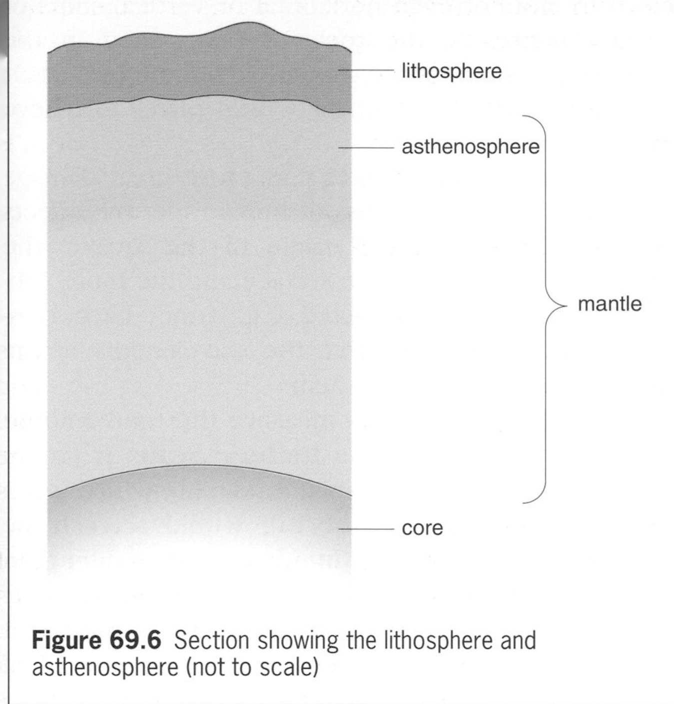

Figure 69.6 Section showing the lithosphere and asthenosphere (not to scale)

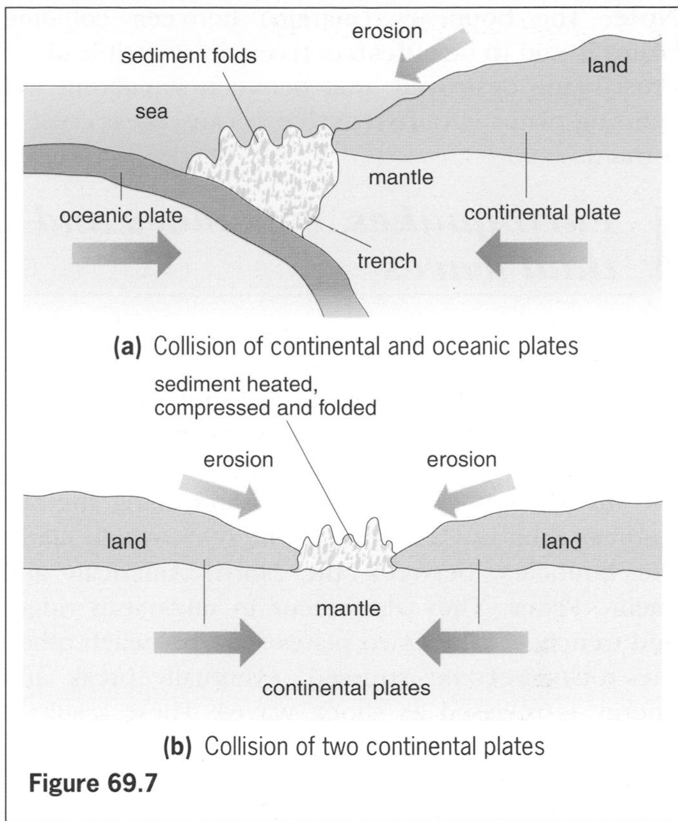

(a) Collision of continental and oceanic plates

(b) Collision of two continental plates

Figure 69.7

The theory of plate tectonics has produced explanations and better understanding of many apparently unrelated geological facts, as we will see in the next section.

Plate boundaries and geological features

Certain important geological features such as mountain ranges, island chains, mid-ocean ridges and rift valleys have their origin along the boundaries between moving tectonic plates.

(a) Colliding plates

When a **continental** plate meets an **oceanic** plate, the continental plate slides up over the denser oceanic plate which is forced down into the asthenosphere where it melts to become part of the mantle, Figure 69.7a. A trench is formed and is gradually filled up with sediment (i.e. rock fragments, gravel, sand, mud) from erosion of the land. As the collision continues, eventually after millions of years the sediment becomes compressed, folded and may be pushed upwards to form a range of mountains or a chain of islands. It is thought the Andes mountains were formed in this way when the oceanic Pacific plate was forced under the continental South American plate.

If two **continental** plates collide, Figure 69.7b, they have the same density and neither sinks. Instead, the sediment produced as they grind into each other is compressed and heated at the boundary to create, in time, large folded mountain ranges. The Himalayas are thought to be the result of this type of action.

(b) Separating plates

When two continental plates move apart on the ocean floor, **magma** (molten rock) rises up from the mantle to fill the resulting trench and produce new oceanic crust, Figure 69.8. As a result mid-oceanic 'mountain' ridges like those in the mid-Atlantic and Pacific oceans are created. The term sea-floor **spreading** is used to describe the separation of the plates. As the new floor is created it 'freezes' the magnetic field and the age of sea-floor material can be determined from this natural 'tape recorder'.

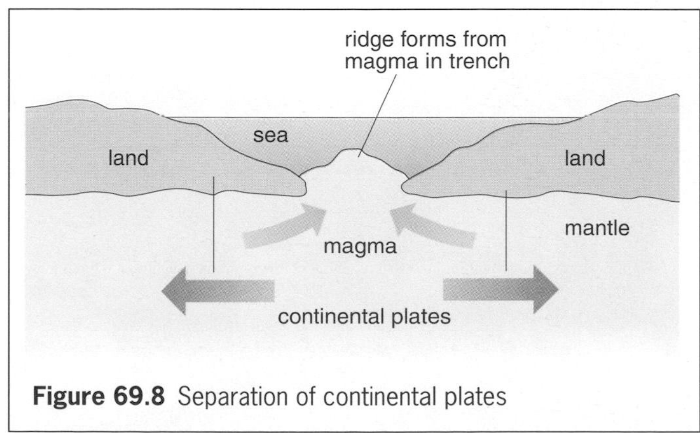

Figure 69.8 Separation of continental plates

(c) Splitting plates

If a continental plate splits, the fault often develops into a rift valley that may become a sea, as may have occurred in the case of the Red Sea.

Note. The boundary (margin) between colliding plates is said to be a **destructive** one since it leads to crust being destroyed; that between separating and splitting plates is **constructive** because new crust is formed.

Earthquakes, volcanoes and tidal waves

(a) Earthquakes

Large earthquakes are due to the movement of tectonic plates along the boundaries where they meet; many are the result of oceanic plates descending into the mantle. For example, they occur along the San Andreas Fault in California, Figure 69.5, which marks the boundary between the North American and Pacific Plates. They also occur in mid-ocean ridges and trenches. When two plates slide past each other, the rocks become stressed, eventually break and energy is released as shock waves. These result in earth tremors or even horizontal or vertical shifts o several metres in the rocks, as happened in th Californian orange grove shown in Figure 69.9a.

Figure 69.9a Lateral displacement of rows of trees due to plate movement along the San Andreas Fault, California

Figure 69.9b Devastation after the Armenian earthquake 1988

Smaller earthquakes occur within plates to reliev local stresses.

While major earthquakes can cause great damag to property and tragic loss of human life, the effect are less the deeper the origin of the 'quake, th firmer the rocks are at the surface and the more sub stantial the buildings affected, e.g. if they have foun dations that are larger than the wavelengths of th tremors that might strike them.

Earthquake magnitudes measure the total amoun of energy released on the **Richter Scale**; it range from 0 to 9. One rated of magnitude 2.0, which is te times greater than one of 1.0, would scarcely b noticed. One of 3.0 is one hundred times greater tha one of 1.0, and so on. The Armenian earthquake o 1988 had magnitude 7.0; in less than 1 minute i reduced two cities to rubble, Figure 69.9b, and kille over 25 000 people. Altogether there are more tha one million earthquakes annually but fortunately onl about one of the Armenian magnitude per year.

Britain is far removed from plate boundaries and s earthquakes today are rare but mild ones occur occa sionally. They arise from slight slipping of rock fault caused by earthquakes that happened millions o years ago. An example is the Great Glen Fault whic separates Scotland into two parts and is now occu pied by three lochs and the Caledonian Canal.

(b) Volcanoes

The fact that many of the main volcanic region occur along tectonic plate boundaries suggests tha volcanic eruptions might arise from the same force that cause earthquakes and the formation of mour tains, i.e. moving plates. Plates rubbing together pro duce heat by friction, some rock melts and travel along cracks in the crust. When eruptions occur, ho magma, being less dense, rises up through the crus and is ejected as **lava**. The eruption is accompanie by a huge explosion due to water near the surfac being turned into steam; clouds of ash and gases, e.g poisonous sulphur dioxide, are also ejected and crater is blown.

There are about 500 active volcanoes on Earth an more under the oceans. They are common round th edges of the Pacific Ocean. The eruption of Mount S Helen's on the north-west coast of the USA in 198C Figure 69.9c, blew away half the mountainside an caused loss of life despite being predicted fron events before the eruption such as earth tremors swelling of the mountainside and the presence c certain gases at its top.

When lava solidifies in the crater of a volcano, th volcano becomes dormant.

igure 69.9c Eruption of Mount St Helen's, Washington State, in 980

Figure 69.10a Granite rocks on Dartmoor (photo: BGS)

Figure 69.10b Basalt rocks in the Giant's Causeway, Northern Ireland

c) Tidal waves (tsunamis)

'hese are huge waves that occur as a result of an arthquake or volcanic eruption under water. They ave enormous energy and travel at speeds up to '00 km/h.

The Earth and its rocks

'he study of rocks is called **geology** and it is a subject vhich has told us much about the history of the Earth. tocks can be put into one of three groups depending n their origin and how they were formed.

a) Igneous rocks

'hese formed when magma from below the Earth's urface cooled and solidified, either in the Earth's rust or on its surface if it was ejected through the rust as lava in a volcanic eruption. Examples are **;ranite** and **basalt**. They are the bedrock of large arts of most continental and oceanic crust and vhen not covered by sediment or the sea, they are xposed at the Earth's surface, Figures 69.10a, b. 'hey contain no fossils.

(b) Sedimentary rocks

These are the result of the weathering (i.e. breaking down) of older rocks over millions of years to form sediments. Sediments consisting of loose pieces of rock, gravel, sand and mud were produced and in many cases carried from their place of origin to the sea-bed. There they were deposited in layers (strata), one on top of the other. The transporting agents responsible for the erosion (i.e. eating away) of the rocks were rain, wind, rivers and glaciers; they moved small-particle sediments farther than large-particle ones.

The lower layers of sediment were pressed together tightly and in time hardened into sedimentary rock. In some instances movements inside the Earth made the layers bend and fold, like those shown in Figure 69.11 (overleaf), often creating faults and fractures in the rock.

Figure 69.11 Folded sedimentary rocks in Dorset (photo: BGS)

Examples of sedimentary rocks are **sandstone**, **limestone**, **clay** and **coal**. They form about three-quarters of the rocks on or very close to the Earth's surface. Many deposits, e.g. of limestone and coal, contain fossils from the remains of animals and plants.

(c) Metamorphic rocks

These were formed when either igneous or sedimentary rocks were subjected to very high temperatures and/or pressures inside the Earth; examples are **slate**, formed from clay, and **marble**, formed from limestone. Metamorphic rocks can be exposed at ground level if weathering or quarrying removes the material covering them, Figure 69.12. They are often found in volcanically active regions. They have no fossils.

Figure 69.12 Slate quarry in the Lake District

Rocks are important in everyday life. They break down to give the **soil** in which our food is grown. Some are used to make **building materials** such as stone, bricks, concrete and glass. Others contain **ores** that provide metals like iron and copper that are essential in industry. All contain **minerals**, i.e. naturally occurring crystalline inorganic substances; for example, granite is mostly quartz, Figure 69.13. There are over 2000 different minerals, each with its characteristic physical and chemical properties. Many minerals are silicates, containing oxygen and silicon; these make up most of the Earth's crust and are called 'rock-making' minerals.

Figure 69.13 Granite showing quartz crystals

Weathering and soil

Weathering is the process which occurs when rocks are broken down into smaller pieces by the action of wind, rain, frost, sun, etc. For example, temperature changes cause the outer layers of rocks to expand during the day and to contract at night. Internal stress is created, leading to cracking and disintegration. Similarly, if rainwater freezes in rock cracks, it expands as it changes to ice, making the cracks wider and deeper and causing bits to break off.

Ultimately the sediment particles produced by weathering become fine enough to be classed as soil in which plants can grow. When the plants die, they decompose to form a dark sticky substance called **humus** which plays a vital part in releasing chemicals from minerals in the soil. These are then absorbed in solution by plant roots.

There are three layers in most soils, as shown by the **soil profile** in Figure 69.14. The top layer contains the smallest particles and humus. The middle layer or sub-soil consists of larger particles and rests on the bottom layer of weathered parent rock.

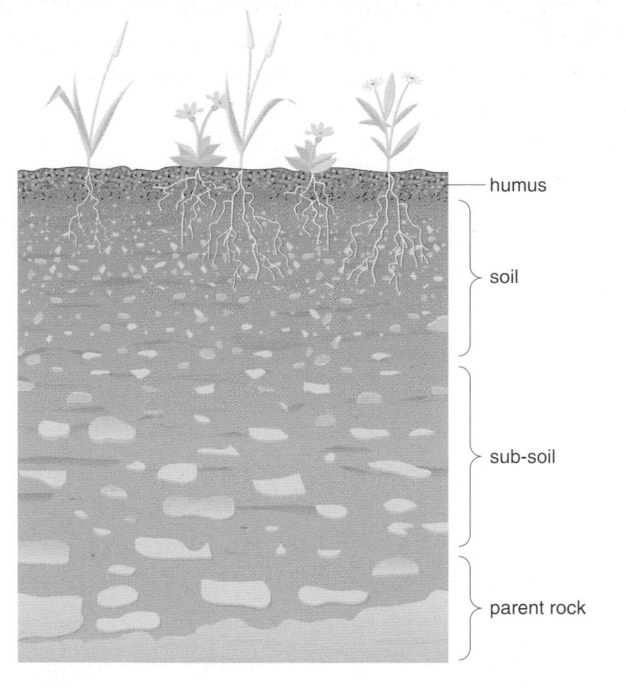

Figure 69.14 Soil profile

Particle sizes vary in different soils. In clay they are very small and the spaces between them soon fill up with water and the soil becomes waterlogged. In sand, particles and spaces are larger and they drain well.

QUESTIONS

1 Name the three main layers of the Earth's structure.

2 What conclusions regarding the density and structure of the Earth can be drawn from the following facts?

a P and S waves travel faster through the mantle than through the crust.

b P and S waves travel faster the deeper they go into the mantle.

c P waves travel more quickly through the inner core than through the mantle.

d S waves cannot pass through the outer core.

3 **a** State *three* pieces of evidence which support the theory of continental drift.

b What causes continental drift?

4 By referring to Figure 69.5 showing tectonic plate boundaries, state which two plates were probably responsible for creating **a** the Himalayan Mountains, **b** the Mid-Atlantic Ridge.

Checklist

After studying this chapter you should be able to

- describe the nature of the Earth's layered structure and state from what the evidence for this structure comes,
- recall the properties of the three types of **seismic wave** and describe how they are transmitted through the Earth,
- recall that the Earth's crust consists of large slabs of rock called **tectonic plates** whose relative motion can account for the drift, shape and rock records of the continents,
- explain the formation of mountain ranges, island chains, mid-ocean ridges and rift valleys in terms of moving tectonic plates,
- use the theory of plate tectonics to explain how earthquakes and volcanoes arise at plate boundaries,
- recall the three main groups of rocks and their properties,
- explain how soil is produced by weathering.

70 THE SOLAR SYSTEM

Members of the Solar System

The **Solar System** consists of the Sun and the nine planets moving round it in elliptical orbits, i.e. slightly flattened circles, Figure 70.1. It also includes the asteroids, comets and the moons that travel round most of the planets.

The **four inner planets**, Mercury, Venus, Earth and Mars are all small, of similar size, solid and rocky, probably with a layered structure.

The **four outer planets**, Jupiter, Saturn, Uranus and Neptune are much larger and colder and consist mainly of gases. The outermost planet, Pluto, is quite small and thought to be made of ice.

The **asteroids** are pieces of rock of various sizes which orbit mostly between Mars and Jupiter.

The **Sun**, being a star, produces its own light which takes 8 minutes to reach us; the planets and our Moon, on the other hand, are seen from the Earth by reflected solar light. The Sun has a surface temperature of about 6000 °C and in its central core, where the temperature must be many millions of °C, nuclear fusion occurs. This results in hydrogen being changed to helium and the whole range of electromagnetic radiation from gamma rays to radio waves is emitted. It is a 'yellow dwarf' star (see Chapter 71) estimated to be about halfway through its lifetime of 10 000 million years.

Figure 70.1 The Solar System (distances from the Sun not to scale)

Motion of the Earth

The occurrence of certain natural events is readily explained by the Earth's motion.

(a) Day and night

These are caused by the Earth spinning on its axis (i.e. about the line through its north and south poles) and making one complete revolution every 24 hours. This creates day for the half of the Earth's surface facing the Sun and night for the other half.

(b) The seasons

Two factors are responsible for these. The first is the motion of the Earth round the Sun once every 365.24 days, i.e. in 1 year, and the second is the tilt of the Earth's axis (at 23.5°) to the plane of its path round the Sun. Figure 70.2 shows the tilted Earth in four different positions of its orbit.

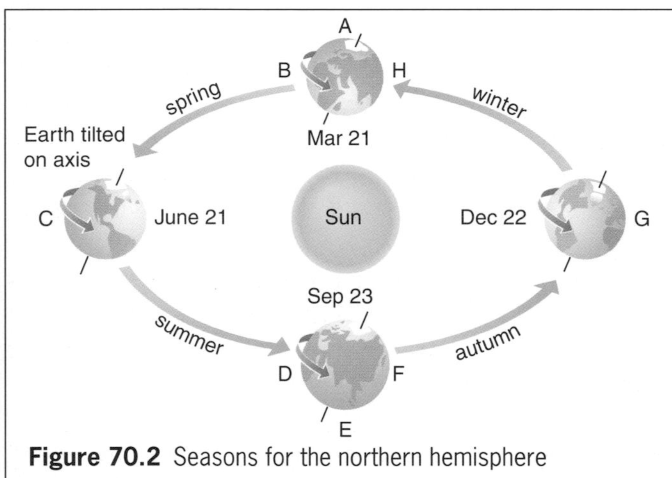

Figure 70.2 Seasons for the northern hemisphere

Over part BCD of the orbit, the northern hemisphere is tilted towards the Sun and so it is spring and summer with the hours of daylight being greater than those of darkness. The southern hemisphere is tilted away from the Sun and is having autumn and winter with shorter days than nights. The northern hemisphere receives more solar radiation and the weather is consequently warmer.

Over FGH the situation is reversed. The southern hemisphere is tilted towards the Sun, while the northern hemisphere is tilted away from it and experiences autumn and winter.

At C the northern hemisphere has its longest day, while the southern hemisphere has its shortest, usually on 21 June. At G the opposite is true and occurs about 22 December.

At A and E night and day are equal in both hemispheres. These are the **equinoxes**, often 21 March and 23 September.

(c) Rising and setting of the Sun

The Earth's rotation on its axis causes the Sun to have an apparent daily journey from east to west. It rises exactly in the east and sets exactly in the west only at the equinoxes. In the northern hemisphere in summer it rises north of east and sets north of west. In winter it rises and sets south of these points.

Each day the Sun is highest above the horizon at noon and directly due south in the northern hemisphere; this height itself is greatest and the daylight hours longest about 21 June. Thereafter the Sun's noon height slowly decreases and near 22 December it is lowest and the number of daylight hours is smallest, Figure 70.3.

In the southern hemisphere the Sun is due north at noon.

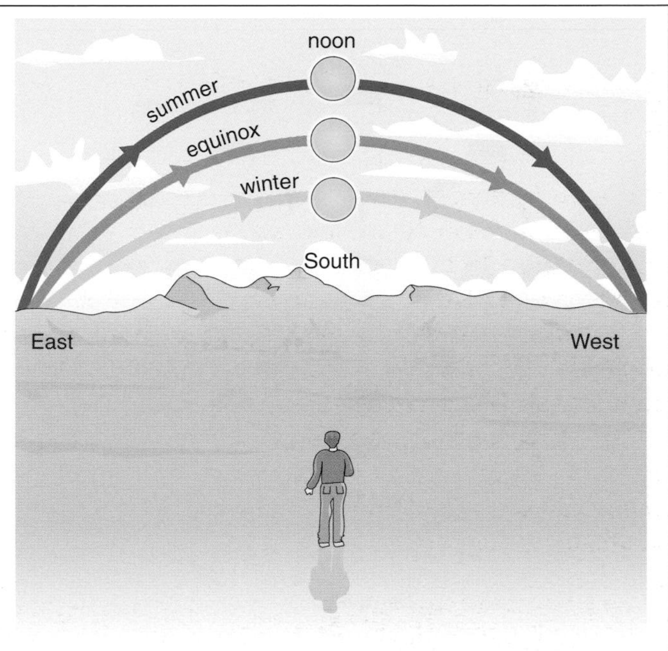

Figure 70.3 Rising and setting of the Sun (in the northern hemisphere)

Motion of the Moon

The Moon is a satellite of the Earth and travels round it in an approximately circular orbit once every 27.3 days at an average distance away of about 400 000 km or 240 000 miles. It also revolves on its own axis in 27.3 days and so always has the same side facing the Earth, hence we never see the 'dark side of the Moon'. We see the Moon by reflected sunlight since it does not produce its own light. It does not have an atmosphere. It does have a gravitational field due to its mass, but the field strength is only one-sixth of that on Earth. Hence moon-walkers move in a 'springy' fashion but do not fly off into space.

(a) Phases

The Moon's appearance from the Earth changes during its monthly journey; it has different phases. In Figure 70.4 the outer circle shows that exactly half of it is always illuminated by the Sun. What it looks like from the Earth in its various positions is shown inside this. In the New phase, the Moon is between the Sun and the Earth and the side facing the Earth, being unlit, is not visible from the Earth. A thin 'new' crescent appears along one edge as it travels in its orbit, gradually increasing until at the First Quarter phase, half of the Moon's face can be seen. At Full Moon it is on the opposite side of the Earth from the Sun and appears as a complete circle. Thereafter it wanes through the Last Quarter until only the 'old' crescent can be seen.

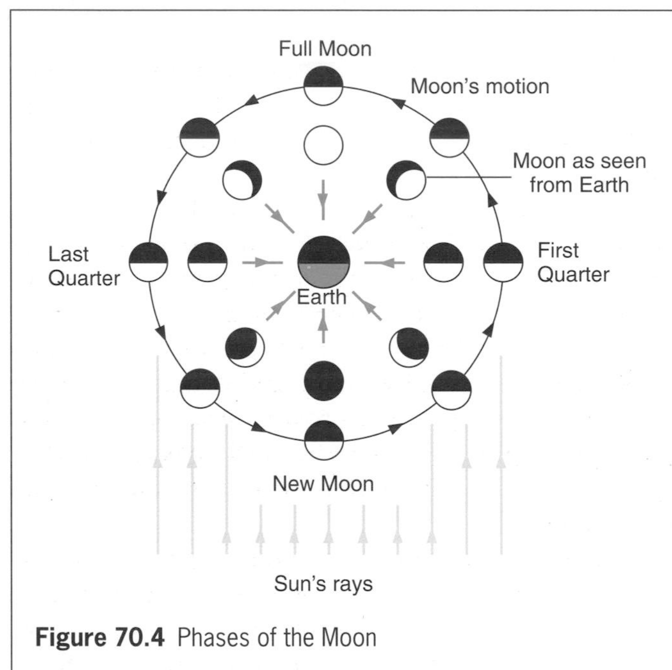

Figure 70.4 Phases of the Moon

(b) Eclipse of the Sun

There is an eclipse of the Sun by the Moon (a **solar eclipse**) when the Sun, Moon and Earth are in a straight line. Solar eclipses can only occur at the New Moon phase. Anyone at B in Figure 70.5a sees a **total** eclipse of the Sun (i.e. they can't see the Sun at all). Anyone at A sees a **partial** eclipse (i.e. part of the Sun is still visible).

Sometimes the Moon is farther from the Earth (since its orbit is not a perfect circle), Figure 70.5b, and then while those at A would still see a partial eclipse, anyone at B sees an **annular** eclipse (i.e. only the central region of the Sun is hidden).

A total eclipse seen from one place may last for up to 7 minutes. During this time, although it is day, the sky is dark, stars are visible, the temperature falls and birds stop singing.

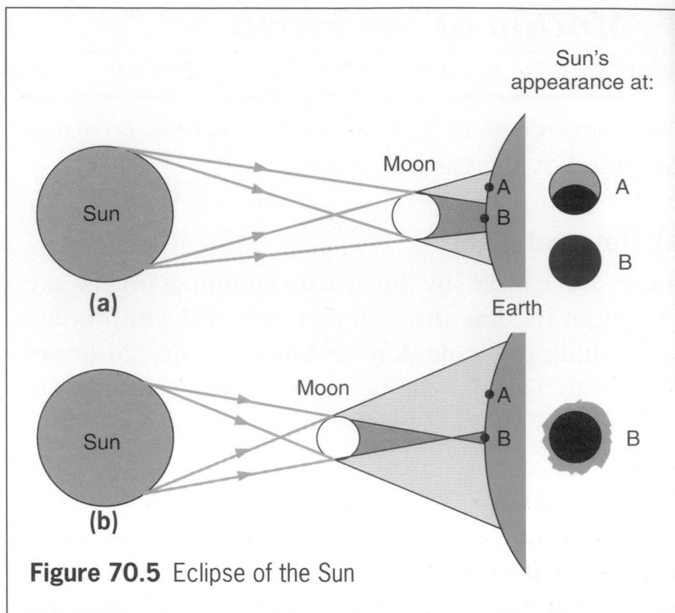

Figure 70.5 Eclipse of the Sun

(c) Eclipse of the Moon

A **lunar eclipse** occurs when the Moon passes into the Earth's shadow, i.e. the Earth comes between the Sun and the Moon, Figure 70.6. Lunar eclipses can only occur at the Full Moon phase.

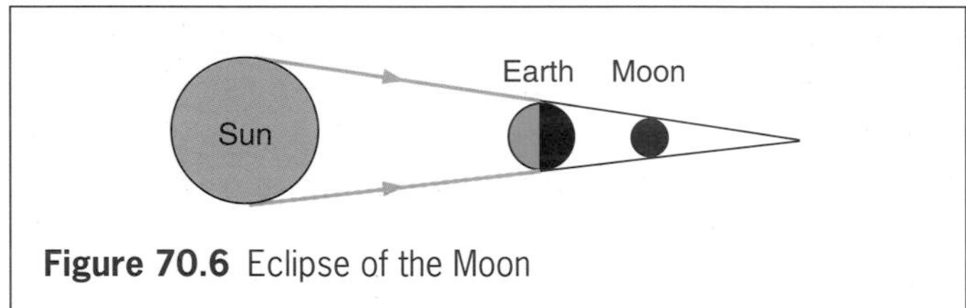

Figure 70.6 Eclipse of the Moon

(d) Rising and setting of the Moon

Like the Sun, the Moon seems to have a daily trip across the sky from the east where it rises to the west where it sets, due to the Earth's rotation on its axis.

More about the planets

(a) Data

Facts and figures about the planets are listed in Table 70.1; times are given in Earth hours (h), days (d) or years (y). As well as being of general interest this data indicates (i) factors that affect conditions on the surface of the planets, and (ii) some of the environmental problems that a visit or attempted colonization would encounter!

Table 70.1

Planet	Av. distance from Sun (million km)	Orbit time round Sun (days/yrs)	Surface temp. (°C)	Spin time about axis (hrs/days)	Diameter (thousand km)	Relative mass (Earth = 1.0)	Surface gravity (N/kg)	No. of moons
Mercury	58.0	88 d	350	58.5 d	4.8	0.05	3.6	0
Venus	108	225 d	460	243 d	12.2	0.81	8.7	0
Earth	150	365 d	20	24.0 h	12.8	1.0	9.8	1
Mars	228	687 d	−23	24.6 h	6.8	0.11	3.7	2
Jupiter	778	11.9 y	−120	10 h	143	318.0	25.9	16
Saturn	1430	29.5 y	−180	10.6 h	120	95.0	11.3	21
Uranus	2870	84 y	−210	17.2 h	49	14.0	10.4	15
Neptune	4500	165 y	−220	16 h	50	17.5	14.0	8
Pluto	5900	248 y	−230	6.3 d	2.3	0.003	?	1

(b) Some features

A planet's year (i.e. orbit time round the Sun) increases with distance from the Sun. The orbital speed decreases with distance; Neptune travels much more slowly than Mercury. Surface temperatures decrease markedly with distance from the Sun, with one exception, that of Venus.

Venus has a high surface temperature (460 °C) due to its dense atmosphere of carbon dioxide acting as a heat trap (i.e. the greenhouse effect). Its very slow 'spin time' of 243 Earth days means its day is longer than its year of 225 Earth days!

Mercury, also with a slow 'spin time' (58.5 days) has practically no atmosphere and so while its noon temperature is 350 °C, at night it falls to −170 °C. Its surfaces are also exposed to long periods of heat or cold because of its slow rotation.

Mars is the Earth's nearest neighbour. It is colder than the Earth, temperatures on its equator seldom exceeding 0 °C even in summer. Its atmosphere is very thin and consists mostly of carbon dioxide with traces of water vapour and oxygen. Its axis is tilted at an angle of 24° and so it has seasons but these are longer than on Earth. There is now no liquid water on the surface but it has polar ice-caps of water ice and solid carbon dioxide ('dry ice'). In some parts there are large extinct volcanoes and evidence such as gorges of torrential floods in the distant past, Figure 70.7a. It is now a comparatively inactive planet though high winds do blow at times causing dust storms.

Jupiter is by far the largest planet in the Solar System. It is a gaseous planet and is noted for its Great Red Spot, Figure 70.7b, which is a massive swirling storm.

Saturn has characteristic rings, Figure 70.7c, which are made of ice particles. Like Jupiter, it has an ever-changing very turbulent atmosphere of hydrogen, helium, ammonia and methane gas.

Uranus and **Neptune** have methane in their atmosphere as well as hydrogen and helium. The four outer 'gas giants' are able to retain these lighter gases in their atmospheres, unlike the Earth, because of the greater gravitational attraction they exert due to their large masses.

The **Voyager** and other unmanned space missions of the 1980s obtained much new information about the outer planets and their many moons as they flew past them. Uranus and Neptune were found to have rings as well as Saturn, but not so large. Signs of

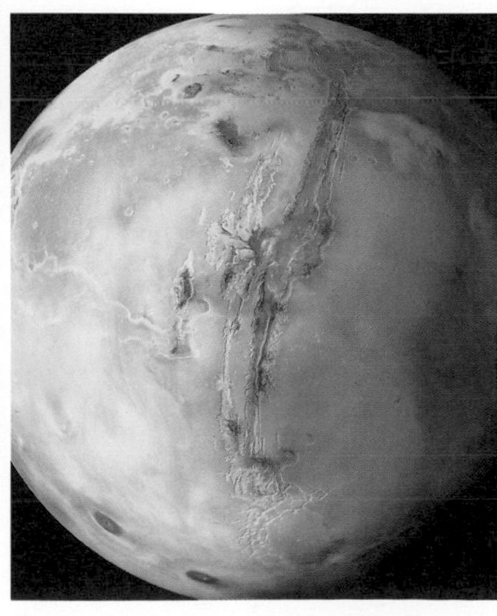

Figure 70.7a The surface of Mars showing gorges and craters

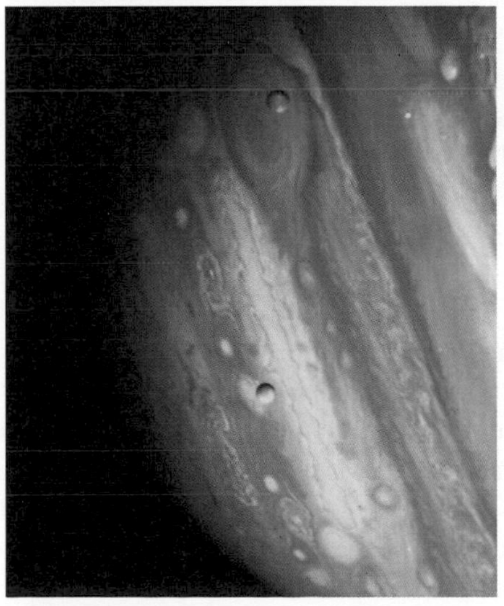

Figure 70.7b Jupiter's Great Red Spot (and note two of the planet's moons)

Figure 70.7c Saturn's rings

geological activity on the planets can be detected by direct observations and, in some cases, from the emission of infrared radiation.

Less is known about **Pluto**. It is thought to be smaller than our Moon and to have an 'atmosphere' of frozen methane.

Observing the planets

In addition to the Sun and Moon, five other fairly bright objects can be seen without a telescope moving among the stars. They are the planets (or 'wanderers') Mercury, Venus, Mars, Jupiter and Saturn.

Like the Sun and Moon, the planets seem to rise in the east and set in the west but sometimes their movements appear to lack the regularity associated with other heavenly bodies and were a puzzle to early astronomers. Their positions against the background of the stars depend on where they and the Earth are as they orbit the Sun.

Venus and Jupiter are the two brightest planets. Venus appears periodically as a very bright evening 'star'. It is first visible just after sunset, close to the Sun, and sets shortly after sunset. It gradually moves eastwards from the Sun on subsequent days and sets later in the evening, Figure 70.8a. Some time later it appears as a morning star, rising at first just before the Sun and then earlier from day to day as its westward motion away from the Sun increases, Figure 70.8b. Jupiter is visible for several months each year.

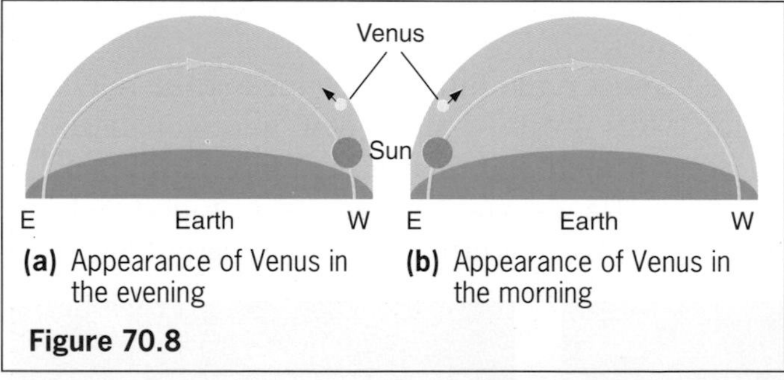

(a) Appearance of Venus in the evening **(b)** Appearance of Venus in the morning

Figure 70.8

Mars is outshone only by Venus and Jupiter and is visible from Earth as an orange-red planet for several months every year.

Mercury is difficult to see from the Earth because it is close to the Sun. It appears briefly, low in the sky after sunset and before dawn.

Maps of the sky are published monthly in national newspapers and show where to look for different planets. Using even a small telescope or binoculars improves the details that can be seen, e.g. Jupiter's moons and Saturn's rings.

Warning. Looking directly at the Sun through a telescope or binoculars can cause blindness. The only safe way to view the Sun is to project its image through a telescope onto a piece of card.

Gravity and satellites

(a) Theory of gravity

Newton proposed that all objects in the Universe having mass attracted each other with a force called gravity. The greater the mass of each object and the smaller their distance apart, the greater is the force. In fact, halving the distance quadruples the force. Earlier we saw that it is the force of gravity arising from the mass of the Earth that causes a body to fall with an acceleration $g = 9.8 \text{ m/s}^2$ (Chapter 33). The Moon has a smaller mass than the Earth and on it $g = 1.6 \text{ m/s}^2$.

To keep a body moving in a circular path requires a centripetal force, as we saw previously (Chapter 37). The centripetal force needed increases if

(i) the mass of the body increases,
(ii) the speed of the body increases, and
(iii) the radius of the orbit decreases.

In the case of the planets orbiting the Sun in near-circular paths, it is the force of gravity between the planet and the Sun which provides the necessary centripetal force. The Moon is similarly kept in a circular orbit round the Earth by the force of gravity between it and the Earth.

(b) Satellites

To put an artificial satellite in orbit at a certain height above the Earth it must enter the orbit at the correct speed. If it does not, the force of gravity, which decreases with height, will not be equal to the centripetal force needed for the orbit.

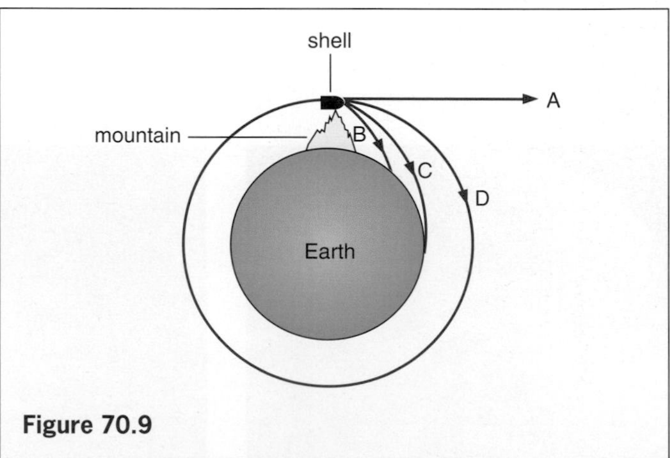

Figure 70.9

This can be seen by imagining a shell fired horizontally from the top of a very high mountain, Figure 70.9. If gravity did not pull it towards the centre of the Earth it would continue to travel horizontally, taking path A. In practice it might take path B. A second shell fired faster might take path C and travel farther. If a third shell is fired even faster, it might never catch up with the rate at which the Earth's surface is falling away. It would remain at the same height above the Earth (path D) and return to the mountain top, behaving like a satellite.

A high-level geostationary communication satellite at 36 000 km experiences a smaller gravitational force than a low-level polar one orbiting at 850 km (Chapter 68) since it is at a greater distance from the Earth. Its speed is less, so it takes longer to make a complete orbit of the Earth, 24 hours compared with 100 minutes.

Space programmes

There have been many beneficial spin-offs from space programmes. They have been the reason for much of the miniaturization of electronic components and systems. They have enabled satellites to be put into orbit to assist air and sea navigation, to improve the reliability of worldwide telecommunications, to make weather forecasting more accurate, to allow astronomical observations to be made unaffected by the Earth's atmosphere and to permit surveillance and monitoring of the Earth's surface for a variety of reasons.

In the realm of space travel one of the most notable feats occurred on 20 July 1969 when the American lunarnauts Edwin Aldrin and Neil Armstrong landed on the Moon with their Lunar Excursion Module (LEM), Figure 70.10.

Figure 70.10 Edwin Aldrin on the Moon with the Lunar Excursion Module

The exploration of space, even by unmanned vehicles, is very costly and the benefits have to be weighed against this. Huge amounts of fuel are needed for rockets and their payloads to leave the Earth and reach the speeds required on their long journeys. Limitations on manned space travel are set by the unfriendliness of the environments of possible destinations and by the difficulties of maintaining the conditions to sustain life, such as the supply of oxygen, water, food, warmth and sanitation. Despite these difficulties and the problems of existing in zero gravity, the Russian cosmonaut Yuri Romanenko spent a record 326 days in the Mir space station during 1987.

Comets

Comets, which consist of lumps of rocky ice surrounded by dust and gas, orbit the Sun in highly elliptical paths, Figure 70.11, and are much closer to it at some times than others. They return to the inner Solar System at regular intervals but in many cases these are so long (due to journeys far beyond Pluto) that they cannot be predicted. On approaching the Sun the dust and gas are blown backwards and the comet develops a bright head and a long tail pointing away from the Sun.

One of the most famous is Halley's comet, Figure 70.12, which visits the inner Solar System about every 76 years, the last occasion being in 1986.

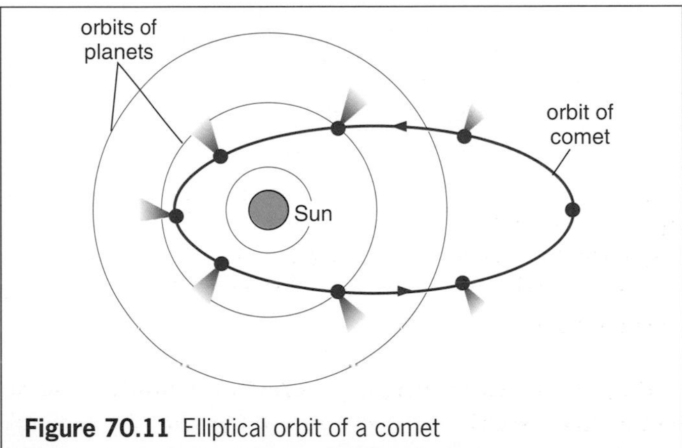

Figure 70.11 Elliptical orbit of a comet

Figure 70.12 Halley's comet

Tides

Tides are caused by the gravitational pull of the Moon, and to a lesser extent the Sun, on the oceans. There is a high tide at places nearest the Moon and *also* opposite on the far side, Figure 70.13a. As the Earth rotates on its axis the positions of high tide move over its surface, giving two high and two low tides daily or, more exactly, every 24 hours 50 minutes. The extra 50 minutes is due to the Moon travelling round the Earth in the same direction as the Earth's daily spin, so the Earth has to make just over one revolution before a given place is again opposite the Moon.

A full explanation of the tides is rather complex. Here it will be enough to say that as a result of the Moon's gravitational pull weakening as the distance from the Moon increases, the oceans nearest the Moon experience a larger pull than the solid Earth as a whole and this raises a high tide there. The Moon's pull on the oceans on the remote side of the Earth is less than that on the solid Earth and allows them to flow away into another high tide. The tide-raising forces thus arise from *differences* of the Moon's attraction for various parts of the Earth.

The tidal range is a maximum when there is a New or a Full Moon. The Sun, Moon and Earth are then in line, Figure 70.13b, with the lunar and solar pulls reinforcing to produce extra high (**spring**) tides. The lowest (**neap**) tides occur when there is a Half Moon and the Sun and Moon are pulling at right angles to each other, Figure 70.13c.

The times of high tide differ from port to port on account of the shape of the shore, friction from the sea-bed and other factors which can only be found from past records.

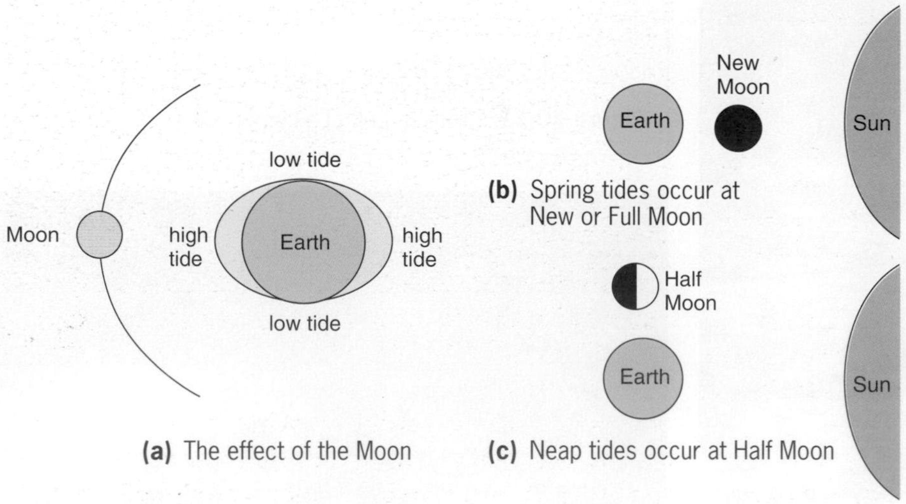

(a) The effect of the Moon

(b) Spring tides occur at New or Full Moon

(c) Neap tides occur at Half Moon

Figure 70.13 Formation of tides

QUESTIONS

1 Which of the following statements are *true*?

A The Sun is a star which produces its own light.
B Planets are seen by reflected solar light.
C The planets revolve round the Earth in elliptical orbits.
D Mercury is the planet closest to the Earth.
E The four outer planets Jupiter, Saturn, Uranus and Neptune consist mainly of gases.

2 Which of the following statements is *not* true?

A The Earth goes round the Sun once each year.
B The Earth spins on its axis once each month.
C The Earth has seasons because its axis is tilted.
D Day and night are due to the Earth spinning on its axis.
E At places on the Earth's hemisphere tilted towards the Sun, the day is shorter than the night.

3 a Why do we never see the other side of the Moon?

b Why does the Moon have phases?

c Why does the Moon rise and set?

d Draw diagrams to show the positions of the Sun, Earth and Moon during (i) a solar eclipse, (ii) a lunar eclipse.

Checklist

After studying this chapter you should be able to

- give a broad description of the Solar System,
- describe how the motion of the Earth explains (a) day and night, (b) the seasons, (c) the rising and setting of the Sun,
- describe how the motion of the Moon explains (a) its phases, (b) eclipses of the Sun and Moon, (c) the rising and setting of the Moon,
- recall some of the main features of the planets and appreciate what factors affect conditions on their surface,
- locate some of the planets in the sky,
- relate the theory of gravity to the motion of artificial satellites,
- discuss the pros and cons of space programmes,
- describe the path of a comet,
- outline how tides occur.

71 STARS AND THE UNIVERSE

The night sky

The night sky has been an object of wonder and study since the earliest times. On a practical level it provided our ancestors with a calendar, a clock and a compass. On a theoretical level it raised questions about the origin and nature of the Universe and its future. It is only in the last 100 years or so that some progress has been made in finding answers and making sense of what we see.

Although the stars may seem close-by on a dark night, in fact the distances involved are mind-boggling. So much so that we need a new unit of length, the **light-year** (l.y.). This is the distance travelled by light in one year and equals about 10 million million kilometres (10^{13} km) or 6 million million miles. The star nearest to the Solar System is Alpha Centauri, 4.3 l.y. away, which means the light arriving from it at the Earth today left 4.3 years ago. The Pole Star is 142 l.y. from Earth. The whole of the night sky we see is past history and we will have to wait a long time to find out what is happening out there right now.

Figure 71.1a The Milky Way from Earth

Figure 71.1b An image of the Milky Way from a space probe (the Cosmic Background Explorer)

Galaxies

A **galaxy** is a large collection of stars which, like our Sun, all produce their own light. Millions of galaxies make up the whole **Universe**.

As well as containing stars, galaxies consist of clouds of gas, mostly hydrogen, and dust. They move in space, many rotating as spiral discs like huge Catherine wheels with a dense central bulge.

The **Milky Way**, part of the spiral galaxy (see Figure 1a, p. vi) to which the Solar System belongs, can be seen on dark nights as a narrow band of light spread across the sky, Figure 71.1a. We are near the outer edge of the galaxy, so what we see when we look at the Milky Way is the galaxy's central bulge. Figure 71.1b is an infrared photograph of the galaxy, taken from beyond the Earth's atmosphere, clearly showing the central bulge.

Galaxies travel in groups and the nearest spiral in our **Local Cluster** is the **Andromeda** galaxy. It is 2.2 million light-years away and visible to the naked eye.

The stars in a galaxy are often millions of times further apart than the planets in the Solar System but millions of times closer than are neighbouring galaxies.

Constellations

On a good night when there is no Moon and the air is clear, although the Earth seems to have an enormous star-spangled dome overhead, only up to 4000 or so stars can be seen by the unaided eye.

From ancient times star-gazers have tried to bring some order into the sky by fitting the stars into patterns, the **constellations**, to which names were given. Some of the major constellations and stars that are seen from the northern hemisphere in winter, looking north and south, are shown in Figures 71.2a and b respectively, The pale bands indicate the Milky Way.

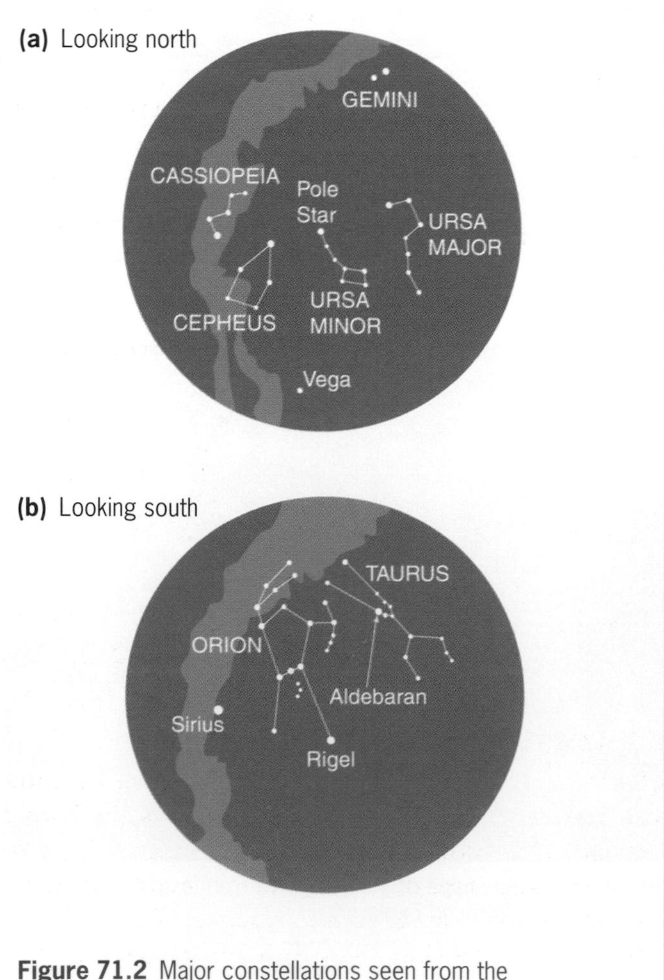

Figure 71.2 Major constellations seen from the northern hemisphere in winter

Ursa Major (or the **Plough**) is one of the most well-known constellations in the northern sky. It consists of seven stars in the shape of a plough, two of these pointing towards the Pole Star which is used in navigation to find north. Opposite the Plough is the W-shaped constellation **Cassiopeia**.

Orion (or the **Hunter**) is one of the most easily recognised constellations in the southern sky on winter nights. He is holding a club (a distorted Y of five stars) aloft in one hand and a shield (a curved line of five stars) in the other hand towards **Taurus** (the **Bull**), situated above and to his right. Orion's belt of three bright stars (from which hangs his dagger of three less bright stars) points towards **Sirius** (the **Dog Star**), the brightest star in the sky. Taurus represents the head of a bull with the large star **Aldebaran** forming one of its eyes.

Motion of the stars

Observation of the night sky at intervals of a few hours shows that in the northern hemisphere the stars appear to revolve anticlockwise about a point near the Pole Star, Figure 71.3a. As they revolve the stars keep the same positions in relation to each other, and in fact the appearance of the constellations has changed very little over the centuries.

Like the Sun and Moon, the stars have a daily journey across the sky, rising upwards in the east and setting in the west – due to the Earth's daily spin about its axis. The **circumpolar** stars, i.e. those near the Pole Star, are an exception to this; they are so close to the pole that they never disappear below the horizon. The trails of the circumpolar stars during a 3-hour period are shown in Figure 71.3b; the brightest trail near the centre is that of the Pole Star.

Although the stars always rise in the same *places* they do not do so at the same *times* every day. They rise about 4 minutes earlier each day (as measured by Sun-time), and in almost but not exactly a year they

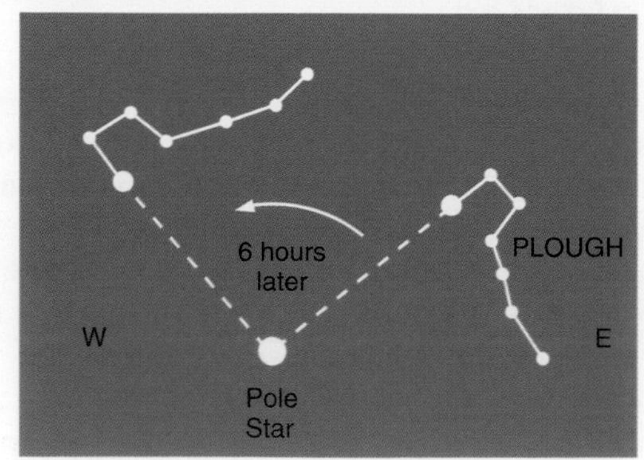

Figure 71.3a Stars appear to revolve about a point close to the Pole Star

Figure 71.3b Trails of the circumpolar stars in a 3-hour period

again rise and set at the same time. If a certain star is seen to rise in the east exactly when the Sun sets on a particular day, a few weeks later it will be seen well above the eastern horizon at sunset. Thus the stars in Orion rise in midwinter just after sunset and set before sunrise; by March, they are high in the heavens at sunset and set about midnight. Or again, a circumpolar constellation such as Cassiopeia seen at midnight directly above the Pole Star, will appear directly below it at the *same time* six months later.

Life cycle of stars

A star is a ball of hot hydrogen and helium gas in which nuclear fusion results in the production and emission of electromagnetic radiation.

Stars vary in age, size, mass, surface temperature, colour and brightness. Colour and brightness both depend on surface temperature which in turn increases with the mass of the star. Stars that are blue or white are hotter and brighter (surface temperatures of 6000 to 25 000 °C) than those that are yellow or red (3000 to 6000 °C).

(a) Origin of stars

Stars are thought to form when gravitational attraction pulls together hydrogen gas and dust (called **nebulae**) in regions of space where their density is greater. As the mass of the star increases its core temperature rises because it is squeezed by gravity (just as the air in a bicycle pump gets warm if it is squeezed). When the hydrogen is hot enough nuclear fusion occurs. If the 'young' star has a very large mass (more than twenty times that of the Sun) it forms a blue or white star. If it has a smaller mass (about equal to that of the Sun) if forms a yellow or red star and this is more common.

(b) Yellow dwarf star

When a star is of this type, as our Sun is at present, the very strong forces of gravity pulling it together are balanced by opposing forces trying to make it expand due to its extremely high temperature. The star is then in a stable state which may last for about 10 000 million years, during which time the hydrogen in its core is changed to helium.

When there is no more hydrogen left, the star becomes unstable, expands, cools and turns into a **red giant**. (In the case of our Sun, which is halfway through its stable state, this stage would be reached in about 5 thousand million years from now. The inner planets Mercury, Venus and Earth would be devoured and all forms of life destroyed.) Later in its history it contracts under its own gravity to become a **white dwarf** in which matter may be millions of times denser than any on Earth. White dwarfs cool to **red dwarfs** and finally, after a very long time, to cold **black dwarfs**, Figure 71.4a.

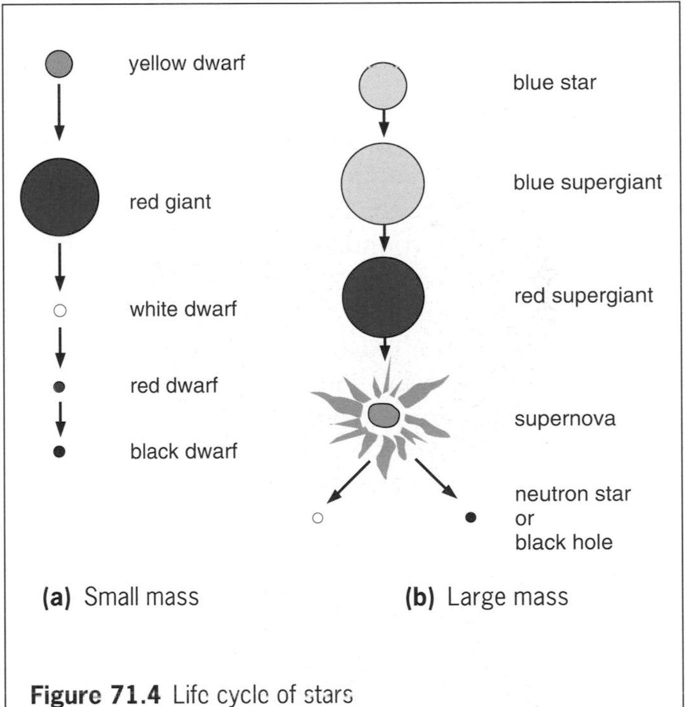

Figure 71.4 Life cycle of stars

(c) Blue star

This type has a much shorter stable stage of about 100 million years. When its hydrogen is used up it expands to a **blue supergiant** and then cools down to a **red supergiant**. It may then become unstable and explode forming a **supernova**, Figure 71.4b. During the explosion there is a huge increase in its brightness, and gas and dust from its outer layers are thrown into space. The Crab Nebula is the remains of the supernova seen by Chinese astronomers on Earth in 1054. It is visible in a telescope as a hazy glow in the constellation Taurus.

The centre of the supernova collapses to a very dense **neutron star**, which spins rapidly and acts as a **pulsar**, sending out pulses of radio waves. If the supergiant is very massive, the remnant at the centre of the supernova has such a large density that its gravitational force stops anything escaping from its surface, even light. This is a **black hole**.

Origin of the Solar System

While the Sun was probably formed like other stars, as described previously, the formation of the planets is not so clear. One theory is that the Solar System was all formed at the same time, about 5 thousand million years ago, and the planets were created when a disc of matter was pushed out from the centre of the cloud of hydrogen gas that produced the Sun. Evidence for this as the approximate age of the Earth comes from the radioactive dating of minerals in rocks (Chapter 63). Further confirmation was provided by the American astronauts who brought back rocks from the Moon which were found to be about 4.5 thousand million years old.

This view that the whole Solar System was formed at the same time is supported by the fact that the orbits of the planets are more or less in the same plane and all revolve round the Sun in the same direction. Neither of these are likely to have happened by chance.

To account for the existence of heavier chemical elements in the Sun and inner planets, it is thought these might have come from exploding supernovae elsewhere in the galaxy. During the lifetime of a star, atoms of hydrogen, helium and other light elements are changed into atoms of heavier elements. It is possible that matter from stellar explosions could have mixed with interstellar hydrogen before the planets formed.

Theories of the Universe

(a) The expanding Universe

In developing a theory about the origin of the Universe two discoveries about galaxies have to be taken into account. The first is that light from other galaxies is 'shifted' to the red end of the spectrum. The second is that the farther away a galaxy is from us, the greater is this **red-shift**. The shift is detected by noting the positions of certain wavelengths due to a known element in the star's spectrum compared with their positions in a laboratory-produced spectrum of the element.

Red-shift can be explained as being due to other galaxies moving away from us very rapidly, and the farther away they are, the faster is their speed of recession. That is, the Universe is expanding. This explanation is based on the Doppler effect, which occurs when a source emitting waves is moving. If the source approaches us the waves are crowded into a smaller space and their wavelength appears to be smaller and their frequency greater; if the source moves away, the wavelength seems larger, Figure 71.5. It occurs with sound waves and explains the rise and fall of pitch of a siren as the vehicle approaches and passes us. The same effect is shown by light. When the light source is receding the wavelength seems longer, i.e. the light is redder. From the size of the red-shift, the speed of recession of a galaxy can be calculated; the most distant ones visible are receding with speeds up to one-third that of light.

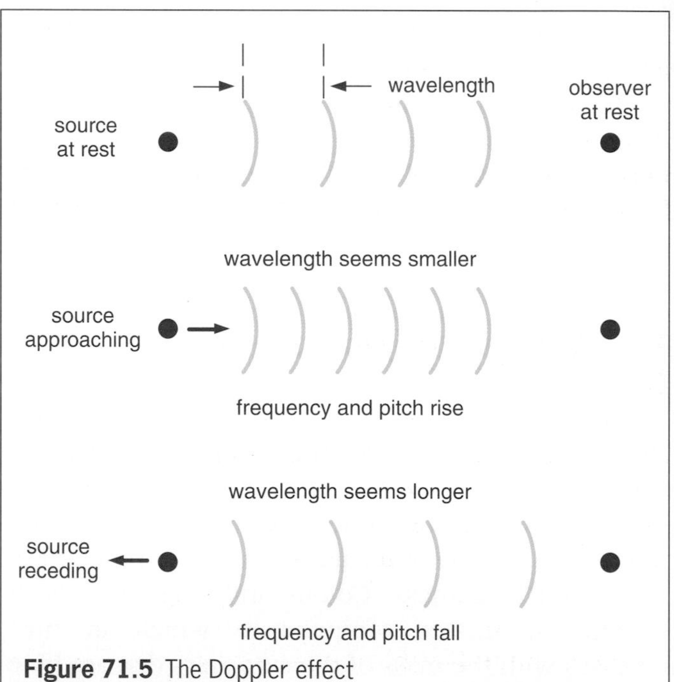

Figure 71.5 The Doppler effect

(b) The Big Bang (or Evolutionary) theory

If the galaxies are receding from each other, it follows that in the past they must have been closer together. It is therefore possible that initially all the matter in the Universe was packed together in an extremely dense state. The Big Bang theory proposes that this was the case and that the Universe started thousands of millions of years ago from one place with a huge explosion – the Big Bang.

The resulting expansion continues today but predictions vary as to what will happen in the future. One is that the expansion will go on. Another is that the expansion will be succeeded by contraction to a dense state, followed by expansion again and so on, leading to a pulsating Universe. If the latter is true then eventually gravitational attraction between galaxies will slow down the observed expansion and cause them to approach, leading to a return to the dense state.

Two further facts are relevant. One is that the rate of expansion is decreasing due to gravitational attraction restraining the galaxies and other material as they speed apart. The other is that most of the mass of the Universe is invisible (the so-called 'dark' matter) and in an unknown form which does not emit radiation but does exert a gravitational force. It may be that the fate of the Universe depends on this hidden mass which could ultimately cause the collapse of the Universe.

(c) Other theories

One of these is the now less-favoured Steady State or Continuous Creation theory. It proposes that new matter (hydrogen) is created all the time to fill up the empty space arising from the expansion of the Universe. In this case the Universe would not change and would always 'look' the same.

The whole question of how the Universe originated and what happened before it formed, is one that may remain an impenetrable mystery, perhaps beyond our understanding. Nevertheless it never fails to excite and fascinate cosmologists and laymen alike.

Age of the Universe

Hubble, an American astronomer, discovered that the speed of recession v of a galaxy is directly proportional to its distance away s. This is called **Hubble's law** and can be written

$$v \propto s \qquad \text{or} \qquad v = H \times s$$

where H is the **Hubble constant**. Its value is found by measuring the speeds of recession of galaxies (from red-shifts) whose distances away are known from other astronomical data. Due to the considerable difficulties involved, H is estimated to be very approximately 20 km per second per million light-years. This means that a galaxy is receding at 20 km/s for every million light-years of distance from us. One that is 2 million light-years away will be receding at 40 km/s.

The age of the Universe can be shown to equal $1/H$ and can be calculated very roughly. We assume that the time for which two galaxies were close together at the Big Bang when they were formed is negligible compared with their present ages, i.e. most of their lifetimes have been spent apart. Their ages are thus approximately the age of the Universe, which equals the time since expansion began.

For the two galaxies, assuming the speed of recession has not changed, we have

$$\text{age} = \frac{\text{distance apart}}{\text{speed of recession}} = \frac{s}{v}$$

From Hubble's law, it follows that

$$\text{age} = \frac{1}{H} = \frac{1}{20\ \text{km/s/million light-years}}$$

$$= \frac{10^{19}\ \text{km}}{20\ \text{km/s}} \quad (\text{since 1 l.y.} = 10^{13}\ \text{km})$$

$$= 5.0 \times 10^{17}\ \text{s}$$

But $\qquad 1\ \text{year} \simeq 3.2 \times 10^{7}\ \text{s}$

$$\therefore \quad \text{age of Universe} \simeq \frac{5.0 \times 10^{17}\ \text{s}}{3.2 \times 10^{7}} \simeq 1.5 \times 10^{10}\ \text{y}$$

$$\simeq 15 \times 10^{9}\ \text{years}$$

$$\simeq 15\,000\ \text{million years}$$

This result has an uncertainty of ± 5000 million years, i.e. it could be between 10 000 and 20 000 million years.

QUESTIONS

1 Which of the following statements are *true*?

A A galaxy is a large collection of stars.
B The Universe is a collection of galaxies.
C The Solar System belongs to the Andromeda galaxy.
D A light-year is the distance travelled by light in 1 year.
E The Pole Star is the brightest star in the sky.

2 a Why is the Pole Star useful for finding direction?

b How does the Plough help in finding the Pole Star?

c If the Plough is in the position shown in Figure 71.6 at 6 p.m. on a certain evening, draw two diagrams to show where it will be (i) at midnight on the same evening, (ii) at 6 p.m. 6 months later.

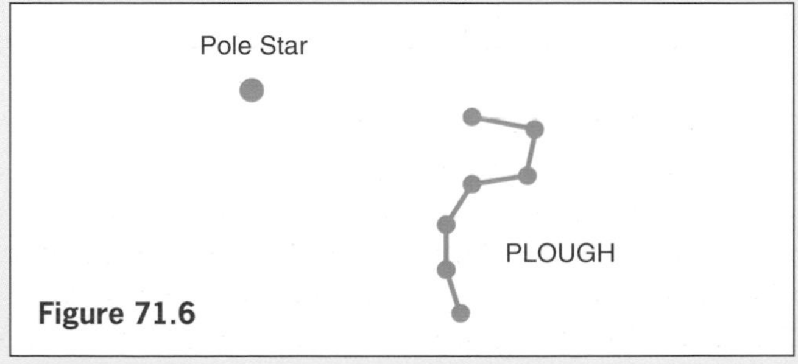

Figure 71.6

3 Name the stages in the life history of a star with mass

a equal to that of the Sun,

b more than 20 times that of the Sun.

Checklist

After studying this chapter you should be able to

- understand the term **light-year** and relate it to star distances,
- recall that a **galaxy** is a large collection of stars and that millions of galaxies make the Universe,
- recognize the major constellations e.g. Ursa Major, Cassiopeia, Orion,
- explain the apparent daily motion of the stars,
- describe how stars may originate and outline the main stages in the life cycles of a yellow dwarf star and a blue star,
- explain how the Solar System may have been formed,
- understand the facts about **red-shift** and appreciate its importance to our understanding of the Universe,
- describe briefly past and present theories about the origin and future of the Universe,
- explain how the **Hubble constant** is used to estimate the age of the Universe.

ADDITIONAL QUESTIONS

The first questions under each topic heading are 'basic' questions intended for all students; the following questions marked with a stripe are 'higher' questions for those seeking grades E to A*.

The atmosphere and weather; weather maps and forecasts

1 Which one of the following statements is *untrue*?

A Land heats up and cools down faster than water.
B When air falls to a lower level its pressure and temperature increase.
C Wind is air in motion caused by pressure differences.
D When air is forced to rise it cools.
E The leeward side of a mountain receives more rain than the windward side.

2 Why does planting trees help to reduce global warming due to the greenhouse effect?

3 **a** What is a cold front?

b Draw a symbol used to show one on a weather map.

c State briefly a common sequence of events that accompanies a cold front.

4 Why does the air temperature decrease with height in the troposphere despite the fact that greater altitudes are closer to the Sun?

5 In the northern hemisphere if you stand with your back to the wind, is low pressure to your left or right? Explain your answer.

Structure of the Earth

6 A simplified map showing the basic geology of the English Lake District is shown.

The **northern area A** contains the oldest rocks, called **Skiddaw Slates**, which were formed about 450 million years ago mainly as sea mud that later underwent changes which turned it into slate. The rocks contain marine fossils.

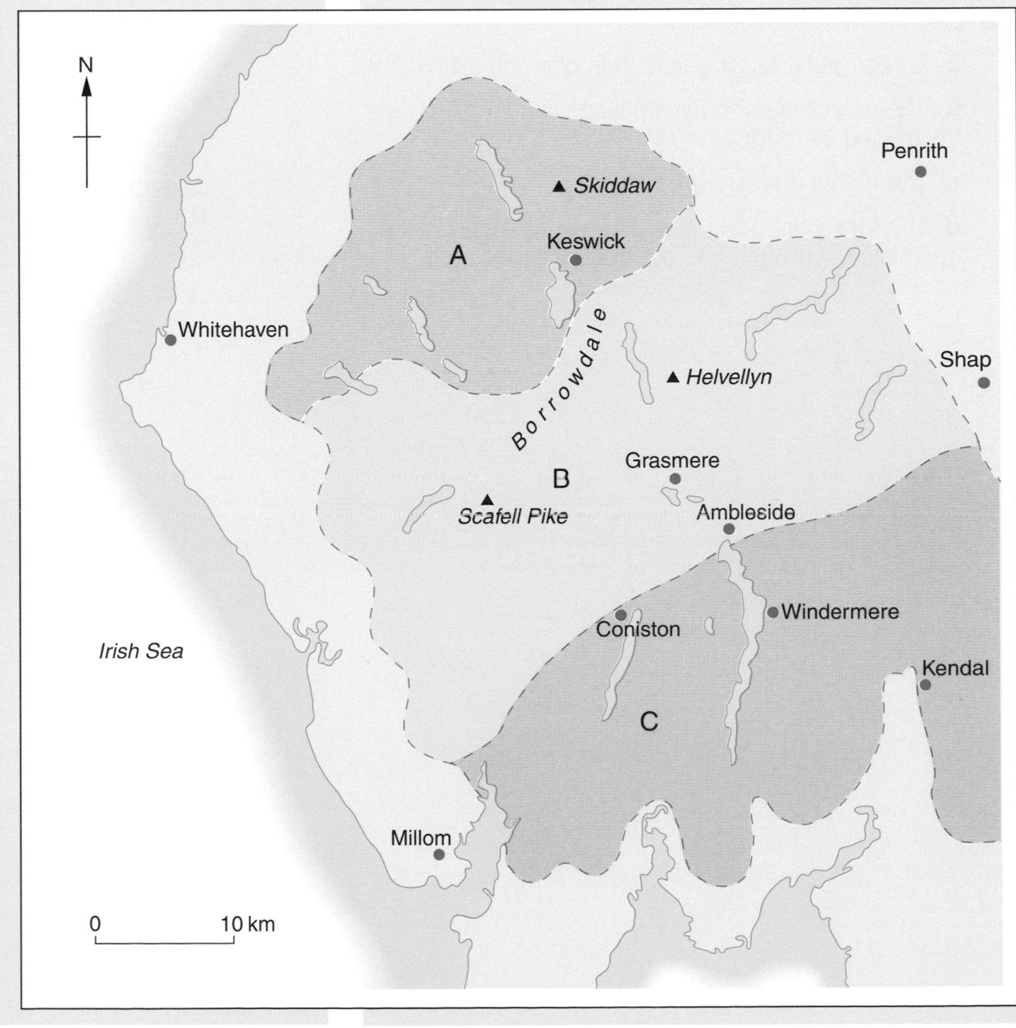

The **central area B** consists of **Borrowdale Volcanic** rocks which were the result of intense volcanic activity when ash and lava were ejected from the Earth. No fossils have been found in this area but certain parts are noted for their attractive green slates.

The **southern area C** is built of rocks called the **Silurian Series** which were formed from sediments that contain marine fossils. Some blue and blue-grey slates are also found in this area.

Describe *briefly*, in terms of the three main rock groups, the rock history of each area.

Find out why the term Silurian is used for the rocks in area C.

7 a Earthquakes and volcanic eruptions often occur together and seem to be connected. What is thought to cause them?

b Name the instrument used to detect earthquakes.

c Name the scale which gives the magnitude of an earthquake.

d How many times more powerful is an earthquake of magnitude 4 compared with one which records (i) 3, (ii) 1?

The Solar System; stars and the Universe

8 Explain the following facts.

a A year on Mercury is less than one on Earth.

b The only planets to have phases like the Moon are Venus and Mercury.

c The nights on Mars are very cold.

d The four giant outer planets have atmospheres containing hydrogen gas but the Earth does not.

9 Outline the evidence for:

a the view that all members of the Solar System were formed about the same time,

b the expanding Universe.

10 Three space vehicles, P (mass *m*), Q (mass *m*) and R (mass 2*m*) are launched into circular orbits above the Earth's atmosphere as shown in the diagram.

a Why must Q be moving more slowly than P, although it has the same mass?

b Why must Q and R have the same speed, although they have different masses?

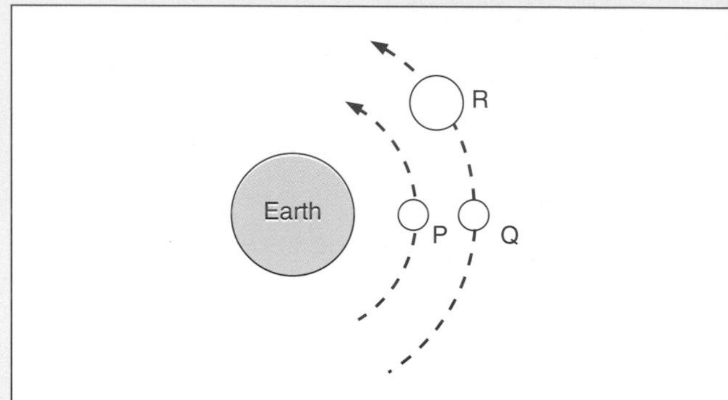

11 a State Hubble's law.

b Write down the equation connecting the age of the Universe to the Hubble constant *H*.

c Some astronomers believe that the true value of *H* is 30 km/s/million light-years. Using this value, calculate the approximate age of the Universe. (1 light-year = 10^{19} km; 1 year = 3.2×10^7 s)

REVISION QUESTIONS

The first questions under each topic heading are 'basic' questions intended for all students; the following questions marked with a stripe are 'higher' questions for those seeking grades E to A*.

Light and sight

1 The image of a window on the screen of a pinhole camera is

A virtual, inverted and larger
B virtual, upright and smaller
C real, upright and larger
D real, inverted and larger
E real, inverted and smaller.

2 In Figure R1 the completely dark region is

A PQ **B** PR **C** QR **D** QS **E** RS

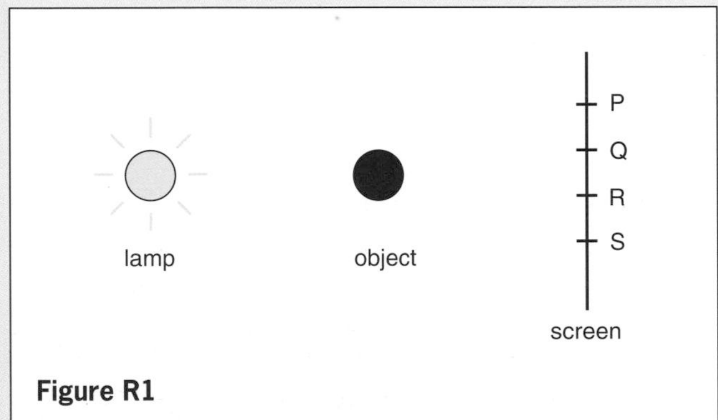

Figure R1

3 In Figure R2 a ray of light is shown reflected at a plane mirror. What is

a the angle of incidence,

b the angle the reflected ray makes *with the mirror*?

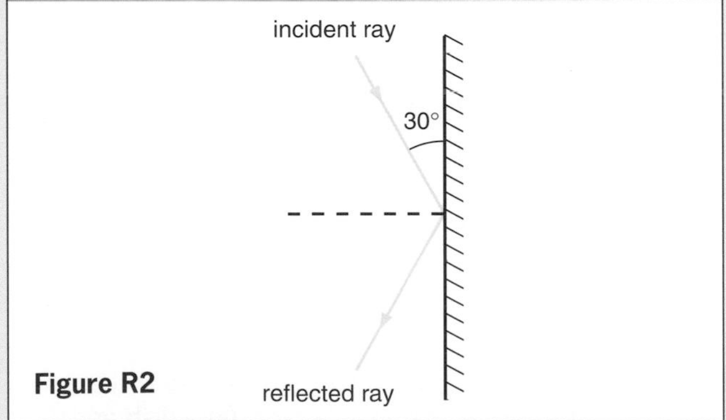

Figure R2

4 In Figure R3 at which of the points **A** to **E** will the observer see the image of the object in the plane mirror?

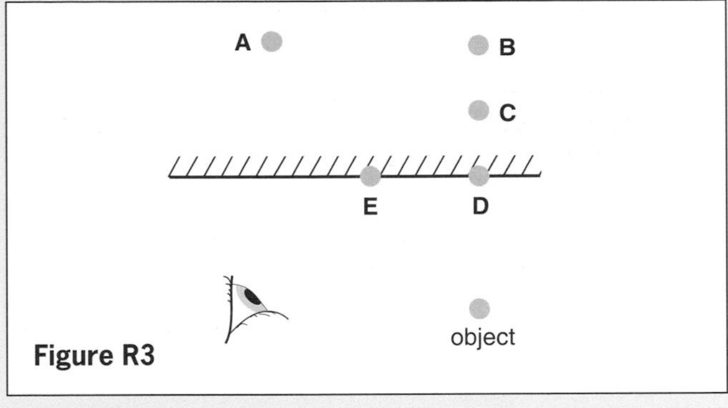

Figure R3

5 In Figure R4 a ray of light IO changes direction as it enters glass from air.

a What name is given to this effect?

b Which line is the normal?

c Is the ray bent towards or away from the normal in the glass?

d What is the value of the angle of incidence in air?

e What is the value of the angle of refraction in glass?

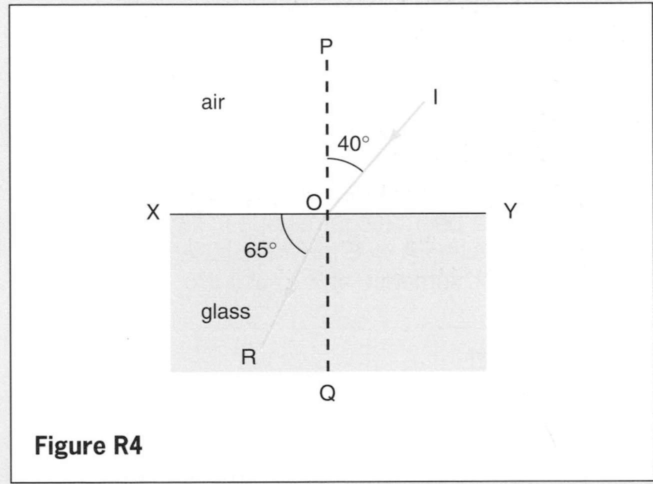

Figure R4

6 In Figure R5 which of the rays **A** to **E** is most likely to represent the ray emerging from the parallel-sided sheet of glass?

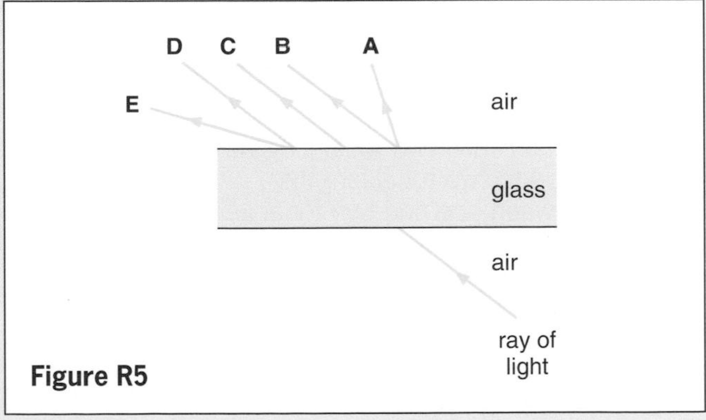

Figure R5

7 In Figure R6 which diagram shows the correct path of the ray through the prism?

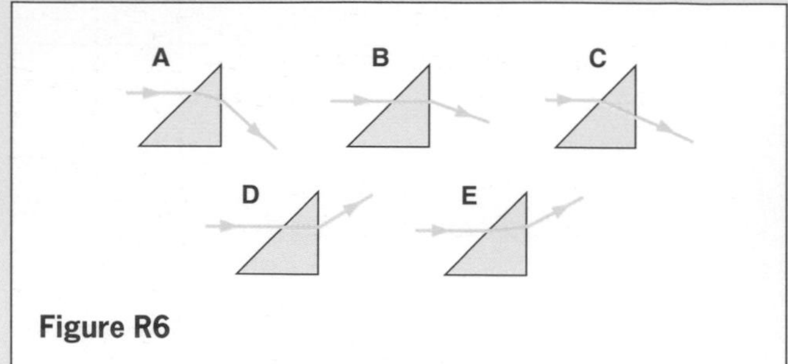

Figure R6

8 If the critical angle for glass is 42°, which diagram **A** to **E** in Figure R7 does *not* represent the behaviour of a ray of light falling on a glass–air boundary?

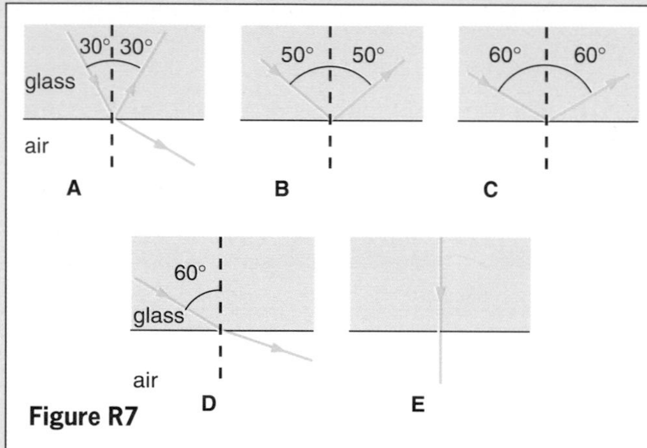

Figure R7

9 A ray of light is shown passing through the upper prism of a prismatic periscope in Figure R8. In which position **A** to **E** must the lower prism be placed for someone at X to use it?

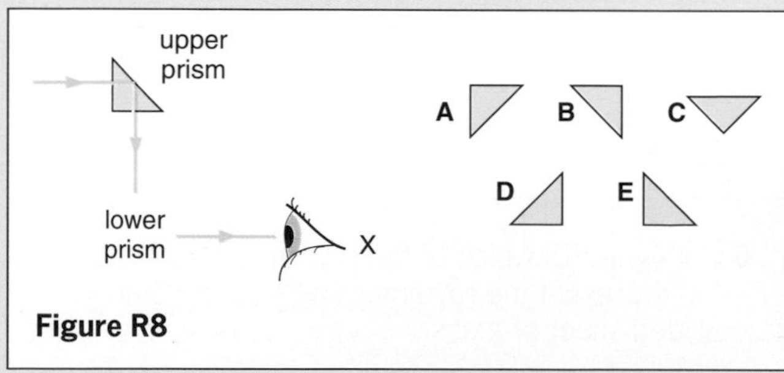

Figure R8

10 A convex lens produces a magnified, inverted image of an object if the distance of the object from the lens is

A greater than two focal lengths
B equal to two focal lengths
C between one and two focal lengths
D equal to one focal length
E less than one focal length.

11 The eye forms an image on the retina which

1 is inverted
2 is focused by the lens changing shape
3 has its brightness controlled by the size of the pupil.

Which statement(s) is (are) correct?

A 1, 2, 3 **B** 1, 2 **C** 2, 3 **D** 1 **E** 3

12 In Figure R9 a narrow beam of white light is shown passing through a glass prism and forming a spectrum on a screen.

a What is the effect called?

b Which colour of light appears at (i) A, (ii) B?

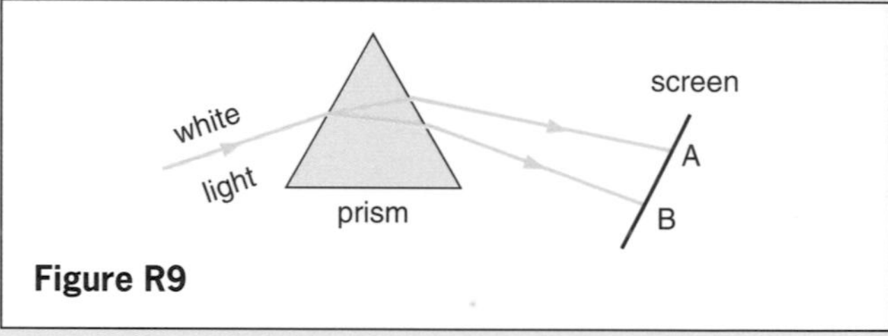

Figure R9

13 Which pair of colours **A** to **E** when added together give white light?

A yellow and blue **B** red and blue
C green and yellow **D** red and green
E green and blue

14 To focus a lens camera on a nearer object is

A the distance between the lens and film increased
B the distance between the lens and film decreased
C the aperture made larger
D the aperture made smaller
E the shutter speed changed?

15 When using a magnifying glass to see a small object

1 an upright image is seen
2 the object should be less than one focal length away
3 a real image is seen.

Which statement(s) is (are) correct?

A 1, 2, 3 **B** 1, 2 **C** 2, 3 **D** 1 **E** 3

16 **a** Why is a concave mirror used as a make-up mirror?

b Why is a convex mirror used as a driving mirror?

c Why is a concave parabolic mirror used as (i) a car headlamp reflector, (ii) a dish aerial?

17 A ray of light is incident on the plane surface of a transparent material at such an angle that the reflected and refracted rays are at right angles to each other. Draw a diagram to illustrate this. *(S. part qn.)*

18 A 45° right-angled prism ABC, made of glass of critical angle 42°, is used to turn light through 90°.

a Copy Figure R10, continue the paths of the two rays until they emerge into the air again.

b Why does light not emerge from face AC?

c Using the two rays, explain what is meant by lateral inversion.

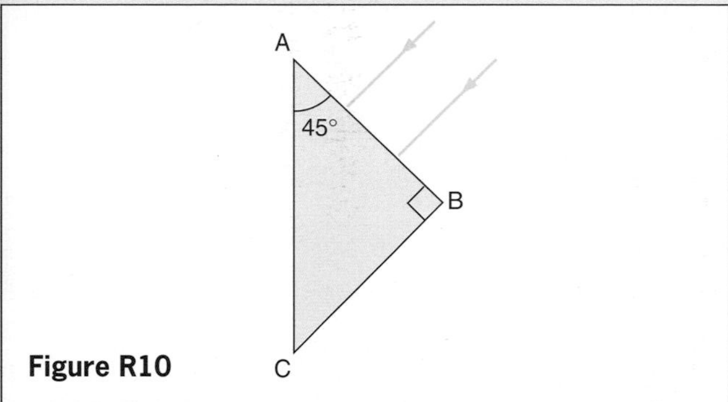

Figure R10

Waves and sound

19 In the transverse wave shown in Figure R11 distances are in centimetres. Which pair of entries **A** to **E** is correct?

	A	**B**	**C**	**D**	**E**
Amplitude	2	4	4	8	8
Wavelength	4	4	8	8	12

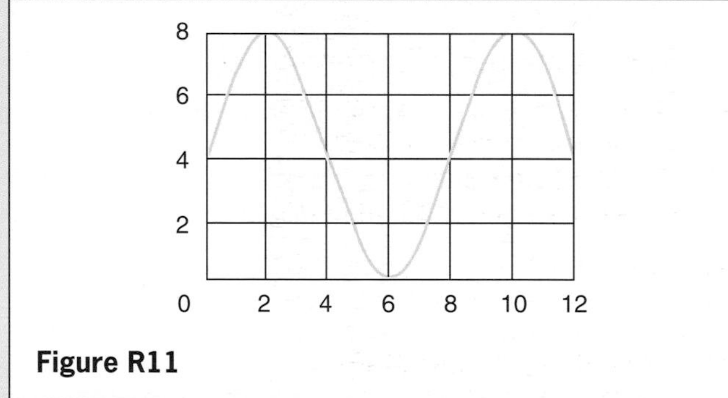

Figure R11

20 The lines in Figure R12 are the crests of straight ripples produced in a ripple tank by a wave generator.

a What is the wavelength of the ripples if there are 6 complete waves in a distance of 30 cm?

b If the wave generator makes 4 vibrations per second what is the frequency of the ripples?

c What is the equation connecting the wavelength (λ), the frequency (f) and the speed (v) of the ripples?

d Calculate the speed of the ripples.

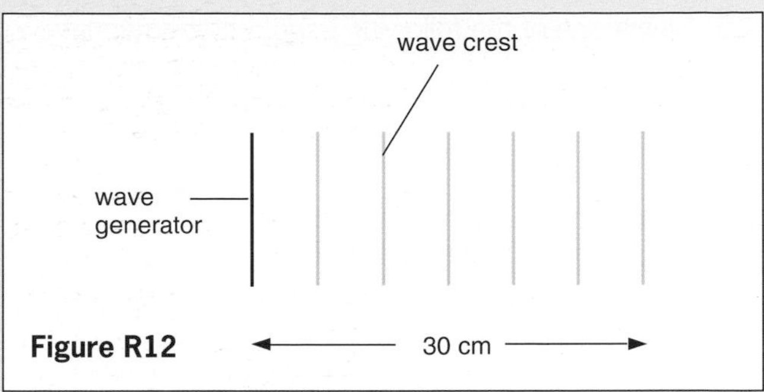

Figure R12

21 When water waves go from deep to shallow water, the changes (if any) in its speed, wavelength and frequency are

	Speed	**Wavelength**	**Frequency**
A	greater	greater	the same
B	greater	less	less
C	the same	less	greater
D	less	the same	less
E	less	less	the same

22 When the straight water waves in Figure R13 pass through the narrow gap in the barrier they are diffracted. What changes (if any) occur in **a** the shape of the waves, **b** the speed of the waves, **c** the wavelength?

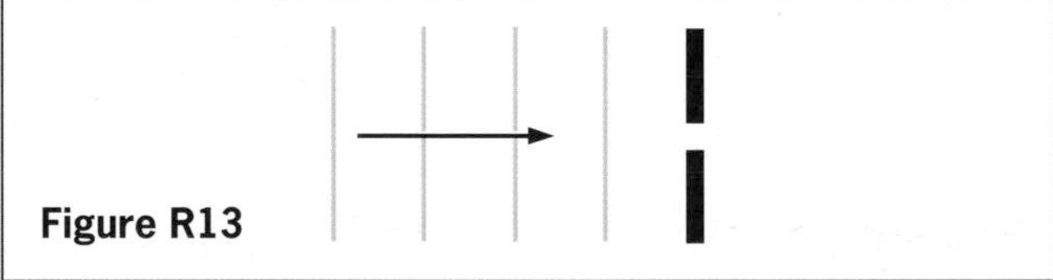
Figure R13

23 a When two sets of waves arrive at the same point at the same time, under what conditions do they (i) completely cancel out, (ii) produce a larger wave?

b If A and B in Figure R14 are two dippers vibrating together (in phase) in a ripple tank, what happens (i) at P if PA = PB, (ii) at Q if QB – QA = half a wavelength, (iii) at R if RB – RA = one wavelength?

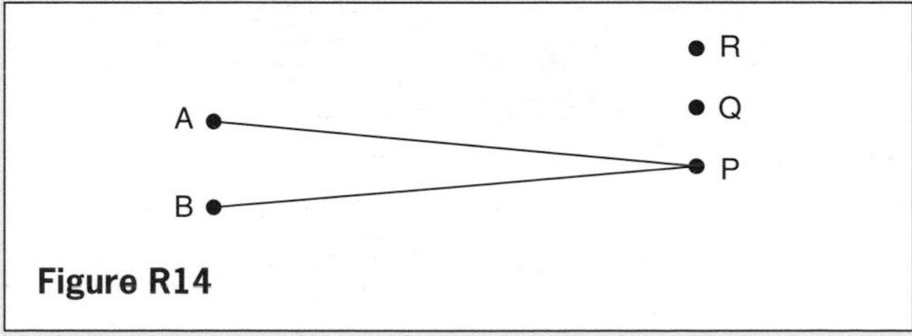

Figure R14

24 Sometimes 'light + light = darkness'. This behaviour of light is called

A reflection **B** refraction **C** dispersion
D interference **E** deviation.

25 Which one of the following is not electromagnetic waves?

A infrared **B** gamma rays **C** ultraviolet
D X-rays **E** sound

26 Compared to radio waves, the wavelength, frequency and speed of light are

	Wavelength	Frequency	Speed
A	greater	smaller	the same
B	greater	smaller	greater
C	greater	greater	the same
D	smaller	greater	smaller
E	smaller	greater	the same

27 The wave travelling along the spring in Figure R15 is produced by someone moving end X of the spring to and fro in the directions shown by the arrows.

a Is the wave longitudinal or transverse?

b What is the region called where the coils of the spring are (i) closer together, (ii) farther apart, than normal?

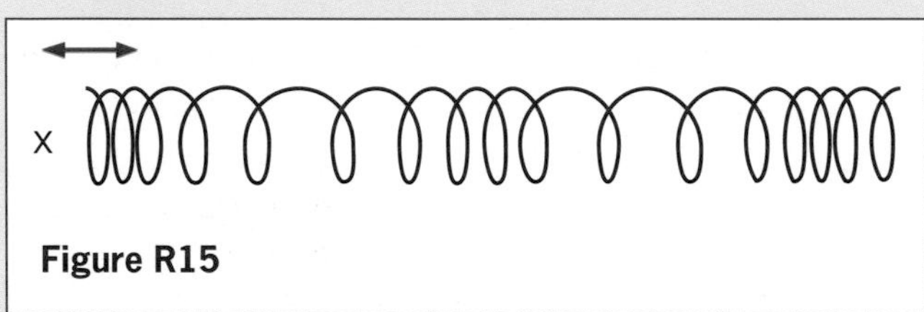

Figure R15

28 Sound and light waves

A travel through a vacuum
B travel as longitudinal waves
C travel with the same speed in air
D can be diffracted
E have similar wavelengths.

29 The signal sent out by a sonar echo sounder in a ship is received back from the sea bed directly below the ship, Figure R16, 2 seconds later. The speed of sound in sea water, in m/s, is

A 750 **B** 1500 **C** 2250 **D** 3000 **E** 3750

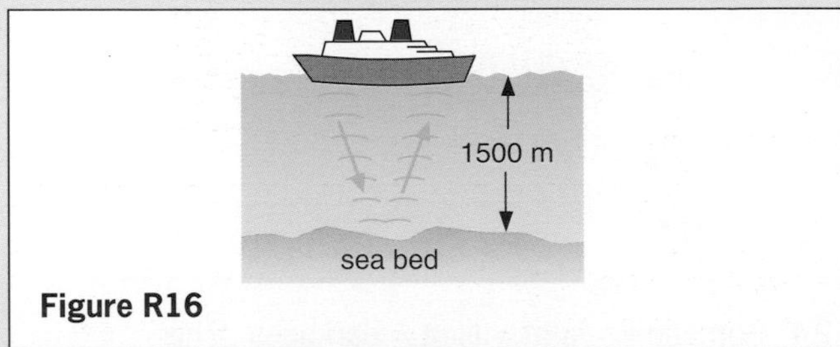

Figure R16

30 If a note played on a piano has the same pitch as one played on a guitar, they have the same

A frequency **B** amplitude **C** quality
D loudness **E** harmonics.

31 The waveforms of two notes P and Q are shown in Figure R17. Which one of the statements **A** to **E** is true?

A P has a higher pitch than Q and is not so loud.
B P has a higher pitch than Q and is louder.
C P and Q have the same pitch and loudness.
D P has a lower pitch than Q and is not so loud.
E P has a lower pitch than Q and is louder.

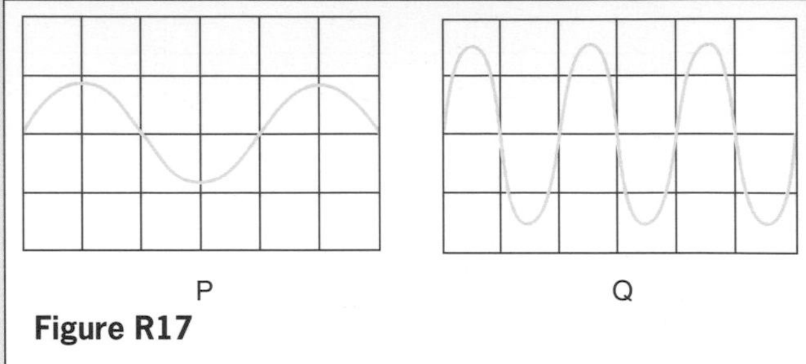

Figure R17

32 The loudness of the note produced by a vibrating wire can be increased by

A increasing the length of the wire
B decreasing the length of the wire
C increasing the amplitude of vibration
D increasing the tension of the wire
E decreasing the tension of the wire.

33 Examples of transverse waves are

1 water waves in a ripple tank
2 all electromagnetic waves
3 sound waves.

Which statement(s) is (are) correct?

A 1, 2, 3 **B** 1, 2 **C** 2, 3 **D** 1 **E** 3

34 Notes of the same pitch played on the guitar and the clarinet sound different because

1 different overtones are present in each case
2 the fundamental frequencies are different
3 a guitar string vibrates transversely and the air column in a clarinet vibrates longitudinally.

Which reason(s) is (are) correct?

A 1, 2, 3 **B** 1, 2 **C** 2, 3 **D** 1 **E** 3

35 State whether the following waves are transverse or longitudinal *and* progressive or stationary:

a the vibrations of a stretched wire (which is fixed at both ends) when plucked,

b the sound from the above wire to an observer,

c a ripple on a large pond,

d the vibrations of an air column in a long test tube. (*S.*)

36 Figure R18 represents a plan view of a horizontal ripple tank. Two dippers, A and B, vibrate in-phase with the same frequency. At a point P constructive interference is observed, and at a nearby point Q destructive interference is observed.

a On the same axes sketch two graphs of displacement against time for the vibration of the water at P, one for the waves from A, the other for the waves from B. On the same axes sketch a third graph showing the displacement of the water at P due to both sets of waves arriving at P together.

b In exactly the same way, sketch three graphs of displacement against time for the water at Q, one for waves from A, another for waves from B and the third for both sets of waves arriving at Q together.

c (i) State a relationship between the distances AP and BP.
(ii) State a relationship between the distances AQ and BQ. *(J.M.B.)*

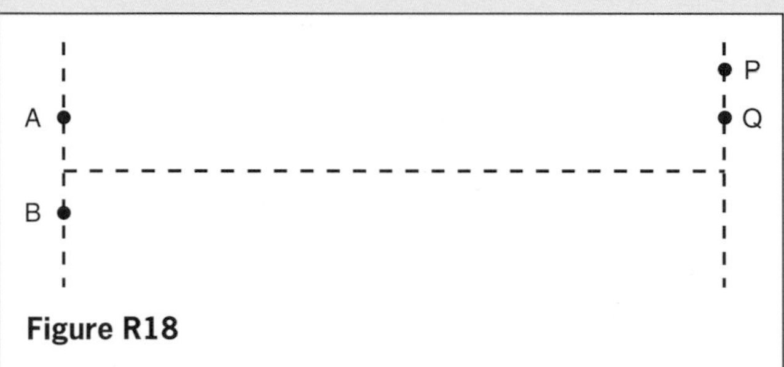

Figure R18

37 Which of the following is (are) true both for light waves and for sound waves?

1 They can travel through a vacuum.
2 Speed of wave = frequency of wave × wavelength.
3 They transfer energy from one place to another.

A 1, 2, 3 **B** 1, 2 **C** 2, 3 **D** 1 **E** 3 *(L.)*

38 Figure R19 shows a ray of sunlight incident upon a triangular glass prism, such that a spectrum is produced on the screen W.

a Copy the diagram, continue the ray to illustrate how the spectrum is formed on a white screen at W. Mark the violet end of the spectrum V, and the red end R.

b If the white screen is replaced with a pure red screen, describe carefully what you would observe on the screen, giving a reason for your answer.

c The refractive index of the glass is 1.50 for red light. Using your answer to **a**, state whether you would expect the refractive index of glass for violet light to be greater than, equal to or less than 1.50. Explain how you deduced your answer.

d The glass prism transmits both some ultraviolet and some infrared radiation.
(i) On the diagram, mark clearly with a U the region on the screen at which you would detect ultraviolet radiation, and with an I the region at which you would detect infrared radiation.
(ii) State one method by which you could detect the presence of ultraviolet radiation.
(iii) State whether the frequency of ultraviolet radiation is greater than, equal to or less than that of infrared radiation. *(N.I. part qn.)*

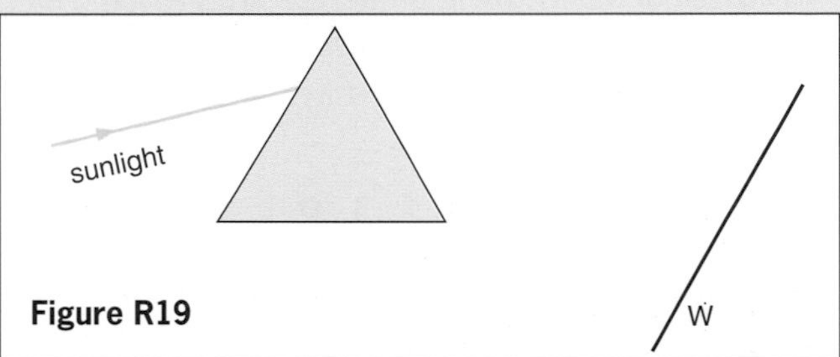

Figure R19

39 When prospecting for oil geologists may cause a small explosion at the surface of the Earth. The distance to possible oil-bearing rocks can then be found by recording how long the sound wave reflected from the rocks takes to return to the surface, Figure R20.

In an exploration 3.6 s is recorded between the explosion and the return of the reflected wave. Taking the average speed of sound through the Earth as 1800 m/s, the rock causing the reflection is at a depth in metres of

A 1800×3.6 **B** $1800/3.6$

C $\dfrac{1800 \times 3.6}{2}$ **D** $\dfrac{1800 \times 2}{3.6}$

E $1800 \times 3.6 \times 2$

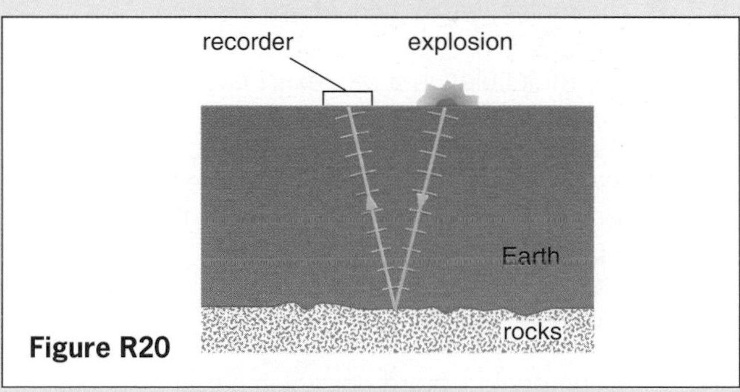

Figure R20

Matter and molecules

40 The basic SI units of mass, length and time are

	Mass	Length	Time
A	kilogram	kilometre	second
B	gram	centimetre	minute
C	kilogram	centimetre	second
D	gram	centimetre	second
E	kilogram	metre	second

41 Density can be calculated from the expression

A mass/volume **B** mass × volume
C volume/mass **D** weight/area
E area × weight

42 A cube of density 1000 kg/m^3 and side of length 2 m has a mass in kg of

A 2000 **B** 4000 **C** 8000 **D** 16 000
E 32 000

43 Which of the following properties are the same for an object on Earth and on the Moon?

1 weight **2** mass **3** density

Use the answer code:

A 1, 2, 3 **B** 1, 2 **C** 2, 3 **D** 1 **E** 3

44 The spring in Figure R21 stretches from 10 cm to 22 cm when a force of 4 N is applied. If it obeys Hooke's law, its total length in cm when a force of 6 N is applied is

A 28 **B** 42 **C** 50 **D** 56 **E** 100

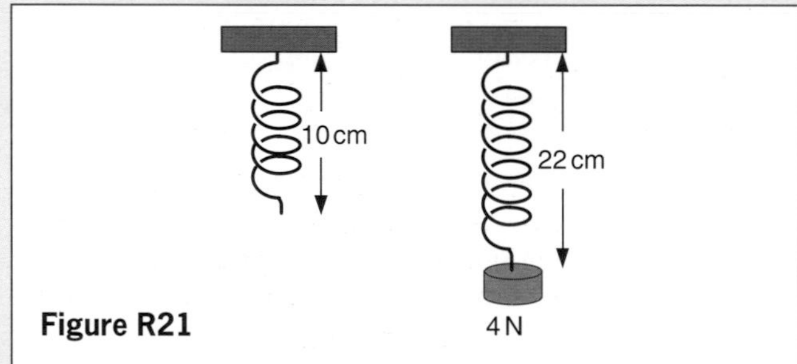

Figure R21

45 Which one of the following statements is *not* true?

A The molecules in a solid vibrate about a fixed position.
B The molecules in a liquid are arranged in a regular pattern.
C The molecules in a gas exert negligibly small forces on each other, except during collisions.
D The densities of most liquids are about 1000 times greater than those of gases because liquid molecules are much closer together than gas molecules.
E The molecules of a gas occupy all the space available.

46 Diffusion occurs more quickly in a gas than in a liquid because

A molecules in a gas have more frequent collisions than molecules in a liquid
B gas molecules are larger
C gas molecules move randomly
D on average, molecules in a gas are further apart than molecules in a liquid
E heat is needed to cause diffusion in a liquid.

47 a The smallest division marked on a metre rule is 1 mm. A student measures a length with the ruler and records it as 0.835 m. Is he justified in giving three significant figures?

b The SI unit of density is

A kg m **B** kg/m^2 **C** kg m^3 **D** kg/m
E kg/m^3

48 A solid block has dimensions 0.1 m × 0.5 m × 0.2 m and is made of material of density 9000 kg/m^3. It rests on a horizontal surface. Calculate:
(i) the mass of the solid,
(ii) the maximum pressure it can exert on the surface. *(W.)*

49 The graph in Figure R22 shows the displacement of a pendulum bob from its rest position as it varies with time. From the graph, determine
(i) the amplitude of the oscillation,
(ii) the time for one complete oscillation,
(iii) the distance of the bob from its rest position after 0.8 seconds.

On a copy of the diagram, draw the graph which represents a pendulum swinging with half the amplitude and twice the frequency. *(W.)*

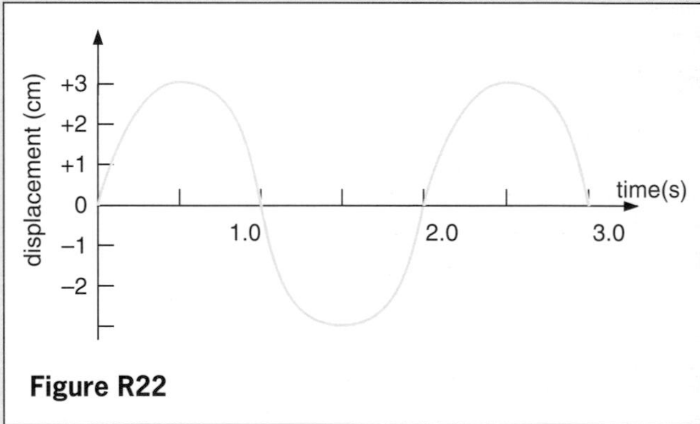

Figure R22

Forces and pressure

50 The metre rule in Figure R23 is pivoted at its centre. If it balances, the mass of *M* is given by the equation

A $M + 50 = 40 + 100$ **B** $M \times 40 = 100 \times 50$
C $M/50 = 100/40$ **D** $M/50 = 40/100$
E $M \times 50 = 100 \times 40$

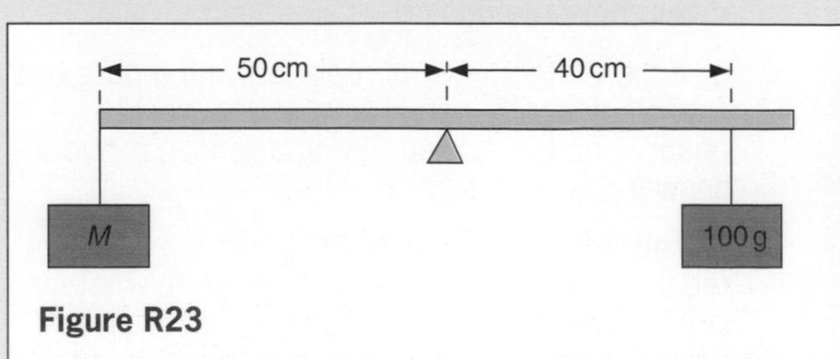

Figure R23

51 The uniform beam of weight 30 N and length 4 m is hinged at end A, as in Figure R24. The force *F*, in N, which must be applied vertically upwards at a distance of 1 m from end B so that the beam is horizontal is

A 60 **B** 50 **C** 40 **D** 30 **E** 20

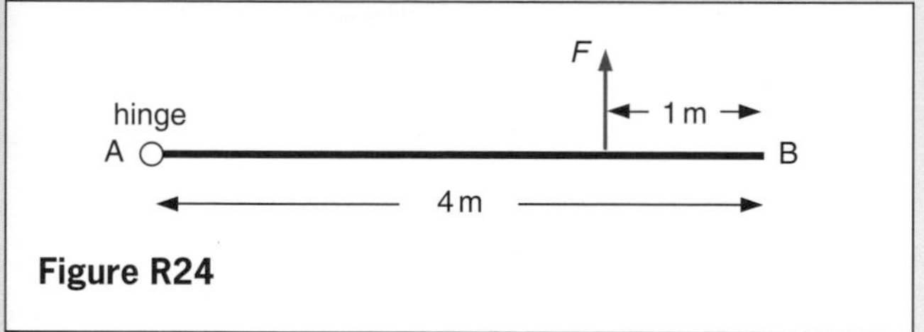

Figure R24

52 The resultant of a force of 5 N acting at right angles to a force of 12 N at a point 0, Figure R25, is

A 5 N **B** 7 N **C** 12 N **D** 13 N **E** 17 N

Figure R25

53 The work done by a force is

1 calculated by multiplying the force by the distance moved in the direction of the force
2 measured in joules
3 the amount of the energy changed.

Which statement(s) is (are) correct?

A 1, 2, 3 **B** 1, 2 **C** 2, 3 **D** 1 **E** 3

54 The main energy change occurring in the device named is

	Device	**Main energy change**
1	electric lamp	electrical to heat and light
2	battery	chemical to electrical
3	pile driver	k.e. to p.e.

Which statement(s) is (are) correct?

A 1, 2, 3 **B** 1, 2 **C** 2, 3 **D** 1 **E** 3

55 If a lift of mass 200 kg is raised by an electric motor through a height of 15 m in 20 s,

1 the weight of the lift is 2000 N
2 the useful work done is 30 000 J
3 the power output of the motor is 1.5 kW.

Which statement(s) is (are) correct?

A 1, 2, 3 **B** 1, 2 **C** 2, 3 **D** 1 **E** 3

56 The efficiency of a machine which raises a load of 200 N through 2 m when an effort of 100 N moves 8 m is

A 0.5% **B** 5% **C** 50% **D** 60% **E** 80%

57 Which one of the following statements is *not* true?

A Pressure is the force acting on unit area.
B Pressure is calculated from force/area.
C The SI unit of pressure is the pascal (Pa) which equals 1 newton per square metre (1 N/m^2).
D The greater the area over which a force acts the greater is the pressure.
E Force = pressure × area.

58 Which of the following will damage a wood-block floor that can withstand a pressure of 2000 kPa (2000 kN/m^2)?

1 A block weighing 2000 kN standing on an area of 2 m^2.
2 An elephant weighing 200 kN standing on an area of 0.2 m^2.
3 A girl of weight 0.5 kN wearing stiletto-heeled shoes standing on an area of 0.0002 m^2.

Use the answer code:

A 1, 2, 3 **B** 1, 2 **C** 2, 3 **D** 1 **E** 3

59 The pressure at a point in a liquid

1 increases as the depth increases
2 increases if the density of the liquid increases
3 is greater vertically than horizontally.

Which statement(s) is (are) correct?

A 1, 2, 3 **B** 1, 2 **C** 2, 3 **D** 1 **E** 3

60 A mercury manometer is connected to a gas supply as shown in Figure R26. The pressure of the gas supply exceeds atmospheric pressure by the pressure exerted by a column of mercury of length

A 15 mm **B** 20 mm **C** 25 mm **D** 30 mm
E 45 mm

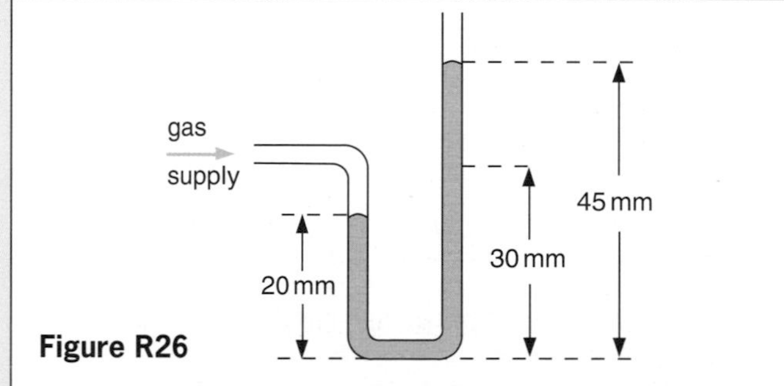

Figure R26

61 If the piston in Figure R27 is pulled out of the cylinder from position X to position Y, without changing the temperature of the air enclosed, the air pressure in the cylinder is

A reduced to a quarter **B** reduced to a third
C the same **D** trebled **E** quadrupled.

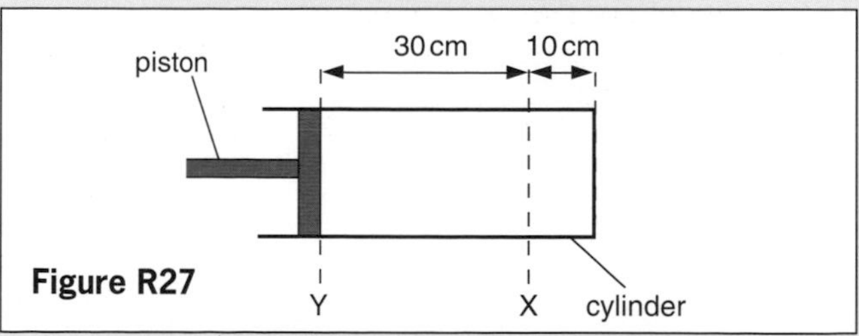

Figure R27

62 An object weighs 200 N in air and 120 N when totally submerged in a liquid of density 800 kg/m^3. What is

a the upthrust on the object,

b the weight of liquid displaced,

c the mass of liquid displaced,

d the volume of liquid displaced,

e the volume of the object?

63 Figure R28 shows the horizontal forces exerted on a tree by two tractors in an attempt to pull it out of the ground.

Draw a diagram to a stated scale and use it to determine the magnitude and direction of the resultant force exerted on the tree by the two tractors. *(C.)*

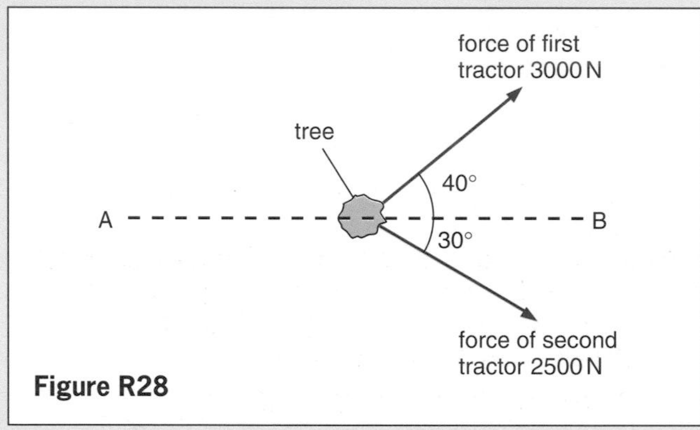

Figure R28

64 An electric motor drives a machine which lifts a mass of 2 kg through the height of 6 m in 4 s at constant speed.

a State how the work done by a force is calculated.

b How much work is done in lifting the 2 kg mass?

c How much power is used to lift the 2 kg mass?

d If the electrical power input is 40 W, what is the efficiency?

e State two causes of loss of efficiency. *(O.L.E.)*

65 Figure R29 illustrates a mercury manometer (using a tube of uniform bore), connected to a gas container through a tap T, which is initially closed. Atmospheric pressure is 750 mm of mercury. On opening tap T, the mercury level in the left-hand arm

A falls through 125 mm
B rises through 125 mm
C falls through 250 mm
D rises through 250 mm
E falls through 500 mm. *(N.I.)*

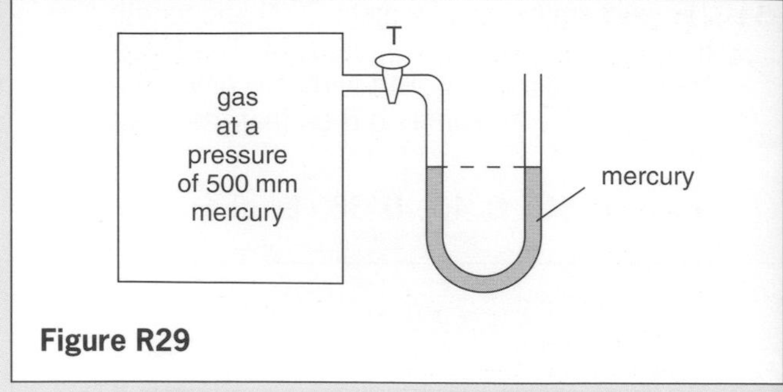

Figure R29

66 Figure R30 represents a simple hydraulic lift. Calculate the maximum load that can be lifted using a downward effort, as shown, of 5 N. *(W.)*

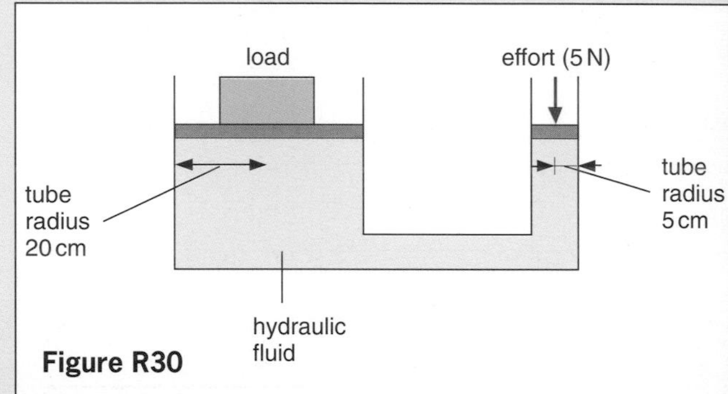

Figure R30

67 An experiment was performed to find the connection between the pressure and volume of a fixed mass of gas when the temperature was kept constant. The graph in Figure R31 shows the pressure (p) plotted against 1/volume ($1/V$).

a (i) What relationship between pressure and volume can you deduce from the graph?
(ii) Give a reason for your answer.

b Using the kinetic theory of gases, explain
(i) how a gas exerts a pressure,
(ii) the variation in pressure when the volume is reduced.

c If the experimental results show that when the pressure of the gas is 160 N/m^2 the corresponding value of $1/V$ is 0.025 m^3, calculate the value of the volume of the gas when the pressure of the gas changes to 4800 N/m^2. *(N.I.)*

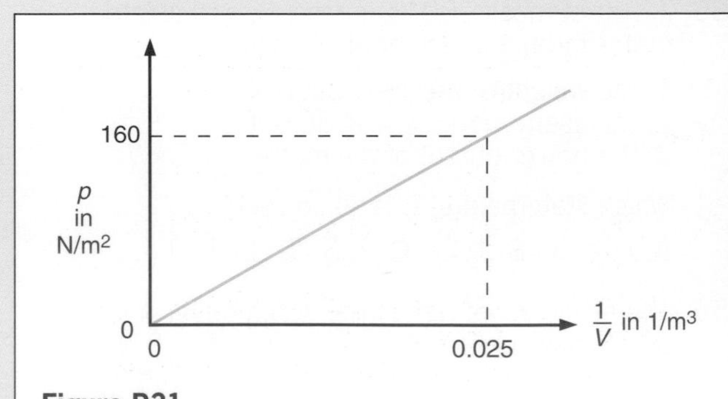

Figure R31

Motion and energy

68 The speeds of a car travelling on a straight road are given below at successive intervals of 1 second.

Time (s)	0	1	2	3	4
Speed (m/s)	0	2	4	6	8

The car travels

1 with an average velocity of 4 m/s
2 16 m in 4 s
3 with a uniform acceleration of 2 m/s^2.

Which statement(s) is (are) correct?

A 1, 2, 3 **B** 1, 2 **C** 2, 3 **D** 1 **E** 3

69 If a train travelling at 10 m/s starts to accelerate at 1 m/s^2 for 15 s on a straight track, its final velocity in m/s is

A 5 **B** 10 **C** 15 **D** 20 **E** 25

70 The equally spaced dots on the ticker tape in Figure R32 are made by a vibrator of frequency 50 Hz. The speed of the tape in m/s is

A 0.04 **B** 0.1 **C** 0.2 **D** 0.4 **E** 1

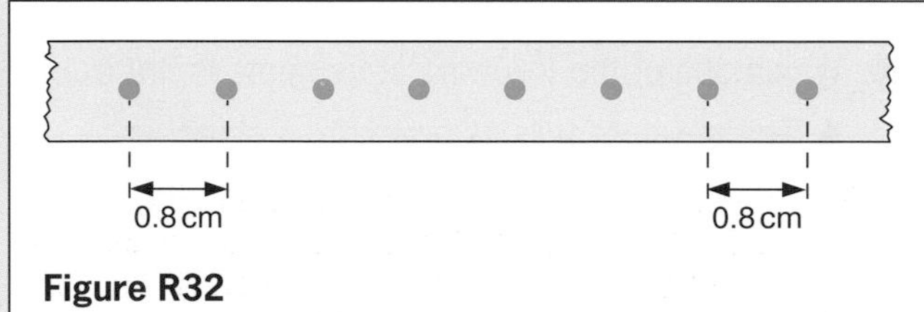

Figure R32

71 The velocity–time graph for a short trip by a girl cyclist is shown in Figure R33.

a What is her initial acceleration?

b What is her average velocity for the trip?

c What is her initial acceleration?

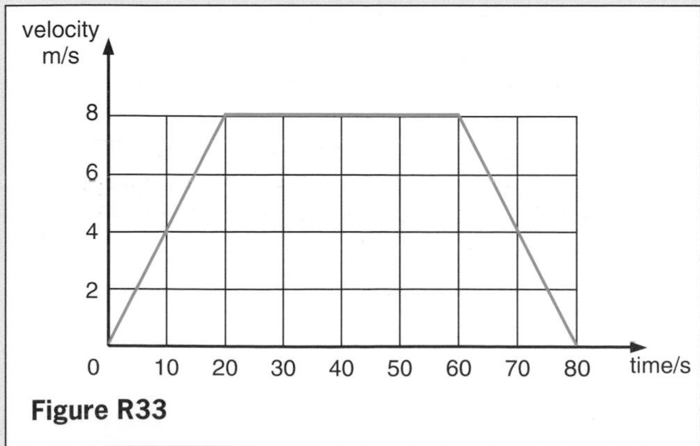

Figure R33

72 If the acceleration due to gravity is 10 m/s^2, an object falling from rest will

1 fall with a constant speed of 10 m/s
2 fall 10 m every second
3 have a speed of 20 m/s after 2 seconds.

Which statement(s) is (are) correct?

A 1, 2, 3 **B** 1, 2 **C** 2, 3 **D** 1 **E** 3

73 Which one of the velocity–time graphs in Figure R34 represents an object falling freely in a vacuum?

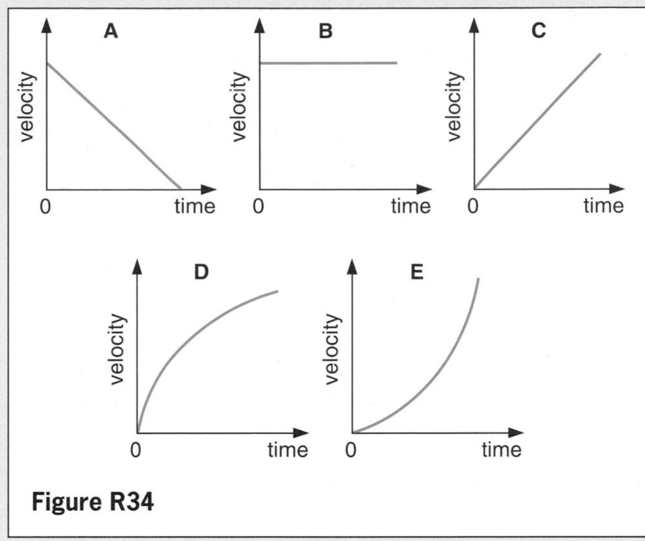

Figure R34

74 An unbalanced force of 50 N acts on a mass of 5 kg. The acceleration of the mass in m/s^2 will be

A 0.1 **B** 10 **C** 45 **D** 50 **E** 250

75 A force of 20 N pulls a block of mass 2 kg along a horizontal bench and is opposed by a constant frictional force of 12 N, Figure R35. The acceleration of the block in m/s^2 is

A 40 **B** 18 **C** 10 **D** 4 **E** 0.1

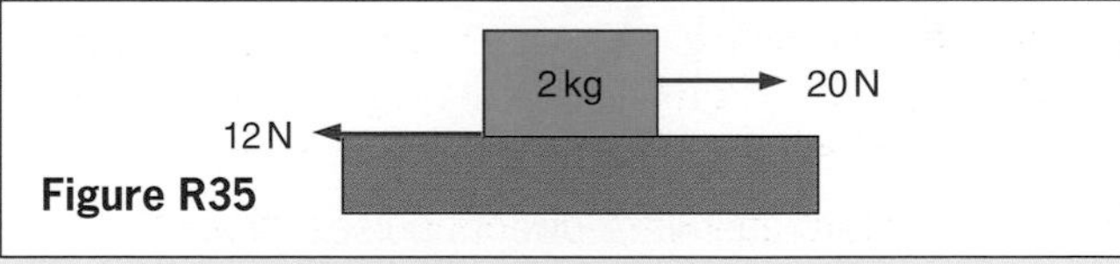

Figure R35

76 A 3 kg mass falls with its terminal velocity. Which combination **A** to **E** gives its weight, the air resistance and the resultant force acting on it?

	Weight	Air resistance	Resultant force
A	0.3 N down	zero	zero
B	3 N down	3 N up	3 N up
C	10 N down	10 N up	10 N down
D	30 N down	30 N up	zero
E	300 N down	zero	300 N down

77 A trolley of mass 3.0 kg moving at 4.0 m/s collides with, and remains attached to, a stationary trolley of mass 1.0 kg, Figure R36. Their combined momentum in kg m/s after the collision is

A 3.0 **B** 4.0 **C** 7.0 **D** 8.0 **E** 12

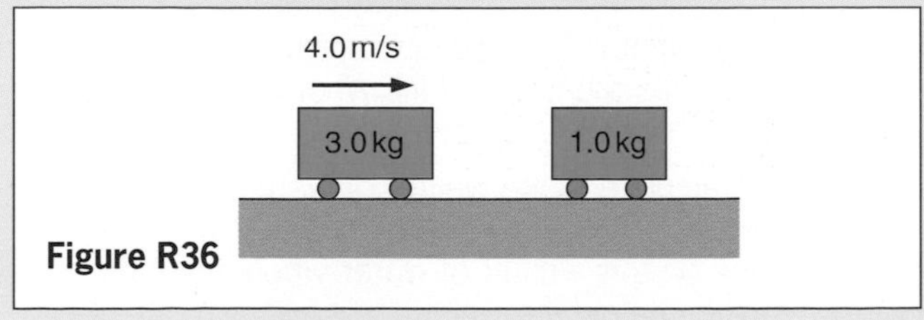

Figure R36

78 A stone of mass 2 kg is dropped from a height of 4 m. Neglecting air resistance, the k.e. of the stone in joules just before it hits the ground is

A 6 **B** 8 **C** 16 **D** 80 **E** 160

79 An object of mass 2 kg is fired vertically upwards with a k.e. of 100 J. Neglecting air resistance, which of the numbers in **A** to **E** below is

a the velocity in m/s with which it is fired,

b the height in m to which it will rise?

A 5 **B** 10 **C** 20 **D** 100 **E** 200

80 An object has k.e. of 10 J at a certain instant. If it is acted on by an opposing force of 5 N, which of the numbers **A** to **E** below is the further distance it travels in metres before coming to rest?

A 2 **B** 5 **C** 10 **D** 20 **E** 50

81 A boy whirls a ball at the end of a string round his head in a horizontal circle, centre O. If he lets go of the string when the ball is at X, Figure R37, the ball flies off in the direction

A 1 **B** 2 **C** 3 **D** 4 **E** 5

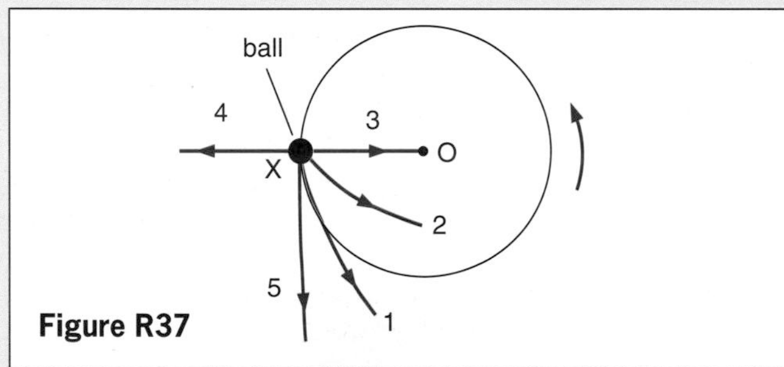

Figure R37

82 A radio transmitter directs pulses of waves towards a satellite from which reflections are received 10 milliseconds after transmission. If the speed of radio waves is 3×10^8 m/s, how far away is the satellite? *(S.)*

83 A helicopter of mass 3000 kg rises vertically at a constant speed of 25 m/s. The acceleration of free fall is 10 m/s^2. The resultant force, in N, acting on the helicopter is

A zero
B 30 000 downwards
C 45 000 upwards
D 75 000 upwards
E 105 000 upwards. *(N.I.)*

84 Water, which flows over a weir at a rate of 900 kg/s, takes 1.5 s to fall vertically into the stream below.

a (i) What is the speed with which the falling water hits the stream? (ii) What is the height through which the water falls?

b Calculate (i) the weight of water falling over the weir in 5.0 s, (ii) the work which has been done on this weight of water when it hits the stream below the weir.

c Calculate the power of the falling water at the instant it hits the stream.

d State the energy transformations which occur as the water falls from the weir into the stream below. *(C.)*

85 Figure R38 shows a drawing pin which rests on the deck of a record player. The deck is horizontal and is rotating at a constant rate. Which two of the following quantities associated with the pin are changed when the drawing pin has been carried from A to B?

speed, velocity, kinetic energy, potential energy, momentum *(O. and C.)*

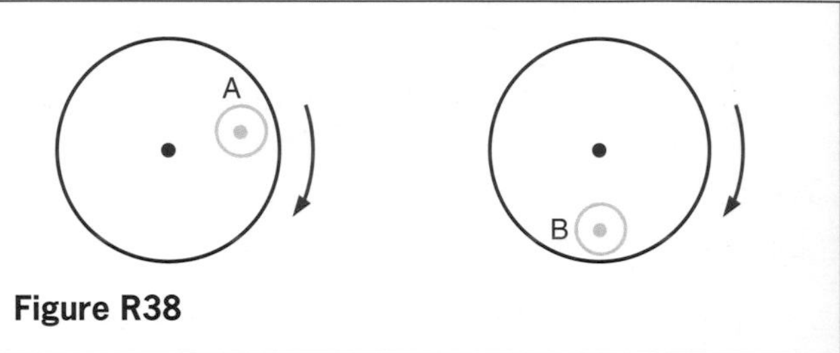

Figure R38

Heat and energy

86 Which one of the following statements is *not* true?

A Temperature tells us how hot an object is.
B Temperature is measured by a thermometer which uses some property of matter (e.g. the expansion of mercury) that changes continuously with temperature.
C Heat flows naturally from an object at a lower temperature to one at a higher temperature.
D The molecules of an object move faster when its temperature rises.
E Temperature is measured in °C, heat is measured in joules.

87 **a** Convert the following temperatures to kelvin: (i) 7 °C, (ii) 25 °C, (iii) −23 °C.

b Convert the following temperatures to °C: (i) 300 K, (ii) 573 K, (iii) 200 K.

88 The pressure exerted by a gas in a container

1 is due to the molecules of the gas bombarding the walls of the container.
2 decreases if the gas is cooled
3 increases if the volume of the container increases.

Which statement(s) is (are) correct?

A 1, 2, 3 **B** 1, 2 **C** 2, 3 **D** 1 **E** 3

89 When equal masses of water and paraffin are supplied with heat at the same rate, the temperature of the paraffin rises faster because paraffin has a

A smaller density **B** lower boiling point
C smaller specific heat capacity
D greater specific heat capacity
E lower melting point.

90 The specific heat capacity (c) of a substance is

1 the quantity of heat needed to raise the temperature of 1 kg by 1 °C
2 calculated using the equation $Q = m \times \Delta\theta \times c$ where $\Delta\theta$ (in °C or K) is the temperature rise when mass m (in kg) is supplied with quantity of heat Q (in J)
3 1000 J/(kg °C) if 4000 J of heat raise the temperature of 0.5 kg from 20 °C to 28 °C.

Which statement(s) is (are) correct?

A 1, 2, 3 **B** 1, 2 **C** 2, 3 **D** 1 **E** 3

91 A drink is cooled more by ice at 0 °C than by the same mass of water at 0 °C because ice

A floats on the drink
B has a smaller specific heat capacity
C gives out latent heat to the drink as it melts
D absorbs latent heat from the drink to melt
E is a solid.

92 Which of the following statements is/are true?

1 In cold weather the wooden handle of a saucepan feels warmer than the metal pan *because* wood is a better conductor of heat.
2 Convection occurs when there is a change of density in the parts of a fluid.
3 Conduction and convection cannot occur in a vacuum.

A 1, 2, 3 **B** 1, 2 **C** 2, 3 **D** 1 **E** 3

93 Which one of the following statements is *not* true?

A Energy from the Sun reaches the Earth by radiation only.
B A dull black surface is a good absorber of radiation.
C A shiny white surface is a good emitter of radiation.
D The best heat insulation is provided by a vacuum.
E A vacuum flask is designed to reduce heat loss or gain by conduction, convection and radiation.

94 Figure R39 shows the variation of the density of water with temperature. It may be deduced from the graph that

1 a given mass of water has a minimum volume at 4 °C
2 between 0 °C and 4 °C convection will not take place in a vessel of water heated at the bottom
3 expansion occurs when water freezes.

Which statement(s) is (are) correct?

A 1, 2, 3 **B** 1, 2 **C** 2, 3 **D** 1 **E** 3 (*L.*)

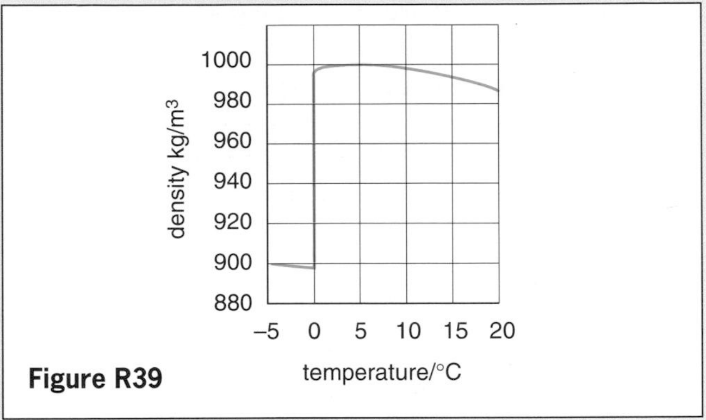

Figure R39

95 A quantity of helium occupies a volume 3.0×10^{-3} m^3 inside a high pressure gas cylinder. The pressure of the helium is 1.5×10^7 Pa (N/m^2) when the cylinder is stored at 7 °C. What volume will the helium occupy if the contents of the cylinder are used to inflate a balloon to a pressure 1.0×10^5 Pa (N/m^2) when the temperature is 20 °C? (*O. and C.*)

96 A small quantity, 0.010 kg, of water at 17 °C is added to a larger mass of ice at 0 °C contained in a vacuum flask. Calculate the greatest mass of ice which can be melted. (Take the specific heat capacity of water as 4200 J/(kg °C) and the specific latent heat of fusion of ice as 340×10^3 J/kg.) (*O. and C.*)

97 Answer each part of this question in terms of the **kinetic theory of matter**.

a A sample of liquid at its boiling point is to be converted to vapour at the same temperature. Why must heat energy be supplied to bring about this change?

b When all the liquid has been converted into vapour at the boiling point, the vapour fills a container at atmospheric pressure. Explain why the volume of the vapour is much greater than the original volume of the liquid.

c Why does the vapour exert a pressure on its container?

d If the temperature of the container is raised, why does the pressure of the vapour inside the container rise? (Assume that the volume of the container does not change.) (*C.*)

Electricity and magnetism

98 If the two uncharged metal spheres R and S on insulating stands, Figure R40, are separated while the negatively charged polythene strip is held near R, the charges on R and S are

	A	**B**	**C**	**D**	**E**
R	+	−	−	+	zero
S	+	−	+	−	zero

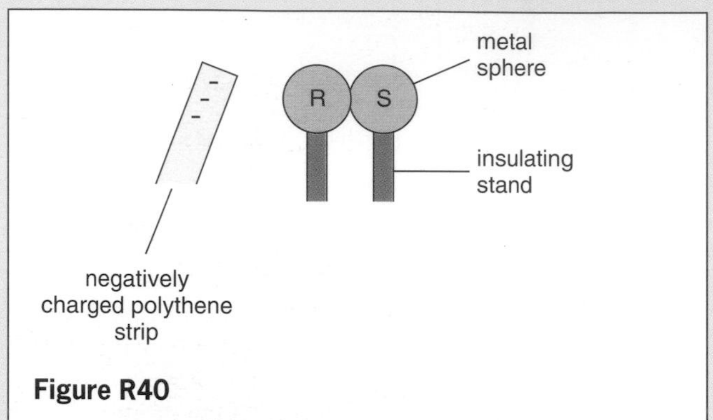

Figure R40

99 For the circuit of Figure R41 calculate

a the total resistance,

b the current in each resistor,

c the p.d. across each resistor.

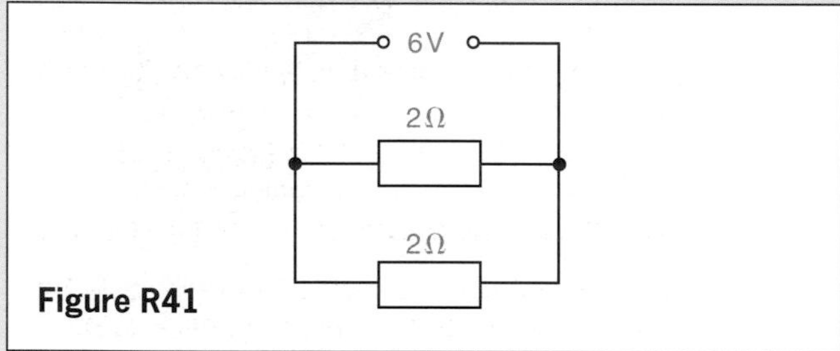

Figure R41

100 Repeat question 99 for the circuit of Figure R42.

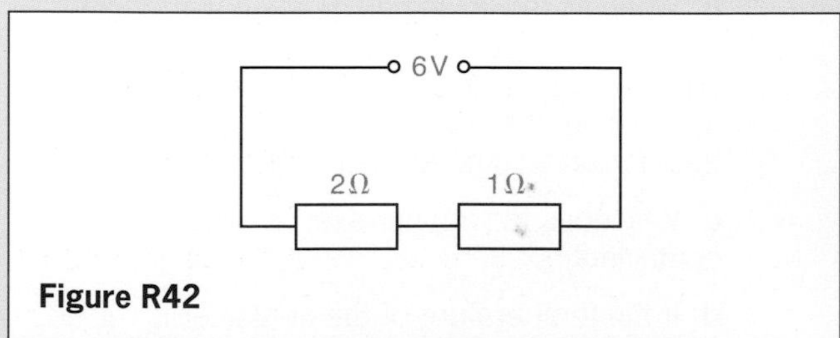

Figure R42

101 An electric kettle for use on a 240 V supply is rated at 3000 W. The minimum current the cable supplying it should be able to carry for safe working is

A 2 A **B** 5 A **C** 10 A **D** 15 A **E** 30 A

102 Which one of the following statements is *not* true?

A In a house circuit lamps are wired in parallel.
B Switches, fuses and circuit breakers should be placed in the neutral wire.
C An electric fire has its earth wire connected to the metal case to prevent the user receiving a shock.
D When connecting a three-core cable to a 13 A three-pin plug the *brown* wire goes to the *live* pin.
E The cost of operating three 100 W lamps for 10 hours at 10p per unit is 30p.

103 The diagrams in Figure R43 represent magnetic fields caused by current-carrying conductors. Which one is due to

a a long straight wire,

b a circular coil,

c a solenoid?

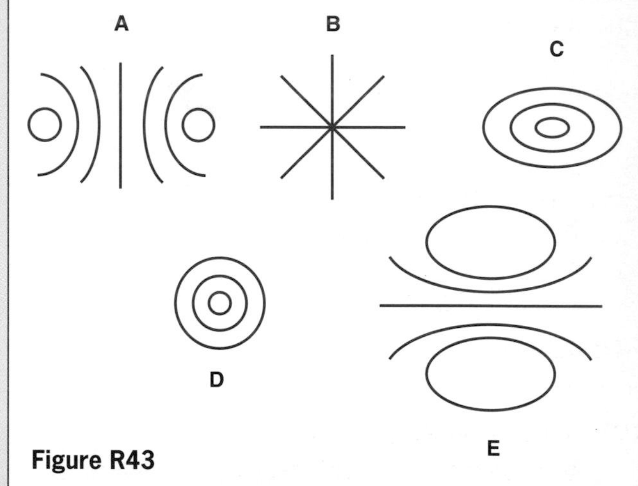

Figure R43

104 Which of the following statements is/are true?

1 An electromagnet consists of a coil of wire wound on a soft iron core.
2 The strength of the magnetic field produced by an electromagnet increases if the strength of the current and/or the number of turns of wire is increased.
3 In Figure R44 when the switch is closed the gap G increases.

A 1, 2, 3 **B** 1, 2 **C** 2, 3 **D** 1 **E** 3

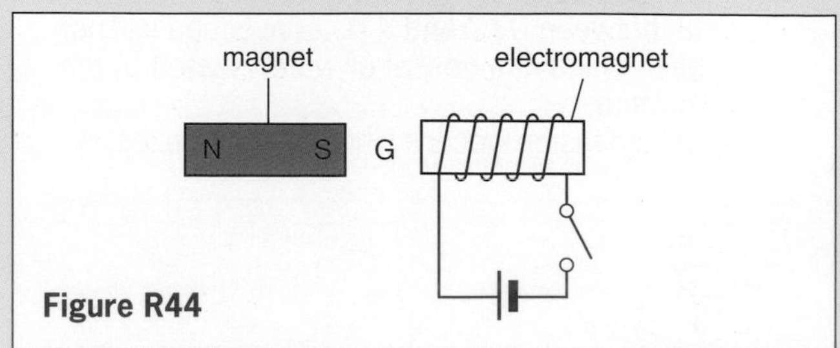

Figure R44

105 Which one of the following statements is *not* true?

A If a current is passed through the wire XY in Figure R45a, a vertically upwards force acts on it.
B If a current is passed through the wire PQ in Figure R45b, it does not experience a force.
C If a current is passed through the coil in Figure R46 it rotates clockwise.
D If the coil in Figure R46 had more turns and carried a larger current, the turning effect would be greater.
E In a moving-coil loudspeaker a coil moves between the poles of a strong magnet.

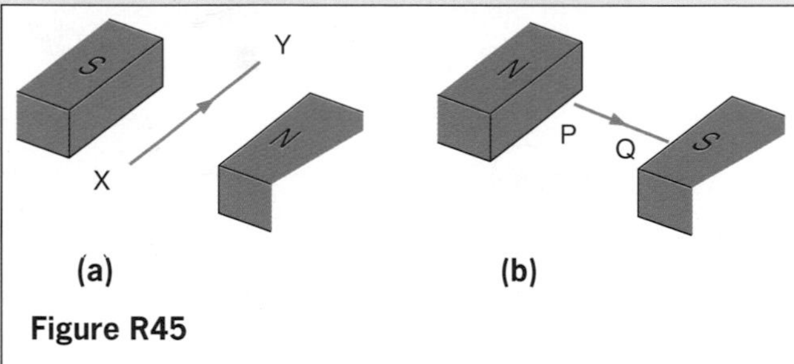

Figure R45

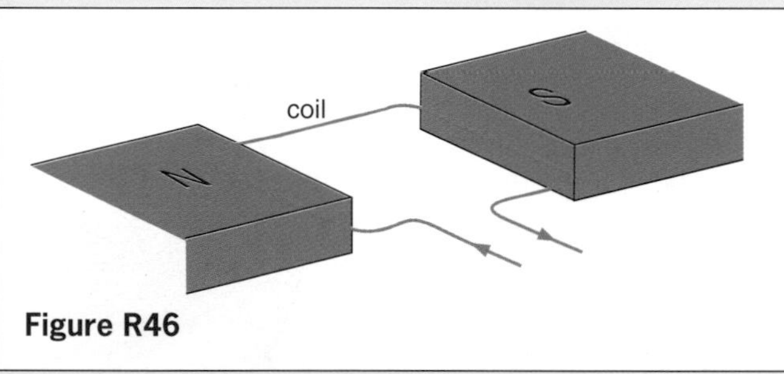

Figure R46

106 Which statement(s) is (are) correct?

1 An ammeter is connected in series in a circuit and a voltmeter in parallel.
2 An ammeter has a high resistance.
3 A voltmeter has a low resistance.

A 1, 2, 3 **B** 1, 2 **C** 2, 3 **D** 1 **E** 3

107 Which one of the following statements is *not* true when a magnet is pushed N pole first into a coil, Figure R47?

A A p.d. is induced in the coil and causes a current through the galvanometer.
B The induced p.d. increases if the magnet is pushed in faster and/or the coil has more turns.
C Mechanical energy is changed to electrical energy.
D The coil tends to move to the right because the induced current makes face X a N pole which is repelled by the N pole of the magnet.
E The effect produced is called electrostatic induction.

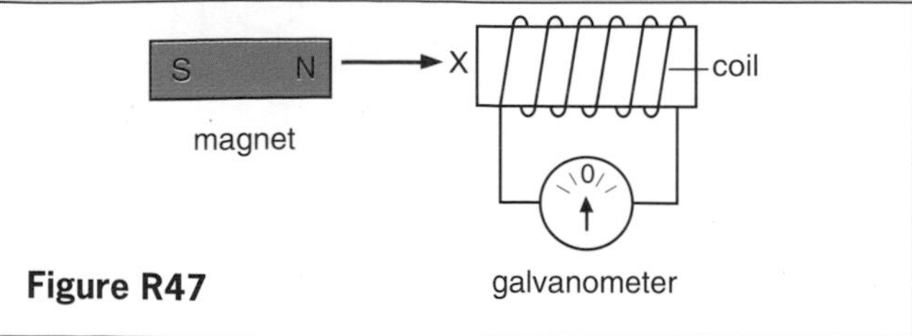

Figure R47

108 Which of the following statements is/are true?

1 A transformer changes an alternating p.d. from one value to another according to the equation $V_s/V_p = N_s/N_p$.
2 Transformers are used to send electrical energy by cable over long distances at high voltage and low current to cut heat loss.
3 The number of turns on the primary of a transformer which has 200 turns on the secondary and is designed to deliver 12 V from the 240 V mains supply, Figure R48, is 5000.

A 1, 2, 3 **B** 1, 2 **C** 2, 3 **D** 1 **E** 3

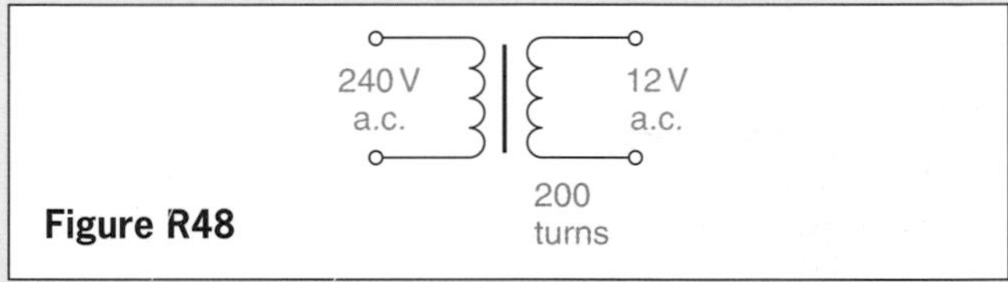

Figure R48

109 Which of the units **A** to **E** could be used to measure **a** electric charge, **b** electric current, **c** p.d., **d** energy, **e** power?

A ampere **B** joule **C** volt **D** watt **E** coulomb

110 a A wire was connected to a battery and it was found that the energy converted into heat was 30 joules when 20 coulombs of charge flowed through the wire in 5 seconds. Calculate
(i) the potential difference between the ends of the wire,
(ii) the current flowing through the wire,
(iii) the resistance of the wire,
(iv) the average power developed in the wire.

b If the current in the wire were doubled and all the energy were released as heat in the wire, how much heat would be produced in the wire in 5 seconds? *(J.M.B.)*

111 The graph in Figure R49 shows how the current *I* through a tungsten filament lamp varies with the voltage *V* across it.

a What is the voltage when the current is 0.2 amperes?

b What is the resistance of the lamp when *I* is 0.2 amperes?

c What is the increase in current when *V* is increased from 2 V to 6 V?

d Does the resistance of the filament increase or decrease as *V* increases from 2 V to 6 V? What causes the change?

e Draw a diagram of the circuit you would use to take readings to plot this graph. *(O.L.E.)*

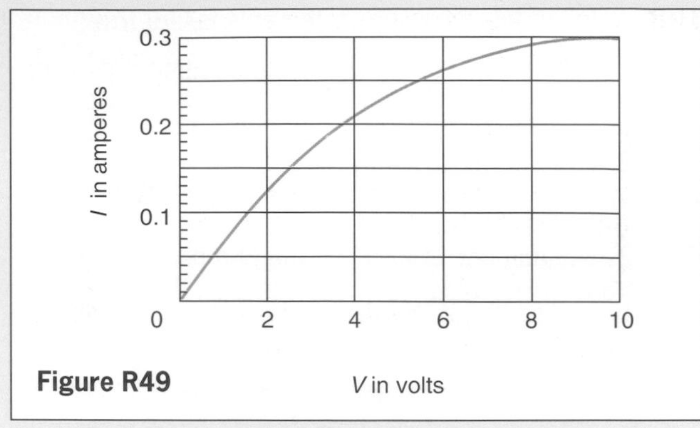

Figure R49

112 An electric motor takes a current of 5.0 A from a 4.0 V supply. Calculate the power input to the motor.

The motor lifts a weight of 50 N through a vertical height of 3.0 m in 10 s. Calculate the average useful work done per second by the motor in lifting this weight.

Suggest a reason for the difference between this quantity and the power input to the motor. *(C.)*

113 Two coils P and Q are wound on a soft iron former as in Figure R50. When a steady current passes through coil P, end N of the coil behaves as a magnetic north pole, while end S behaves as a magnetic south pole. Key K is initially open, and is then closed. Which one of the following statements concerning the circuit with coil Q is correct after the key is closed?

A A current will flow momentarily from X to Y through G.
B A current will flow momentarily from Y to X through G.
C A steady current will flow from X to Y through G.
D A steady current will flow from Y to X through G.
E No current will flow through G. *(N.I.)*

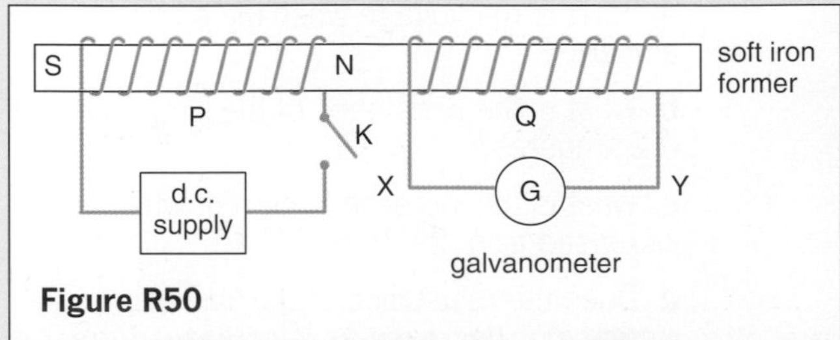

Figure R50

114 Figure R51 shows how the output voltage from an a.c. generator varies with time. The frequency of the output, in Hz, is

A 0.01 **B** 0.10 **C** 10.00 **D** 100.00
E 200.00 *(N.I.)*

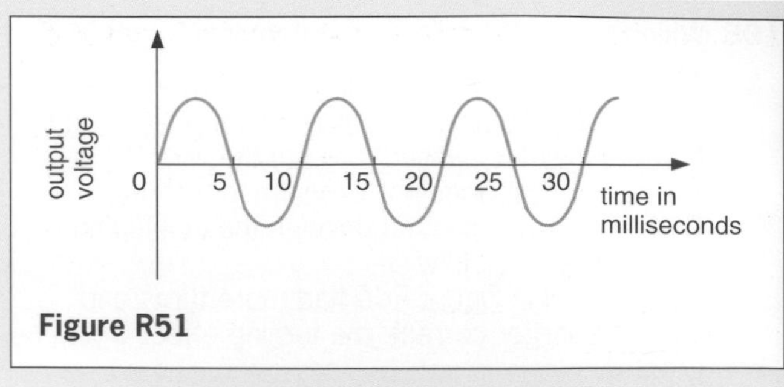

Figure R51

115 The circuit in Figure R52 is for a transformer supplying a lamp. If the transformer is 100% efficient the resistance of the lamp is

A 1.5 Ω **B** 3 Ω **C** 6 Ω **D** 12 Ω **E** 24 Ω

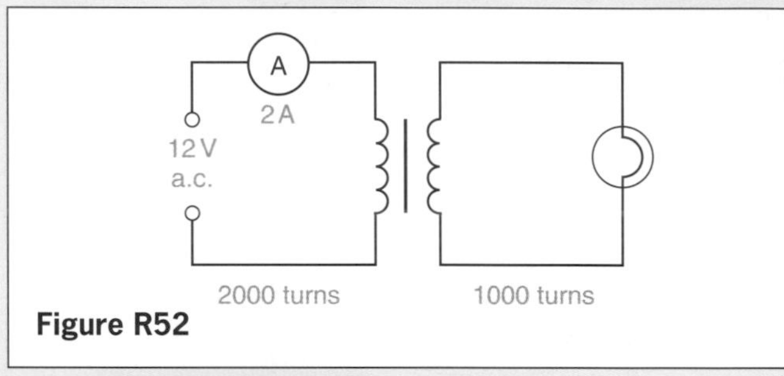

Figure R52

Electrons and atoms

116 Which of the following statements about the deflection of the beam of electrons by the p.d. between the plates P and Q in Figure R53 is/are true?

1 Plate P is negative.
2 The deflection would be greater if the p.d. was greater.
3 The deflection would be greater if the electrons were moving slower.

A 1, 2, 3 **B** 1, 2 **C** 2, 3 **D** 1 **E** 3

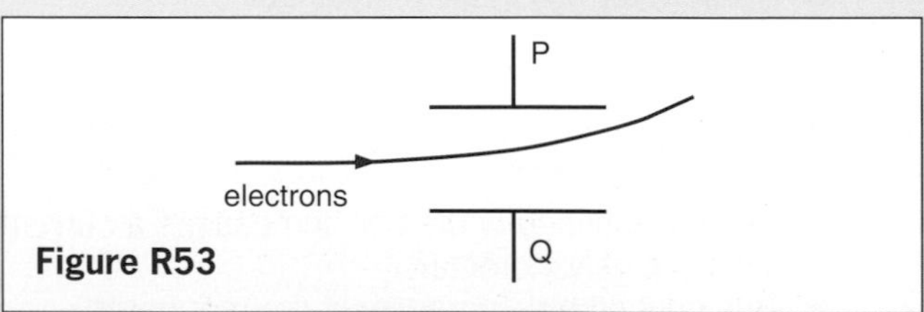

Figure R53

117 Which of the patterns in Figure R54 could appear on the screen of a CRO if an alternating p.d. is applied to the Y-plates with the time base **a** off, **b** on?

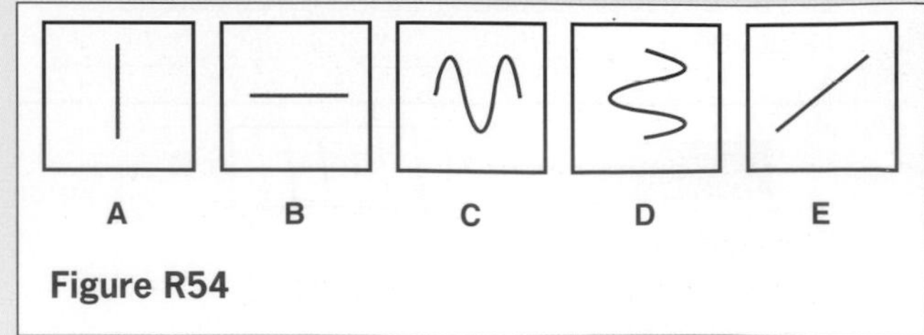

Figure R54

118 Identify the radiations X, Y and Z from Figure R55.

	A	B	C	D	E
X	alpha	beta	gamma	gamma	beta
Y	beta	alpha	alpha	beta	gamma
Z	gamma	gamma	beta	alpha	alpha

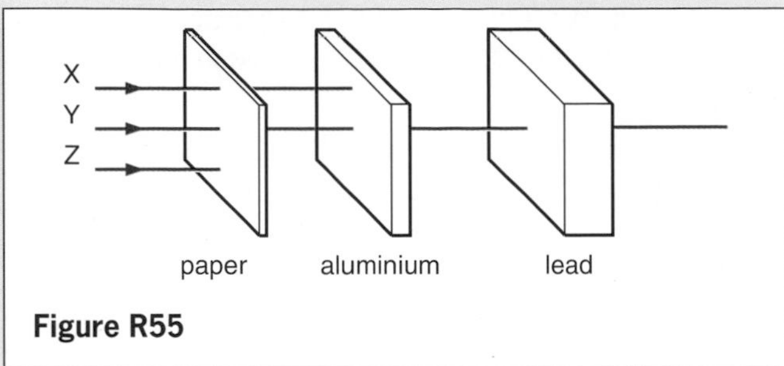

Figure R55

119 The radioactive substance whose decay curve is given in Figure R56 has a half-life in minutes of

A 1 **B** 2 **C** 3 **D** 4 **E** 5

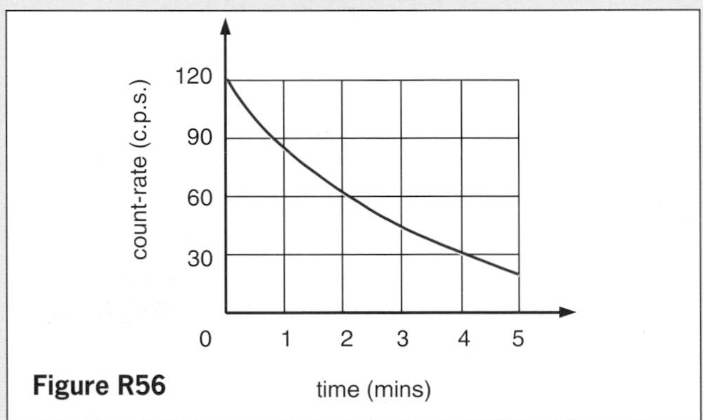

Figure R56

120 A radioactive source which has a half-life of 1 hour gives a count-rate of 100 c.p.s. at the start of an experiment and 25 c.p.s. at the end. The time taken by the experiment was, in hours,

A 1 **B** 2 **C** 3 **D** 4 **E** 5

121 Which one of the following statements is *not* true?

A An atom consists of a tiny nucleus surrounded by orbiting electrons.
B The nucleus contains protons and neutrons, called nucleons, in equal numbers.
C A proton has a positive charge, a neutron is uncharged and their mass is about the same.
D An electron has a negative charge of the same size as the charge on a proton but it has a much smaller mass.
E The number of electrons equals the number of protons in a normal atom.

122 A lithium atom has a nucleon (mass) number of 7 and a proton (atomic) number of 3.

1 Its symbol is $^7_4$Li.
2 It contains 3 protons, 4 neutrons and 3 electrons.
3 One of its isotopes has 3 protons, 3 neutrons and 3 electrons.

Which statement(s) is (are) correct?

A 1, 2, 3 **B** 1, 2 **C** 2, 3 **D** 1 **E** 3

123 Which symbol **A** to **E** is used in equations for nuclear reactions to represent **a** an alpha particle, **b** a beta particle, **c** a neutron, **d** an electron?

A $_{-1}^{0}$e **B** $_0^1$n **C** $_2^4$He **D** $_{-1}^{1}$e **E** $_1^1$n

124 **a** Radon $^{220}_{86}$Rn decays by emitting an alpha particle to form an element whose symbol is

A $^{216}_{85}$At **B** $^{216}_{86}$Rn **C** $^{218}_{84}$Po **D** $^{216}_{84}$Po **E** $^{217}_{85}$At

b Thorium $^{234}_{90}$Th decays by emitting a beta particle to form an element whose symbol is

A $^{235}_{90}$Th **B** $^{230}_{89}$Ac **C** $^{234}_{89}$Ac **D** $^{232}_{88}$Ra **E** $^{234}_{91}$Pa

125 Which one of the following statements about the transistor circuit in Figure R57 is *not* true?

A The collector current I_C is zero until base current I_B flows.
B I_B is zero until the base-emitter p.d. V_{BE} is +0.6 V.
C A small I_B can switch on and control a large I_C.
D When used as an amplifier the input is connected across B and E.
E X must be connected to supply − and Y to the +.

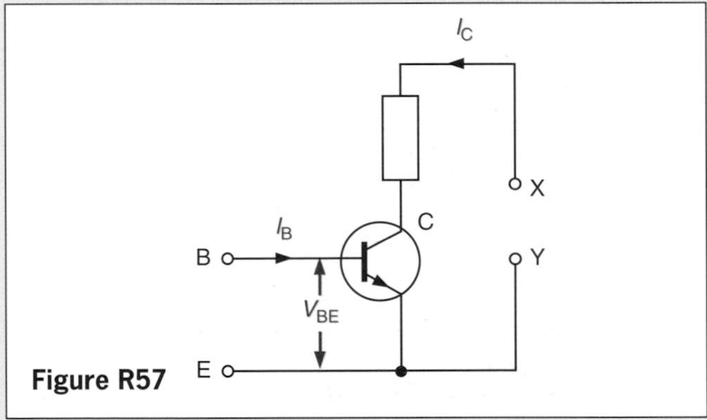

Figure R57

Questions 126 to 128

The diagrams in Figure R58 represent five possible paths taken by atomic particles (initially travelling to the right) after they enter a magnetic field. Crosses (+) indicate a magnetic field directed into the page, while arrows pointing up indicate a magnetic field in that direction.

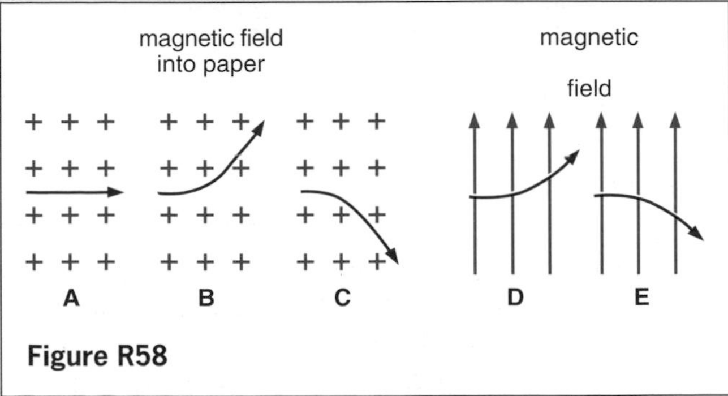

Figure R58

For each of the following particles, select the diagram which corresponds most closely to the expected path taken. Each diagram may be used once, more than once, or not at all.

126 electron

127 α-particle

128 neutron (*N.I.*)

129 The waveform in Figure R59a is displayed on a CRO. A student then alters *two* controls and obtains the waveform in Figure R59b. The two controls adjusted were

A time base and Y-shift **B** X-shift and Y-gain
C Y-gain and time base **D** focus and X-gain
E X-gain and Y-gain.

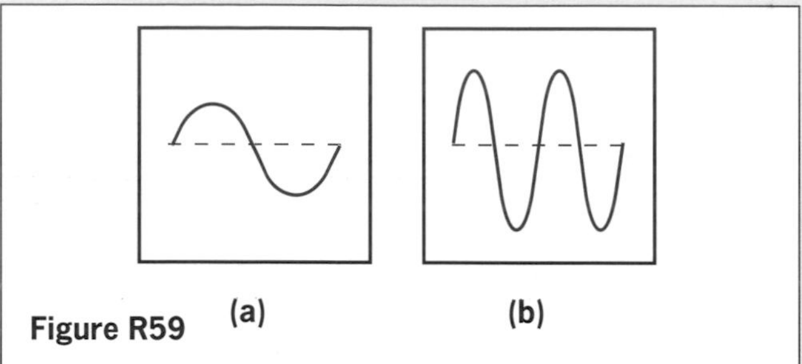

Figure R59

130 A radioactive source emits α, β and γ radiations. A suitable detector is first placed very close to the source and then moved about 10 cm away. Explain any differences in count-rate that you would expect. What further effects would be obtained by your placing

a a very thin sheet of lead,

b a thick sheet of lead,

between source and detector? (*S.*)

131 a Explain what is meant by (i) atomic mass number, (ii) atomic number.

b (i) The isotope $^{238}_{92}U$ decays by alpha-emission to an isotope of thorium (Th). Compare the $^{238}_{92}U$ and thorium nuclei, explaining the changes which have occurred in the uranium nucleus.
(ii) The thorium nucleus decays by beta-emission to an isotope of protactinium (Pa). Compare the thorium and Pa nuclei, accounting for the changes you describe. (*J.M.B.*)

132 A radioactive sample has a half-life of 20 minutes, and at a certain time a detector records 120 counts per second. Calculate the count-rate recorded by the detector one hour later.

The sample emits α-particles. State briefly the nature of these particles. (*C.*)

133 Figure R60 represents a certain atom which is made up of protons (P), neutrons (N), and electrons (e).

a What is meant by the **mass number** of an atom and what is its value in this case?

b What is meant by the **atomic number** of an atom and what is its value in this case?

c What is the charge: (i) on the nucleus of this atom, (ii) on the atom as whole?

d How do protons, neutrons and electrons compare in mass?

e What happens to the atomic number of an atom when it emits: (i) an α-particle, (ii) a β-particle? (*O.L.E.*)

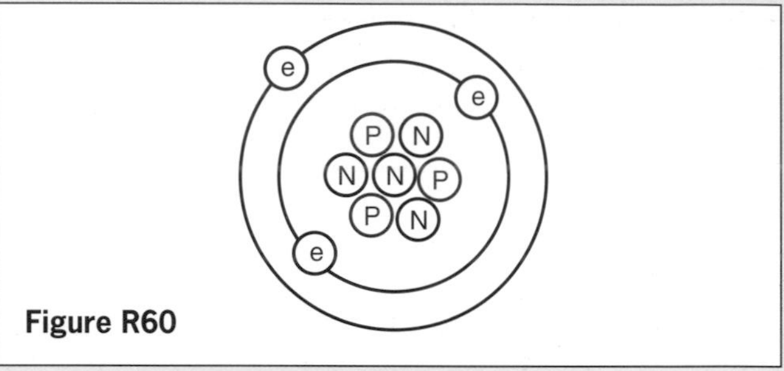

Figure R60

Earth and space physics

134 Refer to the map of Figure 68.13 to answer this question about Britain's weather. Which air stream is most likely to

a give the south-west its warm humid weather,

b cause cool, very dry weather in summer and very cold weather in winter, and

c deposit fine sand on cars parked outdoors?

135 a Draw a diagram to show the Earth's structure.

b On the diagram you have drawn, sketch in lines showing the paths taken by P, S and L waves from an earthquake.

c Why are seismic waves refracted in a *curved* path when they pass through the Earth's mantle?

136 A space shuttle is in orbit round the Earth at a certain height.

a What keeps it in orbit?

b If the shuttle reduces its mass by launching a communications satellite, how does this affect the centripetal force?

c If the shuttle moves to a higher orbit does there have to be an increase or a decrease in (i) the centripetal force, (ii) the orbital speed?

137 The Sun radiates energy E at a rate of 4.0×10^{26} watts by converting some of its mass m according to the equation $E = mc^2$. If $c = 3 \times 10^8$ m/s, calculate how much matter the Sun loses per second.

MATHEMATICS FOR PHYSICS

USE THIS SECTION AS THE NEED ARISES

Solving physics problems

When tackling physics problems using mathematical equations it is suggested that you **do not substitute numerical values until you have obtained the expression in symbols which gives the answer**. That is, work in symbols until you have solved the problem and only then insert the numbers in the expression to get the final result.

This has two advantages. First, it reduces the chance of errors in the arithmetic (and in copying down). Second, you write less since a symbol is usually a single letter whereas a numerical value is often a string of figures.

Adopting this 'symbolic' procedure frequently requires you to change round an equation first. The next two sections and the questions that follow them are intended to give you practice in doing this and then substituting numerical values to get the answer.

Equations – type 1

In the equation $x = a/b$, the subject is x. To change it we **multiply or divide both sides** of the equation by the same quantity.

To change the subject to a

We have

$$x = \frac{a}{b}$$

If we multiply both sides by b the equation will still be true.

$$\therefore \quad x \times b = \frac{a}{b} \times b$$

The b's on the right-hand side cancel

$$\therefore \quad b \times x = \frac{a}{\cancel{b}} \times \cancel{b} = a$$

$$\therefore \quad a = b \times x$$

To change the subject to b

We have

$$x = \frac{a}{b}$$

Multiplying both sides by b as before, we get

$$a = b \times x$$

Dividing both sides by x

$$\frac{a}{x} = \frac{b \times x}{x} = \frac{b \times \cancel{x}}{\cancel{x}} = b$$

$$\therefore \quad b = \frac{a}{x}$$

Now try the following questions using these ideas.

QUESTIONS

1 What is the value of x if

a $2x = 6$ **b** $3x = 15$ **c** $3x = 8$

d $\frac{x}{2} = 10$ **e** $\frac{x}{3} = 4$ **f** $\frac{2x}{3} = 4$

g $\frac{4}{x} = 2$ **h** $\frac{9}{x} = 3$ **i** $\frac{x}{6} = \frac{4}{3}$

2 Change the subject to

a f in $v = f\lambda$ **b** λ in $v = f\lambda$

c I in $V = IR$ **d** R in $V = IR$

e m in $d = \frac{m}{V}$ **f** V in $d = \frac{m}{V}$

g s in $v = \frac{s}{t}$ **h** t in $v = \frac{s}{t}$

3 Change the subject to

a I^2 in $P = I^2R$ **b** I in $P = I^2R$

c a in $s = \frac{1}{2}at^2$ **d** t^2 in $s = \frac{1}{2}at^2$

e t in $s = \frac{1}{2}at^2$ **f** v in $\frac{1}{2}mv^2 = mgh$

g y in $\lambda = \frac{ay}{D}$ **h** ρ in $R = \frac{\rho l}{A}$

4 By replacing (substituting) find the value of $v = f\lambda$ if

a $f = 5$ and $\lambda = 2$ **b** $f = 3.4$ and $\lambda = 10$

c $f = 1/4$ and $\lambda = 8/3$ **d** $f = 3/5$ and $\lambda = 1/6$

e $f = 100$ and $\lambda = 0.1$ **f** $f = 3 \times 10^5$ and $\lambda = 10^3$

5 By changing the subject and replacing find

a f in $v = f\lambda$ if $v = 3.0 \times 10^8$ and $\lambda = 1.5 \times 10^3$

b h in $p = 10hd$ if $p = 10^5$ and $d = 10^3$

c a in $n = a/b$ if $n = 4/3$ and $b = 6$

d b in $n = a/b$ if $n = 1.5$ and $a = 3.0 \times 10^8$

e F in $p = F/A$ if $p = 100$ and $A = 0.2$

f s in $v = s/t$ if $v = 1500$ and $t = 0.2$

Equations – type 2

To change the subject in the equation $x = a + by$ we **add or subtract the same quantity from each side**. We may also have to divide or multiply as in type 1. Suppose we wish to change the subject to y in

$$x = a + by$$

Subtracting a from both sides,

$$x - a = a + by - a = by$$

Dividing both sides by b,

$$\frac{x-a}{b} = \frac{by}{b} = y$$

$$\therefore \qquad y = \frac{x-a}{b}$$

QUESTIONS

6 What is the value of x if

a $x + 1 = 5$ **b** $2x + 3 = 7$ **c** $x - 2 = 3$

d $2(x - 3) = 10$ **e** $\frac{x}{2} - \frac{1}{3} = 0$ **f** $\frac{x}{3} + \frac{1}{4} = 0$

g $2x + \frac{5}{3} = 6$ **h** $7 - \frac{x}{4} = 11$ **i** $\frac{3}{x} + 2 = 5$

7 By changing the subject and replacing, find the value of a in $v = u + at$ if

a $v = 20$, $u = 10$ and $t = 2$

b $v = 50$, $u = 20$ and $t = 0.5$

c $v = 5/0.2$, $u = 2/0.2$ and $t = 0.2$

8 Change the subject in $v^2 = u^2 + 2as$ to a.

Proportion (or variation)

One of the most important mathematical operations in physics is finding the relation between two sets of measurements.

(a) Direct proportion

Suppose that in an experiment two sets of readings are obtained for the quantities x and y as in Table 1 (units omitted).

Table 1

x	1	2	3	4
y	2	4	6	8

We see that when x is doubled, y doubles; when x is trebled, y trebles; when x is halved, y halves; and so on. There is a one-to-one correspondence between each value of x and the corresponding value of y.

We say that y is **directly proportional** to x, or y **varies directly** as x. In symbols

$$y \propto x$$

Also, the **ratio** of one to the other, e.g. y to x, is always the same, i.e. it has a constant value which in this case is 2. Hence

$$\frac{y}{x} = \text{a constant} = 2$$

The constant, called the **constant of proportionality** or **variation**, is given a symbol, e.g. k, and the relation (or law) between y and x is then summed up by the equation

$$\frac{y}{x} = k \quad \text{or} \quad y = kx$$

Notes:

1 In practice, because of inevitable experimental errors, the readings seldom show the relation so clearly as here.

2 If instead of using numerical values for x and y we use letters, e.g. x_1, x_2, x_3, etc., and y_1, y_2, y_3, etc., then we can also say

$$\frac{y_1}{x_1} = \frac{y_2}{x_2} = \frac{y_3}{x_3} = \ldots\ldots\ldots = k$$

or

$$y_1 = kx_1, y_2 = kx_2, y_3 = kx_3, \ldots\ldots$$

(b) Inverse proportion

Two sets of readings for the quantities p and V are given in Table 2.

Table 2

p	3	4	6	12
V	4	3	2	1

There is again a one-to-one correspondence between each value of p and the corresponding value of V but when p is doubled V is halved; when p is trebled, V has one-third its previous value; and so on.

We say that V is **inversely proportional** to p, or V **varies inversely** as p, i.e.

$$V \propto \frac{1}{p}$$

Also, the **product** $p \times V$ is always the same (=12) and we write

$$V = \frac{k}{p} \quad \text{or} \quad pV = k$$

where k is the constant of proportionality or variation and equals 12 in this case.

Using letters for values of p and V we can also say

$$p_1V_1 = p_2V_2 = p_3V_3 = \ldots\ldots\ldots = k$$

Graphs

Another useful way of finding the relation between two quantities is by a graph.

(a) Straight line graphs

When the readings in Table 1 are used to plot a graph of y against x, a **continuous** line joining the points is **a straight line passing through the origin**, Figure M1. Such a graph shows there is direct proportionality between the quantities plotted, i.e. $y \propto x$. But note that the line must go through the origin.

A graph of p against V using the readings in Table 2 is a curve, Figure M2. However if we plot p against $1/V$ (or V against $1/p$) we get a straight line through the origin, showing that $p \propto 1/V$, Figure M3 (or $V \propto 1/p$).

p	V	1/V
3	4	0.25
4	3	0.33
6	2	0.50
12	1	1.00

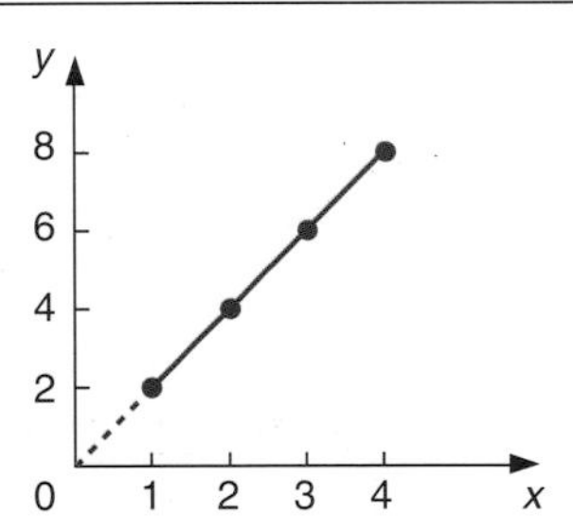

Figure M1

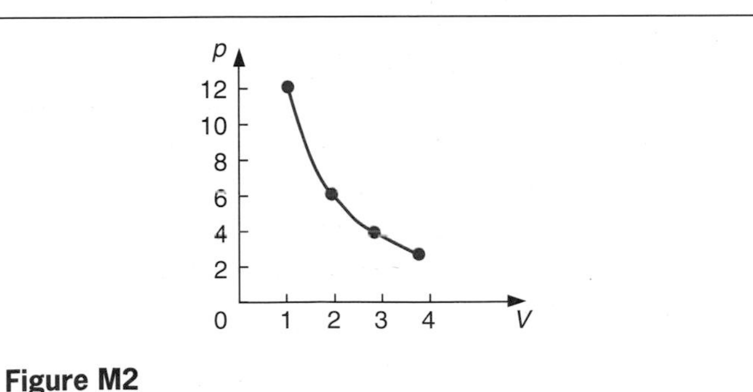

Figure M2

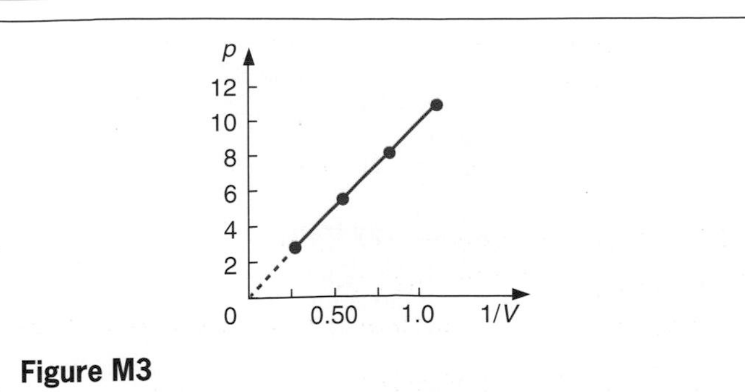

Figure M3

(b) Slope or gradient

The slope or gradient of a straight line graph equals the constant of proportionality. In Figure M1, the slope is $y/x = 2$; in Figure M3 it is $p/(1/V) = 12$.

In practice points plotted from actual measurements may not lie exactly on a straight line due to experimental errors. The 'best straight line' is then drawn 'through' them so that they are equally distributed about it. This automatically averages the results. Any points that are well off the line stand out and may be investigated further.

(c) Practical points

(i) The axes should be labelled giving the quantities being plotted and their units, e.g. I/A meaning current in amperes.

(ii) If possible the origin of both scales should be on the paper and the scales chosen so that the points are spread out along the graph.

(iii) Mark the points ⊙ or ×.

QUESTIONS

9 In an experiment different masses were hung from the end of a spring held in a stand and the extensions produced were as shown below.

Mass (g)	100	150	200	300	350	500	600
Extension (cm)	1.9	3.1	4.0	6.1	6.9	10.0	12.2

a Plot a graph of **extension** along the vertical (y) axis against **mass** along the horizontal (x) axis.

b What is the relation between **extension** and **mass**? Give a reason for your answer.

10 Pairs of readings of the quantities m and v are given below.

m	0.25	1.5	2.5	3.5
v	20	40	56	72

a Plot a graph of m along the vertical axis and v along the horizontal axis.

b Is m directly proportional to v? Explain your answer.

c Use the graph to find v when $m = 1$.

11 The distance s (in metres) travelled by a car at various times t (in seconds) are shown below.

s	0	2	8	18	32	50
t	0	1	2	3	4	5

Draw graphs of **a** s against t, **b** s against t^2. What can you conclude?

ANSWERS

1 Light rays

1 Larger, less bright
2 a 4 images **b** Brighter but blurred
3 Shadow to left is darker
4 When he sees the flash

2 Reflection of light

1 a 50° **c** 50°, 40°, 40° **d** Parallel
2 B
3 Top half

3 Plane mirrors

2 D
3 4 m towards mirror

4 Curved mirrors

2 7.2 cm from mirror, 1.6 cm high
3 Concave, parabolic

5 Refraction of light

3 250 000 km/s
4 D

7 Lenses

1 A
2 c Image 9 cm from lens, 3 cm high

8 The eye

2 E
3 a (i) No (ii) Yes (iii) No
b Concave
4 a (i) No (ii) Yes **b** Convex

9 Colour

2 D
3 a (i) White (ii) White **b** Red
c Red

10 Simple optical instruments

1 Close object
2 (i) Larger, blurred, less bright
(ii) Moved closer to the slide
3 a 4 cm
b 8 cm behind lens, virtual, $m = 2$

11 Microscopes and telescopes

1 a D **b** B
2 a Objective, 100 cm **b** 100 cm
c Eyepiece **d** Real image formed by A; 5 cm **e** 105 cm

Additional questions

1 B
2 B
3 A
4 3.4×10^3 km
6 14 m
7 a Ray passes into air
b Total internal reflection occurs in water
8 E
9 30°
10 C
11 a 4 **b** 3.2 cm
13 a Long sight **b** Convex lens
14 A: convex, $f = 20$ cm;
B: convex, $f = 10$ cm
18 40 cm; 40 cm

12 Mechanical waves

1 a 1 cm **b** 1 Hz **c** 1 cm/s
3 a Speed of ripple depends on depth of water
b AB since ripples travel more slowly towards it, therefore water shallower in this direction
4 b (i) 3 mm (ii) 15 mm/s (iii) 5 Hz

13 Light waves

1 a $S_1O = S_2O$ **b** $S_1O_1 - S_2O_1 = \lambda/2$
c $S_1O_2 - S_2O_2 = \lambda$

14 Electromagnetic radiation

2 D
4 a 3 m **b** 2×10^{-4} s

15 Sound waves

1 D
2 a When he sees flash of light from pistol
b Time taken by light to travel 110 m is negligible
c Error = 110/330 = 1/3 s; this should be *added* to his timing
3 a $v = f\lambda$ **b** 300 m **c** 25 cm
4 1650 m (about 1 mile)
5 a $2 \times 160 = 320$ m/s
b $240/(\frac{3}{4}) = 320$ m/s **c** 320 m

16 Musical notes

1 b (i) 1.0 m (ii) 2.0 m

Additional questions

1 $v = 33/1.5 = 22$ cm/s, $\lambda = 4$ cm,
$\therefore f = v/\lambda = 22/4 = 5.5$ Hz
2 b Refraction **c** 0.75 cm; 0.5 cm
d Equal **e** Speed in A greater than in B **f** A
4 X down, Y and Z up
5 E
9 a B **b** A **c** D
10 a 3×10^8 m/s **b** 15/4 m
c 0.2 MHz (200 kHz)
12 E
13 B
14 c 340 m/s

17 Measurements

1 a 10 **b** 40 **c** 5 **d** 67
e 1000
2 a 3.00 **b** 5.50 **c** 8.70 **d** 0.43
e 0.1
3 a 1×10^5; 3.5×10^3; 4.28×10^8; 5.04×10^2; 2.7056×10^4
b 1000; 2 000 000; 69 000; 134; 1 000 000 000
4 a 1×10^{-3}; 7×10^{-5}; 1×10^{-7}; 5×10^{-5} **b** 5×10^{-1}; 8.4×10^{-2}; 3.6×10^{-4}; 1.04×10^{-3}
5 10 mm
6 a two **b** three **c** four **d** two
7 24 cm^3
8 40 cm^3; 5
9 80
10 a 250 cm^3 **b** 72 cm^3

18 Density

1 a (i) 0.5 g (ii) 1 g (iii) 5 g
b (i) 10 g/cm^3 (ii) 3 kg/m^3
c (i) 2.0 cm^3 (ii) 5.0 cm^3
2 a 8.0 g/cm^3 **b** 8.0×10^3 kg/m^3
3 15 000 kg
4 130 kg
5 a 200 cm^3 **b** 200 cm^3
c 3.0 g/cm^3

19 Weight and springs

1 a 1 N **b** 50 N **c** 0.50 N
2 a 120 N **b** 20 N
3 a 2000 N/m **b** 50 N/m
4 a 19 cm **b** 35 g

20 Molecules

1 E
4 The forces between the layers are weak

Additional questions

1 0.88 g/cm^3
2 b Density of block + lead = (7.5 + 11)/16 = 18.5/16 g/cm^3; it sinks in water since density greater than 1.0 g/cm^3
c Milk
3 a 70.5 cm^3 **b** 20 cm^3
c 1.6 g/cm^3
5 b 8 cm **c** 2 cm
6 50 g
7 c Yes **d** 5 N

22 Moments and levers

1 a Turns anticlockwise
b Remains horizontal
2 B
3 X = 20 N; X = 5 N
4 (i) (c) (ii) (a) (iii) (b)

23 Centres of gravity

2 a B **b** A **c** C
3 Tips to right
4 a (i) 50 cm (iii) 1.2 N

24 Adding forces

1 a 50 N making 37° with 40 N
b 60 N making 25° with 40 N
2 D
3 200 N

25 Energy transfer

1 a Electrical to sound
b Sound to electrical
c K.e. to p.e. **d** Electrical to light (and heat) **e** Chemical to electrical to light and heat

2 A chemical; B heat; C k.e.; D electrical
3 180 J
4 1.5×10^5 J
5 **a** 150 J **b** 150 J **c** 10 W
6 500 W
7 **a** $\frac{300}{1000} \times 100 = 30\%$ **b** Heat
c Warms surroundings

26 Machines

1 **a** 100 J **b** 150 J **c** 67%
2 **a** 5000 J **b** 7500 J **c** 67%
3 **a** (i) 10/7 (ii) 2 (iii) 71%
b (i) 2 (ii) 3 (iii) 67%
c (i) 10/3 (ii) 6 (iii) 56%
4 Distance multiplier

27 Pressure in liquids

2 **a** (i) 25 Pa (ii) 0.50 Pa (iii) 100 Pa **b** 30 N
3 230 Pa
4 **a** 200 m^3 **b** 400 000 kg
c 4 000 000 N **d** 200 000 Pa
e (i) None (ii) Less
5 **a** 100 Pa **b** 200 N
7 1 150 000 Pa (1.15×10^6 Pa) (ignoring air pressure)
8 20 m

28 Atmospheric pressure

1 **a** (i) 860 mm or 100 mm over atmospheric (ii) 860 mm
2 **c** 74 cm
3 10 m

29 Floating, sinking and flying

1 **a** 10 N **b** 8 N **c** 6 N **d** 5 N
2 **a** 480 kg **b** 4800 N
c 4800 – 1600 = 3200 N
3 25 000 – 20 000 = 5000 N
4 **a** 30 g **b** 30 g **c** 30 cm^3

30 Fluid flow

1 **a** River flows faster in middle
b River gets wider so flows slower

Additional questions

1 **a** (ii) 500 N
2 **a** Moments about A:
of $P = P \times 0 = 0$;
of 500 N = 500 × 1 N m (clockwise);
of 200 N = 200 × 2 N m (clockwise);
of $Q = Q \times 3$ (anticlockwise)
b 300 N **c** 700 N **d** 400 N
3 Resultant = 46 N; P = 23 N
4 Tension in AC = 140 N; W = 193 N
6 **a** 1300 J **b** 1300 J **c** 13 W
e 8 J **f** 8 J
g Changed to heat and sound energy
h None **i** 100 N
7 3.5 kW
8 **a** 80% **b** (i) 2400 J (ii) 1.0 m
10 2×10^5 Pa
11 **a** 3 N **b** 0.3 kg **c** 0.0003 m^3
12 D

31 Velocity and acceleration

1 **a** 20 m/s **b** $6\frac{1}{4}$m/s
2 **a** 15 m/s **b** 900 m
3 2 m/s^2
4 **a** 1 s
b (i) 10 cm/tentick^2 (ii) 50 cm/s per tentick (iii) 250 cm/s^2
c 0
5 50 s
6 **a** 6 m/s **b** 14 m/s
7 4 s
8 **a** Uniform acceleration
b 75 cm/s^2

32 Graphs and equations

1 **a** 60 km **b** 5 hours **c** 12 km/h
d 2 **e** $1\frac{1}{2}$ hours
f 60 km/$3\frac{1}{2}$ h = 17 km/h
g Steepest line: EF
3 **a** 100 m **b** 20 m/s
c Slows down
4 **a** 5/4 m/s^2 **b** (i) 10 m (ii) 45 m
c 22 s
5 2.5 m/s^2
6 **a** 10 m/s **b** 25 m **c** 10 s
7 96 m
8 **a** 4 s **b** 24 m

33 Falling bodies

1 **a** (i) 10 m/s (ii) 20 m/s (iii) 30 m/s (iv) 50 m/s
b (i) 5 m (ii) 20 m (iii) 45 m (iv) 125 m
2 5 s; 31 m
3 B
4 3 s; 30 m/s
5 **a** 10 s **b** 2000 m
6 B

34 Newton's laws of motion

1 D
2 20 N
3 **a** 5000 N **b** 15 m/s^2
4 **a** 2 m/s^2 **b** 1000 N
5 **a** 4 m/s^2 **b** 2 N
6 (i) 0.5 m/s^2 (ii) 2.5 m/s (iii) 25 m
7 (1650 – 400) N/1000 kg = 1.25 m/s^2
8 **a** 1000 N **b** 160 N
9 **a** 5000 N **b** 20 000 N; 40 m/s^2

35 Momentum

1 **a** 50 kg m/s **b** 2 kg m/s
c 100 kg m/s
2 2 m/s
3 4 m/s
4 0.5 m/s
5 2.5 m/s
6 **a** 40 kg m/s **b** 80 kg m/s
c 20 kg m/s^2 **d** 20 N
7 2.5 m/s
8 **a** 10 000 N **b** 10 m/s^2

36 Kinetic and potential energy

1 **a** 2 J **b** 160 J
c 100 000 = 10^5 J
2 **a** 20 m/s **b** (i) 150 J (ii) 300 J
3 **a** 1.8 J **b** 1.8 J **c** 6 m/s
d 1.25 J **e** 5 m/s
4 **a** (i) 1000 J (ii) 20 m
5 B
6 3.5×10^9 W = 3500 MW

37 Circular motion

2 **a** Sideways friction between tyres and road
b (i) Larger (ii) Smaller (iii) Larger

Additional questions

1 **a** 14 m/s **b** 26 m/s **c** 1200 m
d –1 m/s^2
2 **a** OA, BC: accelerating; CD: uniform velocity; DE: decelerating
b OA: a = +80 km/h^2; AB: v = 80 km/h; BC: a = +40 km/h^2; CD: v = 100 km/h; DE: a = –100 km/h^2
c OA 40 km; AB 160 km; BC (5 + 40) = 45 km; CD 100 km; DE 25 km
d 370 km **e** 74 km/h
3 **a** Uniform velocity **b** 600 m
c 20 m/s
4 C
5 **b** (i) 0.5 m/s^2 (ii) 1.3×10^4 (12 775) m (iii) 20 m/s
6 **a** 4.0 s **b** 80 m **c** 40 m/s
d 20 m/s
7 A
8 30 N northwards; 21 m
9 C
10 10 000 m/s
11 75 km/h; yes
12 A 80; B 180; C 160; D 360; E 75; F 120; G 240; H 120; I 200; J 420
13 A
14 **a** 5 m/s **b** 0.4 kg m/s **c** 1 J
d 0.2 J

38 Thermometers

1 **a** 1530 °C **b** 19 °C **c** 0 °C
d –12 °C **e** 37 °C
2 C

39 Expansion of solids and liquids

2 Aluminium
3 **a** Aluminium **b** 1.009 m
4 **a** 0.1 m **b** 0.0004 m **c** 0.3 m
5 **a** Water **b** 4 °C

40 The gas laws

1 **a** 4 m^3 **b** 1 m^3
2 546 K (273 °C)
3 **a** Nothing **b** $\frac{1}{3}$: previous volume
c 3 times previous pressure
4 **a** 15 cm^3 **b** 6 cm^3
5 (i) 100 cm^3 (ii) 400 cm^3 (iii) 180 (178) cm^3

41 Specific heat capacity

1 15 000 J
2 A = 2000 J/(kg °C); B = 200 J/(kg °C); C = 1000 J/(kg °C)
3 **a** 6000 J **b** 6000 J/(kg °C)

42 Latent heat

1 **a** 3400 J **b** 6800 J
2 **a** 5 × 340 + 5 × 4.2 × 50 = 2750 J **b** 1700 J
3 680 s
4 0°C **b** 45 g

5 **a** 9200 J **b** 25 100 J
6 157 g

43 Vapours

7 **a** 686 mmHg **b** 624 mmHg

44 Conduction and convection

4 **a** (i) 2.2 MJ (ii) 1.1 MJ **b** 300 W

45 Radiation

1 E
6 D

46 Energy sources

1 **a** 2% **b** Water
c Cannot be used up
d Solar, wind
e All energy ends up as heat which is difficult to use and there is only a limited supply of non-renewable sources
2 Renewable, non-polluting (i.e. no CO_2, SO_2 or dangerous waste), low initial building cost of station to house energy converters, low running costs, high energy density, reliable, allows output to be readily adjusted to varying energy demands

Additional questions

2 B
3 A
4 B
5 C
6 363 K (90 °C)
7 (i) 182 mm (ii) 264 K (–9 °C)
8 B
9 **a** AB and CD **b** BC **c** E
10 B
11 D
12 D
15 6 kW

48 Static electricity

1 D
3 a, b, c
4 **a** B
b Electrons repelled to earth leaving sphere positively charged

49 Electric current

1 **a** 5 C **b** 50 C **c** 1500 C
2 **a** 5 A **b** 0.5 A **c** 2 A
4 B
5 b
6 All read 0.25 A

50 Potential difference

1 **a** 12 J **b** 60 J **c** 240 J
2 **a** 6 V **b** (i) 2 J (ii) 6 J
3 A
4 B
5 **b** Very bright **c** Normal brightness
d No light **e** Brighter than normal
f Normal brightness
6 **a** 6 V **b** 360 J
7 $x = 18$, $y = 2$, $z = 8$

51 Resistance

1 3 Ω
2 20 V
3 C
4 D
5 A = 3 V; B = 3 V; C = 6 V
6 3/2 = 1.5 Ω
7 **a** 15 Ω **b** 1.5 Ω
8 D

54 Electric power

1 **a** 100 J **b** 500 J **c** 6000 J
2 **a** 24 W **b** 3 J/s
3 D
4 3.1 kW
5 **a** 8 V **b** 16 W
6 **a** 3 A **b** 4 Ω
7 **a** 12 A **b** 250/12 = 21 Ω

55 Electricity in the home

1 Fuse is in live wire in (a) but not in (b)
3 **a** 3 A **b** 13 A **c** 13 A
4 **a** Parallel **b** Yes, current = 25 A
c 30 p
5 **a** 20 units **b** 200 p
6 40 p
7 **b** (i) 20 (ii) 1200 W **c** 840 p

Additional questions

1 C
2 B
3 B
4 **a** $\frac{1}{4}$ A **b** 1 A **c** $\frac{1}{2}$ A
5 **a** (ii) 2 Ω
6 B
7 B
8 **a** 3 Ω **b** 0.5 A **c** 0.25 A
9 4/3 Ω; 2A
10 B, D, F = 1.5 A; C, E = 3 V; G = 6 A; H, J = 3 A; I, K = 6 V; L = 2 A; M = 4 V; N, P = 1 A; O = 2 V
11 **a** 6 Ω **b** 2 A **c** 12/7 A
d 4/3 A
12 30 Ω
13 86 Ω
14 **a** (i) 3 kW (ii) 60 W (iii) 750 W
b 4 A
15 **a** 3 A **b** 36 W **c** 10 800 J
d 26 °C
16 27 kWh; 270 p

56 Magnetic fields

1 C
3 A N, B N, C S, D S, E N, F S

57 Electromagnets

2 S
5 **a** B **b** Relay switch **c** B, C

58 Electric motors

1 E
3 Anticlockwise
4 E

61 Transformers

2 B
3 **a** 24 **b** 10 A
4 **b** 12 000

Additional questions

1 E
3 **a** To complete the circuits to the battery negative **b** One contains the starter switch and relay coil; the other contains the relay contacts and starter motor
c Carries much larger current to starter motor **d** Allows wires to starter switch to be thin since they only carry the small current needed to energize the relay
6 **b** 53%
8 **a** 5 A **b** 400 W **c** 80 V
10 B
11 **b** (i) 60 (ii) 0.1 A
12 3.3 A

62 Electrons

1 A
2 **a** A – ve, B +ve **b** Down

63 Radioactivity

4 25 minutes
5 D

64 Atomic structure

2 **a** 27 protons and 32 neutrons in nucleus, 27 electrons outside

65 Electronics and control

2 **a** $V_1 = V_2 = 3$ V
b $V_1 = 1$ V, $V_2 = 5$ V
c $V_1 = 4$ V, $V_2 = 2$ V
5 A: AND; B: OR; C: NAND; D: NOR
7 **a** AND **b** OR
9 **a** Reset **b** Q = 1 **c** Q = 1
d R = 0

66 Telecommunications

2 **a** (ii) 3 m

Additional questions

2 **a** 10 V **b** 50 Hz
3 6 minutes
4 1/16
5 E
6 (i) β (ii) β (iii) α: A, D
8 **a** (i) 6 V (ii) 0 V
b (i) 0 V (ii) 6 V
11 **a** 1.6×10^{-16} J **b** 1.8×10^7 m/s
13 5000 s (83 min)

67 The atmosphere and weather

1 **a** Reflected back into space by atmosphere
b Radiation, convection, evaporation
c Altitude, latitude, nature of surface
2 Land and sea breezes, clouds, thunderstorms
3 Creates winds, e.g. Trade winds

68 Weather maps and forecasts

3 **b** A high; B low
c B, isobars closer
e (i) Easterly (ii) Northerly

69 Structure of the Earth

2 **a** Mantle denser than crust

b Density of mantle increases with depth
c Inner core denser than mantle
d Outer core is liquid

70 The Solar System
A, B, E
A, C, D

1 Stars and the Universe
1 A, B, D

Additional questions
1 E
4 Air thinner
5 Left
6 A: originally sedimentary, later parts became metamorphic
B: originally igneous, later parts became metamorphic
C: originally sedimentary, later parts became metamorphic
7 **d** (i) 10 (ii) 1000
11 **b** Age = $1/H$
c 10 000 million years

Revision questions
1 E
2 C
3 **a** 60° **b** 30°
4 B
5 **a** Refraction **b** POQ **c** Towards
d 40° **e** 90 – 65 = 25°
6 C **7** A **8** D **9** B
10 C **11** A
12 **a** Dispersion **b** (i) Red (ii) Violet
13 A **14** A **15** B **19** C
20 **a** 30/6 = 5 cm **b** 4 Hz **c** $v = f\lambda$
d 20 cm/s
21 E
22 **a** Circular **b, c** No change
23 **a** (i) Troughs from one arrive at same time as crests from other
(ii) Crests from one arrive at same time as crests from other
b (i) Larger wave (ii) Cancel out
(iii) Larger wave
24 D **25** E **26** E
27 **a** Longitudinal
b (i) Compression (ii) Rarefaction
28 D **29** B **30** A **31** D
32 C **33** B **34** D
35 **a** Transverse, stationary
b Longitudinal, progressive
c Transverse, progressive
d Longitudinal, stationary
36 **c** (i) BP – AP = even number of half-wavelengths
(ii) BQ – AQ = odd number of half-wavelengths
37 C **39** C **40** E **41** A
42 C **43** C **44** A **45** B
46 D
47 **a** Yes, 1 mm = 0.001 m **b** E
48 (i) 90 kg (ii) 4.5×10^4 Pa (N/m^2)
49 (i) 3 cm (ii) 2 s (iii) 2.3 cm
50 E **51** E **52** D **53** A
54 B **55** A **56** C **57** D
58 E **59** B **60** C **61** A
62 **a** 80 N **b** 80 N **c** 8 kg
d 0.01 m^3 **e** 0.01 m^3
63 4500 N at 9° to AB
64 **b** 120 J **c** 30 W **d** 75%
65 B
66 80 N
67 **c** 1.33 m^3
68 A **69** E **70** D
71 **a** 480 m **b** 6 m/s
c 8/20 = 0.4 m/s^2
72 E **73** C **74** B
75 D (F = 20 – 12 = 8 N; m = 2 kg;
$a = F/m = 8/2 = 4$ m/s^2)
76 D **77** E **78** D
79 **a** B **b** A
80 A
81 E
82 1.5×10^6 m
83 A
84 **a** (i) 15 m/s (ii) 11 m
b (i) 45 000 N (ii) 506 000 J
85 Velocity, momentum
86 C
87 **a** (i) 280 K (ii) 298 K (iii) 250 K
b (i) 27 °C (ii) 300 °C (iii) –73 °C
88 B **89** C **90** A **91** D
92 C **93** C **94** A
95 0.47 m^3
96 2.1×10^{-3} kg
98 D
99 **a** 3 Ω **b** 2 A
c 4 V across 2 Ω and 2 V across 1 Ω
100 **a** 1 Ω **b** 3 A **c** 6 V
101 D
102 B
103 **a** D **b** A **c** E
104 B **105** C **106** D **107** E
108 B (answer to 3 is 4000)
109 **a** E **b** A **c** C **d** B **e** D
110 **a** (i) 1.5 V (ii) 4 A (iii) 0.4 Ω
(iv) 6 W
b 120 J
111 **a** 4 V **b** 20 Ω **c** 0.13 A
112 20 W, 15 J/s
113 B **114** D **115** A **116** C
117 **a** A **b** C
118 E **119** B **120** B **121** B
122 C (symbol is $^7_3$Li)
123 **a** C **b** A **c** B **d** A
124 **a** D **b** E
125 E **126** C **127** B **128** A
129 C
132 15 counts/second
133 **a** 7 **b** 3 **c** (i) +3e (ii) zero
134 **a** Tropical maritime
b Polar continental
c Tropical continental
135 **c** Because the density of the mantle changes gradually, not abruptly
136 **a** Gravity **b** Reduces it
c (i) Decreases (ii) Decreases
137 4.4×10^9 kg

Mathematics for physics
1 **a** 3 **b** 5 **c** 8/3 **d** 20 **e** 12
f 6 **g** 2 **h** 3 **i** 8

2 **a** $f = \frac{v}{\lambda}$ **b** $\lambda = \frac{v}{f}$ **c** $I = \frac{V}{R}$
d $R = \frac{V}{I}$ **e** $m = d \times V$ **f** $V = \frac{m}{d}$
g $s = vt$ **h** $t = \frac{s}{v}$

3 **a** $I^2 = \frac{P}{R}$ **b** $I = \sqrt{\frac{P}{R}}$ **c** $a = \frac{2s}{t^2}$
d $t^2 = \frac{2s}{a}$ **e** $t = \sqrt{\frac{2s}{a}}$
f $v = \sqrt{2gh}$ **g** $y = \frac{D\lambda}{a}$
h $\rho = \frac{AR}{l}$

4 **a** 10 **b** 34 **c** 2/3 **d** 1/10
e 10 **f** 3×10^8
5 **a** 2.0×10^5 **b** 10 **c** 8
d 2.0×10^8 **e** 20 **f** 300
6 **a** 4 **b** 2 **c** 5 **d** 8 **e** 2/3
f $-\frac{3}{4}$ **g** 13/6 **h** –16 **i** 1
7 $a = \frac{v - u}{t}$ **a** 5 **b** 60 **c** 75

8 $a = \frac{v^2 - u^2}{2s}$
9 **b** Extension ∝ mass since the graph is a straight line through the origin
10 **b** No: graph is straight line but does not pass through the origin
c 32
11 **a** is a curve
b is a straight line through the origin, therefore $s \propto t^2$ or s/t^2 = a constant = 2

INDEX